ISBN 978-1-5285-1914-4
PIBN 10904447

# 1 MONTH OF
# FREE
# READING

## at

## www.ForgottenBooks.com

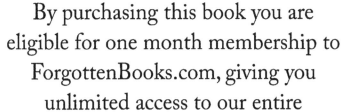

By purchasing this book you are eligible for one month membership to ForgottenBooks.com, giving you unlimited access to our entire collection of over 1,000,000 titles via our web site and mobile apps.

To claim your free month visit:
www.forgottenbooks.com/free904447

English
Français
Deutsche
Italiano
Español
Português

# www.forgottenbooks.com

**Mythology** Photography **Fiction**
Fishing Christianity **Art** Cooking
Essays Buddhism Freemasonry
Medicine **Biology** Music **Ancient**
**Egypt** Evolution Carpentry Physics
Dance Geology **Mathematics** Fitness
Shakespeare **Folklore** Yoga Marketing
**Confidence** Immortality Biographies
Poetry **Psychology** Witchcraft
Electronics Chemistry History **Law**
Accounting **Philosophy** Anthropology
Alchemy Drama Quantum Mechanics
Atheism Sexual Health **Ancient History**
**Entrepreneurship** Languages Sport
Paleontology Needlework Islam
**Metaphysics** Investment Archaeology
Parenting Statistics Criminology
**Motivational**

# Fourth International Congress for Stereology

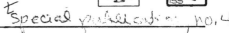

PROCEEDINGS OF THE
Fourth International
Congress for Stereology

HELD AT THE
National Bureau of Standards
Gaithersburg, Maryland, U.S.A.
September 4-9, 1975

EDITED BY:

Ervin E. Underwood

Georgia Institute of Technology
Atlanta, Georgia 30332

Roland de Wit and George A. Moore

Institute for Materials Research
National Bureau of Standards
Washington, D.C. 20234

SPONSORED BY

International Society for Stereology
Société internationale pour la Stéréologie
Internationale Gesellschaft für Stereologie
The National Bureau of Standards
and the
National Science Foundation

---

**U.S. DEPARTMENT OF COMMERCE,** Rogers C. B. Morton, *Secretary*

James A. Baker, III, *Under Secretary*

Dr. Betsy Ancker-Johnson, *Assistant Secretary for Science and Technology*

U.S. NATIONAL BUREAU OF STANDARDS, Ernest Ambler, *Acting Director*

Issued January 1976

Library of Congress Catalog Card Number: 76-8668


National Bureau of Standards Special Publication 431

Nat. Bur. Stand. (U.S.), Spec. Publ. 431, 547 pages (Jan. 1976)

CODEN: XNBSAV


U.S. GOVERNMENT PRINTING OFFICE
WASHINGTON: 1976

For sale by the Superintendent of Documents, U.S. Government Printing Office, Washington, D.C. 204
Price $11 (Add 25 percent additional for other than U.S. mailing).
Order by SD Catalog No. C13.10:431/Stock No. 003-003-01608-0

ABSTRACT

These proceedings contain most of the papers presented at the technical sessions
and workshops of the Fourth International Congress for Stereology, held at the
National Bureau of Standards, Gaithersburg, Maryland, from 4-9 September 1975.
Of the 113 papers recorded here, 10 are invited lectures, 66 are submitted
papers, and 37 come from the six workshops.  Three of these workshops containing 17 papers were integrated into the technical sections for these proceedings.
Topics covered in the Technical Sections are Principles and Mathematical
Developments, Pattern Recognition, Instrumentation, Three-dimensional Recon-
struction, and Stereological Applications in Materials and Biology.  Additional
topics covered in the Workshops are Mathematical Foundations of Stereology,
Particle Science, and Size Distributions.

Key Words:  Anatomy, automated microscopy, biology, convex particles, curvature,
geometrical probability, image analysis, materials science, mathematics, microscopy, microstructure, morphometry, oriented structures, particles, pattern
recognition, probability theory, quantitative microscopy, reconstructions,
sections, serial sectioning, shape parameters, size distributions, spacing
distributions, statistical analysis, statistics, stereology, stereometry,
stereoscopy, stochastic models, structures, three-dimensional reconstruction.

These Proceedings represent, in compact form, a record of all talks delivered
at the Fourth International Congress for Stereology, held at NBS near Washington, D. C., from 4 to 9 September 1975. Previous proceedings of the congresses
of the International Society for Stereology (ISS) have appeared every four
years, starting with the First International Congress for Stereology in Vienna
in 1963, followed by Chicago in 1967 and Berne in 1971.

The ISS is a truly international and interdisciplinary society. It was
founded in 1961 by a small group of anatomists and mathematicians who found
they were working on the same problems of expressing microstructures in
quantitative terms. Now-a-days, the ISS represents equally experimentalists,
theoreticians and instrumentalists, and its growth is due largely to the inter-
change of information among workers in many different fields. Thus, one finds
anatomists and ceramicists, metallurgists and biologists, petrographers and
mathematicians, using the basic equations of stereology and striving to solve
problems that have applicability to all.

Stereology may be defined as a body of methods for relating the microstructure
as seen in the random section or projection plane to the true spatial quantities. It means, in a broad sense, the study of structure in three-dimensions.
Thus it includes the narrower and more restrictive terms, such as stereometry
(measurement in three-dimensions) and morphology (the study of shape). The
basic equations of stereology are well-known by now. Consequently, the main
efforts of the ISS are devoted to advancing stereological theory, solving new
problems, improving the efficiency of measurement methods, determining
accuracy and error, and contributing to the development of automatic image
analysis. The usefulness and the success of the Society depends on its
ability to attack these problems on a broad front, whereby the different
talents of many people contribute to the solution of the problem.

A continuing objective of the ISS is educational in nature, whereby the
organization of meetings, symposia, lectures and the publication of survey
and specialized papers on stereology are constantly encouraged and supported.
Major contributions to the literature of stereology are the quadrennial
congress proceedings. In order to keep the size of the 1975 Proceedings down
to manageable proportions, it has been necessary to severely limit the length
of the papers. Even so, to collect and publish all these papers in one volume
is a tribute to the Managing Editor of the Proceedings, Dr. Roland deWit, and
to the support provided by NBS. The Congress Program Committee, headed by the
President of the ISS, has also contributed to this volume through the initial
selection and continuing editorial supervision of the submitted papers.

I am confident that the readers of these Proceedings will find something of
interest, whether basic or advanced, old or new, routine or stimulating, and
that all will profit from a more extensive use of the powerful, yet elegantly
simple, relationships of stereology.

Ervin E. Underwood, Sc. D.
President, 1971-1975
International Society for Stereology

The International Society for Stereology (ISS) was founded in 1961 as an interdisciplinary society, gathering among its members scientists of disciplines seemingly as disparate as biology, medicine, mineralogy, metallurgy and mathematics. What brought them together was the common problem of quantitatively interpreting the microstructure in three dimensions from two-dimensional sections which reveal only part of the true spatial geometry of the material. Measurements performed on these sections are not necessarily representative of the dimensions of the spatial structures. But the relationships between measurements on sections and the spatial microstructural features can be established by means of appropriate mathematical treatment in the form of stereological principles.

What is "Stereology"?

This term was coined at the Foundation Meeting of the ISS on the Feldberg in 1961. The founder and first president of the ISS, Hans Elias, Chicago, has proposed the following definition:

"Stereology, sensu stricto, deals with a body of methods for the exploration of three-dimensional space, when only two-dimensional sections through solid bodies or their projections on a surface are available. Thus, stereology could also be called extrapolation from two- to three-dimensional space".

The general problem is evident: the interior of opaque solids can only be investigated if the aggregate is somehow "opened". A common way of "opening" is to produce sections or thin slices, thus sacrificing the bulk of the material, but preserving an "image" of the relationship between the components at least in the plane of the section.

This image is distorted: solids become reduced to plane profiles; contact surfaces between component phases appear as linear contours or boundaries of the profiles; linear features of the material (filaments, intersection lines between surfaces) are seen on the section as points. In short, on the section any particular feature of the solid is represented by another, related feature whose dimension is reduced by 1. Furthermore, the diameter of a given profile is not representative of the particle from which it was derived: sectioning of spheres of equal diameters gives rise to a population of circular profiles of varying diameter.

We expect stereology to provide answers to the following fundamental questions: What is the quantitative relationship between the area of the profile and the volume of the solid from which it was derived, or of the profile boundary to the solid's surface, etc.? What is the true shape and size distribution of particles if only profiles can be studied? These are problems for mathematics to solve, particularly for integral geometry and geometric probability theory, as the sections will, most of the times, cut through the solids at random.

Many of these problems await solution; many have been solved and it remains to put them to fruitful use. Improvements in optical and electron microscopes have appreciably improved the resolving power of the most important instruments used to penetrate into the depth of inorganic materials and living organisms; stereological methods supplement these tools by indicating the proper measurements. The development of adequate technical means and the introduction of electronic data analysis help to render these methods efficient in application.

The use of Stereology

Metallurgists are applying stereological principles to study the composition
of metal alloys, the size distribution of grains and particles, the contact
surface area between the different phases, and the anisotropy of the grains,
i.e., microstructural properties which are related to the physical and mechanical properties of materials. The problems investigated by these methods in
mineralogy, petrography and geology are quite similar; it is in these fields
that many of the foundations of basic stereologic principles have been laid,
starting in 1847 with the fundamental principle of Delesse who demonstrated
that the volume fraction of rocks can be directly estimated from the areal
profile fraction of sections. In biology and medicine the internal structure
of cells, tissues and organs is now being quantitatively investigated in many
fields of stereological methods. Besides providing new insights into the
internal geometrical properties of living organisms, such data can be successfully applied to quantitative correlations between structure and function - -
applications that have previously been virtually impossible.

Periodicals

The ISS publishes an annual Newsletter (around 200 pages), under the Editor-
ship of Dr. Gerhard Ondracek, which contains outstanding stereological papers,
discussions, meeting notices and literature surveys. In between, there are
messages of interest to the membership from the President or other officers.

During the past year, special arrangements were made to provide free to
interested members a copy of the "Seattle" conference, edited by Dr. W. L.
Nicholson (260 pages), where statisticians and materials experts met and
discussed stereological topics of quite general interest, and the Handbook
on Powder Science and Technology, by Dr. Brian Kaye (about 150 pages).

The ISS has also arranged with the Royal Microscopical Society for ISS members
to subscribe at a reduced rate to the Journal of Microscopy, and to publish
their papers on stereology therein, if acceptable. The current Editor for the
ISS is Dr. E. R. Weibel, with Drs. E. E. Underwood and G. C. Amstutz serving
on the Editorial Advisory Board.

Membership in the ISS

Membership is open to anyone who wishes to help to promote stereology, make use
of its methods, and exchange ideas with colleagues of related interests in
various disciplines. Membership dues are presently US $5.00 per year, or US
$15.00 for 4 consecutive years. Information on student rates or industrial
affiliates can be obtained from the Secretary-Treasurer:

> Dr. Anna-Mary Carpenter
> University of Minnesota
> Department of Anatomy
> Minneapolis, Minn. 55455, USA

Activities of the ISS

One of the major activities of the ISS is the organization of scientific
meetings. The ISS International Congresses have been held every four years,
starting in 1963 at Vienna, and followed by Chicago in 1967 and Berne in 1971.
The Fourth International Congress for Stereology was held from 4-9 Sept. 1975,
at the National Bureau of Standards near Washington, D. C. Another major objective of the Society is educational in nature. The ISS has organized numerous local seminars and has helped to introduce stereology into many larger
meetings. For example, a Stereology Seminar was organized and presented at the
8th International Congress on Electron Microscopy, at Canberra, Australia, in
August 1974, and at the 4th International Metallography Congress held at Leoben,
Austria in October, 1974. Also, smaller local meetings and symposia are held.

# HOW STEREOLOGY CAME ABOUT

by Hans Elias, *Honorary President*
*International Society for Stereology*

In the process of teaching and producing teaching films for veterinary
and human micro-anatomy, I drew stereograms, i.e., generalized pictures having
a three-dimensional appearance, based on verbal textbook descriptions. However, it was impossible to take photomicrographs that would confirm those
stereograms. I noticed (1945) that alleged tubular glands in the chicken's
stomach were instead concentric trenches, and that the so-called liver cords,
assumed to be connected cylinders, were in reality interconnected walls forming a continuum (1948). These corrections of traditional textbook descriptions
were arrived at by common sense reasoning without the aid of reconstruction and
of mathematics.

Three years after the discovery of the liver wall continuum it dawned on me
that all this was a matter of geometrical probability and that such common sense
determination of structure could be verified mathematically, if one measured and
classified sectional profiles. I wrote a number of papers (1951, 1954) which
dealt with the geometry of sectioning. August Hennig showed me a few minor
mistakes I had made in two of those papers, and we began corresponding. From
him I learned that others before me had used geometrical probability for the
spatial interpretation of flat sections. The method of determining three-dimensional structure from single random sections by mere logical thinking is
described in Stereology 3 (1972).

It should be noted that I used quantitative methods chiefly for shape
determination and problems of continuity versus discreteness. My paper (1954)
on size distribution of spheres was written just for fun, but I used the
results for determining the size distribution of fat drops in fatty liver (1956).

By a chance meeting with Dr. Herbert Haug on a pleasure boat sailing around
Manhattan Island while we were attending the International Congress of Anatomists
in New York (1960), we discovered that we were both using statistico-geometrical
methods to obtain information on three-dimensional structure from sections. We
decided to call a gathering of persons interested in three-dimensional interpretation of flat images.

An announcement in a few journals brought 11 scientists together on the
Feldberg Mountain in the Black Forest (11-12 May, 1961). The participants of
this meeting and hence charter members were: G. Bach (first secretary), A.
Bohle, H. Elias (first president), H. Haselmann, H. Haug (first vice president),
A. Hennig, A. Hossmann (first treasurer), M. Gihr, R. Lorenz, H. Sitte, and
W. Treff (first assistant secretary). The word stereology was then coined and
the International Society for Stereology was founded and incorporated as a non-
profit organization in Neustadt, Schwarzwald. Dr. Ervin E. Underwood had read
the announcement and helped organize the Feldberg meeting, although he could not
attend. Later when Dr. Hossmann resigned as treasurer, Dr. Underwood took his
office and also became editor of our first periodical publication STEREOLOGIA.
The Society was subsequently incorporated as a non-profit organization in
Stuttgart (1962); Vienna (1963); and Deerfield, Illinois (1964).

On my way home from the Feldberg, I stopped in New York and, while walking
along Central Park, met my friend Dr. Ewald R. Weibel. We entered a little
cafe to talk and he told me he had just written on the morphometry of the lung.
He suggested we organize a symposium on the use of geometry in the evaluation
of sections. I then related to him what had transpired at the Black Forest and
that the official word now was stereology.

ELECTED OFFICERS OF THE INTERNATIONAL SOCIETY FOR STEREOLOGY

|  | 1972-1975 | 1976-1979 |
|---|---|---|
| President: | Ervin E. Underwood<br>Atlanta, Georgia<br>U.S.A. | Herbert Haug<br>Lübeck, Germany |
| Vice-Presidents,<br>Regional:<br>for America: | John E. Hilliard<br>Evanston, Illinois<br>U.S.A. | Robert T. DeHoff<br>Gainesville, Florida<br>U.S.A. |
| for Australasia: | -- | Robert S. Anderssen<br>Canberra, Australia |
| for Europe: | Herbert Haug<br>Kiel, Germany | H. Eckart Exner<br>Stuttgart, Germany |
| Vice-President,<br>Corporate: | -- | John E. Hilliard,<br>Evanston, Illinois<br>U.S.A. |
| Secretary-<br>Treasurer: | Anna-Mary Carpenter<br>Minneapolis, Minnesota<br>U.S.A. | Anna-Mary Carpenter<br>Minneapolis, Minnesota<br>U.S.A. |
| Secretary for<br>Publications: | G. Christian Amstutz<br>Heidelberg, Germany | Gerhard Ondracek<br>Karlsruhe, Germany |

HONORARY MEMBERS OF ISS

Prof. Dr. Hans Elias
(Honorary President)
San Francisco, U.S.A.

Dr. R. Buckminster Fuller
Philadelphia, Pennsylvania
U.S.A.

Dr. Otto Röhm
Lausanne, Switzerland

Prof. Dr. S. A. Saltikov
Erevan, Armenia
U.S.S.R.

Dr. med. August Hennig
München, Germany

Dr. Harold W. Chalkley
(Deceased)

CONGRESS ORGANIZING COMMITTEE

| | |
|---|---|
| Chairman: | George A. Moore |
| Deputy: | Roland de Wit |
| Arrangements: | Sara R. Torrence |
| Support Staff: | Robert F. Martin |
| Treasurer: | Ronald B. Johnson |
| International Relations: | Michael B. McNeil |
| Public Relations: | Madeleine Jacobs |

CONGRESS PROGRAM COMMITTEE

Ervin E. Underwood (Chairman)
Atlanta, Georgia, U.S.A.

G. Christian Amstutz
Heidelberg, Germany

Anna-Mary Carpenter
Minneapolis, Minnesota, U.S.A.

Robert T. DeHoff
Gainesville, Florida, U.S.A.

H. Eckart Exner
Stuttgart, Germany

Herbert Haug
Lübeck, Germany

John E. Hilliard
Evanston, Illinois, U.S.A.

George A. Moore
Washington, D.C., U.S.A.

## WELCOMING REMARKS

by John D. Hoffman, *Director*
*Institute for Materials Research, National Bureau of Standards*

Good evening ladies and gentlemen. Welcome to the National Bureau of Standards and to the 4th International Congress for Stereology cosponsored by the International Society for Stereology and the National Bureau of Standards. I would like to add a special welcome to our guests from abroad and wish you a pleasant visit in the United States.

Because this is the first congress of this society to be held at the Bureau of Standards, some of you are visiting here for the first time and may not be familiar with the range of activities at the Bureau. I hope that the tour which many of you took today gave you a first-hand view of the breadth and depth of the work performed here. I would like to take a few minutes now to describe some things about the Bureau that you might not know.

The Bureau of Standards was created in 1901 to develop and disseminate the national standards of measurement, to determine physical constants and properties of important materials, to develop test methods, to aid in the establishment of standard practices, and to provide technical services to other government agencies. More recently, legislation has added new responsibilities in the areas of computer technology, standard reference data, and fire research. Measurements, standard practices and data are the common themes in all our work, and we are perhaps best described as the Nation's measurement laboratory.

The work of the Bureau is carried out in four institutes: The Institute for Basic Standards, The Institute for Applied Technology, The Institute for Computer Sciences and Technology, and The Institute for Materials Research.

The Institute for Basic Standards is responsible for the custody, maintenance and development of the national standards, for methods of physical measurement and for the provision of means for making measurements consistent with those standards. This work includes the basic units of measurement such as mass, length, and time; and NBS is perhaps best known to some of you for its basic standards work. Many of you, I am sure, are also familiar with the NBS work in pattern recognition and computer analysis of images begun about 15 years ago. This work is carried out in the applied mathematics group of The Institute for Basic Standards.

The Institute for Applied Technology is concerned with engineering standards, codes, and test methods. It works to promote the development and use of technology in important areas such as building construction, fire prevention, electronics, and consumer product safety.

The Institute for Computer Sciences and Technology has the responsibility to improve federal management and utilization of computer technology, and to lead federal efforts to stimulate the use of computer technology to meet national needs. With the growing use of remote terminals and computer-computer communications, problems of access, privacy, and rapid turn-around inevitably arise. The computer networking laboratory you visited today is one in which experiments are conducted to develop guidelines for the efficient use of computer networks.

The Institute for Materials Research is concerned with the development of methods for measuring the key properties of materials, the development of methods for relating these properties to the performance of materials, and with the development of standard reference materials. Three of the facilities you visited today - the image analysis laboratory, the reactor, and the dental research laboratory - are concerned with different facets of materials research and illustrate the breadth of our activities. Incidentally, the image analysis laboratory is one of several institute central facilities that we have established during the last few years. The work being carried out in this laboratory includes metallographic work in support of our metallurgy program, particulate analysis in support of our environmental program and analysis of blood cells in support of our clinical standards program. ·The materials research program at NBS also includes work in nondestructive evaluation of materials, corrosion, test methods development ·for materials to be used in advanced energy generation systems, and polymer research.

I note that many of you, perhaps a majority of you, are interested in the stereology of biological systems. While NBS is not an agency concerned with biological problems per se, understanding the behavior of biological systems and delivering health care increasingly require accurate measurement and characterization. Beginning with calibrating clinical thermometers almost 75 years ago, NBS has cooperated with the medical community to develop better methods·of measurement for medical applications. The dental research program described to you earlier today has been contributing to improved dental care for more than 50 years. This program, jointly supported by the American Dental Association, the National Institute for Dental Research and NBS has resulted in new dental restoratives, the high speed drill.and the panoramic x-ray. The dental program has recently been expanded to include the development of techniques for characterizing polymeric implant materials.

Earlier, I alluded to our clinical standards program. For the last six years, NBS has cooperated with a number of groups including the College of American Pathologists and the American Association for Clinical Chemists to develop ·standard reference materials and analytical reference methods needed to improve the accuracy of measurements in the clinical laboratory. This year, about 13000 private and hospital affiliated clinical laboratories will perform about four billion tests. The general trend toward early detection of diseases indicates that measurement and standards will play an ever greater role in the total health care system of the future.

Another area of medical diagnosis, closer perhaps to the interests of this congress, involves the use of ultrasound to characterize human tissue and to detect abnormalities such as malignant tumors. NBS has recently undertaken work in this area in collaboration with the National Institutes of Health. Work on image reconstruction in connection with ultrasonic testing is under consideration. Interestingly, the NBS ultrasonic diagnostic work is an outgrowth of our program in nondestructive evaluation of materials. The NBS work is intended to improve the methods for characterizing flaws in materials or abnormalities in tissue. This work is one.of the many examples where developments in one area stimulate developments in another·and for which the same techniques are adopted for diverse purposes.

The fact that this conference brings together scientists from many disciplines − metallurgy, medicine, earth resources, mathematics − for the common purpose of utilizing the science of stereology indicates to me that your society is well aware of the benefits that may be derived from such sharing of common methodology and mathematical analysis.

I would like to take this opportunity to thank Dr. George Moore of NBS who is serving as chairman of the organizing committee for this congress. For many years, George has worked to advance the science of stereology, most recently, in connection with the analysis of inclusions in metals. To facilitate his metallographic work, George designed, built and applied a system for automated image analysis. He has also been active in ASTM affairs where he has devoted much time to the development of standard practices for metallographic analysis.

Once again, welcome to·the National Bureau of Standards and have a good meeting.

## ACKNOWLEDGMENTS

Many people have contributed to the technical, organizational, and social
success of this Congress and its Proceedings, and we are grateful to all of
them. We wish to single out a few for special mention.


The highlight of the Thursday Night Opening Session was the light-hearted
talk "Toronto Daze" by J. H. L. Watson from the Edsel B. Ford Institute for
Medical Research in Detroit. He covered the gamut and liberally sprinkled
his talk with song and dance. For this he was ably accompanied by pianist
William R. Ott from the Optical Physics Division at NBS, who did a superlative
job at the keyboard on one day notice, when the previously engaged pianist fell
ill. He also provided background piano music through the rest of the Opening
Session. Another fascinating feature of that evening was E. E. Underwood's
talk on "Art in Microscopy."

The organizational and social program would not have been possible
without the unstinting efforts of Mrs. Sara K. Torrence from the Office of
Information Activities at NBS. She kept us moving in the right direction at
all times and also arranged the Saturday Night Bar-B-Que and Old Fashioned
Hoe-Down and the Sunday Tours with the help of Gretchen Poston. We are also
indebted to the firms of Bausch and Lomb, IMANCO, and E. Leitz for their contribution to the success of the Thursday Reception.

For help with the editorial work on the Proceedings we particularly want
to thank Mrs. Betty L. Burdle and Mrs. Mildred C. Reid from the Office of
Technical Publications at NBS. Their meticulous efforts, in close cooperation
with the Visual Arts Section at NBS, led to the attractive physical appearance
of this volume.

ARNOLD LAZAROW

1916 – 1975


Arnold Lazarow's contributions to science are documented in some
250 publications which include original articles, reviews, abstracts and
lectures both to students and at scientific meetings.  His influence on
colleagues, medical students, graduate students and post-doctoral fellows
was so personal that each responded with loyalty and increased productivity.
    Dr. Lazarow was born in Detroit, Michigan in 1916.  He studied at the
University of Chicago where he was granted the Bachelor of Science (1937)
as well as both Ph.D. and M.D. (1941).  He was on the staff of the Department
of Anatomy at Western Reserve University for a decade when he was invited
to the University of Minnesota where he served as professor and head of the
department of Anatomy for two decades.  Honors included:  Phi Beta Kappa,
Sigma Xi, Alpha Omega Alpha, the Capp Prize and the Banting medal of the
American Diabetes Association.  A member of 23 scientific societies, Arnold
was elected to office in many of them; he was secretary of the International
Society for Stereology from 1968 to 1971.  Dr. Lazarow died June 25, 1975
after a brief illness.
    Many of his contributions to science were in advance of their time:  he
pioneered in establishing the concept that intracellular glycogen is in particulate form, that mitochondria contain respiratory enzymes and that sul-
phhydryl groups protect against alloxan.  His interest in mathematical expression of morphologic data and his attraction to instrumentation resulted
in the development of the first quantitators for the application of stereologic methods to light and electron microscopy (1958).  These tools were
used in establishing the percentage of islet tissue in the total pancreas,
the volume per cent of the cellular components of the islets, the cytotoxic
effects of alloxan on the beta cells and the precise increment in beta cell
mass during embryogenesis.  Using these parameters as a basis, sufficient
numbers of beta cells grown in organ culture were transplanted to alloxan
diabetic rats; the animals became normoglycemic and remained normal for as
long as two years.
    Challenged by the computer, Arnold applied information storage and
retrieval methods to the development of the Diabetes Literature Index, a
monthly publication, which NIH distributes to members of the American
Diabetes Association.
    Dr. Lazarow had clarity of thought, drive, and an innate inquisitiveness
coupled with a perfectionism that enabled him to achieve his goals and to
inspire others.  Smilingly he interrupted his own activities to help those
who queried, "Do you have a minute?"; these minutes are precious to all who
knew him.

A-M Carpenter

HAROLD W. CHALKLEY

1886 – 1975


Harold W. Chalkley's principal research efforts were in the field of protozoan cell division. He was the author of approximately 100 papers concerned with the mechanics of cell division in Amoeba proteus, its chemistry and other aspects of protozoan physiology. His contributions to stereology came late in his research career.

Dr. Chalkley was born in London, England, in 1886. He moved to Lawrence, Kansas, in 1905 and entered Kansas University. Two years later, finding university life less fascinating than the still expanding western frontier, he left school and went to work on a Rock Island railway crew surveying across Indian country in the Texas panhandle. He worked later on canal surveys in Louisiana and as a railroad hand, enginewright, and stationary engineer.

After completing World War I service, he entered the Mississippi Agricultural and Mechanical College. He graduated with honors in 1925 and entered the Johns Hopkins University as a Johnston Fellow. He received his Ph.D. in zoology in 1929 and was elected a member of both Phi Beta Kappa and Sigma Xi. He entered government service the same year as a physiologist of the old Hygenic Laboratory, now the National Institutes of Health of the United States Health Service.

His early experience in engineering was frequently reflected in his research. This included the design of a split-field microscope that permitted simultaneous visualization of an object from both the top and side, the adaptation of planimetric methods for the determination of amoeboid volume and a variety of other devices to assist, not only in his own work, but in the work of colleagues. One of these efforts led to his adaptation (1943) of the "flying spot" method to the quantitative analysis of sectioned tissues. He often remarked that this technique, which he developed in less than an hour, would probably be the only finding for which he would be remembered. Later contributions to the armamentarium of stereology included a method for determining numbers of cellular nuclei (1945) and for determining surface/volume ratios (1949). He collaborated in several studies involving the application of these techniques to biological investigation. He retired from government service in 1952.

Dr. Chalkley retained a vigorous interest in the world about him, an all-encompassing sense of humor, and a willingness to respond to challenges that lasted until the very minute of his death on the evening of September 25, 1975, sixteen days after his election as an Honorary Member of the I.S.S.

Donald T. Chalkley

CONTENTS

PART I.  TECHNICAL SECTIONS

1.  PRINCIPLES AND MATHEMATICAL DEVELOPMENTS

Contents

*Contents*

*Contents*

## 9. SIZE DISTRIBUTIONS

PART I.  TECHNICAL SECTIONS

1.  PRINCIPLES AND MATHEMATICAL DEVELOPMENTS

*Chairpersons:*

R. E. Miles
W. L. Nicholson
R. T. DeHoff
R. de Wit
J. E. Hilliard
R. S. Anderssen
J. C. Serra

NOTE

In order to assist the reader with the symbolism
in the more mathematical papers of Miles (p. 3),
DeHoff and Gehl (p. 29), and Rhines (p. 233), an
abbreviated glossary of equivalent symbols has
been assembled on p. 513.

*National Bureau of Standards Special Publication 431*
*Proceedings of the* FOURTH INTERNATIONAL CONGRESS FOR STEREOLOGY
*held at* NBS, Gaithersburg, Md., September 4-9, 1975 (Issued January 1976)

ON ESTIMATING AGGREGATE AND OVERALL CHARACTERISTICS FROM THICK SECTIONS BY TRANSMISSION MICROSCOPY

by R. E. Miles
*Department of Statistics, Institute of Advanced Studies, Australian National University, P.O. Box 4, Canberra, A.C.T. 2600, Australia*

ABSTRACT

The random spatial structure considered is the union $X$ of an aggregate of random opaque convex particles in transparent space, the particle centroids being sited at the points of a homogeneous Poisson process. The model is thus the union of an aggregate of not–necessarily–convex disjoint regions. Expressions are derived for the fundamental stereological quantities $V_V$, $S_V$, $K_V$ (integrated mean curvature density) and $G_V$ (integrated gaussian curvature density) for $X$ ; and for $A_A^I$, $L_A^I$ and $C_A^I$ (integrated curvature density) for the projection of a slice of thickness $t$ of this model onto a plane parallel to the slice, where $t$ may take any non–negative value. The estimates relevant to transmission microscopy suggested by this theory are presented.

The stochastic model considered is an aggregate of random 'opaque' convex particles (with $\overline{V}$, $\overline{S}$ and $\overline{M}$ denoting their mean volume, mean surface area and mean mean (*sic*) caliper diameter, respectively) distributed randomly throughout a 'transparent' medium as follows. Their centroids (or any other well–defined particle point) are sited at the points of a Poisson process with constant density $D$ , and their orientations are (independently) isotropic about these points. Thus the particles may interpenetrate, a prerequisite for any effective mathematical theory; and so the model is essentially one of disjoint opaque regions, each of which is the union of one or more overlapping particles. General properties of this rather natural stochastic model have been given by Giger and Hadwiger [3]. The sets of points contained in at least one particle, and not contained in any particle, are denoted by $X$ and $Y$ , respectively.

As usual in transmission microscopy, a foil (or section) bounded by two parallel planar faces a distance $t$ apart is taken, and the content of this section is projected perpendicularly onto an observation plane parallel to the faces. At each point of the observation plane it may only be discerned whether it is covered by at least one particle, or is uncovered. With a dash relating to the observation plane, we write $X'$ and $Y'$ for the covered and uncovered areas, respectively.

The stereological problem is to estimate $D$, $\bar{V}$, $\bar{S}$ and $\bar{M}$ from measurements solely in the observation plane; these estimates may in turn be used to furnish estimates for the following fundamental 'overall' quantities: the fraction

(1a) $$V_V = 1 - \exp(-D\bar{V})$$

of the volume which is occupied by $X$ ; the interface area

(1b) $$S_V = D\bar{S} \exp(-D\bar{V})$$

between $X$ and $Y$ per unit volume; the total (or integrated) mean curvature

(1c) $$K_V = \left(2\pi D\bar{M} - \frac{1}{32}\pi^2 D^2 \bar{S}^2\right) \exp(-D\bar{V})\cdot$$

of this interface per unit volume (at each point of surfaces under surface tension, the pressure difference between the two sides divided by the surface tension equals twice the mean curvature); and the integrated gaussian curvature of this interface per unit volume

(1d) $$G_V = \left(4\pi D - 2\pi D^2 \bar{M} \, \bar{S} + \frac{1}{96}\pi^2 D^3 \bar{S}^3\right) \exp(-D\bar{V})$$

(for aggregates of non-overlapping particles, not necessarily convex, $G_V$ is $4\pi$ times their number per unit volume). It is of interest that (1a) - (1d) remain true when the particles are no longer convex - in that case $\bar{M}$ is no longer the mean mean caliper diameter, but represents $(2\pi)^{-1}$ times the mean value of the integral of mean curvature of the particles.

In the observation plane, define $A_A'$ to be the covered area fraction, $L_A'$ to be the length of interface curve between $X'$ and $Y'$ per unit area, and $C_A'$ to be the total curvature of this interface curve per unit area. Then

(2a) $$A_A' = 1 - \exp[-D\{\bar{V}+(t\bar{S}/4)\}] \ ,$$

(2b) $$L_A' = \pi D\{(\bar{S}/4)+t\bar{M}\} \exp[-D\{\bar{V}+(t\bar{S}/4)\}] \ ,$$

and

(2c)    $C_A' = \left[2\pi D\,(\overline{M}+t) - 2\pi^2 D^2\{(\overline{S}/8)+(t\overline{M}/2)\}^2\right]\,\exp[-D\{\overline{V}+(t\overline{S}/4)\}]$ .

The derivations of formulae (1) and (2), by the methods of geometrical probability, are given in the Appendix. With $D$ small, (1) and (2) reduce to the relations (20)-(22) in DeHoff [1]. Standard stereological relations for a planar section result from (1) and (2) upon setting $t = 0$ .

Using (2), we may estimate $D\{\overline{V}+(t\overline{S}/4)\}$, $D\{(\overline{S}/4)+t\overline{M}\}$ and $D(\overline{M}+t)$ . Thus, if $t$ (see [6; pp. 188-190]) and any one of $D$, $\overline{V}$, $\overline{S}$ and $\overline{M}$ is known, the other three may be estimated. Alternatively, if sections of differing but unknown thicknesses $t_1$, $t_2$ are available, we have the estimates

(3a)    $V_V = 1 - e^{-u}$ ,

(3b)    $S_V = 4va_{21}e^{-u}$ ,

(3c)    $K_V = \left[2\pi v b_{21} - \dfrac{1}{2}\,\pi^2 v^2 a_{21}^2\right]\,e^{-u}$ ,

(3d)    $G_V = \left[4\pi v c_{21} - 8\pi v^2 a_{21}b_{21} + \dfrac{2}{3}\,\pi^2 v^3 a_{21}^3\right]\,e^{-u}$ ,

(4)    $D : \overline{V} : \overline{S} : \overline{M} : 1 = v^2 c_{21}^2 : u : 4va_{21} : vb_{21} : vc_{21}$ ,

and

(5)    $t_1 : t_2 : 1 = b_1 b_{21} - c_1 a_{21} : b_2 b_{21} - c_2 a_{21} : c_1 b_2 - c_2 b_1$ ,

*where* $a_{21} = a_2 - a_1$ , *etc.*,

(6a)    $a = -\ln\left(1-A_A'\right)$ , $\quad b = \dfrac{1}{\pi}\,\dfrac{L_A'}{1-A_A'}$ , $\quad c = \dfrac{1}{2\pi}\,\dfrac{C_A'}{1-A_A'} + \dfrac{\pi b^2}{4}$ ,

(6b)    $u = \{a_{21}(c_1 a_2 - c_2 a_1) + b_{21}(a_1 b_2 - a_2 b_1)\}/(b_{21}^2 - a_{21}c_{21})$ ,

and

(6c)    $v = (c_1 b_2 - c_2 b_1)/(b_{21}^2 - a_{21}c_{21})$ .

If at least one of $t_1$, $t_2$ be known, or sections of three or more thicknesses (known or unknown) are at hand, estimates of a more statistical nature may be obtained from (1) and (2).

A feature of our theory is its generality: $t$ is not necessarily small, and the convex particles may have any shape distribution (surprisingly, even with this generality, no approximations are made). As observed above, since the particles may interpenetrate, our model is really one of disjoint opaque regions, each of which is the union of one or more overlapping convex particles. When $D(2\overline{V}+\overline{M}\ \overline{S})$ is small, most of these are isolated and therefore convex. For larger values of $D(2\overline{V}+\overline{M}\ \overline{S})$, estimates of $D$, $\overline{V}$, $\overline{S}$ and $\overline{M}$ lose their practical interpretation, unlike those of $V_V$, $S_V$, $K_V$ and $G_V$. In fact, it seems likely that (3) (and (5)) will often give useful estimates for a thick section from a general homogeneous and isotropic two-phase (one opaque, one transparent) structure, e.g. a porous medium.

The only work similar to this appears to be some of Giger [2], but the stereological implications of his paper seem a little obscure.


APPENDIX

In this appendix the derivations of the basic relations (1a)-(1d) and (2a)-(2c) are presented. The remaining formulae given above follow from these, essentially by simple algebra.

The model may be thought of as being constructed in two stages:

(i) The centroids of the particles are located at the points of a Poisson point process in space having constant density $D$ . Thus the number of process points in disjoint volumes $V_1$, ..., $V_k$ are mutually independent and Poisson distributed, with mean values $DV_1$, ..., $DV_k$ , respectively.

(ii) For each process point $x$ , a random convex set $K$ is sampled from some isotropic distribution of convex sets having the origin as centroid. Then the particle associated with $x$ is $x + K$ , i.e., $K$ translated so as to have $x$ as centroid. This procedure is carried out independently for each process point, using the same convex set distribution. The random set $X$ of interest is the union of the particles so formed.

First we have the key

*Theorem* (Hitting Numbers). The number of particles hitting any fixed convex domain $K_0$ is Poisson distributed with mean value

(7) $$\delta = D\left[V_0 + \frac{1}{2} S_0\overline{M} + \frac{1}{2} M_0\overline{S} + \overline{V}\right] .$$

In particular the probability that $K_0$ is not hit is $e^{-\delta}$ .

*Proof of Theorem.* Write $dP$ for the proportion of particles with dimensions in the range $(M, M+dM; S, S+dS; V, V+dV)$. Then by the three dimensional form of the principal formula of integral geometry [5; equation (66)] the number of particles in this range hitting $K_0$ is Poisson distributed with mean value $\mu = D \, dP \left[ V_0 + \frac{1}{2} S_0 M + \frac{1}{2} M_0 S + V \right]$. It follows from the 'complete independence' of Poisson processes and the model that the number hitting $K_0$ is Poisson $\left( \int \mu \right)$ distributed.     Q.E.D.

Although not mentioned, the following theory depends heavily upon this Poisson independence, and would scarcely be possible without it.

Taking $K_0$ to be a point, (1a) follows. The particle surface area per unit volume is $D\overline{S}$. For each such surface point, the probability it is uncovered by other particles is $e^{-D\overline{V}}$. (1b) follows at once from these two observations. (We shall repeatedly use 'uncovered' in this sense below.)

Since the mean caliper diameter of a convex set is $(2\pi)^-$ times its integral of mean curvature, the integral of mean curvature of the particle surfaces per unit volume is $2\pi D\overline{M}$. Hence the contribution of these surfaces to $K_V$ is

(8) $$K_V^{(2)} = 2\pi D\overline{M} \, e^{-D\overline{V}} .$$

To simplify the theory, we suppose the particles have smooth surfaces, so that pairs of surfaces intersect in smooth curves. These we call I-*curves*; likewise, points of intersection of triples of particle surfaces we call I-*points*. In fact, our results extend without change to general convex particles. It only remains to find the contribution $K_V^{(1)}$ to $K_V$ from the uncovered portions of I-curves, since it is readily verified that the contribution from uncovered I-points is zero.

But first we determine the length $L_V$ of I-curve and the number $N_V$ of I-points per unit volume. Consider a small fixed sphere $Q$ of radius $\epsilon \ll 1$. The following statements are true in the limit $\epsilon \downarrow 0$. Intersections of particle surfaces with $Q$ are planar, and their number $n$ is Poisson $(2\epsilon D\overline{S})$ distributed. Further, given $n = n_0$, these $n_0$ hitting planes are

independent isotropic uniform random (IUR) planes in $Q$ . We require the
following specializations of the results (2.13), (1.35P) and (2.31T) of [4]:

(9)    $P$ (2 independent IUR planes in $Q$ intersect in a line

hitting $Q$) $= \pi^2/16$ ;

(10)    $P$ (3 independent IUR planes in $Q$ intersect in a point of $Q$)

$= \pi^2/48$ ;

and given that $2$ independent IUR planes in $Q$ intersect in a line hitting
$Q$ , this line is IUR in $Q_\setminus$, so that the mean length of its intersection with
$Q$ is $4\epsilon/3$ . It follows from the above that

(11)                    $N_V = E\binom{n}{3}\ \left(\pi^2/48\right)/\left(4\pi\epsilon^3/3\right) = \pi D^3 \overline{S}^3/48$

and

(12)                $L_V = E\binom{n}{2}\ \left(\pi^2/16\right)(4\epsilon/3)/\left(4\pi\epsilon^3/3\right) = \pi D^2 \overline{S}^2/8$ .

Now, since the integral of mean curvature for unit length of a wedge of
dihedral angle $\phi$ is $(\pi-\phi)/2$ , we have

(13)                        $K_V^{(1)} = L_V\ E\left[\frac{\pi-\phi}{2}\right]\ e^{-D\overline{V}}$ .

Here $\phi$ has a distribution symmetric about $E(\phi) = 3\pi/2$ , and (1c) follows
from $K_V = K_V^{(1)} + K_V^{(2)}$ .

We now derive (1d). Since the integral of gaussian curvature for any
convex domain is $4\pi$ , the contribution to $G_V$ from particle surfaces is

(14)                            $G_V^{(2)} = 4\pi\ D\ e^{-D\overline{V}}$ .

The contribution from each uncovered I-point is the same as the integral of
gaussian curvature for the complementary cone $C$ at the I-point, which itself
equals the exterior solid angle $\omega$ [4; p. 216] of the cone dual to $C$ .
Hence, since $E(\omega) = \pi/2$ , we have

(15)                            $G_V^{(0)} = N_V\ (\pi/2)\ e^{-D\overline{V}}$ .

It remains to find the contribution $G_V^{(1)}$ from the uncovered portions of
I-curves. We suppose the particles are polyhedral. This is permissible, since
a convex set may be approximated arbitrarily closely within and without by
convex polyhedra. Then the I-curves are equivalently (i) line segments of
intersection of pairs of polyhedral faces; and (ii) points of intersection of

polyhedral faces and edges. We need only consider (ii), since the contribution of (i) to $G_V^{(1)}$ is zero. Consider in particular the contribution from a polyhedral edge of length $l_i$ , dihedral angle $\phi_i$ $\left(0 < \phi_i < \pi\right)$ (and hence exterior angle $\pi - \phi_i$ [4; p. 216]). By considering the intersections of a half-plane with a 'ripple surface' in which the dihedral angles are alternately $\phi_i$ and $2\pi - \phi_i$ , and noting that the sum contribution of the vertices to the integral of gaussian curvature is zero, we see that the contribution from the intersection of a polyhedral face with our edge of dihedral angle $\phi_i$ is *minus* that from the intersection of the face with an edge with dihedral angle $2\pi - \phi_i$ . Averaging over mutual orientations, the average value of the latter contribution is seen to be $\pi - \phi_i$ . Now, by the classical stereological formula "$P_L = \frac{1}{2} A_V$" , the mean number of intersections of the $\left(l_i, \phi_i\right)$ edge with polyhedral faces is $\frac{1}{2} l_i \, \overline{DS}$ . Hence the average contribution of a polyhedron is

$$(16) \qquad\qquad - \sum \frac{1}{2} l_i \; \overline{DS} \; \left(\pi - \phi_i\right) \, e^{-\overline{DV}} \; .$$

That is, since $M = (4\pi)^{-1} \sum l_i \left(\pi - \phi_i\right)$ for polyhedra,

$$(17) \qquad\qquad G_V^{(1)} = - 2\pi \, D^2 \, \overline{M} \, \overline{S} \, e^{-\overline{DV}} \; .$$

(1d) follows from $G_V = \sum_0^2 G_V^{(i)}$ .

Now we turn to the transmission microscopy theory for our model, and derive (2a)-(2c). Taking $K_0$ in the Theorem to be a line segment of length $t$ , for which $M = t/2$ , we obtain

$$(2a) \qquad\qquad 1 - A_A' = \exp[-D\{\overline{V} + (t\overline{S}/4)\}] \equiv \theta \; , \text{ say.}$$

Next we derive (2b). We find $L_A'$ by application of the classical stereological formula "$L_A' = (\pi/2) \, N_L'$ ", where $N_L'$ relates to a line section of $X'$ . To determine $N_L'$ , we consider the number $m$ of particles hitting a rectangle $R$ of width $t$ and length $T \gg 1$ , representing an orthogonal section of the foil of length $T$ . Since $T \gg 1$ , we may assume each hitting particle projects onto each rectangle side of length $T$ as a segment. Since each segment end point contributes towards $N_L'$ with probability $\theta$ , we have

(18)      $$2\theta \ E(m)/T \to N_L' \quad \text{as} \quad T \to \infty \ .$$

For $R$ , $V = 0$ , $S = 2tT$ and $M = (t+T)/2$ [4; (2.29T)] and so, by the Theorem,

(19)      $$E(m) \sim D\{(\overline{S}/4)+t\overline{M}\}T \quad \text{as} \quad T \to \infty \ ,$$

and (2b) follows.

Finally we derive (2c), adopting the following terminology: *portion* for that part of a particle within the foil, *image* for the projection of a portion onto the observation plane, and *vertex* for the intersection point of the boundaries of a pair of images when this point is uncovered by any other image. Now

(20)      $$C_A' = C_A^{(0)} + C_A^{(1)} \ ,$$

where the addends are the contributions to $C_A'$ from the vertices and image boundaries, respectively. Clearly

(21)      $$C_A^{(0)} = - \ (\pi/2) \ N_A' \ ,$$

where $N_A'$ is the number of vertices per unit area, and

(22)      $$C_A^{(1)} = E(\kappa) \ L_A' \ ,$$

where $\kappa$ is the curvature at a uniform random point of the boundary of $X'$ .

First we find $N_A'$ . In the following, the symbols $\pm$ and $o$ refer to the two foil faces and the foil interior, respectively. We describe the intersection of a particle boundary with the $\pm$ faces as a $\pm$ arc (which it should be noted is closed). Again, for a particle intersecting the foil we describe a point on the boundary of the corresponding portion at which the tangent plane is perpendicular to the foil as a tangent point, and describe the arc of projection of the set of tangent points as the associated $o$ arc (which is not necessarily closed). The aggregates of $\pm$ and $o$ arcs are (dependent) planar Poisson aggregates of essentially the same type as the basic spatial particle aggregate, and we now investigate their densities $\rho$ (corresponding to $D$ ) and mean lengths $\overline{L}$ . We have

(23)      $$\rho_+ = \rho_- = D\overline{M} \equiv \rho_\pm \ .$$

The mean perimeter of the section of a convex set by an IUR plane is $\pi S/4M$

[4; (2.31T)]. Since the 'probability' each foil face hits a particle is $\propto M$, we have

(24) $$\bar{L}_{+} = \bar{L}_{-} = E\left(\frac{\pi S}{4M} \cdot M\right)/E(M) = \pi\bar{S}/4\bar{M} \equiv \bar{L}_{\pm} .$$

The determination of $\rho_o$ and $\bar{L}_o$ is not so simple; however, fortunately all we need is their product — the arc length per unit area. Since the mean perimeter of the projection of a convex set onto an isotropic random plane is $\pi M$, we have

(25) $$\rho_o \bar{L}_o = tD \pi \bar{M} .$$

Now each vertex is of one of the six types ++, --, oo; -o, o+ and +- ; corresponding to the two associated particles. It is easily shown that the corresponding densities of arc-arc intersections per unit area are $\pi^{-1} \rho_+^2 \bar{L}_+^2, \dots, \dots; 2\pi^{-1} \rho_-\rho_o \bar{L}_-\bar{L}_o, \dots, \dots$ . Whether such an arc-arc intersection is a vertex depends on

(i) whether, for the relevant point on a $\pm$ arc, it is within or without the projection of the associated portion (each case has probability $\frac{1}{2}$); and

(ii) whether the relevant segment of length $t$ is hit by any other particle.

Taking all this into account, we have

(26) $$N_A' = \pi^{-1} \left(\rho_\pm \bar{L}_\pm + \rho_o \bar{L}_o\right)^2 \theta .$$

Finally we determine $E(\kappa)$ . For a single fixed closed arc,

(27) $$E(\kappa) = \oint \kappa \, ds/\oint ds = 2\pi/L .$$

Hence for an aggregate of such arcs

(28) $$E(\kappa) = \sum \left(2\pi/L_i\right) L_i/\left(\sum L_i\right) = 2\pi/\bar{L} .$$

As $t$ increases, the $o$ arcs become closed and so (28) applies, yielding $E(\kappa) = 2/\bar{M}$ for the $o$ arcs. It follows, taking $\pm$ and $o$ arcs with weights $\frac{1}{2}$ and $1$, respectively, that

(29)
$$E'(\kappa) = \frac{\rho_{\pm} \bar{L}_{\pm} \left(2\pi/\bar{L}_{\pm}\right) + \rho_o \bar{L}_o (2/\bar{M})}{\rho_{\pm} \bar{L}_{\pm} + \rho_o \bar{L}_o}$$

$$= 8(t+\bar{M})/(\bar{S}+4t\bar{M}) \ .$$

(2c) now follows from (20)-(29).

REFERENCES

[1]  DeHoff, R.T., 1968. Quantitative microstructural analysis. *ASTM STP.* 430, *Am. Soc. Testing Mats.*: 63-95.

[2]  Giger, H., 1972. Grundgleichungen der Stereologie II. *Metrika* 18: 84-93.

[3]  Giger, H. and Hadwiger, H., 1968. Über Treffzahlwahrscheinlichkeiten in Eikörperfeld. *Z. Wahrscheinlichkeitstheorie verw. Geb.* 10: 329-334.

[4]  Miles, R.E., 1969. Poisson flats in Euclidean spaces. Part I: A finite number of random uniform flats. *Adv. Appl. Prob.* 1: 211-237.

[5]  Santaló, L.A., 1936. *Integralgeometrie 5: Über das kinematische Mass im Raum.* Hermann, Paris (Act. Sci. Indust. No. 357).

[6]  Underwood, E.E., 1970. *Quantitative Stereology.* Addison-Wesley (Reading, Mass.).

*National Bureau of Standards Special Publication 431*
*Proceedings of the* FOURTH INTERNATIONAL CONGRESS FOR STEREOLOGY
*held at* NBS, Gaithersburg, Md., September 4-9, 1975 (Issued January 1976)

COMPUTATIONAL METHODS IN STEREOLOGY

by R. S. Anderssen and A. J. Jakeman

*Computer Centre, Australian National University, Canberra, ACT, 2600, Australia*

INTRODUCTION

A classical problem in stereology is the determination of three-dimensional
(structural) properties of an α-phase in a β-matrix from two-dimensional observations of the α-phase. Although numerous numerical methods have been proposed
for the approximate solution of the resulting mathematical formulations (such as
the random spheres and thin section models), when they exist and are known, none
appear to be completely satisfactory. Little attention appears to have been
paid to the actual classification of the mathematical formulations involved, and
the use of appropriate results from the numerical analysis literature for the
numerical solution of the resulting classes. A reason for this was the lack,
until recently, of suitable results in the literature.

This situation has now changed since many of the questions concerning the
construction of basic methods in computational mathematics have been resolved.
For this reason, a systematic and comparative study of the different classes of
methods available for the different types of stereological estimations (invers-
ions) is now possible. For the estimation of the size distribution of spheres
from planar or thin section observations of their circular contours, the following classes of methods can be identified: (1) Degradation of Problem Formulation; (2) Non-parametric Methods; (3) Parametric Methods; (4) Evaluation of
Inversion Formulas; (5) Regularization Methods; and (6) Filtering Methods.

There is little justification for the degradation of the problem formulation (when available) such as the use of the Random Spheres Model {see, for
example, Baudhuin and Berthet (1967)} to determine the size distribution of
spheres from thin sectional observations of circles. The use of degraded
problem formulation leads to the necessity to adjust for its effect which can be
avoided, if the correct formulation is used. Thus, the necessity to correct for
the Holmes effect is avoided, if the Thin Section Model is used for thin section
data {see, for example, Jakeman and Anderssen (1975a)}.

It follows from the systematic study of methods by Anderssen and Jakeman
(1974) and Jakeman and Anderssen (1974) that, except under special circumstances, the use of non-parametric and parametric methods as well as the evaluation of inversion formulas can behave unsatisfactorily as a consequence of the
following factors: stereological data is observational and therefore non-exact,
and the mathematical formulations involved are improperly posed (small perturbations of the data can correspond to arbitrarily large perturbations of the
solution). The non-parametric methods do not take account of the improperly
posed nature of the formulation and were developed for exact rather than non-
exact data. They can be used to yield results, but at the expense of accuracy
in order to control error growth. For example, when using finite difference
methods, the choice of the class width (grid spacing) must be a trade-off of
accuracy against error growth by ensuring that it is not taken to be too small.
As indicated by Anderssen (1975), the same comment applies to the use of direct
numerical methods to evaluate inversion formulas.

The difficulty posed by the parametric methods, and the use of pseudo-
analytic methods to evaluate an inversion formula, is the choice of functions
(distributions) which model correctly (not approximately) the structure in the
data. In general, this can only be done approximately, since the level of
information about the underlying structure is virtually non-existent. Even in
the exceptional situation where it is possible, this does not necessarily cir-
cumvent the difficulty associated with the improperly posed nature of the

13

formulation.

   Although regularization methods {see, for example, Lavrentiev (1967)}
stabilize the solution of an improperly posed problem formulation, they were not
designed to filter the error from the signal in non-exact data, whereas the
filtering methods were designed to do the latter and not the former. It is
clear then that satisfactory numerical methods must combine the advantages of
both the regularization and filtering methods at the expense of their disadvant-
ages. One possibility is the application of a filtering method to the data
followed by the use of a regularization method to invert the appropriate problem
formulation for the filtered data. It is not necessary however to limit attention to such a simplistic implementation of these two methods. Other forms for
the problem formulations, such as inversion formulas, could be used when known.
Methods which implicitly combine the advantages of both regularization and
filtering should be sought.

   The methods proposed recently by Anderssen and Jakeman (1974) and Jakeman
and Anderssen (1974) aim to achieve this through the evaluation of known inversion formulas using product integration and spectral differentiation. In order
to illustrate, consider the thin section model for spherical particles

$$c(y) = \frac{1}{2a + t} \left\{ 2y \int_y^\infty s(x) [x^2 - y^2]^{-\frac{1}{2}} dx + t\, s(y) \right\} , \qquad (1)$$

where $s(x)$ denotes the probability density for the radii of the spheres, $c(y)$
$= dC(y)/dy$ the probability density for the radii of the circles in thin sections
of thickness $t$, and $a = \int_0^\infty x\, s(x)\, dx$ is the expectation of the radii of the
spheres. For the model, explicit inversion formulas are known, including

$$s(x) = - [2/\pi]^{\frac{1}{2}} \frac{(2a + t)}{t} \frac{d}{dt} \int_x^\infty f\{[2\pi(y^2 - x^2)]^{\frac{1}{2}}/t\}\, C(dy) , \qquad (2)$$

where

$$f(w) = \sqrt{2\pi} \exp(w^2/2) \left\{1 - \frac{1}{\sqrt{2\pi}} \int_{-\infty}^w \exp(-u^2/2) du\right\} .$$

   Using product integration and spectral differentiation, the method proposed
in Anderssen and Jakeman (1974) was:

   (a)   Evaluate, using product integration

$$\hat{V}(x_k) = \int_{x_k}^{y_n} f\{[2\pi(y^2 - x_k^2)]^{\frac{1}{2}}/t\}\, (dC_L/dy) dy \qquad (3)$$

where $\{y_k\}$ $(k=0,1,2,...,n)$ denotes the non-uniform grid on which the data
is given, $\{x_k\}$ $(k=0,1,2,...,N)$ the uniform grid onto which the data is
integrated, and $C_L(y)$ a localized Lagrange interpolation smoothing of the data
$\{C(y_k)\}$ {see, Anderssen and Jakeman (1975), for details}.

   (b)   Use spectral differentiation to determine

$$\hat{s}(x_k) = - [2/\pi]^{\frac{1}{2}} \frac{(2a + t)}{t} \frac{d}{dx} [\{\hat{V}(x_k)\}]_{x=x_k} . \qquad (4)$$

## COMPARISON OF METHODS

   Using the exact   (effect of sampling not included) synthetic data of
Anderssen and Jakeman (1974), Jakeman and Anderssen (1974) have compared the
above method with the non-parametric method of Bach (1967) and the parametric
method of Keiding et al. (1972). Their graphs clearly illustrate the superior-
ity of the above method over these two and exemplify the above general comments
about parametric and non-parametric methods. In Figure 1, we compare the above
method with a pseudo-analytic method {of the type discussed by Piessens and
Verbaeten (1973) for an Abel equation in a non-stereological context} for (2)
based on an exact inversion result of Bach (1966), §§ VI, VII, : viz., if

$$(2a + t) \; c(y) = p_\theta(y) \sum_{i=0}^{m} a_i \; \theta^i \; y^{2i} \; , \quad \theta > 0 \; , \tag{5}$$

then

$$s(x) = p_\theta(x) \sum_{i=0}^{m} A_i \; \theta^i \; x^{2i} \; . \tag{6}$$

Here $p_\theta(y) = 2\theta y \exp(-\theta y^2)$ with the parameter $\theta$ estimated (using a maximum likelihood argument) by

$$\theta = n / \sum_{i=0}^{n} y_i^2 \; . \tag{7}$$

The constants $a_i$ and $A_i$ are related by

$$A_i = [\theta^{\frac{1}{2}} a_i - (1 - \delta_i^m) \sum_{j=i+1}^{m} \binom{j}{i} \Gamma(j - i + \tfrac{1}{2}) A_j] / (t\theta^{\frac{1}{2}} + \pi^{\frac{1}{2}}) \; , \tag{8}$$

$$i = m, m-1, \dots, 1, 0 \; ,$$

with $\Gamma(x)$ denoting the gamma function with argument $x$ , and $\delta_i^m$ equal to one when $m = i$ and zero otherwise.

The actual computational implementation was: (a') For given $m$ , fit the representation (5) to the data $\{(2a + t) \; c(y_k)\}$ $(k=0,1,2,\dots,n)$ using linear least squares. (b') Substitute in (8) the $\{a_i\}$ $(i=0,1,2,\dots,m)$ obtained to find the corresponding $\{A_i\}$ , and hence the representation (6) for $s(x)$ .

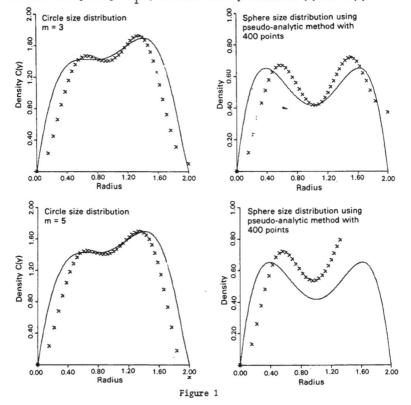

Figure 1

It is clear from Figure 1 that there exist data for which this pseudo-analytic method is unable to yield a satisfactory solution. Potentially, it will only work for data which have the structure of (5) with m finite. The solution obtained from the proposed method (using the same exact data) was quite close to the exact solution {see Figure 1(b) of Jakeman and Anderssen (1975a)}.

In the comparisons mentioned above, exact synthetic data (with sampling effects neglected) were used. When synthetic sample cumulative data are used, methods will not perform as well as when exact synthetic data are used. This is a consequence of the properties of the sample cumulative rather than the methods. Thus, methods which fail for exact data will also fail for the sample cumulative data, while methods which yield satisfactory results for the exact will exhibit degraded performance on the sample cumulative. For the last mentioned class of methods, however, the degraded performance can be improved by increasing the number of data points. This is again a consequence of the properties of the sample cumulative data (when the method under consideration is stable). It is best explained by observing that the sample cumulative data $H_n(y) = \{y_i \rightarrow i/n ;$ $i=1,2,3,...,n\}$ sampled from $H(y)$ have variance satisfying

$$\mathrm{var}[H_n(y)] = \frac{1}{n} H(y) (1 - H(y)) .\tag{9}$$

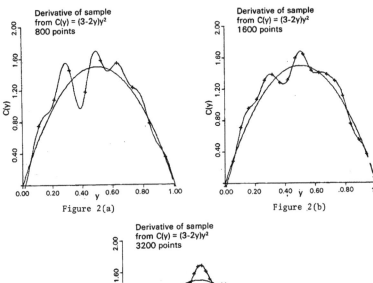

Figure 2(a)  Figure 2(b)

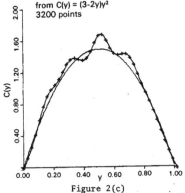

Figure 2(c)

Consequently, the error between $H_n(y)$ and $H(y)$ behaves like $1/\sqrt{n}$ .

The nature of the problem can be fully illustrated by considering the numerical differentiation of sample cumulative data sequences collected from the cumulative distribution function $C(y) = (3 - 2y)y^2$ on the interval $[0,1]$. The spectral derivative of one such sample cumulative is shown in Figures 2(a), (b) and (c). For the same amounts of exact synthetic data, the spectral derivatives were graphically indistinguishable from the exact solution shown. The improvement resulting in the accuracy of the spectral derivative through increasing the number of points in the sample cumulative can be seen from a comparison of Figure 2(a) with 2(b) and 2(c).

In implementing the spectral methods, other points which require future attention are the use of the even ordinate grid of size distribution data, and the grouping of size distribution data onto an even abscissae grid.

## TRUNCATED MOMENTS FOR THIN SECTION DATA

Some of the numerical difficulties posed by the estimation of three dimensional particle size distribution functions can often be avoided when only linear properties of the size distributions are required. A general procedure has been developed by Nicholson (1970) for estimating such linear properties from given truncated data. It requires that the particles be completely specifiable by one parameter. It has been applied to spheres as well as to a certain class of cylinders with both sampled by planar sections. The theory also extends to observations in a thin section, though its specific application to thin section data has yet to be published.

When an inversion formula is known, which is the case for the thin section model for spherical planar data, the approach used by Jakeman and Anderssen (1975b) for deriving linear estimators of the Nicholson type from planar sections of random spheres can be applied. Using the inversion formula, they first transform the linear property into a linear functional defined on the size distribution function of the observations. Next, they stabilize the actual evaluation of the functional on the observational (cumulative) data $\{C(y_k)\}$ by evaluating it, using product integration, on a localized Lagrange interpolation smoothing $C_L(y)$ of data. For example, when using thin section data for spherical particles, the truncated zeroth moment

$$M_0^* = \int_{y_0}^{\infty} s(y)dy \qquad (10)$$

transforms using the inversion formula (2) to the functional

$$M_0^* = [2/\pi]^{\frac{1}{2}} \frac{(2a + t)}{t} \int_{y_0}^{\infty} f\{[2\pi(y^2 - y_0^2)]^{\frac{1}{2}}/t\} \, c(y)dy \ . \qquad (11)$$

Thus, using an appropriate form for $C_L(y)$ {see, for example, Jakeman and Anderssen (1975b)}, the estimation of $M_0^*$ reduces to the evaluation of

$$\hat{M}_0^* = [2/\pi]^{\frac{1}{2}} \frac{(2a + t)}{t} \int_{y_0}^{y_n} f\{[2\pi(y^2 - y_0^2)]^{\frac{1}{2}}/t\} \, C_L'(y)dy \qquad (12)$$

using product integration, where $C_L'(y) = d \, C_L(y)/dy$ .

These stabilized estimators have been shown to be better than those previously proposed {see Anderssen and Jakeman (1975)}.

## REFERENCES

R.S. Anderssen and A.J. Jakeman (1974) On computational stereology, *Proc. 6th. Aust. Comp. Conf.*, Sydney, Vol. 2, 353-362.

R.S. Anderssen and A.J. Jakeman (1975) Product integration for functionals of particle size distributions, *Utilitas Mathematica* (in press).

R.S. Anderssen (1975) A stable procedure for the inversion of Abel's equation, *JIMA* (to appear).

G. Bach (1966) *Zuffallsschnitte durch Haufwerke von Korpen*, (Habilitationsschrift), Braunschweig.

G. Bach (1967) Size distribution of particles derived from the size distribution of their sections. In *Stereology*, Elias, H. (ed.), Springer Inc., New York, 174-186.

P. Baudhuin and J. Berthet (1967) Electron microscopic examination of subcellular fractions II: Quantitative analysis of the mitochondrial population isolated from rat liver, *J. Cell. Biol.* 35, 631-648.

A.J. Jakeman and R.S. Anderssen (1974) A note on numerical methods for the thin section model, *Electron Microscopy* 1974, Vol. 2, 4-5.

A.J. Jakeman and R.S. Anderssen (1975a) On optimal forms for stereological data, *Fourth International Congress for Stereology*, September 1975, National Bureau of Standards, Gaithersburg, Maryland.

A.J. Jakeman and R.S. Anderssen (1975b) Abel type integral equations in stereology I: General Discussion, *J. Microsc.* (to appear).

N. Keiding, S. Tolver Jensen, and L. Ranek (1972) Maximum likelihood estimation of the size distribution of liver cell nuclei from the observed distribution in a plane section, *Biometrics* 28, 813-829.

M.M. Lavrentiev (1967) *Some Improperly Posed Problems of Mathematical Physics*, Springer Tracts in Natural Philosophy, Vol. 11, Springer, Berlin.

W.L. Nicholson (1970) Estimation of linear properties of particle size distributions, *Biometrika* 57, 273-298.

R. Piessens and P. Verbaeten (1973) Numerical solution of the Abel integral equation, *BIT* 13, 451-457.

*National Bureau of Standards Special Publication 431*
*Proceedings of the* FOURTH INTERNATIONAL CONGRESS FOR STEREOLOGY
*held at* NBS, Gaithersburg, Md., September 4-9, 1975 (Issued January 1976)

ESTIMATION OF LINEAR FUNCTIONALS BY MAXIMUM LIKELIHOOD

by W. L. Nicholson
*Pacific Northwest Laboratories, A Division of Battelle Memorial Institute, Richland, Washington 99352,*
*U.S.A.* .

## 1. INTRODUCTION

A population of spherical particles, with cumulative size distribution $G(x)$, is randomly distributed in an infinite three-dimensional matrix. The position of particle centers is defined by a Poisson field. $N_V$ is the volume number density of sphere centers (centers per unit volume). Thus, $N_V G(x)$ is the volume number density contribution from spheres of diameter $\leq x$. $\mu$ is the mean sphere diameter, assumed to be finite. This model is an idealization for physical situations where the sphere density is low (so there is small chance for crowding) and spatially homogeneous.

A classical problem of stereology is the estimation of linear functionals of the form

$$\theta = N_V \int_0^\infty \ell(x) G(dx) \qquad (1)$$

from linear or planar probe data. Nicholson [5] describes an estimation theory for a more general situation of geometrically similar particles distributed in a finite matrix. The theory is a mathematical statement of an entire class of stereology problems, treated more or less independently in the literature. Results are extended to the Poisson field infinite matrix in [6]. For simplicity we consider here spheres and planar probes. More general treatment is possible, but mathematically tedious.

A planar probe of finite area A is inserted into the matrix to obtain a sample of n circular intersections. The measured diameters of these intersections $y_1 < y_2 < \ldots < y_n$ are the event points for a realization of a generalized Poisson process $\{n(y) | 0 \leq y\}$. The process is defined by: i) $n(0) = 0$ with probability one; ii) non-overlapping increments are independent; iii) and for $0 \leq y' < y''$ the increment $n(y'') - n(y')$ is Poisson distributed with mean

$$E\{n(y'') - n(y')\} = N_A[H(y'') - H(y')]. \qquad (2)$$

In (2) $N_A = \mu N_V$ is the area number density (circle centers per unit area) on the probe and

$$H(y) = 1 - \mu^{-1} \int_0^\infty (x^2 - y^2)^{1/2} G(dx). \qquad (3)$$

The contribution to area number density from circles of diameter $\leq y$ is $N_A H(y)$. The measured diameters are an ordered random sample with cumulative distribution function $H(y)$. The sample size is $n = n(+\infty)$. Differentiation of (3) gives the familiar Abel's integral equation relationship between the circle frequency function $h(y) = H'(y)$ and the sphere frequency function $g(x) = G'(x)$ first derived by Wicksell [9].

---
*Research sponsored by U.S. Energy Research and Development Administration under contract E(45-1):1830.

From [5] and [6] an unbiased estimate of the functional (1) is

$$\hat{\theta} = A^{-1} \int_0^\infty m(y)n(dy),$$ (4)

where $m(y)$ is a solution to the integral equation

$$\int_0^x (x^2-y^2)^{-1/2} ym(y)y dy = \ell(x).$$ (5)

Uniqueness (non-existence of the solution) implies that (4) is the only (there is no) linear estimate which is unbiased over the class of all discrete and absolutely continuous $G(x)$. Watson [8] points out that in many applications the variance of (4) is infinite, even though the estimator is consistent. The estimator of $\rho(x_0) = N_V[1-G(x_0)]$ is such a case. Here the solution to (1) is unique. The estimator is

$$\hat{\rho}(x_0) = (2/A\pi) \sum_{y_i > x_0} (y_i^2 - x_0^2)^{-1/2}.$$ (6)

With $x_0 = 0$, (6) is Fullman's formula [4] for estimating $N_V$.

Anderssen and Jakeman [1] consider a family of product integration formulas which are smoothed versions of (4). Their simplest formula, a trapezoidal rule, is

$$\hat{\theta}_{AJ} = A^{-1} \sum_{i=1}^{n-1} (y_{i+1} - y_i)^{-1} \int_{y_i}^{y_{i+1}} m(y)dy.$$ (7)

A factor $1/n$ is replaced by $1/A$ in (7) since Anderssen and Jakeman estimate functionals of $G(x)$ not $N_V G(x)$. With appropriate regularity conditions, Anderssen and Jakeman show that their estimators have finite variance and are consistent. Tallis [7] employs a special case of product integration formulas to grouped data to get finite variance estimates of $N_V$ and $1-G(x)$.

An alternative form of smoothing the estimate (4) is to construct a smooth estimate $N_A \hat{H}(y)$ from the sample function $n(y)$. The estimate of (1) is (4) with $A N_A \hat{H}(dy)$ substituted for $n(dy)$. One method of construction is to use the probability structure of the data and the theory of maximum likelihood estimation. Section 2 is an outline of the method. Section 3 is a discussion of relative merits of the several approaches. Section 4 is an example of maximum likelihood applied to estimation of $N_V[1-G(x)]$. Data are collected at several magnifications, a complicating feature for other approaches, but not for maximum likelihood.

2. MAXIMUM LIKELIHOOD ESTIMATION

Consider $k$ fixed cells with boundaries $0 = z_0 < z_1 < \ldots < z_k$. The number of circle diameters in the $j^{th}$ cell is $\Delta n_j = n(z_j) - n(z_{j-1})$. Because of the generalized Poisson process character of the data, $\Delta n_1, \Delta n_2, \ldots, \Delta n_k$ are mutually independent Poisson random variables with means, respectively,

$$E\Delta n_j = AN_A \int_{z_{j-1}}^{z_j} H(dy) = AN_A h(\overset{\vee}{z}_j)\Delta z_j.$$ (8)

In (8) $\overset{\sim}{z}_j$ is a specific point satisfying $z_{j-1} < \overset{\sim}{z}_j < z_j$ and $\Delta z_j = z_j - z_{j-1}$. The observed frequencies in the form $\Delta n_j / A\Delta z_j$ (number of circles per unit area per diameter unit) are estimates of the areal density function, say,

$$f(\overset{\sim}{z}_j) = N_A h(\overset{\sim}{z}_j). \tag{9}$$

Suppose that $f(z) = f(z|\phi)$ is known up to a p-dimensional parameter $\phi = (\phi_1, \phi_2, \ldots, \phi_p)$. Then maximum likelihood theory provides a method for estimating $\phi$. The likelihood function for the observed Poisson frequencies is

$$L(n|\phi) = \prod_{j=1}^{k'} e^{-B_j(\phi)} [B_j(\phi)]^{\Delta n_j} / \Delta n_j!. \tag{10}$$

In (10) $B_j(\phi) = A\Delta z_j f(z_j'|\phi)$. Because $\overset{\sim}{z}_j$ is unknown, we approximate by the cell midpoint $z_j' = (z_j + z_{j-1})/2$. The maximum likelihood estimate of $\phi$ is that value $\hat{\phi}$ which maximizes (10). In practice $\hat{\phi}$ is taken as a solution to the likelihood equations

$$S_\alpha(n|\phi) = \frac{\partial \log_e L(n|\phi)}{\partial \phi_\alpha} = \sum_{i=1}^{N} \frac{\Delta n_i - B_i(\phi)}{B_i(\phi)} \frac{\partial B_i(\phi)}{\partial \phi_\alpha} = 0 \tag{11}$$

with $\alpha = 1, 2, \ldots, p$.

A maximum likelihood smoothed estimate of the linear functional (1) is $\hat{\theta}_M = \theta_M(\hat{\phi})$ where

$$\theta_M(\phi) = \int_0^{+\infty} m(y)f(y|\phi)dy. \tag{12}$$

With appropriate regularity conditions on $f(z|\phi)$ (see e.g., [10]) $\hat{\phi}$ is asymptotically distributed as multivariate normal with mean $\phi_0$, the true value of $\phi$, and covariance matrix $\{AN_A C_{pq}(\phi_0)\}^{-1}$ where

$$C_{pq}(\phi) = \int_0^{+\infty} \cdots \int_0^{+\infty} S_p(n|\phi)S_q(n|\phi)L(n|\phi)d\Delta n. \tag{13}$$

Further,

$$\chi^2(\phi_0) = -2 \log_e[L(n|\phi_0)/L(n|\hat{\phi})] \tag{14}$$

is asymptotically distributed as chi-square with p degrees of freedom. In these limiting results $AN_A$ plays the role of the large sample size. That is, the distributional results are approximately true for $N_A$ fixed and A sufficiently large. Let $T(\hat{\phi}|p,\alpha)$ be the set of $\phi$ values which satisfy the inequality, $\chi^2(\phi) < \chi^2(p,\alpha)$. Here, $\chi^2(p,\alpha)$ is the 100 $\alpha^{th}$ percentile for the chi-square distribution with p degrees of freedom. Let $\underline{\theta}_M$ and $\overline{\theta}_M$ be the minimum and maximum value of (14) over the set $T(\hat{\phi}|p,\alpha)$. Then $(\underline{\theta}_M, \overline{\theta}_M)$ is a $100\beta\%$ confidence interval for the linear functional $\theta$ of (1) where $\alpha \leqq \beta \leqq 1$ whenever A is sufficiently large.

## 3. DISCUSSION

The present status of linear functional estimation in stereology is that many classical procedures exist (that is, those that reappear periodically in the many application oriented journals of the diverse specific fields where stereological methods are used) which are in the linear category of (4). In actual practice the properties of these estimators are not as bad as mathematical theory predicts, though admittedly poor in some instances. Natural truncation of small features by observational thresholds removes infinite variance stigma from reciprocal type estimators, like Fullman's density formula. Variances are known specifically and can be estimated from the data with the Poisson assumption. But, now there is a negative bias because the contribution from small features is omitted.

Grouping of data into cells by observational instruments and use of a single cell point as representative gives finite known variance, again at the expense of bias. With many observations so cells can be narrow, the bias is probably tolerable but, without investigation, specifically not known.

The Anderssen and Jakeman formulas are an attractive alternative which in many situations give more stable estimates. These formulas are non-linear in the y's and hence, probabilistically speaking, complicated. Exact variances, and hence, variance estimates are not yet available. Maximum likelihood includes a theory for asymptotic variance estimation, which is a plus. However, the dependence on a family of smooth frequency functions $f(z|\phi)$ to represent the reality of areal density is a minus. In practical situations the classical distributions don't seem to apply. Even lognormal, when appropriate for actual particle size, is not correct for observed intersections. Often competing evolutionary processes give rise to compound multi-modal distributions. Work is in progress on low order spline functions which may offer a general solution.

A tremendous plus for maximum likelihood is its proper weighting of information in the cell frequencies. Other methods lack this sophistication. Weighting is critical when frequencies range from zero to several powers of ten, for example, when data are a composite of observations from several microscopes using distinct magnifications. Composite observation is necessary when feature sizes range over several orders of magnitude. In an overlap region there are a few large features on the right tail of higher magnification data and many small features on the left tail of lower magnification data. With insufficient experimental planning there may be voids in the middle of the data. Large right tail features completely disappear while left tail features are still too small to be observed. Another complicating aspect of composite observation is that the observation area for high magnification is often a sub area for that of low magnification. Maximum likelihood can be made to handle all of these situations if a family $f(z|\phi)$ is available.

4.  AN EXAMPLE

In connection with an experimental characterization of $UO_2$ fuel [3], planar probe data were collected on near-spherical pores distributed in as-fabricated fuel pellets using four distinct magnifications. Higher magnification (21,000X and 6440X) fields were observed with a scanning electron microscope over a size range of 0.05 to 5.0 micrometers. Lower magnification (850X and 61X) fields were observed with an optical microscope over a size range of 1.0 to 100.0 micrometers. The features contained on 76 micrographs were measured using a Zeiss Particle Size Analyzer. For simplicity of presentation, data from the Zeiss 48 exponential scale cells were combined into 24 cells. Pore frequency vs. size data are given in Table 1 as a function of magnification. Arrows in the first three columns indicate the beginning points for the next lower magnification data. Figure 1 is a plot of the area density estimates ($\Delta n/A\Delta z$) of Section 2 versus cell midpoint $z'$.

These data are used to illustrate maximum likelihood estimation of $N_V[1-G(3.0)]$, the contribution to the density from pores of diameter $> 3.0$ micrometers. To obtain this estimate, it is sufficient to model the right tail of the size distribution, say starting at 1.0 micrometer. For that region, a quadratic relationship between log density and log diameter seems appropriate from Figure 1. Thus, the parametric family of Section 2 was taken as

$$f(z|\phi) = \alpha z^{\beta+\gamma \ln z} \qquad 1.0 < z < 100.0 \qquad (15)$$

22

where $\phi = (\alpha, \beta, \gamma)$ is the unknown parameter and z is intersection diameter.

TABLE 1. PORE FREQUENCY VS SIZE DATA
ON PLANAR SECTIONS OF UO$_2$ FUEL PELLETS

| | MICROSCOPE | | | |
| --- | --- | --- | --- | --- |
| | SEM | SEM | OM | OM |
| | MAGNIFICATION | | | |
| | 21000X | 6440X | 850X | 61X |
| CELL MIDPOINT (mm)* | MICROGRAPH AREA MEASURED (mmxmm) | | | |
| | 1008640 | 504320 | 504320 | 378240 |
| 1.18 | 2 | 360 | 1415 | 232 |
| 1.37 | 4 | 384 | 701 | 50 |
| 1.57 | 11 | 499 | 484 | 41 |
| 1.81 | 23 | 599 | 334 | 24 |
| 2.08 | 38 | 648 | 194 | 17 |
| 2.41 | 64 | 652 | 151 | 6 |
| 2.77 | 58 | 633 | 78 | 5 |
| 3.20 | 81 | 510 | 63 | 0 |
| 3.63 | 89 | 403 | 38 | 2 |
| 4.18 | 90 | 320 | 26 | 0 |
| 4.77 | 122 | 209 | 13 | 0 |
| 5.43 | 131 | 132 | 7 | 0 |
| 6.19 | 138 | 69 | 6 | 0 |
| 7.05 | 151 | 37 | 4 | 1 |
| 8.03 | 151 | 18 | 1 | 0 |
| 9.15 | 125 | 15 | 1 | 0 |
| 10.42 | 100 | 13 | 0 | 0 |
| 11.87 | 92 | 6 | 0 | 0 |
| 13.53 | 56 | 4 | 0 | 0 |
| 15.41 | 39 | 3 | 0 | 0 |
| 17.56 | 22 | 1 | 0 | 0 |
| 20.00 | 9 | 0 | 0 | 0 |
| 22.39 | 4 | 1 | 0 | 0 |
| 25.96 | 2 | 0 | 0 | 0 |

\* AS MEASURED ON MICROGRAPH USING
ZEISS PARTICLE SIZE ANALYZER

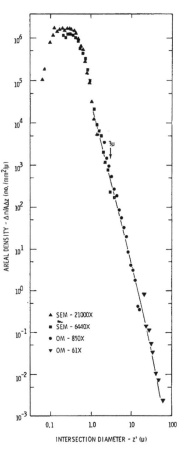

FIGURE 1. AREAL DENSITY OF PORES ON PLANAR SECTIONS OF UO$_2$ FUEL PELLETS

The products in the likelihood function (10) and the likelihood equations (11) involve a double indexing to handle the composite nature of the data. Let $\Delta n_{ij}$ (i=1,2,3,...,24; j=1, 2,3,4) be the frequency in the $i^{th}$ cell for the $j^{th}$ magnification. The products are taken over all index pairs (i,j) with cell midpoints $z'_{ij}$ satisfying $1.0 \leqq z'_{ij} \leqq 100.0$. Also,

$$B_{ij}(\phi) = A_j \Delta z_{ij} \, f(z'_{ij}|\phi) \quad \text{where} \quad z_{ij} = d_i/M_j.$$

Here, $A_j$ is the total area observed for the $j^{th}$ magnification. $M_j$ is the $j^{th}$ magnification factor, and $d_i$ is the upper boundary of the $i^{th}$ Zeiss cell.

The maximum likelihood equations were solved with LEARN program [2]. LEARN is a general purpose non-linear estimation program which uses second degree Taylor expansion logic. Initial parameter guesstimates came from least squares regression of the logarithm of non-zero areal density data on the quadratic polynomial in log-diameter. The specific maximum likelihood estimate of $\phi$ is

$$\hat{\alpha} = 27930.0 \pm 3060.0$$
$$\hat{\beta} = -3.43 \pm 0.14$$
$$\hat{\gamma} = -0.147 \pm 0.037.$$

The precisions are standard deviation estimates from asymptotic theory. The smoothed function $f(z|\hat{\phi})$ is plotted as the curve in Figure 1.

Substitution of $f(z|\hat{\phi})$ into (12) gives the maximum likelihood smoothed estimate of the density contribution from pores of diameter > 3.0 micrometers as

$$N_V[1-G(x_0)] = (2\hat{\alpha}/\pi) \int_{x_0}^{x_1} (y^2-x_0^2)^{-1/2} y^{\hat{\beta}+\hat{\gamma}\ln y} \, dy, \qquad (16)$$

with $x_0 = 3.0$. Since the data end at $x_1 = 100.0$ micrometers, the estimate is truncated at that point. With the contribution near $x_1$ down a factor of $10^7$ from that near $x_0$, and since the observed areal density has a negative slope near $x_1$, the density estimate is clearly not dependent on any detail at or beyond $x_1$.

The final estimate is

$$N_V[1-G(3.0)] = 1.46 \times 10^8 \ (1.35 \times 10^8 - 1.58 \times 10^8) \ \text{pores/cc.}$$

The approximate 95% confidence interval from likelihood contours and asymptotic theory is given in parenthesis.

## 5. ACKNOWLEDGMENTS

The author is indebted to J. L. Daniél and H. A. Treibs for discussion which helped formulate the mathematical model for microscopy data. B. H. Duane obtained the solution to the maximum likelihood equations.

## REFERENCES

[1] Anderssen, R. S. and A. J. Jakeman, "Product Integration for Functionals of Particle Size Distributions", to be published in Utilitas Mathematica.

[2] Duane, B. H., Maximum Likelihood Non-linear Correlated Fields (BNW Program LIKELY), September, 1967, BNWL-390, 46-49.

[3] Freshley, M. C., P. E. Hart, J. L. Daniel, D. W. Brite, T. D. Chikalla, Proceedings of Joint Topical Meeting on Commercial Nuclear Fuel Technology Today, Toronto, Canada, April 28-30, 1975, CNS ISNN 0068-8517, 2-106--2-126.

[4] Fullman, R. L., (1953), J. Metals, 5, 447-52.

[5] Nicholson, W. L., (1970), Biometrika, 57, 273-297.

[6] Nicholson, W. L., (1970), "Sampling and Estimation Theory for Thinly Distributed Particulate Materials", presented at the Symposium on Sampling Theory in Characterization of Bulk Materials Under Japan-U.S. Cooperative Science Program, Honolulu, Hawaii, October, 1970.

[7] Tallis, G. M., (1970), Biometrics, 26, 87-103.

[8] Watson, G. S., (1971), Biometrika, 58, 483-490.

[9] Wicksell, S. D., (1925), Biometrika, 17, 84-99.

[10] Wilks, S. S., (1962), Mathematical Statistics, John Wiley and Sons, Inc., New York.

*National Bureau of Standards Special Publication 431*
*Proceedings of the* FOURTH INTERNATIONAL CONGRESS FOR STEREOLOGY
*held at* NBS, Gaithersburg, Md., September 4-9, 1975 (Issued January 1976)

THE GEOMETRY AND STEREOLOGY OF 'CUBO-SPHERICAL' PARTICLES

by R. Warren

*Department of Engineering Metals, Chalmers University of Technology, Gothenburg, Sweden*

SUMMARY

A class of three-dimensional shapes is defined, consisting of rounded cubes and orthogonally-faceted spheres. Two linear parameters, namely the radius of the spherical surfaces and the distance between parallel faces, are sufficient to define the shape and size of the bodies in this class, while the shape alone is defined by their ratio. The surface areas and volumes, and consequently the mean intercept lengths, of the bodies have been derived as a function of the parameters. Conditions and possible methods of measurement are presented for the determination of the shape, size and number per unit volume of particles in the shape class when observed on two-dimensional sections. Some implications and applications of the analysis are discussed briefly.

1. INTRODUCTION

A particle shape often observed in metallic and other crystalline systems is the rounded cube. In fact, there exists a continuous class of shapes, ranging from the sharp-edged cube to the sphere. These can be described as orthogonally faceted spheres and as rounded cubes, formed by the intersection of a cube and a sphere having a common centre. Two representative members of this shape class are shown in Figures 1 and 2. The shape and size of any body in the class is defined uniquely by two parameters, e.g. the half distance between two parallel flat faces, $c$, and the radius of the spherical surfaces, $r$. The shape alone can be defined by a single parameter, e.g. $\beta$, the ratio $c/r$.

It is convenient to divide the class into two distinct groups, namely (i) faceted spheres, for which $1 > \beta > 1/\sqrt{2}$; and (ii) rounded cubes for which $1/\sqrt{2} > \beta > 1/3$. The boundary between the two groups occurs at $\beta = 1/\sqrt{2}$, i.e. when the circular flat surfaces on adjacent sides just touch.

Fig. 1. Orthogonally faceted
sphere (group 1)

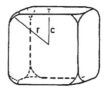

Fig. 2. Rounded cube
(group 2)

2. GEOMETRIC ANALYSIS OF THE SHAPES

*Facetted spheres* $(1 > \beta > 1/\sqrt{2})$

For shapes in this group, the surface area of the flat surfaces is given by

$$S_F = 6\pi \ (r^2 - c^2) \tag{2.1}$$

The total surface area of the body is found simply by replacing the spherical surface area of the six spherical caps by $S_F$ giving:

$$S = 2\pi(6cr - r^2 - 3c^2) \tag{2.2}$$

A similar approach allows derivation of the volume of the body:

$$V = 2\pi(9cr^2 - 4r^3 - 3c^3)/3 \qquad (2.3$$

A general equation for convex bodies (Underwood, 1970) gives the mean intercep length of the body:

$$\bar{L}_3 = 4V/S = 4(9\dot{c}r^2 - 4r^3 - 3c^3)/3(6cr - r^2 - 3\dot{c}^2) \qquad (2.4$$

Since for a given shape,

$$\bar{L}_3 = 4r(9\beta - 3\beta^3 - 4)/3(6\beta - 3\beta^2 - 1) \qquad (2.5$$

Values of $S$, $V$ and $\bar{L}_3$ for different shapes, i.e. different values of $\beta$, are included in Table 1.

*Rounded Cubes* $(1/\sqrt{2} > \beta > 1/\sqrt{3})$
    The geometry of this group is somewhat more complicated. The area of the flat surfaces is:

$$S_F = 6\pi(r^2 - c^2) - 24[A(r^2 - c^2) - c(r^2 - 2c^2)^{1/2}]$$

where:

$$A = \sin^{-1}[(r^2 - 2c^2)^{1/2}/\{r^2 - c^2\}^{1/2}] \qquad (2.6$$

Derivation of the total surface area involves double integration (Warren,1972a and is given by:

$$S = S_F + 4\pi r^2 + 12\pi cr - 24B$$

where:

$$B = r^2\tan^{-1}[c^2/r(r^2 - c^2)^{1/2}] + 2cr\tan^{-1}[(r^2 - 2c^2)/c^2]^{1/2} \qquad (2.7$$

Derivation of $V$ and $\bar{L}_3$ for this group is omitted here. It will be seen that the omission does not seriously affect the subsequent analysis.

Table 1. Geometric Properties of Cubo-Spherical Particles.

| Particle shape  $\beta = c/r$ | Surface area $S$ (for $r=1$) | Volume $V$ (for $r=1$) | Mean intercept length $\bar{L}_3$ (general) | Surface area of unit volume particle | $\bar{L}_3$ for unit volume particle | $\dfrac{S_F}{S}$ |
|---|---|---|---|---|---|---|
| 1 (sphere) | $4\pi$ | $1.333\pi$ | $1.333r$ | 4.835 | 0.827 | 0 |
| 0.9 | $3.94\pi$ | $1.275\pi$ | $1.295r$ | 4.908 | 0.815 | 0.29 |
| 0.8 | $3.76\pi$ | $1.109\pi$ | $1.180r$ | 5.140 | 0.778 | 0.59 |
| 0.707 (i.e. $1/\sqrt{2}$) | $3.484\pi$ | $0.868\pi$ | $0.997r$ | 5.604 | 0.713 | 0.86 |
| 9.577 (i.e. $1/\sqrt{3}$) (cube) | $2.546\pi$ (i.e. 8) | $0.490\pi$ | $0.770r$ | 6 | 0.667 (i.e. $2/3$) | 1 |

## 3. MEASUREMENTS ON SECTIONS

Combination of equations 2.1 with 2.2 and 2.6 with 2.7 shows that, for given shape in the class, the fraction $S_F/S$ is a function solely of $\beta$ i.e. i is independent of size. The fraction is plotted as a function of $\beta$ for the whole shape range in Figure 3. Thus, for an assembly of randomly-oriented particle of a single shape, $\beta$ can be determined from a two-dimensional section b measurement of $S_F/S$. This involves counting the number of intersections mad by test lines with the curved and flat surfaces of the particle section (Warren, 1972b).

If the particles are of *uniform size and shape*, measurement of the mea intercept length will, together with $\beta$, allow determination of the particle size using, for example, equation 2.5. Similarly, application of genera stereological equations for convex particles allows the determination of othe properties of the dispersed system. For example, the number of particles per unit volume, $N_V = 4 N_L/S$ (Underwood, 1970) and so for the present class of shapes $N_V$ can be determined by measurement of $N_L$, $\beta$ and $\bar{L}_3$.

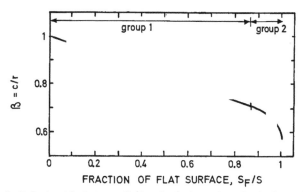

Fig. 3. Relationship between $S_F/S$ and $c/r$ for cubo-spherical particles

## 4. SOME IMPLICATIONS AND APPLICATIONS

### *The dependence of mean intercept length on shape*

Figure 4 shows the dependence of mean intercept length on shape for particles of fixed volume. This clearly has implications for studies of particle size in systems in which a shape change occurs. For example, in particle growth studies, account must be taken of shape changes if growth is measured in terms of $\bar{L}_3$, particularly at low growth rates. Similarly, in studies of the pressing or sintering of spherical powders, the development of interparticular contact areas will lead to an apparent reduction in particle size if measurement is made in terms of $\bar{L}_3$ and shape is not accounted for.

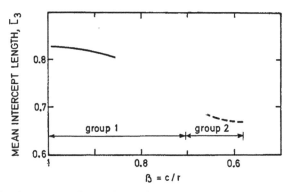

Fig. 4. The dependence of mean intercept length on the shape of cubo-spherical particles of unit volume.

*Anisotropy of surface energy*
In certain circumstances crystalline particles develop an equilibrium cubic or cubo-spherical shape because the surface or interfacial energy of their (100) planes is lower than that on the other crystallographic planes. The equilibrium shape is given by the Wulff construction, in which the distance from the centre of the particle to a given plane on the surface is proportional to the energy of that plane (Winterbottom, 1967). For 'cubo-spherical' particles therefore, the surface energy ratio, $\gamma_{100}/\gamma_{hkl}$, is given by $c/r$. Thus in a system of particles, all having the equilibrium shape, the energy anisotropy can be determined by determination of $\beta$ (e.g. by measurement of $S_F/S$).

REFERENCES
Underwood, E.E. (1970) *Quantitative Stereology*, Chapter 4. Addison-Wesley, Reading, Mass.
Warren, R. (1972 a) *The Microstructure of Cemented Carbides Produced by Liquid-Phase Sintering* (Ph.D. Thesis) University of Surrey, Guildford.
Warren, R. (1972 b) Carbide Grain Shape in Cemented Carbide Alloys of Cubic Refractory Carbides. *J. Inst. Metals*, 100, 176.
Winterbottom, W.L. (1967) Crystallographic Anisotropy in the Surface Energy of Solids. In: *Surfaces and Interfaces* (Ed. by J.J. Burke, N.L. Reed, and V. Weiss), p. 33. Syracuse University Press, New York.

*National Bureau of Standards Special Publication 431*
*Proceedings of the* FOURTH INTERNATIONAL CONGRESS FOR STEREOLOGY
*held at* NBS, Gaithersburg, Md., September 4-9, 1975 (Issued January 1976)

QUANTITATIVE MICROSCOPY OF LINEAL FEATURES IN THREE DIMENSIONS

by R. T. DeHoff and S. M. Gehl,
*Department of Materials Science and Engineering, University of Florida, Gainesville, Florida 32611, U.S.A.*

## SUMMARY

Microstructural features that may be characterized as lines in three dimensional space possess, at every point along their length, two properties: <u>curvature,</u> which is the rate of change of direction of the tangent to the curve, and <u>torsion,</u> which is the rate of change of direction of the binormal to the curve. If, in addition, the lineal feature is an edge, i.e., is defined by the meeting of two surfaces, the feature also possess a third characteristic; the <u>dihedral angle,</u> defined to be the angle between the surface normals at the edge.

A set of three new fundamental relations in stereology are presented which permit the estimation of the total and average curvature and torsion of a set of lineal features, and the average dihedral angle of a set of edges in a microstructure.

## INTRODUCTION

Lineal features are common in the three dimensional structures that are explored in microscopy. In materials science, the archtype of lineal features is the dislocation; a multitude of these one dimensional, tortuously curved crystal imperfections pervade the three dimensional lattice of any crystalline solid. More generally across the spectrum of microstructural analysis, any structure composed of contiguous cells contains cell edges of a variety of classes; these triple lines form partially or wholly connected networks that frame interfaces in the structure. Finally, there is the class of features that are actually three dimensional, but have one dimension that is very much larger than the other two, so that the approximation that the feature is one dimensional is valid for some purposes. Capillaries, nerve fibres, or slag stringers fall into this class of essentially one dimensional features in three dimensional space.

Methods for characterizing the <u>extent</u> of lineal features were first proposed by Saltykov (1958) and Smith and Guttman (1953). The description of the anisotropy of such an array was elegantly developed by Hilliard (1962). These results are now part of standard texts in stereology (DeHoff and Rhines, 1968; Underwood, 1970; and Weibel and Elias, 1967). In addition to their length, lineal features also possess other unambiguously definable geometric properties, associated with their local curvature and torsion. Further, if the lineal feature is an edge, it has an additional property that derives from the implicit fact that an edge is defined by two surfaces that meet in space. This property is the dihedral angle, defined as the angle between the normals to the mating surfaces at the edge. This paper explores the relations between these additional properties of space curves and observations that may be made in two dimensions. The result is three new stereological relations, presented here for the first time.

## THE GEOMETRY OF SPACE CURVES

The concept of curvature at a point P on a plane curve is usually defined by establishing two other points on the curve, Figure 1a, and constructing a circle through the three points

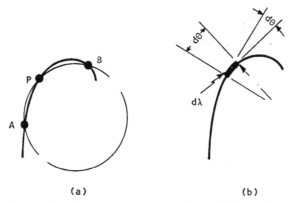

(a)                                        (b)

Figure 1.   Alternate constructions defining the
            local curvature on a plane curve.

(Schwartz, 1967).  The outlying points are allowed to approach P,
and in the limit, the circle approaches the osculating circle.
Its center is the center of curvature at P; its radius is the
radius of curvature of the curve at P, and the reciprocal of the
radius is called the curvature of the curve at the point P.

Alternatively, the curvature may be formulated by considering
an element of arc, $d\lambda$, Figure 1b.   The circular image of that
element is defined as the angle, $d\theta$, swept by the normals to $d\lambda$.
The angle swept out by the tangent as it moves along $d\lambda$ is also
$d\theta$.   Simple geometric considerations yield $d\lambda = r\ d\theta$, or
(Schwartz, 1967)

$$k = 1/r = d\theta/d\lambda \tag{1}$$

This construction of the concept of curvature may be extended
to an element of arc of a space curve, Figure 2a (Struik, 1950).
As the points A and B approach P, succeeding constructions of the
circle through A, P, and B do not generally lie in the same plane.
Nonetheless, there exists a limiting osculating circle that
defines the center and radius of curvature of the space curve at
P.   If $d\theta$ is considered to be the rotation of the tangent vector
to the space curve as an element, $d\lambda$, is traversed, then equation
(1) holds for space curves.

The trajectory of a space curve has an additional degree of
freedom, not associated with plane curves, that is described as
the rotation of the osculating plane as P moves along the curve,
Figure 2b.   Locally, this rotation may be described as the angle
$d\psi$ traced out by the normal to the osculating plane as the element
$d\lambda$ is traversed.   A local property, analogous to the curvature at
P may be defined:

$$\tau \equiv d\psi/d\lambda \tag{2}$$

where $\tau$ is called the torsion of the curve at P (Struik, 1950).
The torsion may be thought of as a measure of the out-of-planeness
of the curve.

It is convenient to define a local reference system at P,
Figure 2c: $\bar{t}$ is the tangent vector; $\bar{n}$, the vector pointing to the

center of curvature is called the <u>normal</u> vector at P; and б, the vector that is perpendicular to both t̄ and n̄, and hence to the osculating plane, is called the <u>binormal</u>. The two vectors, n̄ and б, taken together define a plane that is perpendicular to the tangent vector, called the <u>normal plane</u>. Rotations of this coordinate system as the curve is traversed define its local curvature and torsion.

## THE GEOMETRY OF EDGES ON SURFACES

The intersection of two curved surfaces, Figure 3, defines a space curve, usually called an <u>edge</u> on the figure defined by the surfaces. (In cell structures, such edges are usually defined by the incidence of three cells and the intersection of three surfaces; in such cases, one must consider that three separate, though not independent, edges are thus defined.) In addition to curvature and torsion, such edges possess, at each point, a <u>dihedral angle</u>, χ, Figure 3a. Note that the dihedral angle is defined as the angle between the surface normals at the edge, and not the angle between the tangent planes, which is ambiguously defined. Edges thus have some of the aspects of surfaces, and some of space curves. For example, an edge may be convex, concave, or have a saddle configuration. At the same time, an edge possesses local curvature and torsion.

## THE STEREOLOGICAL RELATIONS

The <u>Length of Lineal Features</u>. When a three dimensional structure containing lineal features is sectioned with a test plane, the features appear as points on the microsection. If the number of points is counted, and the result divided by the total area scanned, the <u>area point count</u>, $P_A$, is obtained. This

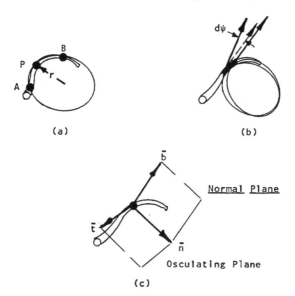

(a)                              (b)

(c)

Figure 2.   Construction illustrating local curvature (a) torsion (b) and the reference trahedron (c) for a segment of a space curve.

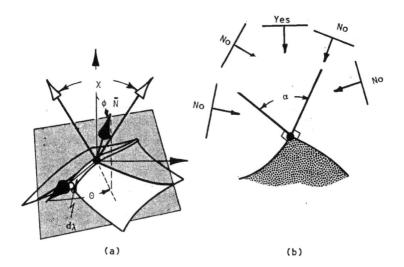

(a)                              (b)

Figure 3.   Configuration of an element of edge, showing the
            dihedral angle ($\chi$) and coordinate system used
            (a) and the formation of tangents with the edge
            trace on a sectioning plane (b).

quantity has been shown to be an unbiased estimator of the total
length of the lineal features thus measured, per unit volume of
microstructure (DeHoff and Rhines, 1968).

$$P_A = \frac{1}{2}L_V \qquad (3)$$

This established counting measurement will be used to normalize
the new stereological measurements presented below in order to
obtain average values of curvature, torsion, and dihedral angle.

   Curvature of Lineal Features.   Focus upon an element of arc,
$d\lambda$, of a collection of space curves, Figure 4. Figure 2 shows the
normal plane at a point P on such an element. The spherical image
of this normal plane is obtained by centering P in a unit sphere
(the orientation sphere) and noting the intersection of the plane
with the sphere. The spherical image of the normal plane is thus
a great circle whose pole is the spherical image of the tangent
direction. As the point P moves along the arc, $d\lambda$, in Figure 4a,
the tangent rotates, as does the normal plane. This rotation
traces out a spherical image that (in the first order of
differentials) has the shape corresponding to the region between
two longitudinal lines on the globe, Figure 4a. The area of this
spherical image $d\Omega_n$ is simply related to the angle $d\theta$ of rotation
of the tangent, $\bar{t}$:

$$d\Omega_n = 4 \, d\theta \qquad (4)$$

   Let this structure be sampled by sweeping planes through its
volume, $V_o$. The event to be counted in this statistical
experiment is the number of tangents formed by these sweeping
planes and the element $d\lambda$. This event occurs when the sweeping

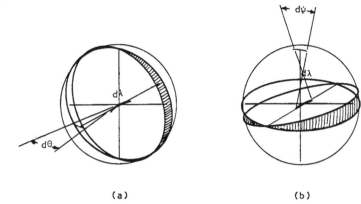

(a)                                        (b)

Figure 4.    Spherical images of (a) the normal plane, and
              (b) the osculating plane for an element of a space
              curve, d$\lambda$.

plane contains the tangent direction, $\bar{t}$, somewhere in the interval
d$\lambda$. An equivalent condition is that the normal to the sweeping
plane must lie within the spherical image d$\Omega_n$.

    If a large number, M, of orientations of sweeping test planes
is chosen at random from the uniform distribution of directions on
the unit sphere, the fraction that form tangents with d$\lambda$ is
d$\Omega_n$/4$\pi$. The number of tangents formed is Md$\Omega_n$/4$\pi$. The total
volume sampled is MV$_0$. The number of tangents formed with d$\lambda$, per
unit volume of structure sampled is:

$$dT_V = (Md\Omega_n/4\pi) \cdot (1/MV_0) = d\Omega_n/4\pi V_0 \qquad (5)$$

Apply equations (1) and (4) to convert the spherical image to the
local curvature:

$$dT_V = (1/4\pi V_0)(4d\theta) = d\theta/\pi V_0 = kd\lambda/\pi V_0 \qquad (6)$$

This result holds for each element of arc in the set of space
curves. The number of tangents formed with all elements is the
integral over the length of these features:

$$T_V = (1/\pi V_0) \int kd\lambda \qquad (7)$$

The quantity $\int kd\lambda$ may be usefully defined to be the <u>total
curvature of lineal features</u> in the structure. The average
curvature may be defined by dividing by the length of arc:

$$\bar{k} \equiv \int kd\lambda / \int d\lambda = \pi\ T_V/L_V \qquad (8)$$

    This volume tangent count may be obtained by sampling the
structure with a series of closely spaced microsections, and
noting appearances and disappearances of the lineal features;
these events correspond to tangencies of the feature with a plane
swept in a direction normal to the sectioning plane.
Alternatively, if the structure can be viewed in transmission, the
operation of sweeping a line across the transmitted image
corresponds to sweeping a plane through the volume of the
structure. Tangencies with this sweeping plane will be observed
as tangencies with the sweeping line on the projected image. If

$T_{Aproj}$ is the number of tangents formed per unit area of projected image traversed, then

$$T_{Aproj} = T_V t$$

and, from equation (7),

$$T_{Aproj} = (t/\pi V_o) \int k d\lambda \qquad (9)$$

where t is the thickness of the thin section being viewed in transmission. Again applying equation (3), and noting that, on a transmitted image observed in projection, the $P_A$ count may be made by a line intercept count, i.e., by counting the number of intersections a test line makes with lineal features, and dividing by the length of test line, one obtains (Underwood, 1972)

$$N_{Lproj} = P_A t = (\tfrac{1}{2}) L_V t \qquad (10)$$

Substituting these results into equation (8) gives an expression for the average curvature of lineal features in terms of counting measurements that may be made upon a projected image:

$$\bar{k} = (\pi T_{Aproj}/t)/(2N_{Lproj}/t) = \frac{\pi}{2} T_{Aproj}/N_{Lproj} \qquad (11)$$

Torsion of Lineal Features. Just as the curvature may be expressed in terms of the rotation of the tangent vector or the normal plane, the torsion may be visualized as the rotation of the binormal vector, or the osculating plane. The spherical image of the osculating plane is a great circle whose pole is the binormal. As the element $d\lambda$ is traversed, Figure 4b, the binormal rotates, and the spherical image of the osculating plane traces out a double-lune shape similar to that obtained for the normal plane in Figure 4a. The area of this spherical image $d\Omega_o$ is simply related to the angle of rotation of the binormal, $d\psi$, and thence to the local torsion at $d\lambda$:

$$d\Omega_o = 4 d\psi \qquad (12)$$

Imagine that the array of lineal features is viewed in projection. Let the projection direction, i.e., the orientation from which the structure is viewed, be randomly chosen from the uniform distribution of orientations on the unit sphere.* Focus upon the element $d\lambda$ as the structure is viewed. The event to be counted is the number of times that the projected image of $d\lambda$ contains an inflection point. This event will occur for a given projection direction if the orientation of the point of view of the observer lies in the osculating plane. Thus, those projection directions that lie within the spherical image of the osculating plane, $d\Omega_o$, will produce inflection points at $d\lambda$ on the projection plane.

Suppose the structure is viewed from a large number, M, of orientations. Each view samples the entire volume $V_o$ of the structure. The fraction of these observations of $d\lambda$ that show an inflection point is $d\Omega_o/4\pi$. The number of inflection points thus

---

*In practice it will not be practical to make such a sequence of observations. Usually only one direction of projection is available. Nonetheless, the thought experiment required to obtain the fundamental relationship requires averaging the observations over the sphere of orientation. This procedure is inherent in essentially all of the basic stereological relationships; the present case is not an exception.

observed is $Md\Omega_o/4\pi$. The total volume sampled is $MV_o$. The number of inflection points counted per unit volume of structure sampled is:

$$dI_V = (Md\Omega_o/4\pi)(1/MV_o) = d\Omega_o/4\pi V_o \qquad (13)$$

Evaluate the spherical image in terms of the local torsion of $d\lambda$ by applying equations (2) and (12)

$$dI_V = 4d\psi/4\pi V_o = \tau d\lambda/\pi V_o \qquad (14)$$

This result is valid for each element of arc in the structure. Thus, for the structure as a whole, the number of inflection points observed in a projected image, per unit volume of structure sampled in the projected image is:

$$I_V = (1/\pi V_o) \int \tau d\lambda \qquad (15)$$

where the integration is carried out over all of the length of space curves in the structure. The average torsion of the set of curves may be defined by dividing by the total length of lineal features in the structure:

$$\bar{\tau} = \int \tau d\lambda / \int d\lambda = \pi \, I_V/L_V \qquad (16)$$

In practice, the inflection point count is performed over an area, A, of a projected image of the structure. The structure itself is a thin film or section of thickness t. Let $I_{Aproj}$ be the number of inflection points counted per unit area of projected image. Then

$$I_{Aproj} = I_V \, t$$

and, from equation (15),

$$I_{Aproj} = (t/\pi V_o) \int \tau d\lambda \qquad (17)$$

The average torsion of the set of space curves becomes

$$\bar{\tau} = (\pi I_{Aproj}/t)/(2N_{Lproj}/t) = \frac{\pi}{2} \, I_{Aproj}/N_{Lproj} \qquad (18)$$

Thus, the average torsion of a collection of space curves in a transparent structure may be obtained from two counting measurements made upon a representative projected image. If the structure is opaque, estimation of the torsion requires serial sectioning, with the counting of projected inflection points as the sectioning planes traverse the three dimensional structure.

   Total Curvature and the Dihedral Angle at Edges. The total curvature of a collection of smoothly curves surfaces is defined by (Cahn, 1967; DeHoff, 1967)

$$M \equiv \int\int H ds \qquad (19)$$

where H is the local mean curvature of the surface,

$$H \equiv 1/2(\kappa_1 + \kappa_2) \qquad (20)$$

$\kappa_1$ and $\kappa_2$ being the local principal normal curvatures (Struik, 1950). One of the established relationships of stereology permits estimation of total curvature per unit volume of surfaces, $M_V$, by means of the area tangent count, $T_A$, applied to a representative

R. T. DeHoff and S. M. Gehl

section through the structure (DeHoff, 1967).*

$$T_{Anet} = M_V/\pi \qquad\qquad (21)$$

In its original formulation, this relationship was limited in application to smooth surfaces, i.e., to surfaces possessing no edges or vertices. However, in a note added in proof to his paper, Cahn (1967) suggested that the notion of total curvature could be extended to include convex polyhedral surfaces.** The present development accomplishes this extension, and in addition removes the limitation that the surfaces must be convex. Thus, the result derived in this section applies to surfaces with edges of arbitrary configuration, excluding as usual the collection of mathematical curiosities classed as topologically "wild" surfaces.

The derivation makes use of the concept of density of features developed in integral geometry, and presented by Santalo (1967) and others (Blaschke, 1955). Figure 3a shows an edge defined by two surfaces, together with the dihedral angle, $\chi$. The structure containing this element of edge is imagined to be sectioned with a random sample of test planes drawn from the uniform distribution of planes in space. Within each plane, the sectioned structure is sampled by selecting randomly from the set of sweeping test lines that traverse the structure. The event of interest is the formation of a tangent between such sweeping test lines and the vertex which is the emergence of the edge on the sectioning plane. This event requires the simultaneous occurrence of two independent events: namely that the plane intersects the element of edge length, $d\lambda$, and that the direction of sweep of the test line in the plane is contained between the normals to the meeting traces, Figure 3b.

The density of planes in space is given by $dp \sin\phi \, d\theta \, d\phi$, (Santalo, 1967) where $dp$ is an element of length along a direction specified by the usual spherical coordinates of an orientation ($\theta$, $\phi$). The density of sweeping test lines in a plane is equal to the density of orientations in two dimensions, which is an angle, $d\alpha$. Thus, the density of sweeping test lines in three dimensional space is

$$dT = d\alpha \, dp \, \sin\phi \, d\theta \, d\phi$$

The measure of the set of sweeping test lines that form tangents at $d\lambda$ is obtained by integrating this density over the set of values of ($\alpha$, p, $\theta$ and $\phi$) that are contained within the event. The measure of the total area swept out by the set of sweeping lines as the volume of the structure, $V_o$, is sampled is $\iiiint A \, dp \, d\alpha \, \sin\phi \, d\theta \, d\phi$ where A is the area of the local sectioning plane.

---

*This tangent count is similar to that applied in equations (9) and (11); it differs in that it is determined upon a <u>section</u> through the structure observed in reflection, as opposed to a projected image obtained in transmission. The tangents counted are those observed to form between a sweeping test line and the traces of the surface whose total curvature is being measured.

---

**It is not clear whether Cahn limited his result to convex polyhedra because he was interested primarily in the relation of total curvature to the caliper diameter of the particle (Minkowski's formula (Minkowski, 1903)) or because he had not explored nonconvex cases. He merely states the equation, without proof or reference.

Thus the number of tangents formed with all edge elements in the volume $V_o$, normalized to unit area of structure swept over, is:

$$T_A = \iiiint_{\text{(Lines form tangents)}} d\alpha dp \sin\phi\, d\theta\, d\phi / \iiiint A dp d\alpha \sin\phi\, d\theta\, d\phi \qquad (22)$$

Since for any orientation $\int Adp$ is the volume of the sample, $V_o$, the denominator integrates to $8\pi^2 V_o$. For any given orientation, $(\theta, \phi)$, the domain over which the plane intersects the edge $d\lambda$ is $d\bar\lambda \cdot \bar N$ where $\bar N$ is the unit vector normal to the plane. For the convention shown in Figure 3a,

$$dp = d\lambda \cos\theta \sin\phi \qquad (23)$$

For a fixed plane that intersects $d\lambda$, the domain of $\alpha$ over which tangents are formed is the dihedral angle in the plane at the edge, $\alpha(\theta, \phi, \chi)$, Figure 3b. Thus, the net number of tangents per unit area of test plane traversed is

$$T_A = \iiint \alpha \quad \cos\theta \sin^2\phi \, d\theta \, d\phi \, d\lambda / 8\pi^2 V_o \qquad (24)$$

The value of the dihedral angle observed on the sectioning plane, $\alpha$, depends upon the dihedral angle $\chi$ in three dimensions at the edge segment and the orientation of the sectioning plane specified by $(\theta, \phi)$. Numerical integration of the function $\iint\alpha(\theta, \phi, \chi)\cos\theta \sin^2\phi \, d\theta \, d\phi$ for a sequence of values of $\chi$ from 0 to $\pi$ gave the result:

$$\iint\alpha(\theta, \phi, \chi)\cos\theta \sin^2\phi \, d\theta \, d\phi = 4\pi\chi \qquad (25)$$

Inserting equation (25) into (24) yields:

$$T_A = 4\pi\int\chi d\lambda / 8\pi^2 V_o = (1/\pi)(1/2V_o) \int\chi d\lambda \qquad (26)$$

Comparison with equation (21) suggests that the quantity $(1/2)\int\chi d\lambda$ is analogous to the concept of total curvature of

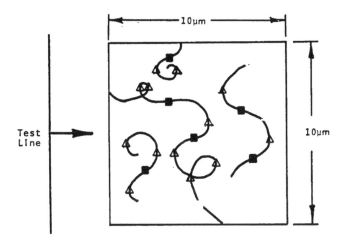

Figure 5. Projected image of a structure containing lineal features. Tangents formed with sweeping test lines are indicated by △ ; inflection points by ■ .

37

surfaces in the volume and may reasonably be defined as the total curvature of edges as suggested by Cahn (1967):

$$M_{edges} \equiv \frac{1}{2}\int \chi d\lambda \qquad (27)$$

where the integration is carried out over the length of the edges in the structure. Then relation (26) becomes:

$$T_{Aedges} = (1/\pi)\, M_{edges}/V_o = M_{Vedges}/\pi \qquad (28)$$

where $M_{Vedges}$ is the total curvature of edges per unit volume of structure. Thus, the application of the area tangent count separately to the edges that exist in a structure permits estimation of the total curvature of edges in unit volume of structure, where the contribution of edges to the total curvature is defined in terms of the dihedral angle at the edge according to equation (27).

The average dihedral angle at edges may be defined as:

$$\bar{\chi} \equiv \int \chi d\lambda / \int d\lambda$$

Substituting from equations (26) and (3) yields:

$$\bar{\chi} = (2\pi V_o\, T_{Aedges})/(2V_o\, P_{Aedges}) = \pi\, T_{Aedges}/P_{Aedges} \qquad (29)$$

An area point count and a tangent count at edges together permit the estimation of the mean dihedral angle at edges in a three dimensional structure.*

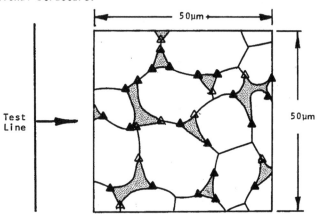

Figure 6. The minor phase (β) has edges (ααβ); the dihedral angle may be estimated by counting ααβ triple points (△+▲) and edge tangents with the test line (▲).

---

*Sections through edges may be convex or concave; the corresponding dihedral angle on the section may thus be positive or negative. Tangents to convex corners are counted as positive, and those at concave corners as negative. In the preceding equations $T_A$ must be interpreted to mean $T_A$(CONVEX) - $T_A$(CONCAVE), just as is the case for smooth surfaces (DeHoff, 1967).

## MECHANICS OF THE MEASUREMENTS

While it is possible to estimate torsion and curvature of lineal features from serial sections through opaque structures, these measurements are more likely to be made upon projected images through transparent structures. Accordingly, the illustrations presented in this section dealing with torsion and curvature are based on analysis of projected images. Estimates of the dihedral angle at edges requires that the observations be made upon a section, as opposed to a projection.

Curvature. Figure 5 is a projected image of a structure containing curved lineal features. The dimensions of the image sampled are shown; the thickness, t, is known to be $10^{-5}$ cm. The number of tangents formed when a vertical line is swept across the structure is found to be 13, with each tangent indicated by a $\triangle$. The area tangent count is obtained by dividing this number by the area swept out: $T_{Aproj} = 1.3 \times 10^7$ (cm$^{-2}$). The total curvature per unit volume is:

$$1/V_o \int k d\lambda = (\pi/t) T_{Aproj} = 4.1 \times 10^{12} \ (1/cm^3)$$

A line intercept ($N_L$) count applied to the structure yields 3200 counts per centimeter. From this, the total length of features per unit volume is estimated, using equation (10):

$$L_V = 2N_{Lproj}/t = 6.4 \times 10^8 \ cm/cm^3$$

The average curvature of these lineal features is obtained by applying equation (11)

$$\bar{k} = 4.1 \times 10^{12}/6.4 \times 10^8 = 6400 \ (1/cm)$$

Torsion. The lineal features in Figure 5 also possess torsion. The total or integral torsion may be estimated by applying the inflection count to the projected image. Each inflection point is designated by a ■; the number in the area shown is 6. The number of inflections per unit area of image is: $I_A = 6 \times 10^6$ (cm$^{-2}$). The total torsion per unit volume is obtained by applying equation (17), and is found to be:

$$(1/V_o)\int \tau d\lambda = \pi \ I_{Aproj}/t = 1.88 \times 10^{12} \ (1/cm^3)$$

The average torsion of these space curves may be determined from equation (18):

$$\bar{\tau} = 1.88 \times 10^{12}/6.4 \times 10^8 = 2900 \ (1/cm)$$

Edge Curvature. Figure 6 shows a two phase structure ($\alpha+\beta$) in which the minor constituent ($\beta$) appears at the boundaries between cells of the major constituent. As a result, the structure contains triple lines of the type $\alpha\alpha\beta$, indicated with a $\triangle$ or $\blacktriangle$. A vertical test line is swept across the structure producing tangents at $\alpha\alpha\beta$ triple points, indicated by $\blacktriangle$. The number of tangents formed is 20. The area swept out is $2.5 \times 10^{-5}$ cm$^2$. The area tangent count is thus $8.0 \times 10^5$ (cm$^{-2}$). The total curvature of $\beta$ edges along $\alpha\alpha\beta$ triple lines is estimated from equation (28) to be:

$$M^{\alpha\alpha\beta}_{V\text{edges}} = \pi(8.0 \times 10^5) = 2.5 \times 10^6 \ (cm/cm^3)$$

The total length of triple lines of the type $\alpha\alpha\beta$ per unit volume may be estimated by counting $\alpha\alpha\beta$ triple points ($\blacktriangle+\triangle$) and applying equation (3). This length is estimated to be:

$$L_V^{\alpha\alpha\beta} = 2P_A = 2(30/2.5\times10^{-5}) = 2.4\times10^6 \text{ cm/cm}^3$$

The average dihedral angle along $\beta$ edges of $\alpha\alpha\beta$ triple lines in the three dimensional structure is thus estimated to be:

$$\bar{\chi} = 2M_{Vedges}/L_V = 2.09\text{rad} = 120^\circ$$

The interior angle is $(180^\circ-\chi)$ or $60^\circ$.

## SUMMARY

The average curvature of a set of lineal features in a three dimensional structure may be estimated by combining a tangent count on a projected image of the structure (which estimates the integral of the curvature over all arc length in the sample) with the line intercept count on the projection (which estimates the length of features):

$$\bar{k} = \frac{\pi}{2} T_{Aproj}/N_{Lproj} \tag{11}$$

The average torsion of a set of lineal features may be estimated from an inflection point count on a projected image:

$$\bar{\tau} = \frac{\pi}{2} I_{Aproj}/N_{Lproj} \tag{18}$$

The contribution of the edges bounding particles in a structure to the total curvature of the particles may be usefully defined in terms of the dihedral angle $\chi$ at the edge:

$$M_{edge} \equiv \frac{1}{2}\int\chi d\lambda \tag{27}$$

The average dihedral angle at an edge may be estimated by counting tangents formed by a sweeping test line with the emergence of edges on the section, and combining the result with the established area point count:

$$\bar{\chi} = \pi \, T_{Aedges}/P_{Aedges} \tag{29}$$

## ACKNOWLEDGEMENTS

This paper was the result of interaction between investigators from two projects; one was sponsored by the National Science Foundation, the other by the United States Atomic Energy Commission. Their support is gratefully acknowledged. The authors also thank Mr. A. Gokhale and Mr. C. Iswaran for their participation in the discussions that led to these results.

#### References

Blaschke, W. (1955), *Vorlesungen uber Integral-geometrie*, 3rd Ed., Berlin.

Cahn, J. W. (1967) *Trans. Met. Soc. AIME*, 239, 610.

DeHoff, R. T. (1967) *Trans. Met. Soc. AIME*, 239, 617.

Hilliard, J. E. (1962) *Trans. Met. Soc. AIME*, 224, 1201.

Minkowski, H. (1903) *Math. Ann.*, 57, 447.

Saltykov, S. A. (1958) *Stereometric Metallography*, 2nd. Ed., Metallurgizdat, Moscow.

Santalo, L. A. (1967) *Studies in Global Geometry and Analysis*, S. S. Chern, Ed., Prentice-Hall, p. 147.

Schwartz, A. (1967) *Calculus and Analytic Geometry*, 2nd, Ed., Holt, Reinhart and Winston, New York, p. 474.

Smith, C. S. & Guttman, L. (1953) *Trans. AIME*, 197, 81.

Strulk, D. J. (1950) *Lectures on Classical Differential Geometry*, Addison-Wesley, Cambridge, Mass., p. 13.

Underwood, E. E. (1970) *Quantitative Stereology*, Addison-Wesley, Reading, Mass.

Underwood, E. E. (1972) *Jnl. Mic.*, 95 - 1, 25.

Weibel, E. R. & Elias, H. (1967) *Quantitative Methods in Morphometry*, Springer Verlag, Inc., Berlin.

National Bureau of Standards Special Publication 431
Proceedings of the FOURTH INTERNATIONAL CONGRESS FOR STEREOLOGY
held at NBS, Gaithersburg, Md., September 4-9, 1975 (Issued January 1976)

CONNECTIVITY OF "DISPERSED" PARTICLES: A PROBABILISTIC COMPUTATION

by George A. Moore
National Bureau of Standards, Washington, D.C., 20234, U.S.A.

The false appearance of total dispersion of particles usually seen on plane sections through an aggregate is a persistent and serious stereological error. This is inherent in the fact that it is topologically impossible to represent more than one continuous phase on an extended plane, although an unlimited number of independent and totally interconnected systems or phases can co-exist in volume. Difficulties in tracing actual connections require a method of estimating contact frequency from simple information.

In ball models of either of the regular close packed structures, each ball has a "hard contact" with 12 neighbors in space. The number of such "hard contacts" per ball will here be used as the measure of connectivity for any aggregate. It is easily seen that only a few specifically oriented and positioned section planes through the close-packed model will show contact-connected circles, while most sections falsely show a dispersion of small circles. Other regular arrangements of balls, as used as models of crystal structures, have 4 to 8 hard contacts. For these models the number of contacts increases quite regularly with the fraction of total volume occupied by the balls, as noted in Figure 3. This suggests that random arrangements of particles in an aggregate should, on the average, make increasing numbers of contacts as the volume fraction of particles increases.

The first key to modeling a random aggregate lies in determining the probability that one added particle comes to rest in a position at which it generates a contact with some other particle. The probability of finding the center of any specified particle within any defined small volume is equal to the volume fraction occupied by the defined small volume. To make the model independent of absolute particle size the unit of measurement of volume is taken equal to the average volume of 1 particle. For spheres of uniform size any one sphere will acquire a contact whenever the center of some other sphere falls within a shell of doubled radius. The sphere in which contact is possible thus has a volume of 8 units. It is, however, more realistic to consider that placing the new center within the original particle is either impossible or forms a single larger particle. Thus a shell volume of 7 units gives a more realistic model. The intrinsic probability of contact, $P$, is thus taken as 7 for this particular uniform sphere model.

We may consider two forms of the uniform sphere model. In the "soft" form spheres in motion deform on contact with final center to center distance of any value between r and 2r. Growth of spheres upon small nuclei by precipitation from solution yields the same configuration. In this soft case each added particle shields against future additions by a factor, s = 7, by occupying one-seventh of the contact shell.

In the "hard" form spheres are assumed to be subject to a mixing motion in a viscous fluidized matrix which will subsequently solidify. Elastic rebound between particles is prevented. Thus whenever the relative motion of two particles would result in a center to center distance less than 2r the particles will roll about each other, while remaining in contact, until they come to rest at 2r spacing. In this condition a shielding factor, s = 12, applies as each contact uses one-twelfth of the possible contact positions. There is no assumption that the new sphere rolls into a pocket or ends in any other specific configuration.

It is also assumed that there are no significant attractive or repulsive forces between the particles, or surface forces from the matrix, and that the densities of the two phases are sufficiently similar to eliminate gravitational effects. If these assumptions do not hold, the intrinsic probability, $P$, will be different from 7. Irregular particle shapes generally will call for $P>7$. Nonuniform particle sizes could be computed in several classes of configurations, using different values of $P$ for each class.

We now proceed toward the construction of an iterative computer program which models the series of microprocesses which occur each time an incremental addition of new particles is made to an existing, but still fluidized, mixture. The quantities $F_0$, $F_1$, $F_2$, . . . $F_{12}$ represent the fractions of all particles currently present which exist in states having 0, 1, 2, . . . 12 contacts. Starting with an originally negligible particle concentration $V_V \simeq 0$, all particles are in the zero contact state where $F_0 = 1$ and $F_1$, . . . $F_{12}$ all = 0. A small incremental particle addition, $\Delta V_V$, is now made. The part of the present zero state particles which are involved in contacts with this addition is:

$$P \cdot F_0 \cdot \Delta V_V$$

Each such contact-forming action forms 2 contacts, one on each particle. While the 2 particles may in fact end in different configurations in later steps, the new particles are not in fact distinguishable from old particles. Therefore, the part of the zero state portion promoted to the one contact state is given by

$$\Delta F_1 = -\Delta F_0 = 2P \cdot F_0 \cdot \Delta V_V \qquad [1]$$

At each small addition of particles, this first action is followed by up to 11 more actions in each of which some portion of the particles in one contact state are promoted to the next higher state by contact with part of the remaining portion of the addition. For calculation, these actions are treated as successive, with each new state formed only from the next lower state. The fraction in the lower state is corrected before each new action. The probability $P$ is multiplied by a shielding factor for all states above zero. The contacting ability of $\Delta V_V$ is decreased after each usage. Each action thus is represented by 4 computations as follows.

$$\Delta F_{(n+1)} = 2P \, F'_n (\frac{s-n}{s}) \, \Delta V_{V(n)} \qquad [2]$$

$$F'_{(n+1)} = F_{(n+1)} + \Delta F_{(n+1)} \qquad [3]$$

$$\Delta V_{V(n+1)} = \Delta V_{V(n)} \, [1 - \frac{F'_{(n+1)}}{s}] \qquad [4]$$

$$F_n = F'_n - \Delta F_{(n+1)} \qquad [5]$$

The index "n" represents the number of contacts in the starting state undergoing reaction with the addition. $F'$ represents the temporary fraction in the state after the new contacts are formed but before this state is again decreased by reaction to the next higher state [eq. 5]. Rather than attempt integration of these operational equations, they have been incorporated as an iterative cycle in a programmable calculator operation which accepts chosen values of $P$, $s$, and $\Delta V_V$ and prints values of the fractions, $F_0$, $F_1$ . . . $F_{12}$, and the mean number of contacts per particle, M, after each $\Delta V_V$ addition. The state fractions first rise to individual maxima, then fall as most are converted to higher states (Figures 1 and 2). The mean number of contacts per particle rises continuously with increasing volume fraction, with moderate downward concavity (Figure 3). The mean contact number reaches a maximum of 6.6 for hard spheres at the highest concentration obtainable in practice (about $V_V = 0.64$) and of 5.1 for soft spheres at the same concentration.

The computed mean contact, or coordination, numbers for the hard sphere random aggregate may be compared with experimentally known combinations of coordination and occupied volume fraction (Figure 3). This curve passes slightly above the regular diamond cubic array (M = 4) and slightly below the simple cubic (M = 6), then cuts through the lower corner of the region covered by reports for "dense random packing" (Ref. 1&2) obtained either from "heaps" of

real balls or computer construction of such heaps. All "dense random" determinations include either an effect of gravity or a condition of fitting new
spheres into 3-sphere "pockets," hence it is expected that the true coordination of spheres in a matrix should be slightly lower than by these models.
Selected distributions from Figures 1 and 2 have been checked for variance and
their standard deviations found to agree well with Poisson statistics. Poorly
mixed aggregates would have an excess number of particles of high coordination
number in the more concentrated regions and therefore a higher average coordination number.

Average contact numbers of 2, 3, and 4 correspond to structures dominated
by lines (filaments), sheets, and 3-d networks of particles respectively.
The compositions at which these structures would be dominant are noted in
Figure 3.

The probabilistic model of an aggregate thus appears to give valid estimates for contact connectivity as a direct function of volume fraction for the
assumed uniform random mixture of noninteracting uniform hard spheres. Depar-.
tures from this ideal model can be treated by the same program by inserting
different constants, $P$ and s. Computation for Bernal's (Ref. 2) model of a
liquid with contacts to $r = 1.05$ gave 3% lower coordination number but 2.4 times
greater state variance than Bernal reported. The present computation will
apparently give a realistic model of a liquid by using $P$ based on an enlarged
sphere of attraction and the appropriately larger value of s.

REFERENCES
1.  C. H. Bennett, "Serially Deposited Amorphous Aggregates of Hard Spheres,"
    J Appl. Phys. 43, 2727-34 (1972).
2.  J. D. Bernal, "The Geometry of the Structure of Liquids" (pp. 25-50) in
    Liquids, Structure Properties and Solid Infractions, ed. T. J. Hugel,
    Elsevier (1965).

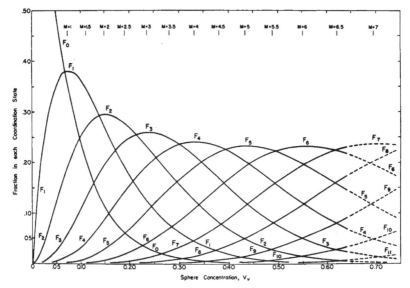

Figure 1 - Distribution of Contact States for "Hard Sphere" Model.

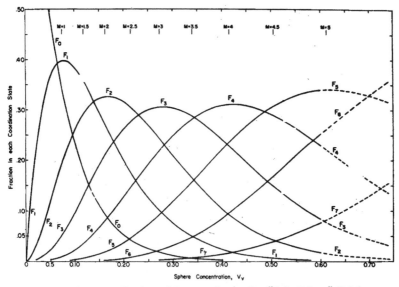

Figure 2 – Distribution of Contact States for "Soft Sphere" Model.

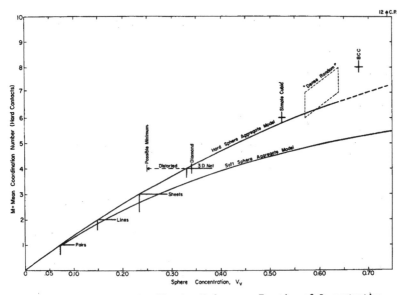

Figure 3 – Mean Contact Coordination Number as a Function of Concentration.

*National Bureau of Standards Special Publication 431*
*Proceedings of the* FOURTH INTERNATIONAL CONGRESS FOR STEREOLOGY
*held at* NBS, Gaithersburg, Md., September 4-9, 1975 (Issued January 1976)

## ANALYTIC RESULTS FOR THE CONNECTIVITY OF DISPERSED PARTICLES

by Roland de Wit
*National Bureau of Standards, Washington, D.C. 20234, U.S.A.*

In the previous paper of these Proceedings G. A. Moore has presented a probabilistic computation of the connectivity of dispersed spheres. He constructed an iterative computer program, based on a set of four operational equations, which he used to generate the curves for the state fractions. We wish to examine whether it is possible to integrate his operational equations and thus find an analytical solution to his problem. The present paper shows that such a solution always exists for the no-contact state and that analytic solutions can be found for the other states if a certain approximation is made.

We use the same notation as Moore and derive the equations *ab initio*. Hence, $F_0$, $F_1$, $F_2$,..., $F_n$,..., $F_s$ are defined as the fractions of all spheres in states having 0, 1, 2,..., n,..., s contacts with neighbouring spheres. Here s is the maximum number of contacts a sphere can have, i.e., the maximum coordination number. Suppose now that we have a certain distribution of contact states at a particular volume fraction $V_V$ of spheres. Then the effect on this distribution of a small increment $\Delta V_V$ is that a portion of spheres in each contact state is promoted to the next higher state as shown in the following scheme:

$$\left.\begin{array}{l} F_0 \\ F_1 \\ F_2 \end{array}\right\} \quad 2P\, F_0 \Delta V_V$$
$$2P\, \frac{s-1}{s}\, F_1\, \Delta V_V\, (1 - \frac{F_1}{s})$$

$$\cdots$$

$$\left.\begin{array}{l} F_{n-1} \\ F_n \\ F_{n+1} \end{array}\right\} \quad 2P\, \frac{s-n+1}{s}\, F_{n-1}\, \Delta V_V\, (1 - \frac{F_1}{s})\, \cdots\, (1 - \frac{F_{n-1}}{s})$$
$$2P\, \frac{s-n}{s}\, F_n\, \Delta V_V\, (1 - \frac{F_1}{s})\, \cdots\, (1 - \frac{F_n}{s})$$

$$\cdots$$

$$\left.\begin{array}{l} F_{s-1} \\ F_s \end{array}\right\} \quad 2P\, \frac{1}{s}\, F_{s-1}\, \Delta V_V\, (1 - \frac{F_1}{s})\, \cdots\, (1 - \frac{F_{s-1}}{s})$$

Here the factor 2 occurs because each contacting action changes the state of 2 spheres. The factor P is the intrinsic local probability of contacting action for a sphere with no contacts. As the sphere acquires n contacts, this probability is decreased by the shielding factor $(s-n)/s$. The portion promoted to a higher contact state $F_{n+1}$ is naturally proportional to the fraction in the contact state $F_n$ and the concentration increment $\Delta V_V$. Finally, the product of the expressions in parentheses represents the decrease in contact forming ability of the concentration change $\Delta V_V$ at each step. This scheme leads to the following set of s differential equations

$$\frac{dF_n}{dV_V} = Q_{n-1} - Q_n, \quad (n=0, 1, \ldots, s), \tag{1}$$

$$Q_n = 2P\, \frac{s-n}{s}\, F_n\, \prod_{i=0}^{n} (1 - \frac{F_i}{s}), \qquad Q_{-1} = 0, \tag{2}$$

which are completely equivalent to Moore's 4 operational equations. Here $Q_n$ represents the fraction of spheres promoted from the contact state $F_n$ to the contact state $F_{n+1}$ as the composition $V_V$ is increased. We note a point of consistency. By definition the complete set of state fractions $F_0$, $F_1$,..., $F_s$ must add up to unity:

$$\sum_{n=0}^{s} F_n = 1. \tag{3}$$

Now it follows from (1) and (2) that

$$\frac{d}{dV_V} \sum_{n=0}^{s} F_n = \sum_{n=0}^{s} \frac{dF_n}{dV_V} = Q_{-1} - Q_s = 0, \tag{4}$$

so that the condition (3) can be maintained at all concentrations. The no-contact state $F_0$ satisfies the simple equation

$$\frac{dF_0}{dV_V} = -2P F_0. \tag{5}$$

With the initial condition that $F_0 = 1$ when $V_V = 0$ the solution is

$$F_0 = \exp(-2P\, V_V). \tag{6}$$

Hence the no-contact state undergoes a simple exponential decay with increase in composition, as is also borne out by Moore's curves. For $n > 1$ the differential equations are nonlinear due to the product factor in (2). As a consequence, an analytic solution does not appear feasible. However, if we linearize the equations by deleting the product factor, i.e., if we take

$$Q_n = 2P \frac{s-n}{s} F_n \tag{7}$$

instead of (2), then there exist simple solutions, which can be obtained by the standard methods of solving differential equations. With the initial conditions that $F_n = 0$ when $V_V = 0$ for $n \geq 1$, they are

$$F_n = \frac{s!}{n!(s-n)!} \sum_{i=0}^{n} \frac{(-1)^i\, n!}{i!(n-i)!} \exp(-2P\, V_V \frac{s-n+i}{s}), \quad (n=0, 1, \ldots, s). \tag{8}$$

It is straightforward to show that these solutions satisfy the consistency condition (3) by a double application of the binomial theorem. Note that the result (6) is included in (8).

The result (8) can also be applied to two dimensions, if $V_V$ is replaced by $A_A$, and to one dimension, if $V_V$ is replaced by $L_L$. The following table summarizes the best values of the parameters:

| Dimension | 1 | 2 | 3 | |
|---|---|---|---|---|
| | | | Hard spheres | Soft spheres |
| s | 2 | 6 | 12 | 7 |
| $P = 2^D - 1$ | 1 | 3 | 7 | 7 |
| Maximum of $L_L$, $A_A$, or $V_V$ | 1 | $\pi/2\sqrt{3}$ 0.907 | $\pi/3\sqrt{2}$ 0.7405 | - |

The maximum values of $L_L$, $A_A$, and $V_V$ are those for the close-packed arrangement. In three dimensions Moore has made a distinction between hard spheres ($s=12$) and soft spheres, for which he has chosen $s=7$. Our results for hard spheres are plotted in Figure 1 and for soft spheres in Figure 2. These curves differ only by an insignificant amount from Moore's results.

Another comparison is afforded by the mean coordination number, which is defined as

$$M = \sum_{n=0}^{s} n\, F_n. \tag{9}$$

The differential equation for M is found as follows

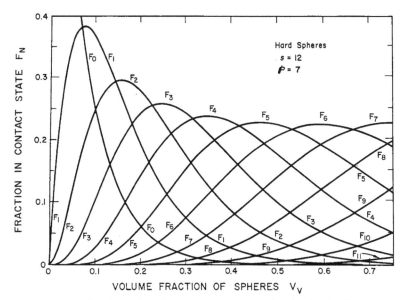

Figure 1. State fractions versus concentration for hard spheres

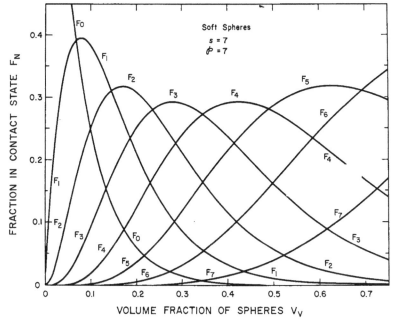

Figure 2. State fractions versus concentration for soft spheres

47

$$\frac{dM}{dV_V} = \sum_{n=0}^{s} n \frac{dF_n}{dV_V} = \sum_{n=0}^{s} n \ (Q_{n-1} - Q_n) = \sum_{n=0}^{s} 2P \frac{s-n}{s} F_n = 2P \ (1 - M/s), \qquad (10)$$

from (9), (1), (7), and (3). With the initial condition that M=0 when $V_V$=0 the solution is

$$M = s \ [1 - \exp \ (-2P \ V_V/s)]. \qquad (11)$$

The same result could also have been obtained from (8) and a double application of the binomial theorem and its derivative. Equation (11) is plotted in Figure 3 for hard spheres (s=12) and soft spheres (s=7), using P=7. The curves are lower than Moore's result by 5% for hard spheres and 1% for soft spheres.

The comparison with Moore's work would seem to indicate that the approximation (7) is not too severe, and that the product factor in (2) supplies only a small correction. We believe that the analytic solutions may sometimes have certain advantages over the numerical ones, even though we had to approximate. One advantage is that equation (8) directly gives the concentration $F_n$ for any value of $V_V$ any choice of P and s, thus showing the qualitative dependence of the solution on those parameters.

Some computer calculations and experimental results for the packing of spheres have been summarized by Norman and others. Their findings for the mean coordination number agree qualitatively with ours, but differ in the details.

REFERENCES
"Computer simulation of particulate systems" by Lindsay D. Norman and others, U.S. Bureau of Mines Bulletin 658 (1971).

"Connectivity of dispersed particles: A probabilistic computation" by George A. Moore. This volume.

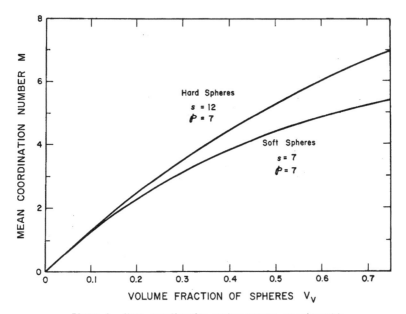

Figure 3. Mean coordination number versus concentration

*National Bureau of Standards Special Publication 431*
*Proceedings of the* FOURTH INTERNATIONAL CONGRESS FOR STEREOLOGY
*held at* NBS, Gaithersburg, Md., September 4-9, 1975 (Issued January 1976)

## TOPOLOGICAL ANALYSIS OF DENDRITIC TREES

by Martin Berry
*Anatomy Department, Medical School, University of Birmingham, England*

The dendritic trees of neurons offer a large surface area for contact with presynaptic elements over wide areas of neuropil. At the same time, the biophysics of the membrane and the impedance properties of the core of dendrites impose constraints on the flow of current such that the characteristics of the branching patterns of dendrites are thought to have profound effects on both the logical and information processing capacity of neurons (Rall, 1970). In the light of this widely held view, that the branching patterns of dendrites are in some way related to the integrative capabilities of neurons, it is surprising that a technique has not been developed for the analysis of dendritic topology. In fact, since the 1940's geographers have been studying the networks created by rivers, roadways and other lines of communication, using the method of network analysis (Haggett and Chorley, 1969) and it has recently been shown that the topological and ordering methods they use are directly applicable to the investigation of both the branching patterns and the growth of the dendritic fields of neurons (Berry et al, 1975; Hollingworth and Berry, 1975; Berry and Bradley, 1975a,b).

The method of network analysis requires the tree to be viewed as a plane graph, made up of points, called vertices, interconnected by lines called arcs. The outermost tips of the tree are termed pendant vertices and the respective arcs, pendant arcs. The point of origin of the tree is called the root vertex. Translated into dendritic terminology arcs are the segments and pendant arcs the terminal segments of the tree. All vertices are nodes, the root vertex is taken to be the axon hillock and the pendant vertices are the tips of the terminal segments. Different forms of branching can be defined by the number of arcs draining into each vertex and the order of magnitude of branching at a vertex may be described as dichotomous if it drains three arcs, trichotomous if it drains four arcs, and so on. Branching patterns can be established by a variety of processes of growth. Terminal growth occurs when arcs are added to pre-existing pendant vertices and segmental growth when arcs are added to pre-existing arcs including pendant arcs. Monochotomous, dichotomous or trichotomous growth patterns are generated by the addition of one, two or three arcs, respectively, at the newly established vertices. The connectivity of dendritic trees is thus described by defining both the mode of interconnection of arcs in the networks and the frequencies of vertices of differing orders of magnitude of branching.

Topological analysis can be used to study branching in both small and large dendritic trees, e.g., the small arrays of the basal dendrites of neocortical pyramids and the massive trees of Purkinje cells in the cerebellum (Hollingworth and Berry, 1975). In the former case the analysis is performed on the entire tree whilst, in the latter case, only the peripheral parts of the network, subtending a small number of pendant arcs, are analysed. The latter strategy is adopted because the numbers of topologically distinct patterns which are possible in networks with only small numbers of pendant arcs is so large that the number of examples needed for analysis becomes unmanageable (Berry et al, 1975).

### Growth of dendrites

The growth of dendrites occurs at specialised sites called growth cones which can be located along segments (Morest, 1969) or at the tips of dendrites (Bray, 1973). This latter location probably predominates in most growing trees and growth of any new branches from segments is only likely to be transitory since growth cones will be carried to the tips of all newly formed dendrites (Berry and Bradley, 1975). Thus terminal growth should predominate in dendritic systems. Vaughn et al (1974) have suggested that synaptic engagement of filopodia may determine branching and Berry and Bradley (1975a) have elaborated this idea predicting that the order of magnitude of branching at nodes, the direction of growth and the length of segments will be dependant on an interplay between the growth cone and synaptic potential in the surrounding axon field.

49

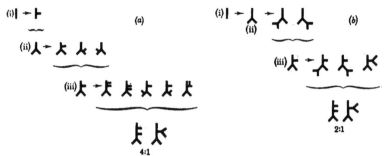

Figure 1.   Topological branching patterns formed by (a) segmental and (b)
terminal growth.   The Roman numerals refer to the consecutive stages of branching and the patterns to the right of each arrow depict where branching can occur
on the pattern to the left of the arrow.   The branching patterns seen under
each bracket represent the topological types drawn in a standard format, used
throughout this paper, into which the patterns above the bracket may be resolved.   The rationale for resolving mirror images and rotational differences are
given by Berry et al (1975).   The ratios against the products of the third
stage of branching refer to the frequency distribution of each topological
branching pattern.

There are, of course, an infinite number of growth hypotheses but perhaps
the most useful formulations are stochastic models which highlight any deviation of growth from a purely random process.   Random models also seem relevant
in terms of growth cone/axon field interaction since filopodial synaptogenesis is
likely to be a chance event as may be the number, length and stability of
filopodia emanating from a growth cone at any one point in time.   To test the
prediction of Berry and Bradley (1975a), that terminal growth predominates in
dendritic systems, two models of growth were formulated (Fig. 1).   In the
first, called the <u>terminal growth hypothesis</u>, two arcs were added randomly to
existing pendant vertices and, in the second, called the <u>segmental growth
hypothesis</u>, a single arc was added randomly to existing arcs including pendant
arcs.   Starting from the root vertex the method of generating specific topological branching patterns according to each hypothesis is illustrated in fig.1.
It can be seen that by the third stage of branching the frequency distribution
of topological branching patterns discriminates between the two growth models.
If the calculation for each hypothesis is continued, the distribution of topologies at subsequent stages of branching can be worked out.   The complete
series of topological types possible at each stage of branching is defined as
an <u>absolute pendant arc series</u> and the frequencies of topologies in the 1st -
9th absolute pendant arc series for random terminal and random segmental growth
models are given in figures 2a and b, respectively.

Hollingworth and Berry (1975) have compared their observed data with those
in figs. 2a and b and shown that the mature tree appears to grow in a manner
indistinguishable from random terminal growth.   In fact, certain constraints
during development do tend to deviate growth from a purely random process
(Berry and Bradley, 1975a), as do the effects of irradiation (Bradley and
Berry, 1975; Berry and Bradley in preparation) and starvation (McConnell and
Berry in preparation) although growth at pendant vertices always predominates.

The lengths of segments can easily be incorporated into the analysis and the
mean probability of branching in the network can be calculated from the frequency distribution of arcs in the mature tree.   Thus, if there are $Y_1$ arcs of
length $X_1$ then the frequency of $Y_2$ arcs, of length $X_2$ will be given by the
expression:-

$$Y_2 = Y_1 (1 - p)^{(X_2 - X_1)}$$

where p = mean probability of branching in the tree.   $Y_1$ and $X_1$, $Y_2$ and $X_2$ can
be obtained from the frequency distribution of segment lengths of the tree and
statistical adjustments between observed and expected distribution verify and

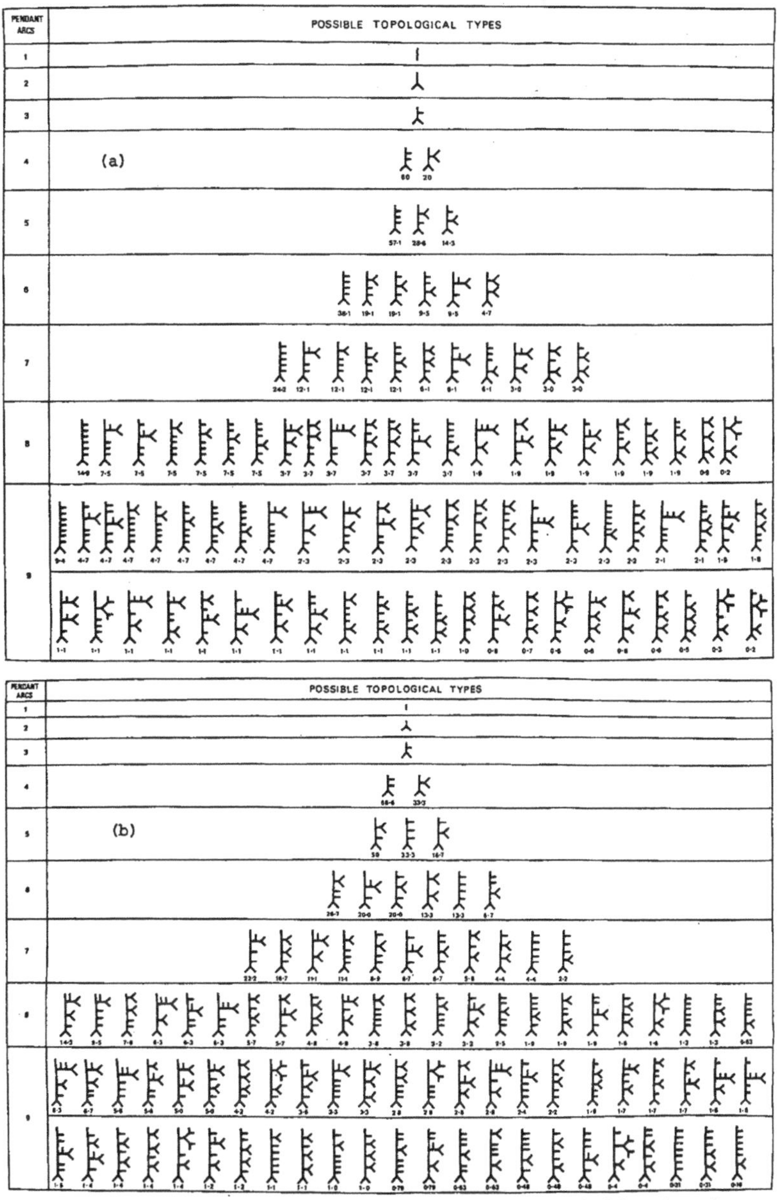

Figure 2. Topological types generated by random segmental (a) and random terminal growth (b) are listed together with the frequency of occurrence (%) of types in networks with from 1 to 9 pendant arcs. (see Berry et al. (1975) and Berry and Bradley (1975a) for sources of data).

51

refine the value of p (Berry and Bradley, 1975a).    This estimate of p is based
on the assumption that segment lengths remain fixed once a vertex, defining the
distal end of a segment, has been formed.    Evidence supporting the validity of
such an assumption has come from cinematographic studies of growing 'neurites'
in tissue culture (Bray, 1970, 1973) and from studies on the microtubular
skeleton within neural processes which appears to stabilise established segments
(Bunge, 1973; Yamada, Spooner and Wessells, 1971; Yamada and Wessells, 1971).
    Estimates of the frequency of different orders of magnitude of branching
are important because they also provide an estimate of the intensity of interaction between axons and dendritic growth cones and thus it is to be expected
that the magnitude of this parameter should be inverseley correlated with the
mean frequency of segment lengths (Berry and Bradley, 1975a; Hollingworth and
Berry, 1975).

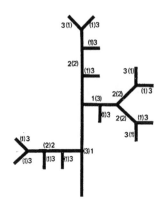

Figure 3.    Illustration of the Strahler
method of ordering.    The tree can be ordered
from the most peripheral branches inwards by
assigning all terminal branches (pendant arcs)
as order 1.    Where two order 1 branches join,
an order 2 branch is formed.    Where two order
2 branches join, an order 3 branch is formed,
etc.    Daughter branches of 'n - 1' Strahler
order form parent branches of Strahler order
'n'.    Collateral branches of 'n - 1' Strahler
order, or less, divide the parent branch into
segments but do not change the order of the
parent branch.    To reverse, the highest
Strahler order branch, draining into the root
vertex, is assigned order 1 and all other orders
are consecutively assigned order 2,3... etc.
In this illustration Strahler order branches
are given in parenthesis and the reversed order
numbers occur to the side of each of these brack-
eted numbers.

## Definition of connectivity

    The topological analysis of small trees defines the mode of interconnection
of arcs in a given network pictorially.    The difficulties of quantifying
connectivity are enormous and centre on choosing a method of assigning a ranking
order to all arcs.    Such methodologies are fraught with problems (Berry et al,
1975; Uylings et al, 1975) but the method we have chosen, called the Strahler
technique can be used to order the tree both centripedally and centrifugally
(Fig. 3) which does give this technique some advantages (Uylings, 1975;
Hollingworth and Berry, 1975; Berry and Bradley, 1975b).
    The Strahler technique defines a branch as a series of arcs of identical
Strahler order.    Daughter branches of 'n - 1' Strahler order create parent
branches of Strahler order 'n' and collaterals draining into the parent branch
are always 'n - 1' Strahler order, or less.    By estimating the relative
frequencies of branches of each order the relationship between orders can be
expressed in terms of the bifurcation ratio i.e. the ratio between adjacent orders.
The number of branches of different Strahler order in a given network tends to
approximate to an inverse geometric series defined by the mean, or overall, bifurcation ratio.    Some examples of mean bifurcation ratios of different branching
systems are given in figure 4.
    The application of ordering methods to the study of dendritic connectivity
is complicated by changes in the order of magnitude of branching at different vertices in the same tree and also by conceptual difficulties in the interpretation
of the meaning of the bifurcation ratio parameter in the context of connectivity.
In this latter respect, the greatest difficulties arise when different networks
give the same overall bifurcation ratio (Berry et al, 1975).    This lack of

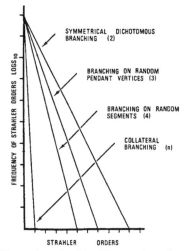

FREQUENCY OF STRAHLER ORDERS LOG$_{10}$

SYMMETRICAL DICHOTOMOUS BRANCHING (2)

BRANCHING ON RANDOM PENDANT VERTICES (3)

BRANCHING ON RANDOM SEGMENTS (4)

COLLATERAL BRANCHING (n)

STRAHLER ORDERS

Figure 4.    Graphic illustration of some of the possible bifurcation ratios (numbers in parentheses) of networks grown by the addition of one arc with the formation of each new vertex (n = total number of pendant arcs).

discrimination by the method comes about for two reasons; firstly, because the overall bifurcation ratio represents the slope of an inverse geometric series and thus is a measure of the size of an arborescence in terms of the maximal Strahler order attained by a given tree and secondly, Strahler ordering defines branches and not segments in the tree and thus gives only general information about connectivity (Berry et al, 1975). These objections can be overcome either by choosing to quantify relationships between all arcs in terms of daughter and parent Strahler orders or of combining the topological analysis with the method of Strahler ordering and calculating <u>absolute bifurcation ratios</u>. This latter measure is obtained by computing the mean bifurcation ratios between successive Strahler orders for each topological branching pattern contained in an absolute pendant arc series for a given hypothesis of growth. Thus, absolute bifurcation ratios give information about both growth and connectivity in absolute and observed networks.

## Conclusions

Topological analysis of dendritic networks together with the application of ordering systems defines the connectivity of trees so that their anatomy and growth can be comprehended in precise quantitative terms. The application of the method is thus likely to yield a fuller understanding of important problems in neurobiology like the degree of interaction between axons and dendrites during the establishment of connections, the plasticity of axonal and dendritic fields and the relative contribution of nurture and nature in determining form and function in the central nervous system. The precise definition of the morphology of dendritic trees afforded by network analysis means that the effects of treatments in experimental studies on growing and adult networks can be assessed quantitatively and that accurate models of dendritic networks can now be built from which their logic can be deduced for a given set of assumptions about the physical properties of their membrane.

## References

Berry, M., Hollingworth, T. Anderson, E.M. and Flinn, R.M. (1975) Application of network analysis to the study of the branching patterns of dendritic fields. In <u>Advances in Neurology</u> Vol. 12 Ed. G.W. Kreutzberg. Raven Press, New York pp. 217-245.

Berry, M.and Bradley, P. (1975a). The application of network analysis to the study of branching patterns of large dendritic fields. <u>Brain Res.</u>(submitted for publication).

Berry, M. and Bradley, P. (1975b). The growth of the dendritic trees of Purkinje cells in the cerebellum of the rat. <u>Brain Res.</u> (submitted for publication).

Bradley, P.and Berry, M. (1975). The effects of reduced climbing and parallel fibre input on Purkinje cell dendritic growth. <u>Brain Res</u>. (Submitted for publication).

Bray, D. (1970)Surface movements during growth of single explanted neurons. <u>Proc. Natn. Acad. Sci. U.S.A.</u> <u>65</u>, 905-910.

Bray, D. (1973) Branching patterns of sympathetic neurons in culture. J. Cell. Biol. 56, 702-712.

Bunge, M.B. (1973) Fine structure of nerve fibres and growth cones of isolated sympathetic neurons in culture. J. Cell. Biol., 56, 713-735.

Haggett, P and Chorley, R.J. (1969) Network analysis in geography, Arnold, London.

Hollingworth, T and Berry, M. (1975). Network analysis of dendritic fields of pyramidal cells in neocortex and Purkinje cells in the cerebellum of the rat. Phil. Trans. Roy. Soc. B. 270 227-264.

Horton, R.W. (1945) Erosional development of streams and their drainage basins, hydro-physical approach to quantitative morphology. Bull. geol. Soc. Am., 56 275-370.

Morest, D.K. (1969). The growth of dendrites in the mammalian brain. Z. Anat. Entwgesch. 128 290-317.

Rall, W. (1970) Cable properties of dendrites and effects of synaptic location. In Excitatory synaptic mechanisms. Proceedings of the Fifth International Meeting of Neurobiologists. Ed. P. Anderson and J.F.S. Jansen. Oslo, Bergen, Tronsö, Universitetsforlaget. pp. 175-188.

Strahler, A.N. (1953) Revisions of Horton's quantitative factors in errosional terrain. Trans. Am. geophys. Un., 34. 345.

Uylings, H.B.M. Smit, G.J. and Veltman, W.A.B. (1975) Ordering methods in quantitative analysis of branching structure of dendritic trees. In Advances in Neurology Vol. 12. Ed. G.W. Kreutzberg Raven Press. New York pp.247-254.

Vaughn, J.E. Henrikson, C.K. and Grieshaber, J.A. (1974). A quantitative study of synapses in motor neuron dendritic growth cones in developing mouse spinal cord. J. cell. Biol. 60, 664-672.

Yamada, K.M. Spooner, B.S. and Wessells, N.K. (1971) Ultrastructure and function of growth cones and axons of cultured nerve cells. J. Cell Biol. 49, 614-635.

Yamada, K.M. and Wessells, N.K. (1971) Axon elongation. The effect of nerve growth factor on microtubular proteins. Exptl. Cell. Res. 66, 346-352.

*National Bureau of Standards Special Publication 431*
*Proceedings of the* FOURTH INTERNATIONAL CONGRESS FOR STEREOLOGY
*held at* NBS, Gaithersburg, Md., September 4-9, 1975 (Issued January 1976)

THEORETICAL AND EXPERIMENTAL STUDIES OF THE STEREOLOGICAL PROPERTIES OF POROSITY, SPECIFIC SURFACE, AND CONNECTIVITY

by R. Bruce Martin
*West Virginia University, Morgantown, W. Va. 26506, U.S.A.*

Theoretical proofs of the equivalence of linear, areal, and volumetric measures of porosity, specific surface, and connectivity are discussed. Most such equivalencies are found to be independent of the orientation of a cross-section or transect line in an anisotropic material. The application of these results to materials of various kinds and degrees of porosity is discussed, and experimental data is presented illustrating the utility of the theory in regard to dimensionally-reduced modeling and connectivity.

## INTRODUCTION

As an engineer who is interested in analyzing bone as a structural material and, furthermore, in studying the relationships between the physiology of this material and its mechanical properties, this author has been compelled to enter the realm of stereology. Specifically, in attempting to analyze the mechanical properties of bone in terms of its histology, and in trying to formulate mathematical representations of physiologic processes, the following questions arise.

(1) Is the areal porosity ($p_A$, void area/total area) of a cross-section numerically equal to the volumetric porosity ($p_V$, void volume/total volume) of the original specimen?

(2) Is the areal specific surface ($c_A$, void perimeter/total area) of a cross-section numerically equal to the volumetric specific surface ($s_V$, void surface/total volume) of the original specimen?

(3) How may one relate the connectivity of various features in a cross-section to three-dimensional connectivity patterns?

In surveying the history of these problems the author found that while geologists and others had for some time accepted the equivalence of areal and volumetric measures of relative composition or porosity, no rigorous mathematical proofs of this equality were available. The writer has therefore attempted to formulate such proofs and to analyze the other questions raised above. This work has been reported elsewhere (1); in this paper some additional results are derived and an application of these results in the problems just outlined is discussed.

*R. B. Martin*

The writer has previously shown that areal measures of porosity can be expected to equal volumetric porosity. Furthermore, it was shown that linear measures of porosity ($p_L$, the fraction of a transect line which falls in inclusions) can be expected to equal $p_A$ and $p_V$, and the specific surface measured in two dimensions can be expected to equal the actual value in the whole specimen if the cross-section is representative. In all these cases the orthogonality of certain geometric relationships requires that the above results are insensitive to the orientation of the transecting plane or line (1). The next result stands in contradistinction to this.

THEORY

Theorem: The perimeter, c, of any simply connected plane figure is equal to the projected length of the figure in a certain direction, $\lambda_y$, multiplied by a geometric constant, $\bar{g}_y$, for that direction. Using this theorem, it can be shown that the frequency, $f_i$, with which a transect line hits inclusions on a cross-section is related to the two-dimensional specific surface measure, $c_A$, as follows:

Fig. 1

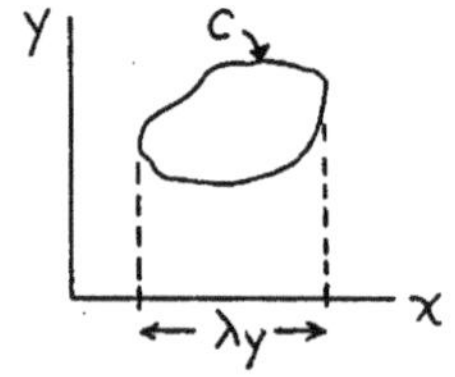

$$f_i \stackrel{e}{=} c_A/\bar{g}_i \quad * \tag{1}$$

where $\bar{g}_i$ is a geometric property of the inclusions associated with the i direction. The linear measure $f_i$ is proportional to but can not be expected to equal the areal measure of internal surface area, $c_A$. We have then,

$$p_L \stackrel{e}{=} p_A \stackrel{e}{=} p_V \ , \tag{2}$$

$$\bar{g}_i f_i \stackrel{e}{=} c_A \stackrel{e}{=} s_V \ . \tag{3}$$

It is seen that linear and areal measurements of porosity on a representative cross-section or transect line of a solid body can be expected to equal the volumetric value, and that this equivalency is entirely independent of the orientation of the section or transect line. A similar situation connects areal and volumetric measures of specific surface area, but the linear measure of this parameter is neither equal to its higher order analogs nor independent of direction. In all cases the inclusions need only be simply connected.

DISCUSSION

The requirement that the cross-section (or transect) be "representative" means that it must reflect the homogeneity of the original specimen; if the specimen contains regions of greater inclusion density, the cross-section which misses such regions will not be representative. If one has specimens which are reasonably homogeneous with regard to inclusion density, only a few cross-sections may be required to give an accurate result.

---

*In this paper the symbol $\stackrel{e}{=}$ is used to mean "can be expected to equal" in the sense that if enough measurements are made on representative sections, the mean value of the equation's left side will approach that of the right side.

Although the equivalence of specific surface and porosity measures in two and three dimensions does not depend on the geometry of the pores or inclusions, the relationship between specific surface and porosity does. Bone of low porosity has cylindrical, tunnel-like voids that run parallel to one another several void-diameters apart. High porosity bone has large, polyhedral voids. Fig. 2 shows the relationship between specific surface and porosity as measured on microradiographs of human bone specimens. The dimensional equivalence of these variables allows one to make two dimensional models of porous bodies. Fig. 2 also shows the $c_A$-$p_A$ relationship for two such models of bone in which the voids are allowed to grow and break into one another in a natural way. Model 2 is better than Model 1 because its features were more nearly scaled to those of real bone. The usefulness of such models in studying dynamic porous materials is enhanced by the insensitivity to anisotropy of the material. If the patterns of change are realistic, one orientation with respect to the actual body is as good as another, but changes in anisotropy experienced during the remodeling will not be apparent unless simultaneous models on orthogonal planes are studied.

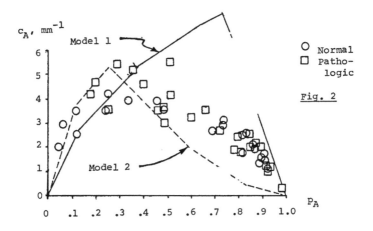

Fig. 2

Consider now the relationship between $f_i$ and "connectivity." Others have used the frequency with which a transect line intercepts an inclusion or other feature as an indication of the degree to which these features are connected across the path of the transect (2) (3). For instance if $f_z = 25$ cm$^{-1}$ and $f_x = 50$ cm$^{-1}$, there would be twice as many cross connections in the z-direction. Equation 3 shows that such a measure of connectivity is directly related to internal surface area. In fact, one might well use specific surface ($s_v$ or $c_A$) as a non-directional measure of connectivity.

Referring again to Fig. 2, it is clear in the light of equation 3 that the maximum value of $c_A$ is associated with the greatest connectivity in the bone, both of the voids and the solid matrix. The connections between voids increase as one moves from left to right and the growing voids break into one another. As even greater porosity values are reached, however, the voids tend to engulf the inter-connections themselves, and the connectivity decreases. The reverse scheme is true for the connectivity of the solid matrix.

*R. B. Martin*

REFERENCES

(1) Martin, R. Bruce, Equivalence of porosity and specific surface measures in one, two, and three dimensions. Pore Structure and Properties of Materials, Prague, Academia, 1973.

(2) Rose, R.M., Dynamic properties of subchondral bone. Proceedings, 5th Annual Biomaterials Symposium, Clemson, S. C., 1973.

(3) DeHoff, R. T. and Rhines, F. N., Quantitative Microscopy. New York, McGraw-Hill, 1968.

National Bureau of Standards Special Publication 431
Proceedings of the FOURTH INTERNATIONAL CONGRESS FOR STEREOLOGY
held at NBS, Gaithersburg, Md., September 4-9, 1975 (Issued January 1976)

ASSESSMENT OF SAMPLING ERRORS IN STEREOLOGICAL ANALYSES

by J. E. Hilliard
Department of Materials Science and Engineering, Northwestern University, Evanston, Illinois 60201, U.S.A.

ABSTRACT

Theoretical and experimental results are presented for the sampling errors in (1) Volume fraction analysis by point counting; (2) The estimation of boundary area per unit volume; and (3) The estimation of the length of lineal features per unit volume.

In all three cases the coefficient of variation of the estimate has the form:

$$\sigma(x)/\bar{x} = k/\sqrt{P_T}$$

in which $P_T$ is the total number of points counted and $k$ is a parameter which may depend on the form of the structure and the test probe. For volume fraction analysis by point counting $k = (1 - V_V)$ provided the point density is not too high. In the estimation of boundary area of contiguous grains $k \rightarrow 1.0$ or 1.4 for straight or circular probes respectively as the probe length tends to zero. For long probes $k \approx 0.6$ for both types. If the boundaries are surfaces of discrete bodies, $k = 1.4$ for long, straight probes. For an analysis of line length $k = 1.0$ if the lineal features are randomly dispersed. A procedure is described for determining $k$ experimentally for structures that do not conform to the ones treated here.

INTRODUCTION

Stereological analyses are subject to various types of errors. These include: (1) Statistical sampling errors which, provided that certain sampling conditions are satisfied, are inversely proportional to the square-root of the number of observations; (2) Errors due to the limited resolution of the microscope and scanning instrument (if the latter is used); (3) Biases introduced by incorrect sampling procedures; (4) Observational errors (for example, miscounts if measurements are made visually or lapses in discrimination with automatic instruments). We shall be considering only sampling errors and we will further limit the treatment to three types of analysis: Volume fraction estimation by point counting; the estimation of boundary area per unit volume; and, the estimation of the length per unit volume of lineal features. These three properties can be determined from measurements on a section without any assumption about the form of the structure. In addition, they require only a counting measurement (as distinct from an analog one). Before considering the individual analyses, we will first derive some statistical relationships that are common to all three.

STATISTICS OF COUNTING MEASUREMENTS

Consider an analysis in which a point, linear, or planar probe is applied n times to a section. If $\bar{P}$ is the average number of point intersections of the probe with structural features then, by definition, the total number of intersections $P_T$ is given by:

$$P_T = n\bar{P}. \tag{1}$$

We next utilize a well known result in statistics that if:

$$X = c_1 x_1 + c_2 x_2 + \ldots \tag{2}$$

where the x's are random variables and the c's are constants, then:

$$\overline{X} = c_1\overline{x}_1 + c_2\overline{x}_2 + \ldots \qquad (3)$$

If, in addition, the x's are independent of one another (we will later examine the implications of this condition in stereological measurements) then:

$$\sigma^2(x) = c_1^2\sigma^2(x_1) + c_2^2\sigma^2(x_2) + \ldots \qquad (4)$$

in which $\sigma^2(\cdot)$ is the variance [i.e., the square of the standard deviation $\sigma(\cdot)$].

Applying Eq. (4) to Eq. (1) we obtain:

$$\sigma^2(P_T) = n^2\sigma^2(\overline{P}). \qquad (5)$$

But since:

$$\overline{P} = (1/n)(P_1 + P_2 + \ldots + P_n),$$

it also follows from Eq. (4) that:

$$\sigma^2(\overline{P}) = (1/n)\sigma^2(P). \qquad (6)$$

[The distinction between $\sigma^2(\overline{P})$ and $\sigma^2(P)$ is that the latter is the variance for a single application of the test probe whereas the former is the variance for the average of n applications.] Substituting into Eq. (5) we find:

$$\sigma^2(P_T) = n\sigma^2(P).$$

Finally, a division by $P_T^2$ $(= n\overline{P}P_T)$ yields:

$$\sigma^2(P_T)/P_T^2 = k^2/P_T, \qquad (7)$$

where

$$k^2 = \sigma^2(P)/\overline{P}. \qquad (8)$$

As we will later see, k depends on the form of the structure and on the shape and size of the test probe. However, once k is known it is possible to estimate a priori the standard deviation of an analysis in terms of the number of observations. Also, since an analysis usually comprises a 100 or more individual observations, it can be assumed from the central limit theorem that the results will be normally distributed. It is therefore possible to attribute a probabilistic interpretation to the standard deviation; for example, that there is approximately a 2/3 probability that the true value of a result x lies in the range x ± σ(x).

Another result that we will be needing is the probability, p(P) that there will be exactly P intersections on a single application of the probe. If the intersections are "randomly" distributed on the probe and if they are independent of one another, then p(P) follows a Poisson distribution; i.e.:

$$p(P) = (1/P!)\overline{P}^P\exp(-\overline{P}). \qquad (9)$$

The variance of this distribution is given by:

$$\sigma^2(P) = \overline{P}. \qquad (10)$$

VOLUME FRACTION ANALYSIS

There are three different procedures for estimating the volume fraction of a constituent: (1) An areal analysis in which the areal fraction of the constituent intercepted by a planar probe is determined. (2) The measurement of the fractional length of a linear probe (applied to a section) that intercepts the constituent. (3) The application of a grid of points to a section and a measurement of the fractional number of points falling within the constituent. In each case the fractional occupancy gives an unbiased estimate of the volume fraction. (This is true even if the measurements are made on a non-randomly oriented section through an anisotropic structure.)

A theoretical analysis made by Hilliard and Cahn (1961) revealed that the standard deviation for all three analyses was determined almost entirely by the number of observations and that the type of observation (estimation of the area of a profile, of the length of an intercept, or the identification of the constituent occupied by a point) had little effect. They therefore concluded that point counting would be the most efficient method of estimation. They also found that if the grid spacing was such that not more than one point can fall on a profile of the constituent being estimated (if the profiles are not discrete the corresponding condition is that there should be little or no correlation between the occupancy of adjacent points) then, in the limit as the volume fraction $V_V \to 0$,

$$\sigma^2(V_V)/V_V^2 = 1/P_\alpha, \tag{11}$$

where $P_\alpha$ is the total number of points falling on the constituent being estimated. It is apparent that Eq. (11) cannot be valid for large volume fractions because it predicts that as $V_V \to 1$, $\sigma^2(V_V)/V_V^2 \to 1/P$ (where $P$ is the total number of points applied) whereas the variance has to go to zero at this limit. An expression for the variance must satisfy the following two conditions if it is to be valid over the whole range of volume fractions. (1) It should be symmetric in $V_V$ and $(1 - V_V)$ [this follows from the requirement that $\sigma^2(V_V)$ be independent of the constituent on which the count is made] and that Eq. (11) should be satisfied in the limit $V_V = 0$. Hilliard (1968) has proposed that the simplest expression* meeting these requirements is:

$$\sigma^2(V_V) = V_V(1 - V_V)/P, \tag{12}$$

or

$$\sigma^2(V_V)/V_V^2 = (1 - V_V)/P_\alpha. \tag{13}$$

Recently, Andersen, James and Hilliard (to be published) have made a study of point-counting sampling errors on a series of two-phase β-brasses. Six different samples were used which were of the same composition but which had been rolled by various amounts to produce elongations varying between 20 and 70 pct. There was thus an appreciable variation in grain shape. Altogether, approximately 8,500 applications of a 9-point grid were made, by two observers, on the specimens. The average volume fraction was 28.1 pct and the average value of $\sigma(V_V)/V_V$ for one application of the grid was 0.532. Setting $V_V = 0.281$ and $P_\alpha = 9 \times 0.281 = 2.53$ in Eq. (13) yields $\sigma(V_V)/V_V = 0.533$. [The agreement between the theoretical and observed values of $\sigma(V_V)/V_V$ is actually better than would be expected on the average.]

From this investigation it can be concluded that Eq. (12) or (13) can be used with some confidence for the prediction of the standard deviation of a volume fraction analysis by point counting. However, it is to be emphasized that they are valid only for low point densities. If, on the average, there are several points per profile these expressions will underestimate the standard deviation.

BOUNDARY-AREA ANALYSIS

The boundary (or surface) area per unit volume, $S_V$, can be estimated by applying a linear probe to a section and counting the number of intersections, $P_L$, per unit length of the probe with boundary profiles. Then, by a well known relationship:

$$S_V = 2\overline{P}_L. \tag{14}$$

The probe need not necessarily be straight but can be of any shape. However,

---

* The same expression has also been given by Gladman and Woodhead (1960) and others. However, the derivations were based on a random point count performed on a single sub-area and they fail to reveal that $\overline{\text{Eq. (12)}}$ is completely invalid if a high point density is used.

if the structure is anisotropic, Eq. (14) is only valid if the probe and the section to which it is applied are randomized with respect to orientation.

It follows from Eqs. (4) and (14) that:

$$\sigma^2(S_V) = 4\sigma^2(P_L) = 4\sigma^2(P_T)/L_T^2, \qquad (15)$$

where $P_T$ is the total number of intersections on a total probe length $L_T$ (i.e., the probe length times the number of applications). Dividing through by $S_V^2$ we obtain:

$$\sigma^2(S_V)/S_V^2 = \sigma^2(P_T)/\bar{P}_T^2, \qquad (16)$$

or from Eq. (8):

$$\sigma^2(S_V)/S_V^2 = k_s^2/\bar{P}_T. \qquad (17)$$

The value of $k_s$ depends on the form of the structure. We will consider two cases: (1) Contiguous grains of the same constituent. (2) Discrete particles embedded in a matrix.

Contiguous Grains

We will assume that three (and only three) grains meet along an edge. (This is usually so if surface tensions are a significant factor.) We first consider the case when the probe is in the form of a circle and will derive a theoretical estimate for $k_s$ for the limit when the expected number of intersections, $\bar{P}$, on the probe approaches zero.

Fig. 1. In an analysis for the boundary area of contiguous grains it is only necessary to consider the outcomes: 0, 2, and 3 intersections for a circular probe and 0, 1, and 2 intersections for a straight probe as the length of the probe tends to zero.

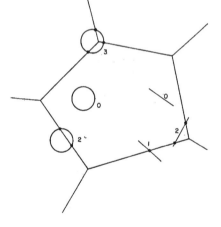

The condition $\bar{P} \rightarrow 0$ is equivalent to the diameter of the probe tending to zero. We therefore need only consider the three possible outcomes: 0, 2 and 3 intersections illustrated in Fig. 1. Let $p(0)$, $p(2)$ and $p(3)$ be the corresponding probabilities of these outcomes, then:

$$\bar{P} = 0p(0) + 2p(2) + 3p(3). \qquad (18)$$

Applying the usual expression for the variance:

$$\sigma^2(P) = -\bar{P}^2 + \Sigma P_i^2 p(P_i),$$

we obtain:

$$\sigma^2(P) = 0p(0) + 4p(2) + 9p(3) - \bar{P}^2. \qquad (19)$$

Multiplying Eq. (18) through by 2 and subtracting from Eq. (19) we find after rearrangement:

$$k_s^2 \equiv \sigma^2(P)/\overline{P} = 2 + \left\{[3p(3)/\overline{P}^2] - 1\right\}\overline{P}, \ [\overline{P} \rightarrow 0], \qquad (20)$$

The next step is to estimate $p(3)$. This probability is equal to the area of the circular probe ($= \pi d^2/4$, where d is the diameter) times the number, $P_A$, of triple junctions per unit area of the section. Thus:

$$p(3) = (\pi d^2/4)P_A. \qquad (21)$$

If $L_V$ is the edge length per unit volume, then:

$$P_A = L_V/2. \qquad (22)$$

It also follows from Eq. (14) that:

$$P = (\pi d)P_L = (\pi d/2)S_V. \qquad (23)$$

Substituting from the last three equations into Eq. (20) we obtain:

$$k_s^2 = 2 + [(3L_V/2\pi S_V^2) - 1]\overline{P}, \ [\overline{P} \rightarrow 0]. \qquad (24)$$

The ratio $L_V/S_V^2$ can be determined for a particular specimen. In order to get a rough estimate of $k_s$ that is generally applicable, we will assume that the specimen is composed of equal size tetrakaidecahedral grains. In this case it is easily shown that $L_V/S_V^2 = 0.7570$ which yields:

$$k_s^2 = 2 - 0.639\overline{P}, \ [\overline{P} \rightarrow 0]. \qquad (25)$$

A plot of this theoretical $k_s$ is shown by the upper dashed curve in Fig. 2.

Fig. 2. The dependence of $k_S$ on the average number of intersections $\overline{P}$ for boundary traces from contiguous grains on a probe for (1) circular probes (upper curve) and (2) straight probes (lower curve). The dashed curves are the theoretical values calculated from Eqs. (25) and (34).

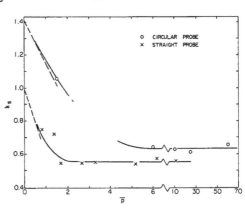

The points are from measurements made by Hilliard (1962) on a sample of silicon iron. It will be seen that both the theoretical and experimental curves display an initial decrease in $k_S$ with increasing $\overline{P}$ (or, equivalently, with increasing size of the probe). However, at $\overline{P} \approx 6$, there is a leveling off and thereafter $k_S$ remains constant at a value of $\sim 0.63$.

We next consider the case of a straight test probe. At the limit $\overline{P} \rightarrow 0$ we can neglect any possibilities other than 0, 1 and 2 intersections (Fig. 1). If $p(0)$, $p(1)$ and $p(2)$ are the corresponding probabilities, then:

$$\bar{P} = 0p(0) + 1p(1) + 2p(2), \tag{26}$$

and

$$\sigma^2(P) = 0p(0) + 1p(1) + 4p(2) - \bar{P}^2. \tag{27}$$

Taking the difference between the two equations and dividing by $\bar{P}$ we obtain:

$$k_s^2 \equiv \sigma^2(P)/\bar{P} = 1 + \left\{[2p(2)/\bar{P}^2] - 1\right\}\bar{P}, \quad [\bar{P} \to 0]. \tag{28}$$

In order to estimate $p(2)$ we will assume that the angle between the boundary traces at a triple junction is $120^\circ$. Consider a probe of length $\ell$ oriented at an angle $\theta$ to one of the boundary traces. Then, as shown in Fig. 3, in order to obtain two intersections, the center of the probe has to fall within

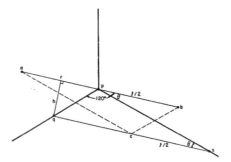

Fig. 3. The center of a straight probe of length $\ell$ (= ab) oriented at an angle $\theta$ to a boundary must fall within the triangle abc in order to generate two intersections.

the triangle abc, where the construction is such that $ap = pb = qc = \ell/2$, and ab is parallel to qs. Thus the probability that a probe randomly oriented over the range $0 \le \theta \le \pi/3$ will yield two intersections is:

$$p(2) = \bar{A}P_A \tag{29}$$

where $\bar{A}$ is the average area of the triangle and $P_A$ is the number of triple junctions per unit area.
Noting that:

$$pq = \ell\sin\theta/\sin(2\pi/3) = (2\ell/\sqrt{3})\sin\theta,$$

and that the height of the triangle is given by:

$$h = pq\sin[(\pi/3) - \theta],$$

we obtain for the area of the triangle abc:

$$A(\theta) = (\ell^2/\sqrt{3})\sin\theta\sin[(\pi/3) - \theta]. \tag{30}$$

We next have to determine the average area of the triangle for a randomly oriented probe. For $0 \le \theta \le \pi/3$ the distribution function, $f(\theta)$, for $\theta$ is $3/\pi$. Hence:

$$\bar{A} \equiv \int_0^{\pi/3} A(\theta)f(\theta)d\theta = (3/\pi)\int_0^{\pi/3} A(\theta)d\theta.$$

Substituting for $A(\theta)$ from Eq. (29) and performing the integration we obtain:

$$\bar{A} = (1/4\pi\sqrt{3})(3\sqrt{3} - \pi)\ell^2 \approx 0.0944\ell^2.$$

Hence from Eq. (29):

$$p(2) = 0.0944\ell^2 P_A. \tag{31}$$

From Eq. (14):

$$\ell = 2\bar{P}/S_V.$$

Substitution of $\ell$ into Eq. (31) together with the value of $P_A$ given by Eq. (22) yields:

$$p(2) = 0.1888(L_V/S_V^2)\bar{P}^2. \tag{32}$$

We thus obtain from Eq. (28):

$$k_s^2 = 1 + [0.3776(L_V/S_V^2) - 1]\bar{P}, \quad [\bar{P} \rightarrow 0]. \tag{33}$$

If we again assume that the grains are equal-size tetrakaidecahedra, then:

$$k_s^2 = 1 - 0.714\bar{P}, \quad [\bar{P} \rightarrow 0]. \tag{34}$$

A plot of this theoretical $k_s$ is shown by the bottom dashed curve in Fig. 2. The experimental points were determined by Philofsky (1966) on the same specimen of silicon iron as was used for the circular probe measurements. The grains in this specimen were somewhat elongated and Philofsky found that $k_s$ for probes parallel to the direction of elongation were about 15 pct lower than those in the transverse direction. (The values plotted in Fig. 3 are the averages for the two directions.) It will be seen that for large $\bar{P}$, $k_s$ for the straight probe is slightly less than that for a circular probe. However, this is probably due to the anisotropy in the grain shape since, in general, one would expect the properties of a circular probe to approach those of a straight probe as the diameter of the former is increased.

### Non-Contiguous Particles

For the case of non-contiguous particles we will restrict the treatment to linear probes that are long compared with the mean free path between particle profiles. With this condition it is permissible to neglect instances when the probe starts or terminates within a profile and assume that all intersections occur in pairs. Thus, if $\bar{N}$ and $\sigma^2(N)$ are the average and variance in the number of intercepts on the probe, we have from Eqs. (3) and (4) with respect to the number of point intersections on profile boundaries:

$$\bar{P} = 2\bar{N}, \tag{35}$$

and

$$\sigma^2(P) = 4\sigma^2(N). \tag{36}$$

We cannot proceed further without some assumption about the dispersion of the profiles on the section. A reasonable hypothesis is that the particles are randomly dispersed. (Some measurements we are currently making indicate that this is a valid approximation for nodular and flake graphite particles in cast irons.) The assumption of a random dispersion requires that the volume fraction of particles be small (say, < 15 pct) as, otherwise, there is an excluded area effect. For a random dispersion, the number, N, of intercepts on the probe will follow a Poisson distribution. Hence, from Eqs. (10), (35), and (36):

$$k_s = \sigma^2(P)/\bar{P} = 2. \tag{37}$$

Thus, for non-contiguous particles, $k_s$ is about twice as large as it is for a long probe on contiguous grains.

### Discussion

The theoretical and experimental results indicate that for contiguous grains $k_s$ decreases with increasing probe length until a limit is reached corresponding to $\bar{P} \approx 2$ for a straight probe and $\bar{P} \approx 6$ for a circular probe. Thereafter $k_s \approx 0.6$ for both types of probe. It thus follows from Eq. (17) that the relative standard deviation of the analysis can be estimated from:

$$\sigma(S_V)/S_V \approx 0.6/\sqrt{P_T}, \qquad (38)$$

in which $P_T$ is the total number of intersections counted. For non-contiguous particles occupying a small volume fraction we have:

$$\sigma(S_V)/S_V \approx 1.4/\sqrt{P_T}. \qquad (39)$$

Two qualifications are necessary in the use of these expressions. The first is that they should not be applied when the structure does not approximate the two cases considered [i.e., a space filling cellular type for Eq. (38) and a low density of discrete bodies for Eq. (39)]. For other types of structure $k_s$ should be estimated experimentally as described in a later section.

The second qualification is that Eq. (7) [and therefore Eqs. (38) and (39)] are valid only if the individual measurements are uncorrelated. As an extreme example consider an estimation of $P_L$ in which a series of traverses are made that exactly overlap one another. It is obvious that, no matter how many traverses are made, the result can never be any more reliable than that obtained on the first traverse. Thus, if parallel traverses (or probes) are used, the spacing between them should be commensurate with the scale of the structure. This condition is not satisfied by automatic image scanners using closely spaced scans. However, an approximate estimate of the sampling error can be obtained in such cases by setting $P_T$ in Eqs. (38) or (39) equal to the actual number of intersections divided by the average number of scanlines intercepting a grain or particle profile.

LINEAL FEATURE ANALYSIS

Lineal features within a specimen yield point intersections on a planar probe. If $L_V$ is the length per unit volume and $P_A$ the number of intersections per unit area, then:

$$L_V = 2P_A. \qquad (40)$$

By analogy with Eq. (39):

$$\sigma(L_V)/L_V = k_1/\sqrt{P_T} \qquad (41)$$

where $P_T$ is the total number of intersections counted. If the features are randomly dispersed then it follows from Eqs. (8) and (10) that:

$$k_1 = 1. \qquad (42)$$

An experimental study by Hilliard (1966) showed that the assumption of a random dispersion was valid for grain edges in a polycrystalline metal specimen and that $k_1$ was independent of the size of the probe. However, these results will not necessarily be true in general and $k_1$ should therefore be determined experimentally.

EXPERIMENTAL DETERMINATION OF k

As an illustration of the experimental determination of k we will consider some of the data collected in the study of grain edge length referred to in the previous section. A circle was applied 200 times to non-overlapping regions on a set of micrographs and the number, P, of triple junctions falling within the circle was counted. The observed frequency of counts is given in the Table.

TABLE

Distribution of Counts in Test Circle

| $P_i$ | Frequency, $n_i$ |
|---|---|
| 0 | 114 |
| 1 | 61 |
| 2 | 18 |
| 3 | 6 |
| 4 | 1 |
| > 5 | 0 |

The average number of counts is calculated from:

$$\bar{P} = \Sigma n_i P_i / n, \tag{43}$$

where $n = \Sigma n_i$. For the data given in the Table, $n = 200$ and $\Sigma n_i P_i = 119$, which yield $\bar{P} = 0.595$. The variance is given by:

$$\sigma^2(P) = [\Sigma n_i P_i^2 - n\bar{P}^2]/(n - 1), \tag{44}$$

which yields $\sigma^2(P) = 0.664$. Thus:

$$k_1 \equiv \sigma(P)/\sqrt{\bar{P}} = 1.057,$$

which differs slightly from the theoretical value of 1. In order to test whether the difference is statistically significant we need to calculate the standard deviation of $k$. It can be shown that:

$$\sigma(k) \approx (k/2\sqrt{n})\left\{[\mu_4(P)/\sigma^4(P)] + (k^2/\bar{P}) - 1\right\}^{1/2}, \tag{45}$$

in which $\mu_4(P)$ is the fourth moment of P about the average; i.e.:

$$\mu_4(P) = (1/n) \Sigma n_i (P_i - \bar{P})^4. \tag{46}$$

The data in the Table yield $\mu_4(P) = 2.106$ which on substitution into Eq. (45) gives $\sigma(k) = 0.09$. We can therefore conclude that observed deviation of 0.06 from unity is not significant.

REFERENCES

Gladman, T. & Woodhead, J. H. (1960) The accuracy of point counting in metallo-
    graphic investigations. J. Iron Steel Inst., 194, 189.
Hilliard, J. E. (1962) Grain-size estimation. General Electric Research
    Laboratory Report 62-RL-3133M.
Hilliard, J. E. (1966) Sampling error in the estimation of grain-edge length.
    Trans. Met. Soc. AIME 236, 589.
Hilliard, J. E. (1968) Measurement of volume in volume. In: Quantitative
    Microscopy, p. 45. (Ed. by R. T. DeHoff and F. N. Rhines) McGraw-Hill,
    New York.
Hilliard, J. E. & Cahn, J. W. (1961) An evaluation of procedures in quanti-
    tative metallography for volume-fraction analysis. Trans. Met. Soc.
    AIME 221, 344.
Philofsky, E. M. (1966) MS Thesis, Northwestern University.

*National Bureau of Standards Special Publication 431*
*Proceedings of the* FOURTH INTERNATIONAL CONGRESS FOR STEREOLOGY
*held at* NBS, Gaithersburg, Md., September 4-9, 1975 (Issued January 1976)

ON OPTIMAL FORMS FOR STEREOLOGICAL DATA

by A. J. Jakeman and R. S. Anderssen
*Computer Centre, Australian National University, Canberra, ACT, 2600, Australia*

INTRODUCTION

A basic aim in stereology is the determination of three-dimensional (structural) properties of some given object from lower-dimensional data ($\ell$-d data) of it. Recent research has concentrated on the construction and inversion of the formulas which define some particular three-dimensional (structural) property in terms of the various forms of $\ell$-d observational data which can be used. See, for example, the work of Santaló (1955) on the size distribution of similarly shaped convex particles, and that of Matheron (1974) on the application of random set theory to the construction of stereological formulas, as well as the examinations of Jakeman and Anderssen (1974) and Anderssen and Jakeman (1974) of the different computational methods proposed for the inversion of the formulas for the thin section and random spheres models.

For any three-dimensional structural property, it seems that a classification of the various forms of $\ell$-d data would be useful, so that *optimal* forms for it can be delineated for the determination of the given property. Other advantages of such a classification are obvious. Among other things, it can be used by experimenters as the basis for the design of stereological experiments to determine specific three-dimensional properties. For each three-dimensional (structural) property, the aim is to highlight, and thereby avoid the use of, suboptimal forms of $\ell$-d data.

It is clear that the actual classification must be based on the accuracy and efficiency with which the required three-dimensional property can be obtained. The implementation of a classification along such lines is not straightforward. The first complication is the fact that the types of lower-dimensional observational data which can be used to estimate the given property depend heavily on underlying assumptions. For example, Hilliard's (1968) conclusion that systematic point counting data is optimal for volume-in-volume estimation rests largely on his definition of efficiency and on his choice of minimum grid spacing.

In trying to produce a classification, the problem of the underlying assumptions has even deeper ramifications than this. For example, let us consider a distribution of ovoids in a field. If no assumption is made about these ovoids other than that they are convex, then it follows from the general Minkowski functional formulation for Crofton's theorem {see, for example, Hadwiger (1957), p.209} that it is not possible to estimate the number density of ovoids from $\ell$-d observations - i.e., no formula relating the number density of the ovoids to $\ell$-d observations exists.

Thus, the essential steps in such a classification must be:

Step 1. For a particular three-dimensional (3-d) property and given assumptions, the determination of the types of $\ell$-d data for which the existence of formulas (which relate this data to the required 3-d property) is guaranteed.

Step 2. The specification of the conditions under which the classification is made.

Step 3. The classification.

Since space does not allow a full classification of all stereological data at this stage, we limit attention to the following points:

(i)    The determination of the size distribution and number density of

69

spheres in some given matrix.

(ii)   Optimal properties of linear versus planar data.

For brevity, we do not pursue the question of the exclusion of $\ell$-d data on the basis of Crofton's theorem or other grounds such as smallness of sample size. In fact, with respect to a given 3-d property, we will only work with $\ell$-d data for which formulas relating the two are known, and will assume that the amount of available data is adequate for the computational methods to be considered.

In this note, the classification will be based on the numerical stability and reliability of the computational methods available for the evaluation of the 3-d property from the known formula when using observational (noisy) data. It is also assumed that the wrong formula is not used, as has been the practice in the past when, for example, the random spheres model has been used to invert thin section data. Use of the thin section model avoids the Holmes' effect (see, for example, Tallis (1970)) and the necessity to correct for it. In fact, computing expertise has reached the point where it is no longer necessary to use expedients such as model degradation and subsequent correction of the abovementioned type. The correct formula should be solved on its own merits using available techniques. If such techniques do not yield satisfactory results, it is preferably to seek new ones which do, rather than appeal to some model degradation strategy.

## THE SIZE DISTRIBUTION AND NUMBER DENSITY OF SPHERES

Under the assumption that the population of spherical particles, with a size distribution (probability density) function $g(x)$ for the diameters, is randomly dispersed throughout space and that their centers are defined by a Poisson process of low intensity such that overlapping of spheres does not occur, it is known that the following formulas exist for defining $g(x)$, and the number density $N_V$ (number per unit volume), in terms of $\ell$-d data:

1.   Plane-of-Polish Data   (The Random Spheres Model)

$$h(y) = \frac{y}{m} \int_y^\infty (x^2 - y^2)^{-\frac{1}{2}} g(x)dx, \quad 0 \leqslant y < \infty , \tag{1a}$$

$$N_V = N_A/m , \tag{1b}$$

where $m$ is the expectation of the spherical diameters, $h(y)$ is the size distribution of the circular diameters observed in the random plane-of-polish sections, and $N_A$ is the expected number density of circles (number per unit area).

2.   Thin Section Planar Data   (The Thin Section Model)

$$c(y) = \frac{1}{m + t} \left\{ y \int_y^\infty (x^2 - y^2)^{-\frac{1}{2}} g(x)dx + t\, g(y) \right\} , \tag{2a}$$

$$N_V = N_A/(m + t) , \tag{2b}$$

where $c(y)$ is the size distribution of circular diameters observed in random planar thin sections of thickness $t$ .

3.   Linear Probe Data In Plane-of-Polish

$$g(x) = -\frac{d}{dx} \left\{ \frac{f(x)}{x} \right\} / f'(0) , \tag{3a}$$

$$N_V = \frac{1}{2} f'(0) N_L , \tag{3b}$$

where $f(x)$ is the observed size distribution of the length of line intercepts between the linear probes and circles in the plane-of-polish, and $N_L$ is the expected number density of intercepts.

Note.   The above formulas for number density exist because the assumption that the ovoids are simply convex has been greatly strengthened by assuming that they are in fact spherical particles with centers defined by a Poisson process. This is much stronger than the minimum assumptions required to yield number density of ovoids from ℓ-d data; viz., similarity of the convex particles distributed homogeneously and isotropically throughout space.

Now, Moran (1972) states in §6 in his review of "the probabilistic basis of stereology" that:

"linear probes should not be used, if the density of
centers and diameter distribution are required because
the use of plane sections is more effective".

The basis for his conclusions is Nicholson's (1970) work on the use of linear probe data to estimate diameter distribution and the density of centers.

Classifying this ℓ-d data on the basis of the numerical stability and reliability of the computational methods available for the evaluation of the above formulas, we aim to confirm and substantiate Moran's conclusion.

For the numerical solution of (1a) and (2a), we use the procedures proposed by Anderssen and Jakeman (1974). For synthetic planar data, which yield a bimodal size distribution of spheres, the results are shown in Figures 1(a) and 1(b). We compare the results with those obtained for the solution of 3(a) by the following two methods.

M1.    Group the given cumulative data $\{F(x_k);\ k=0,1,2,\ldots,n\}$ onto an even grid $\{t_k\}$ in the abscissae direction to obtain $\{F(t_k);\ k=0,1,2,\ldots,N, N <$ $n/2\}$ . Then estimate $g(t_k)$ of (3a) as

$$\tilde{g}(t_k) = -\left\{\frac{1}{t_k}\left[\frac{d^2F(t)}{dt^2}\right]^*_{t=t_k} - \frac{1}{t_k^2}\left[\frac{dF(t)}{dt}\right]^*_{t=t_k}\right\} \Big/ \left[\frac{d^2F(t)}{dt^2}\right]^*_{t=0}$$

for $k=0,1,2,\ldots,N$ . Here $[d^mF(t)/dt^m]^*_{t=t_k}$ denotes the FFT spectral derivative of order $m$ of Anderssen and Bloomfield (1974) evaluated at the point $t=t_k$ for the data $\{F(t_k)\}$.

M2.    Using the even grid of cumulative size distribution data in the ordinate direction, estimate $g(x_k)$ of (3a) as

$$\tilde{\tilde{g}}(x_k) = -\left\{\frac{1}{x_k}\left[\frac{d^2x}{dF^2}\right]^*_{x=x_k} \Big/ \left(\left[\frac{dx}{dF}\right]^*_{x=x_k}\right) \right. +$$

$$\left. 1\Big/\left(x_k^2\left[\frac{dx}{dF}\right]^*_{x=x_k}\right)\right\} \cdot \left\{\left[\frac{dx}{dF}\right]^*_{x=0}\right\}^3 \Big/ \left[\frac{d^2x}{dF^2}\right]^*_{x=0}$$

For synthetic linear probe data, which yield the abovementioned biomodal size distribution of spheres, the results obtained using these two method are shown in Figures 2(a) and 2(b).

It is clear that a comparison of Figures 1 and 2 justifies Moran's conclusion. In addition, we see that the inversion of planar and thin section data

71

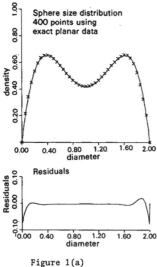

Figure 1(a)

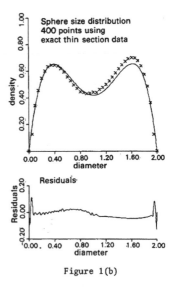

Figure 1(b)

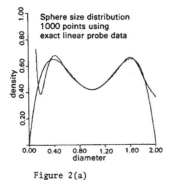

Figure 2(a)

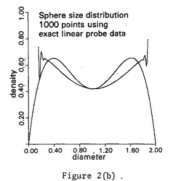

Figure 2(b) .

does not suffer from such severe numerical problems when methods of the proposed
type are used. As shown in the Figures, even the use of exact data does not
alleviate the problem posed by linear probe data. Thus, we conclude that, for
size distribution estimation, linear probe data are suboptimal. The two factors
which contribute most to yield this conclusion are the effect of the $1/x$ factor
in the neighbourhood of the origin and the near impossibility of being able to
accurately estimate the constant of proportionality $f'(0) = F''(0)$ from truncated data. Because of the latter problem it follows that the best which can be
achieved with linear probe data is the determination of the shape of $g(x)$ but
not its true-scaled value.

## VOLUME IN VOLUME

Even though we have concluded above that there exists a context in which
planar data are superior to linear probe, this does not establish a general rule
of thumb. In fact, Hilliard (1968) has shown that there do exist situations
where the linear probe are superior to the planar.

When estimating volume in volume, the computational problem (viz., evaluation of a fraction) is trivial, and so, the classification of data on the basis
of the numerical stability and reliability of the resulting solutions is inappro-
priate, and some more suitable criterion is required. Using efficiency, defined
in terms of the least effort required to yield a given accuracy, as the basis
for classification, and excluding automatic measuring processes from consideration, Hilliard (1968) concludes that point counting data are optimal for the
measurement of volume in volume. The conclusion hinges on the assumption that it
is easier and simpler to decide whether a point falls on a region than it is to
specify the boundaries of the region.

This corroborates the above point that the optimality of $\ell$-d data for the
determination of a given 3-dimensional structural property depends heavily on
underlying assumptions and the criterion of optimality.

## CAVEAT

A conclusion that, with respect to a given optimality criterion, a particular
type of $\ell$-d data is suboptimal for the determination of some 3-d structural property does not necessarily exclude its use. However, it does indicate that there
exist other $\ell$-d data for which the required structural property can be obtained
with greater reliability and/or efficiency. For example, in concluding above
that, compared with planar data, linear probe data are suboptimal when determining
a size distribution for spheres, it is not implied that linear probe data cannot
be used; but rather than they should be avoided unless there exists for their use
a strong independent argument which is outside the context of the present considerations.

## REFERENCES

R.S. Anderssen and P. Bloomfield (1974) A time series approach to numerical
differentiation, *Technometrics* 16, 69-75.

R.S. Anderssen and A.J. Jakeman (1974) On computational stereology, *Proc. 6th.
Aust. Comp. Conf.*, Sydney, Vol. 2, 353-362.

H. Hadwiger (1957) *Vorlesungen über Inhalt, Oberfläche und Isoperimetrie,*
Springer, Berlin.

J.E. Hilliard (1968) Measurement of volume in volume, in *Quantitative Metallography* (Eds. R.T. DeHoff and F.N. Rhines), McGraw Hill, New York.

A.J. Jakeman and R.S. Anderssen (1974) A note on numerical methods for the thin
section model, *Electron Microscopy 1974*, Vol. 2, 4-5.

G. Matheron (1974) *Random Sets and Integral Geometry*, J. Wiley Interscience, New York.

P.A.P. Moran (1972) The probabilistic basis of stereology, *Suppl. Adv. Appl. Prob.* (1972), 69-86.

W.L. Nicholson (1970) Estimation of linear properties of particle size distributions, *Biometrika* 57, 273-298.

L.A. Santaló (1955) Sobre la distribución de los tamaños de corpúsculos contenidos en un cuerpo a partir de la distribución en sus secciones ó proyecciones, *Trabajos Estadist* 6, 181-196.

G.M. Tallis (1970) Estimating the distribution of spherical and elliptical bodies in conglomerates from plane sections, *Biometrics* 26, 87-103.

National Bureau of Standards Special Publication 431
Proceedings of the FOURTH INTERNATIONAL CONGRESS FOR STEREOLOGY
held at NBS, Gaithersburg, Md., September 4-9, 1975 (Issued January 1976)

MULTIVARIATE DATA ANALYSIS TO DESCRIBE INTRA-AND INTERGRANULAR RELATIONS IN THIN SECTIONS

by W. Good

Swiss Federal Institute for Snow and Avalanche Research, CH-7260 Weissfluhjoch/Davos, Switzerland

## Introduction

Structure and texture are only part of the physical reality. The physical and mechanical properties of matter are other important qualities. The description of structural properties therefore should aim at a physical understanding of the material being investigated.

The scope of this paper is to present a semi-quantitative correlation between structural elements and mechanical properties.

## Experimental part

Samples of homogeneous snow were altered under controlled conditions. The resulting samples were tested according to ram hardness $R(N)$ and tensile strength $\sigma_T(N/m^2)$.

In parallel experiments, thin sections were prepared and structural elements investigated.

## Numerical treatment of thin section information

The primary information for each thin section is a point grid of light intensities. Each observation point is implicitely characterized by the orientation of the crystallographic c-axis.

The computer program MSTER.FTN either generates coherent areas joining points with the same crystallographic information (Fig 1), or links contiguous areas even if crystallographically inhomogeneous (Fig 2).

A parametrization of these simple or complex areas yields 28 structural parameters (1) divided into:
- Parameters of point information (Point density e.g.).
- Stereological parameters (Grain surface per unit volume, mean grain diameter, mean void diameter etc.).
- Geometrical parameters (Grain area and circumference, mean radius of convex and concave borders, mass distribution (Principal moments of inertia tensor) etc.).
- Crystallographic information (Azimuthal and zenithal angle of c-axis).

Out of this redundant set of parameters, 21 are kept for a subsequent treatment.

## Factorial analysis

Each thin section, being a sample of the corresponding snow, is therefore characterized by its p parameters (or 2p parameters for inter- and intra-grain conditions).

N samples with p parameters form a nxp matrix. Substituting each value in this matrix by its centered and reduced one, a pxp dimensional correlation matrix may be obtained. It can be shown (2) that the eigenvalue problem for this quadratic and symmetric matrix yields an optimum, orthogonal coordinate frame. The axes are linear combinations of the parameters. Parameters and samples may both be located in this coordinate system and the euclidian

W. Good

distance between them is a measure of their differences.

*Discussion of results*

For each sample two sets of p parameters have been generated.

Fig 1 corresponds to the crystallographically homogeneous
areas generated(intragrain representation)whereas Fig 2 is an
intergrain representation of the same sample with connected
structures.

The second one is better fitted for the discussion of
mechanical properties. Fig. 3 and 4 are projections of samples
and parameters for Fig 3 in the first two principal planes. It
is seen from Fig. 3 that $F_1$ is mainly a linear combination of
parameter groups III and IV (see further discussion).

In addition, sample groups with similar tensile strength
have been delimited (from left following the arrows $\sigma_T = 12000$,
40750, 56500 and 217000 $N/m^2$). This evolution in physical phase
space corresponds more or less to the path from parameters groups
I, II, ... to VI. Briefly they are characterized as follows:
- I     Important variances of domain area and circumference as
        well as crystallographic index.
- II    Large voids and few grains.
- III   Well defined border conditions i.e. lengths  and radius of
        convex and concave circumferences.
- IV    High density and large grain areas.
- V     Well defined mass distribution and orientation of the grains.
- VI    Large Ellipticity
        ($a^2/4\pi\sqrt{I_1 I_2}$ , a = area, $I_1$, $I_2$ principal moments of
        inertia).

*Conclusion*

In addition to this description, it is possible for a given
sample to compute by discriminant analysis its probability to
belong to one of the different subgroups.

This kind of semi-quantitative comparison is a first step
in the finding of models without assumptions, connecting structural and physical parameters.

*Acknowledgment*

I am indebted to Dr. H.U. Gubler (Swiss Federal Institute
for Snow and Avalanche Research, Weissfluhjoch/Davos) for preparing
the snow samples and performing the mechanical experiments as well
as for valuable discussions.

*Bibliography*

1 Good, W. 1975. Numerical Parameters to Identify Snow Structure.
  To be published in:
  IAHS Publ. No. 114 (International Association of Hydrological
  Sciences (Proceedings of the Grindelwald meeting 1974).

2 Lebart, L. and Fenelon, J.P. 1973. Statistique et Informatique
  appliquées, Paris, Dunod 2nd Edition.

## Fig.1 Homogeneous Areas

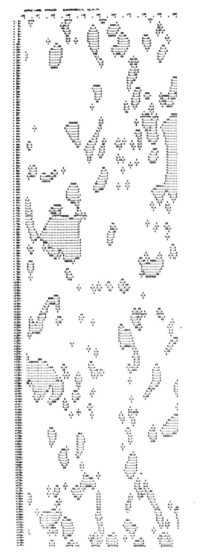

## Fig.2 Contiguous Areas

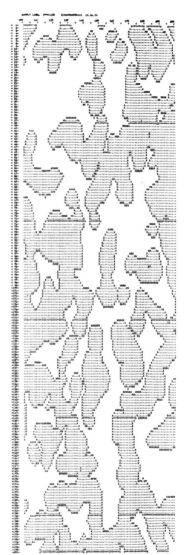

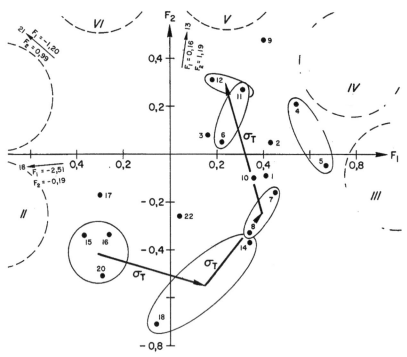

Fig. 3. Intergrain relations (samples 1-22, Gu, Kr, Ho) and tensile strength in plane of first two factors (68,7% of total information).

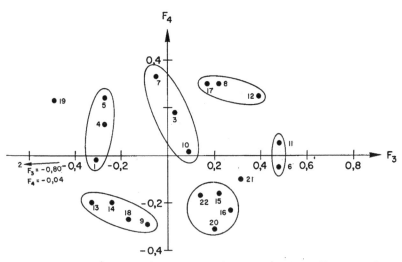

Fig. 4. Intergrain relations (samples 1-22, Gu, Kr, Ho) and tensile strength in plane of factors 3 and 4 (84,4% of total information).

*National Bureau of Standards Special Publication 431*
*Proceedings of the* FOURTH INTERNATIONAL CONGRESS FOR STEREOLOGY
*held at* NBS, Gaithersburg, Md., September 4-9, 1975 (Issued January 1976)

CORRECTION OF STEREOLOGICAL PARAMETERS FROM BIASED SAMPLES ON NUCLEATED PARTICLE PHASES

by Luis-M. Cruz Orive,
*Department of Human Biology & Anatomy, The University, Sheffield S10 2LA, and Department of Probability*
*& Statistics, The University, Sheffield S3 7RH, England*

INTRODUCTION

A system of 'free' cells in a medium can be modelled by a 3-d particulate phase $\alpha$ dilutely embedded in a matrix $\beta$. The particles, however, are not homogeneous; in general, each of them consists of a body $K_2$ contained in a 'bigger' one $K_1$ (i.e., $K_1 \supset K_2$). $K_1$ may be regarded as a cell, and $K_2$ its nucleus. The matrix may consist of many distinct phases.

It is required to estimate, (i) the true ratio $V_{VT} = V_2/V_1$ of the total nuclear volume to the total cellular volume, and (ii) the specific surface area $(S_1/V_1)_T$ of the cells, from measurements made on random plane sections through the aggregate. In normal circumstances, the corresponding frequency-free consistent estimates of the above parameters are well known, namely,

$$V_{VT} = (V_2/V_1)_T \sim (\Sigma_1^n A_2/\Sigma_1^n A_1) \quad ; \quad (S_1/V_1)_T \sim \frac{4}{\pi}(\Sigma_1^n B_1/\Sigma_1^n A_1), \qquad (1)$$

where $A_1$, $A_2$ denote cellular and nuclear profile areas, respectively, whereas $B_1$ is the perimeter length of a cell profile. The symbol $\sim$ becomes $=$ as the number $n$ of profiles monitored becomes infinity.

Suppose, however, that the cell profiles cannot be identified unless a nuclear fraction is present in them. In other words, only transnuclear profiles are analyzed, whereas the paranuclear ones - arising when the plane of section T hits $K_1$ but not $K_2$ - are discarded. Under this sampling regime, the observed ratios

$$V_{VO} \sim (\Sigma_1^n A_2/\Sigma_1^n A_1 \mid T \dagger K_2) \quad ; \quad (S_1/V_1)_O \sim \frac{4}{\pi}(\Sigma_1^n B_1/\Sigma_1^n A_1 \mid T \dagger K_2). \qquad (2)$$

are no longer consistent estimates of the required parameters (1), and correction functions which relate $V_{VO}$ to $V_{VT}$ on the one hand, and $(S_1/V_1)_O$ to $(S_1/V_1)_T$ on the other, are necessary. This is, in essence, the statement of the problem. (The bars in (2) denote 'conditional to', and '$T \dagger K_2$' means 'T hits $K_2$').

THE MATHEMATICAL MODEL: GENERAL CORRECTION FUNCTIONS

Let us assume that the cells $K_1$ have a fixed shape and a variable size determined by a single parameter X; likewise, $K_2$ has a fixed shape (different from that of $K_1$, in general) and a variable size, described by another parameter Y. In general, (X,Y) will be a 2-d random variable with a distribution function $G(x,y)$. We then speak of a 'polydispersed phase of nucleated particles', and $G(x,y)$ represents the size distribution function of the nucleated particles in the system. When X and Y are constant, we speak of a 'monodispersed phase of nucleated particles'.

In the general case, the stereological parameters sought are,

$$V_{VT} = k_2 \cdot E(Y^3)/ k_1 \cdot E(X^3) \quad ; \quad (S_1/V_1)_T = k_3 \cdot E(X^2)/ E(X^3), \qquad (1a)$$

respectively, where $k_1$, $k_2$ and $k_3$ are constant shape coefficients, and $E(.)$ denotes 'expectation' with respect to G.

The correction functions are obtained by evaluating the right hand sides of (2) when $n \to \infty$, and combining the results with the respective equations (1a). For simplicity, it can be assumed that the relative orientation of $K_2$ with respect to the containing $K_1$ does not change, so that their relative position is fully determined by a random vector $\vec{h} = \overrightarrow{O_2 O_1}$ joining their respective centroids. With the above requirements and notation, the equations (2) become

$$V_{VO} = \frac{2\pi k_2 \cdot E(Y^3)}{E\int_{h}^{} E\int_{\Omega_2}^{} A_1(T|x,h)\ dT} \quad ; \quad (S_1/V_1)_O = \frac{4}{\pi}\frac{E\int_{h}^{} E\int_{\Omega_2}^{} B_1(T|x,h)\ dT}{E\int_{h}^{} E\int_{\Omega_2}^{} A_1(T|x,h)\ dT} \qquad (2a)$$

79

which combined with (1a) give rise to the correction functions of biased nuclear volume fraction and specific surface area estimates, respectively, on a polydispersed phase of nucleated particles. $E$ and $\underset{g}{E}$ denote expectations with respect to the distribution of $(X,Y)$ and of $h$, respectively, whereas $dT = \sin\theta \, dp \, d\phi \, d\theta$ is the element of invariant measure for random planes in 3-d (see, e.g., Santaló 1953, p. 121; Kendall&Moran, 1963). $\Omega_2$ denotes the set $\{T : T \uparrow K_2\}$.

The corresponding correction formulae for monodispersed phases of nucleated particles are obtained by deleting $\underset{g}{E}$ in the above expressions.

CORRECTION FORMULAE FOR MONODISPERSED PHASES CONSISTING OF A CERTAIN CLASS OF NUCLEATED PARTICLES.

Suppose that the bodies $K_1(X) \supset K_2(Y)$, $(X, Y, \text{constant})$, constituting each particle are convex and satisfy the additional conditions,

(i) $K_2$ is a contraction of $K_1$ with ratio $0 < \lambda < 1$, and the centroid 0 of $K_1$ as invariant point.

(ii) $K_1$ and $K_2$ belong to a class of bodies in which the area and boundary length of a plane section of coordinates $(p, \phi, \theta)$ factorize as

$$A(p,\phi,\theta) = f\{p/H(\phi,\theta)\}\cdot g(\phi,\theta) \; ; \; B(p,\phi,\theta) = u\{p/H(\phi,\theta)\}\cdot v(\phi,\theta), \qquad (3)$$

where $H(\phi,\theta)$ is the function of support of the boundary of the body, and represents the distance of the origin 0 from the tangent plane perpendicular to the direction $(\phi,\theta)$ (see, e.g., Bonnesen&Fenchel, 1934, p.23).

If $K_1$ and $K_2$ satisfy the above conditions, the equations (1a) and (2a) yield the following correction formulae,

$$V_{VO} = F(1)\cdot\frac{V_{VT}}{F(V_{VT}^{\frac{1}{3}})} \; ; \qquad (S_1/V_1)_0 = \frac{F(1)}{U(1)}\cdot\frac{U(V_{VT}^{\frac{1}{3}})}{F(V_{VT}^{\frac{1}{3}})}\cdot(S_1/V_1)_T, \qquad (4)$$

where $F(Z) = \int_0^Z f(z)\,dz$, and $U(Z) = \int_0^Z u(z)\,dz$.

EXAMPLE. Let the particles consist of two triaxial ellipsoids $K_1 \supset K_2$ with the same eccentricities, a constant size and the three principal axes in common. It can be seen that the conditions set out above are fulfilled; the relevant functions are found to be $f(z) = 1-z^2$, $u(z) = (1-z^2)^{\frac{1}{2}}$, and the corrections (4) become (after appropriate inversion).,

$$V_{VT} = \left(\frac{3V_{VO}}{V_{VO} + 2}\right)^{\frac{3}{2}} \; ; \qquad (S_1/V_1)_T = \frac{\pi}{4}\cdot\frac{3\,v_{VT}^{\frac{1}{3}} - V_{VT}}{\sin^{-1}V_{VT}^{\frac{1}{3}} + V_{VT}^{\frac{1}{3}}\{1-v_{VT}^{\frac{2}{3}}\}^{\frac{1}{2}}}\cdot(S_1/V_1)_0.\, (5)$$

The above formulae coincide with that of a 'concentric spheres' model; the result is logical, since this cell results after applying a linear transformation to the ellipsoid-ellipsoid cell described above, and conversely. The first (5) was first given by Konwinski&Kozlowski(1972), and rederived by Mayhew&Cruz (1973).

CORRECTION FORMULAE FOR MONODISPERSED PHASES WHERE NUCLEUS AND CELL ARE DISSIMILAR SPHEROIDS.

When $K_1$, $K_2$ are either dissimilar or eccentric with respect to each other, the corrections (4), (5) do not hold, and the general formulations (1a) and (2a) have to be recalled. This we have done for monodispersed phases where $K_1$ and $K_2$ are ellipsoids of revolution (spheroids) of the same kind – either prolate or oblate – since the nucleus usually becomes prolate as the cell elongates, and oblate as the cell flattens. The correction formulae are collected in Tables 1 and 2, together with the description of the models.

In Table 1, formula (6a) should be applied when the nucleus $K_2$ can be tangential at the poles of $K_1$ (i.e., when $K_2$ is rather unequiaxed, or small), whereas (6b) is more appropriate when these circumstances are not met. Formula (6a) generalizes (17) of Mayhew&Cruz (1973) for eccentric spheres.

General $(S_1/V_1)$ corrections for eccentric spheroids can readily be derived, but unfortunately they are far too complicated. By assuming that $x_1^2 = x_2^2 = 0$ or that $h = 0$, (eccentric spheres and concentric spheroids, respectively), the corrections become more tractable (Table 2).

TABLE 1. THE SPHEROID-SPHEROID NUCLEAR VOLUME FRACTION CORRECTIONS WHEN THE DISTANCE BETWEEN THE CENTRES OF NUCLEUS AND CELL IS UNIFORM. $V_{VO}$: Volume fraction estimated from nuclear-biased samples. $V_{VT}$: Corrected estimate.

*MODEL (P): Nucleus and cell are prolate spheroids with different eccentricities and a common axis of revolution.*
*MODEL (O): Nucleus and cell are oblate spheroids with different eccentricities and sharing the equatorial plane.*

MODEL (P): PROLATE-IN-PROLATE,
$K_1(a_1,b_1,b_1) \supset K_2(a_2,b_2,b_2)$

MODEL (O): OBLATE-IN-OBLATE,
$K_1(a_1,a_1,b_1) \supset K_2(a_2,a_2,b_2)$

$$x_1^2 = 1 - (b_1/a_1)^2,$$
$$x_2^2 = 1 - (b_2/a_2)^2,$$
$$q = (1-x_1^2)/(1-x_2^2).$$

CASE 1. $x_1^2 > x_2^2$, $0 < V_{VT} < q^3$,

or $x_1^2 < x_2^2$, $0 < V_{VT} < 1/q$.

CASE 1. $x_1^2 > x_2^2$, $0 < V_{VT} < q^3$,

or $x_1^2 < x_2^2$, $0 < V_{VT} < q^{-\frac{3}{2}}$.

$$V_{VT} = \left[\frac{G_2 V_{VO} + \sqrt{6G_1 V_{VO} + (G_2^2 + 3G_1 G_3)V_{VO}^2}}{G_3 V_{VO} + 2}\right]^3 , \quad \dots (6a)$$

CASE 2. $x_1^2 > x_2^2$, $q^2 < V_{VT} < q^{\frac{3}{2}}$.

CASE 2. $x_1^2 > x_2^2$, $q^2 < V_{VT} < q$. $\quad \dots (6b)$

$$V_{VT} = \left[\frac{3G_4 V_{VO}}{G_5 V_{VO} + 2}\right]^{3/2}$$

$$P_1 = \int_0^1 \sqrt{\frac{1-x_2^2(1-t^2)}{1-x_1^2(1-t^2)}}\left[\frac{1-x_2^2 t^2}{1-x_1^2 t^2}\right]^{\frac{1}{2}} dt ; \qquad O_1 = \int_0^1 \left[\frac{1-x_2^2 t^2}{1-x_1^2 t^2}\right]^{\frac{1}{2}} dt ;$$

$$P_2 = \int_0^1 \frac{\{1-x_2^2(1-t^2)\}^{\frac{1}{2}}}{\{1-x_1^2(1-t^2)\}^{\frac{3}{2}}} dt ; \qquad O_2 = \int_0^1 \frac{\{1-x_2^2 t^2\}^{\frac{1}{2}}}{\{1-x_1^2 t^2\}^{\frac{3}{2}}} dt ;$$

CORRECTION COEFFICIENTS:

$$G_1 = q^{\frac{1}{3}}\{(1-\tfrac{1}{3x_1^2})P_1 - \tfrac{1}{3}(1-\tfrac{1}{x_1^2})P_2\} ; \qquad G_1 = q^{\frac{1}{6}}\{(1-\tfrac{1}{6x_1^2})O_1 - \tfrac{1}{6}(1-\tfrac{1}{x_1^2})O_2\} ;$$

$$G_2 = q^{\frac{2}{3}}\{\tfrac{1}{x_1^2}P_1 + (1-\tfrac{1}{x_1^2})P_2\} ; \qquad G_2 = \tfrac{1}{2}q^{\frac{1}{3}}\{\tfrac{1}{x_1^2}O_1 + (1-\tfrac{1}{x_1^2})O_2\} ;$$

$$G_3 = q^{-\frac{1}{3}}G_1 + \tfrac{2}{3}q^{\frac{1}{3}}(3-\tfrac{1}{q})G_2 ; \qquad G_3 = q^{-\frac{1}{3}}G_1 + \tfrac{1}{3}q^{\frac{1}{6}}(9-\tfrac{5}{q})G_2 ;$$

$$G_4 = G_1 + \tfrac{1}{3}q^{\frac{1}{3}}G_2 ; \qquad G_4 = G_1 + \tfrac{1}{3}q^{\frac{1}{6}}G_2 ;$$

$$G_5 = G_3 - q^{-\frac{2}{3}}G_2 . \qquad G_5 = G_3 - q^{-\frac{5}{6}}G_2 .$$

(Tables of the coefficients $G_1, \dots, G_5$ for $x_1^2, x_2^2 = .00, .40, .60, .70, .80, .90, .95, .97, .99$, are available from the author, on request).

81

TABLE 2. THE SPHEROID-SPHEROID CORRECTIONS OF BIASED SPECIFIC SURFACE AREA.
$(S_1/V_1)_0$: Estimate obtained from nuclear-biased profiles.
$(S_1/V_1)_T$: 'True' (corrected) estimate.
$V_{yT}$ : True nuclear volume fraction (see Table 1).

*MODEL (S): Nucleus and cell are eccentric spheres, the distance between their centers being uniformly distributed.*

Correction formula:

$$(S_1/V_1)_T = C(V_{VT}) \cdot (S_1/V_1)_0,$$ (7)

where, putting $V_{VT}^{\frac{1}{3}} = k$,

$$C(V_{VT}) = \frac{\pi}{3} k(1-k)(4+k-2k^2)\{k \int_0^1 \frac{1}{t}\{\sin^{-1}(k+(1-k)t) - \sin^{-1}(k-(1-k)t)\}dt +$$

$$+ \frac{\pi}{12}(6-3k-2k^3) - \frac{1}{9}(30-11k+20k^2)\sqrt{k(1-k)} - \frac{1}{6}(6-9k+2k^3)\sin^{-1}(1-2k) +$$

$$+ \frac{1}{3}(2+k^2)\sqrt{1-k^2}\log(1+2k+2\sqrt{k(1+k)})\}^{-1}$$ (7a)

*MODEL (P): Nucleus and cell are concentric prolate spheroids.*
*MODEL (0): Nucleus and cell are concentric oblate spheroids.*

MODEL (P): $K_1(a_1,b_1,b_1) \supset K_2(a_2,b_2,b_2)$. MODEL (0): $K_1(a_1,a_1,b_1) \supset K_2(a_2,a_2,b_2)$:

Correction formula:

$$(S_1/V_1)_T = C(V_{VT},x_1^2,x_2^2) \cdot (S_1/V_1)_0,$$ (8)

where $$C(V_{VT},x_1^2,x_2^2) = \frac{3\pi}{8} Q(x_1) \frac{kI_1 - \frac{1}{3}k^3\{(x_2/x_1)^2 I_1 + (1-(x_2/x_1)^2)I_2\}}{M_0 - \sum_{n=1}^{\infty}\left[\frac{(2n-1)!!}{(2n)!!}\right]^2 \cdot \frac{x_1^{2n}}{2n-1} \cdot M_n}$$ (8a)

$k = q^{\frac{1}{3}}V_{VT}^{\frac{1}{3}}$ ;  $k = q^{\frac{1}{6}}V_{VT}^{\frac{1}{3}}$;

$Q(x_1) = (1/x_1)\sin^{-1}x_1 + (1-x_1^2)^{\frac{1}{2}}$;  $Q(x_1)=1+\{(1-x_1^2)/2x_1\}\log\{(1+x_1)/(1-x_1)\}$;

$M_n = \int_0^1 (1-t^2)^n\{\sin^{-1}(ku)+ku\sqrt{1-(ku)^2}\}dt$; $M_n = \int_0^1 \frac{(1-t^2)^n}{(1-x_1^2t^2)^{n-\frac{1}{2}}}\{\sin^{-1}(ku)+ku\sqrt{1-(ku)^2}\}dt$;

$u = \{(1-x_2^2(1-t^2))/(1-x_1^2(1-t^2))\}^{\frac{1}{2}}$;  $u = \{(1-x_2^2t^2)/(1-x_1^2t^2)\}^{\frac{1}{2}}$;

$I_1 = \int_0^1 u \, dt$ ;  $I_1 = \int_0^1 u \, dt$ ;

$I_2 = \int_0^1 u/\{1-x_1^2(1-t^2)\} \, dt$ ;  $I_2 = \int_0^1 u/(1-x_1^2t^2) \, dt$.

(Curves for a direct obtention of (7a) and of (8a) when either $x_1^2 = 0$ or $x_2^2 = 0$, are available from the author, on request.)

REFERENCES

Bonnesen, T. and Fenchel, W. (1934) *Theorie der konvexen Körper.* Springer, Berlin.
Kendall, M.G. & Moran, P.A.P. (1963) *Geometrical Probability.* Griffin, London.
Konwinski, M. & Kozlowski, T. (1972) Zellforsch. mikrosk. Anat., 129, 500.
Mayhew, T.M. & Cruz, L. M. (1973) J. Microsc., 99, 287.
Santaló, L.A. (1953) *Introduction to Integral Geometry.* Hermann, Paris.

*National Bureau of Standards Special Publication 431*
*Proceedings of the* FOURTH INTERNATIONAL CONGRESS FOR STEREOLOGY
*held at* NBS, Gaithersburg, Md., September 4-9, 1975 (Issued January 1976)

STOCHASTIC MODELS IN STEREOLOGY: STRENGTH AND WEAKNESSES

by Jean P. Serra
*Ecole Nationale Supérieure des Mines de Paris, Centre de Morphologie Mathématique, 77305-Fontainebleau,*
*France*

SUMMARY
       In stereology, the necessity of a model appears as soon as one wants to
extrapolate, predict or interpret the geometrical properties of a structure.
One can imagine questions being asked at several levels when modelling. For
statistical extrapolation (estimation variance of a volume percentage, for instance) it is sufficient to fit a punctual covariance to a mathematical function.
To predict the mill degradation of ore, one needs axioms for the granulometries
"in situ". To define and characterize clusters of features, one has to compare
the experimental results with moments of random functions. At the present time,
the main handicap in modelling is the lack of statistical theory when using the
notion of a random function.

INTRODUCTION.
       In stereology, we have two main aims, firstly the aim of describing the
geometric structure structural analysis, secondly the linking of this description to physical properties. On one hand, we want to extract several significant parameters, or functions, from complicated geometrical sets. On the other
hand, we hope to emerge from pure geometrical considerations and explain, or
simply correlate, physical phenomena with morphological data. For example, we
may want to relate mechanical strengths to sizes of metallic grains, or, in
the case of a rock, petrographic arrangements to ease of milling (1), or, in
biology, morphological changes in neurons to hypoxia (2), etc., etc..
       From a practical point of view, the second of these objectives is the most
important. The people who use stereology place little importance in the manner
in which the correlation was found, as long as they are satisfied that it truly
exists. Nevertheless, the first objective is fundamental, for only it gives an
overall view of the features of interest. In fact, in practical studies it becomes clear that these two objectives are two faces of the same reality and
complement each other.

LEVEL OF SOPHISTICATION.
       Let us for the moment focus our attention on the structural analysis ap-
proach. Classically, we define several basic parameters, such as the phase areas,
perimeters, connectivity numbers, size distributions, etc.. By doing this, we
implicitly assume that the features studied are the realisations of a random
function. We use random functions here as a means to an end. We do not want to
say whether the features observed are indeed random or deterministic phenomena,
but only that at the scale of observation, they appear chaotic. Our interest
however, is not in local description, but in mean values, which carry physical
meanings. For example, let us consider the porosity of a rock. We are not interested in the fact that a given point x belongs to the grains or to the pores,
but only in the proportion of such points : this is exactly a changing of scale,
and the theoretical tool for approaching it is the concept of a random function.
       Along this line of thought, the classical stereological parameters appear
as the various moments of the random function. The proportion of a phase (classically denoted by $A_A$ or $V_V$) is an estimate of the first order moment. In the
same way, one can show that the other basic stereological parameters (surface,
mean and total curvature) are the moments associated with groups of two, three
or four points. In many applications we need no more than this first set of
moments.
       But one can easily imagine some problems for which we need to go to higher
order moments. Let us give two examples. First the estimation variance of a
volume percentage : it is known (3) that this variance estimation depends only

on the punctual covariance, which is the moment of order two (probability of finding two points simultaneously in the chosen phase). Second example : the size measurement of one particular medium (chord distribution for example). G. Matheron proved that this kind of measurement depends only on the probability that on an arbitrary point x, we can put a given convex set that is contained totally within the medium (infinite moment).

The two steps above (first moments, and more complicated ones) are not sufficient for solving all the problems. Let us consider for example the following question : "Does the structure that we see, exhibit clustering?". The concept of a cluster is not defined on the same level as the perimeter, and needs, for its description, the reference to a random function with clusters (or without).

Another reason for fitting a model comes from the very large number of data provided by the moment measurements. Each size and shape of the structuring element corresponds to a different probability. It seems often that the key parameters are marked by too much data. The model allows us to predict the theoretical value of every moment, and, as soon as it is correctly fitted, substitutes its own two or three parameters into the moments.

The last reason is that the stereological inference from $R^2$ to $R^3$ becomes easier when using models.

Today, the library of models in stereology is not very large, and we do not pretend to give here an exhaustive list of them. The most developed are the point processes (4). Starting from the Poisson Point process, one can sophisticate it by varying the density, and introducing clusters around the Poisson Points.

A second group of models consists of the tessellations of the space $R^n$ (5). The basic one is the Poisson flat process, studied by R.E. Miles and later by G. Matheron. In the same group of models, one can quote the doublet of Poissonnian planes of Cauwe, and the Voronoï polyhedra (Meijering, Gilbert, Miles).

A third group consists of the boolean schemes. We will present later the basic prototype.

The three groups probably represent the building blocks in model construction, since the Poisson point processes, the Poisson flats and the boolean scheme are indefinitely divisible.

THE BOOLEAN SCHEME.

In order to illustrate the previous considerations, let us present now, with more details, the boolean scheme.

The definition of the Poisson point process in $R^n$ is well known. It consists of a random set of points which satisfies the two following properties :

a) if B and B' are two sets such as $B \cap B' = \emptyset$ , the numbers N(B) and N(B') of points falling in B and B' are two independent random variables.

b) the elementary volume dv contains one point with the probability $\theta(dv)$ and none with the probability $1-\theta(dv)$.

The measure $\theta$ is called density of the process. Here we will take $\theta$ = constant, because it leads to more geometrical results.

By integration, the two properties imply that for any bounded B the random integer N(B) satisfies the Poisson law with parameter

$$\theta = \int_B \theta(dv)$$

$$P\{N(B)=n\} = \frac{\theta^n}{n!} e^{-\theta} \tag{1}$$

Let us take a realization of a Poisson process of constant density . This realization has points which may be indexed by the set I. We consider each point as the germ of a crystalline growth. If two crystals

meet each other, they are not disturbed in their growth, which stop independently for each component. Let us transpose this description in terms of random
sets. The points of the Poisson realization are in $x_i$ ($i \in I$). The elementary
grain is a nonstationary random set A'. Successively, we pick out various rea-
lizations $A'_i$ of A' from its space $\Omega$ of definition and implant each $A'_i$ at the
corresponding point $x_i$. The different $A'_i$ are thus independent of each other. We
will call the realization A of a Boolean scheme the union of the $A'_i$ associated
with a given realization of the points $x_i$.

$$A' = \bigcup_{i \in I} A'_i$$

The Boolean scheme is extremely fertile. It represents one of the first steps,
in modelling, when one admits negligible interactions between the particles $A_i$.
We had the opportunity to use it in sedimentary petrography, in dendritic
crystalline growth, in studies of forests, and all its fields of application
are probably not yet discovered.

The basic study of the Boolean scheme has been made by Matheron.

We center the primary grain A' at the origin. This random set is known by
the datum of the two families of functionals $\varpi(B)$ and $\chi(B)$.

$$\varpi(B) = P\left\{B \subset A'\right\} \text{ and } \chi(B) = P\left\{B \cap A' \neq \emptyset\right\} \qquad (2)$$

If A' is translated from the origin to the point z, it admits the new
functionals $\varpi_z(B)$ and $\chi_z(B)$ easily deducible from (2)

$$\varpi_z(B) = P\left\{B \subset A'_z\right\} = P\left\{B_{-z} \subset A'\right\} = \varpi(B_{-z}) \qquad (3)$$

$$\chi_z(B) = \chi(B_{-z})$$

Let us consider now the Boolean scheme itself, namely the union of all
the $A'_i$, and compute the probability of B being included in the pores $\complement A$
of the scheme.

According to the property (a) of the Poisson process, each element of
volume dz of the space makes its contribution independently of the others. In
dz, centered in z, two incompatible favourable events may happen :

1 - no germ in dz : probability $1-\theta dz$.

2 - one germ in dz, but the grain $A'_z$ does not reach B : probability
$\theta dz \cdot \chi(B_{-z})$.

By composition of these two probabilities and summation in z extended to
the space $R^n$, we find

$$Q(B) = P\left\{B \subset \complement A\right\} = e^{-\theta \int_{R^n} \left[1 - \chi(B_{-z})\right] dz} \qquad (4)$$

To interpret the geometrical meaning of the integral in (4), let k(z) be
the indicator function of the random set $A' \oplus \check{B}$, that is

$$k(z) = 1 \quad \text{when} \quad A' \cap B_z \neq \emptyset$$

$$k(z) = 0 \quad \text{when} \quad A' \cap B_z = \emptyset$$

and mes $A' \oplus \check{B} = \int_{R^n} k(z)\,dz$.

By taking the mathematical expectation, and inverting the operations E and $\int$, we find

$$E\left[\text{mes } A \oplus \check{B}\right] = \int_{R^n} \left[1 - \chi(B_{-z})\right]dz$$

Thus formula (4) may be written

$$Q(B) = e^{-\theta E\left[\text{mes}(A' \oplus \check{B})\right]} \tag{5}$$

which is the fundamental formula for the Boolean scheme. It links the functionals after randomisation to those of the primary grains. In other words, the proportion $1-Q(B)$ of the dilated $A \oplus B$ (final state) is related to the measure $E\left[\text{mes } A' \oplus \check{B}\right]$ of the dilated primary grain by an exponentiation.

Unfortunately, one cannot have such a beautiful result for the erosions. The principle of the construction of the scheme gives priority to the logic of intersection of the grains.

Formula (5) leads directly to the computation of the usual functionals

(a) The covariance :

$$C(h) = q^2 \left[e^{\theta K(h)} - 1\right] \qquad q = \text{porosity} = e^{-\theta E\left[\text{Mes } A'\right]}$$

$$K(h) = E\left[\text{Mes}(A' \cap A'_h)\right]$$

(b) Intercepts distribution function (dir. $\alpha$) :

$$1 - F_\alpha(\ell) = e^{-\theta D_\alpha \ell} \qquad D_\alpha = \text{diametral variation of A} \quad (\text{dir. } \alpha)$$

The intercepts law is a negative exponential. This important property appears as a particular case of the general notion of semi Markov processes (Matheron (3)). In practice, it may be used for testing the Boolean hypothesis.

REFERENCES.

1. Ph. Cauwe — See paper in the same Congress.

2. O. Hunziker et Al.— See paper in the same Congress.

3. G. Matheron — Random Sets and Integral Geometry.— John Wiley and Sons, New York, 1975.

4. Stochastic Point Processes.— Edited by Peter A. Lewis, John Wiley and Sons, New York, 1972.

5. R.E. Miles — in Proceedings of the Symposium on Statistical and Probabilistic Problems in Metal.— Adv. in Appl. Prob.— 1972.

6. J. Serra — Mathematical Morphology and Image Analysis.— SIAM, Washington, D.C.— To be published.

*National Bureau of Standards Special Publication 431*
*Proceedings of the* FOURTH INTERNATIONAL CONGRESS FOR STEREOLOGY
*held at* NBS, Gaithersburg, Md., September 4-9, 1975 (Issued January 1976)

A STEREOLOGICAL VIEW OF DATA ANALYSIS

by George A. Moore and George T. Eden
*National Bureau of Standards, Washington, D.C. 20234, U.S.A.*

The basic stereological reasoning by which structure in a two-dimensional image is interpreted as structure of a solid can be usefully applied to rational deduction of meaning in collections of numerical data. Data collections from the real world often involve many variables and fail to conform to theoretical ideals of independence of variables or rectangularization of design. Customary statistical tests often fail to support logical or apparently obvious relationships while at the same time developing "statistically significant" associations which have no logical basis for acceptance as functional relations. Visualization of simultaneous numerical observations as points plotted in data space or hyperspace is a substantial aid in rational sorting of functional relationships from accidental or indirect associations.

In a stereological investigation of a material problem, parameters of a three-dimensional substance are determined with the expectation that they will predict, explain, or control some subsequent behavior. A large number of redundant parameters may be invented; hence functional meaning may be expected only for a limited subset of parameters which properly define a Gestalt (1) controlling the behavior considered. A proper description of this Gestalt, together with a tracing of the effect of treatment on the descriptive parameters, is usually necessary before these treatments can be effectively modified to improve behavior.

In the geometrical visualization each specimen supplies the position of one point in a data hyperspace having behavior as the dependent dimension. Some geometrical figure descriptive of the functional relationships may be discovered either by computational processes or by visual study of 2-d or 3-d abstracts. Projections into 2-d are subject to the same types of stereological illusions encountered when real 3-d structures are examined from a single viewpoint. Complex illusions in interpreting higher dimensionality must also be expected. Among the usual analytical operations simple correlation and ordinary regression computations treat only 2-d projections of the data. Multivariate analysis methods and in particular simultaneous regression computations attempt to interpret the n-dimensional form of the data structure.

*Total* regression is in bad repute for producing false relations when redundant dimensions are present in the data. We tested the actual computer program by deliberately introducing a column of pseudo-data derived by a linear relationship from a column of real data. High slope coefficients resulted but even higher standard deviations of these coefficients clearly showed them to have zero significance. This implementation thus may produce no solution when an excess number of dimensions are used. The original representation of all available data as a single hyperspace normally offers excess dimensions. Analysis is facilitated by logically dividing this hyperspace into lower dimensional spaces with each lower space limited to a set of parameters which might be functionally related. A rule that cause must be present immediately before effect can occur enables data on a material problem to be divided into one space representing possible effects of structural parameters on behavior and other spaces representing effects of preparation and treatment on each structural parameter separately.

Limited data can support determination of only a limited number of equation terms. Thus it is usually necessary to forego the luxury of estimating curvature and synergistic relationships and confine attention to the approximation given by the best fit plane or hyperplane. For this the "statistical significance" of each parameter can be conveniently indicated by the ratio t = slope

87

coefficient ÷ standard deviation of this coefficient. A high statistical association may be found when a parameter has a small overall effect on the dependent variable. Hence, "functional significance" also requires a slope coefficient of sufficient magnitude that the total range of the "causative" parameter
accounts for a substantial fraction of the range of the dependent variable.
The product of t with this fraction appears to be useful in judging relative
functional significance.

An illustration of the highly variable and often contradictory results
obtainable by various methods of analyzing the same experimental data will be
presented using data previously obtained by one of us (GTE) on brittle fracture
of dental amalgams (2). Eight screen separated fractions of spray-formed
"spherical" silver-tin alloy powder were used, along with commercial alloy in
the form of turnings as a control. Following customary dental practices, 30
cylindrical amalgam restorations were made from each powder. Six cylinders
from each particle size were compacted at each of five specific pressures.
After setting, five specimens from each of the 40 sets were broken in diametri-
cal compression and the nominal (tensile) fracture strength determined. A full
cross section micrograph of one specimen of each set was analyzed by scanner to
determine parameters of the void structure. Mercury analysis was obtained from
the broken pieces. One determination of unreacted silver-tin core size was
made for each particle size, with the finding that the unreacted core size was
directly related to the original nominal particle size. Very early scanner
data have been discarded as unreliable. Data Set B uses void measurements made
at an arbitrary threshold setting and includes amalgams made from commercial
alloy turnings. For data Set C the scanner data were recomputed at objectively
determined thresholds and additional parameters computed. Only the spherical
alloys are included in this set.

Simple (2-d) correlation coefficients for all data pairings are shown in
the lower triangle of Table 1. Partial correlation coefficients for the overall
n-dimensional systems are shown in the upper triangle of this table. Of the 45
correlation coefficients for which full observations are available in Data C,
25 are statistically significant at the 95% confidence level (*). The matching
partial correlation coefficients show reversed sign in 13 cases, of which 6 are
above the 95% confidence level for 1 or both coefficients. Thus it is apparent
that no conclusions regarding functional relationships can be drawn from these
correlations. It is, however, useful to note that the cross-correlation between void content and void width is slightly lower than the correlation of fracture strength to void content for Data B, but higher than the correlation of
fracture strength to either void content or void width for Data C.

Two-dimensional linear regressions, Figure 1A & B, were computed for fracture strength against each of the six potential causative parameters in Data B
and against the nine parameters in Data C. Ranking of the potential significance of these parameters by the ratio t was essentially the same as by simple
correlation and unrelated to the partial correlation coefficients. The fraction
of the total *variance* of fracture strength ascribed to the "causative" parameter
was determined for each regression and summed for each set. *Variance* was found
to be overaccounted for by a factor of 5.3 for Data B and of 7.9 for Data C.
Such excessive explanation of the *variance* may be taken to indicate that each
two- dimensional view is more dependent on the unstated hidden factors than on
the parameter supposedly represented. In subsequent multidimensional regressions, it was usually found that most of the *variance* was assigned to whichever
parameter had been listed first in the programmer's order, although the coefficients produced and their t values did not change with this order. Thus for
these data at least, *assignment of variance* is of no value in determining the
proper dimensions of the functional relationships.

Being unable to discover the functional causes of lowered fracture strength
by two-dimensional methods, simultaneous regressions were tried in various combinations of dimensions. A method of sorting dimensions is illustrated graphically in Figure 1C & D for the ambiguity between void content and void size as

| TABLE 1 | Correlation Coefficients Between Amalgam Data | | | | | | | | | |
|---|---|---|---|---|---|---|---|---|---|---|
| Parameter | F.S. | V.C. | V.W. | MFP | S_V/S_Vo | P.D. | P.D.R. | Hg | Pk. P. | Tr. T. |
| Fracture Strength | 1 | +.210 | -.301 | +.192 | -.330 | -.123 | -.264 | -.538* | -.161 | -.232 |
| | | -.648* | +.012 | ---- | ---- | -.454* | +.296 | -.428* | -.221 | ---- |
| Void Content | -.827* | | +.807* | -.783* | +.544* | -.293 | +.098 | +.270 | +.097 | +.179 |
| | -.840* | | +.529* | ---- | ---- | -.184 | +.480* | -.402* | -.328* | ---- |
| Void Width | -.823* | +.864* | | +.570* | -.677* | +.593* | +.060 | -.083 | -.030 | +.119 |
| | -.729* | +.750* | | ---- | ---- | +.295 | -.544* | +.371* | +.287 | ---- |
| M.F.P., Solid | +.475* | -.739* | -.382* | | -.178 | -.028 | -.175 | +.493* | +.429* | -.317 |
| | | | | | ---- | ---- | ---- | ---- | ---- | ---- |
| Surface Coverage, S_V/S_Vo | -.725* | +.669* | +.557* | -.534* | | +.799* | -.118 | -.001 | +.072 | +.134 |
| | | | | | | ---- | ---- | ---- | ---- | ---- |
| Ag-Sn Particle Diam. | -.683* | +.605* | +.558* | -.438* | +.971* | | -.277 | -.408* | -.225 | -.288 |
| | -.613* | +.588* | +.548* | ---- | ---- | | +.297 | -.719* | -.449* | ---- |
| Part. Diam. Range | +.008 | +.143 | +.148 | -.057 | -.538* | -.598* | | -.110 | +.126 | +.350* |
| | +.218 | -.106 | -.401* | ---- | ---- | -.268 | | +.542* | +.399* | ---- |
| Mercury Content | +.044 | -.141 | -.120 | +.251 | -.526* | -.588* | +.449* | | -.725* | -.263 |
| | +.038 | -.130 | -.098 | ---- | ---- | -.541* | +.342* | | -.738* | ---- |
| Packing Pressure | +.204 | -.198 | -.021 | +.311 | -.068 | 0 | 0 | -.546* | | -.142 |
| | +.221 | -.195 | -.138 | ---- | ---- | 0 | 0 | -.587* | | ---- |
| Trituration Time | -.389* | +.462* | +.536* | -.111 | -.125 | -.179 | +.730* | +.287 | 0 | 1 |
| | | | | | | ---- | ---- | ---- | 0 | |

Southwest Triangle = Simple Correlation Coefficients; Data C/Data B
Northeast Triangle = Partial Correlation Coefficients; Data C/Data B
*Indicates significance above 95% confidence level.

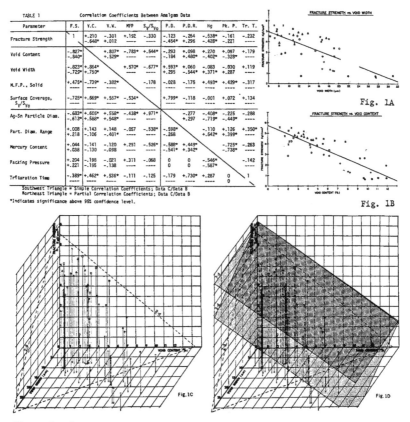

Fig. 1A

Fig. 1B

Fig.1C  Fig.1D

Figure 1: Fracture Strength vs. Void Parameters in 2, 3, and 7 Dimensions.

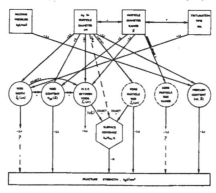

Figure 2
Influences of Preparation on Microstructural Parameters and on Fracture Strength of Spherical Powder Amalgams (Data C)

89

the major cause of decreased fracture strength within the choices of Data B.
Figure 1C represents a three-dimensional box in which an observed fracture
strength is plotted simultaneously against void content·and void width. Data
for each amalgam are represented by a ball mounted on a stalk whose height indicates the fracture strength and whose base is planted at the combination of void
content and width measured for this specimen. The substantial cross-correlation
of void parameters can be seen in the pattern of the base points. The two-
dimensional regression lines (1A & 1B) of the projections of the cloud of balls
against the two vertical planes are shown as dashed lines in these planes. If
another plane is visualized as defined by these two regression lines, it may
easily be seen that this plane lies well below most of the balls. Thus one or
both regression lines must be an illusion of the projection into two dimensions.
The best fit three-dimensional regression plane is represented by the lower of
the two shaded planes in Figure 1D. This plane has nearly the same slope in
the void content direction as the linear regression. However, the slope of the
3-d plane in the width direction is much less than that of the linear regression, indicating a lowered functional significance for width. In a final step,
the best fit hyperplane regression was computed in seven-dimensional hyperspace.
The three-dimensional abstract of this hyperplane is shown as the upper shaded
plane in Figure 1D. The slope of this plane in the void content direction is
little changed from that of the 3-d plane, but the slope in the void width
direction has been reduced nearly to zero. It is thus demonstrated that void
content is the functionally significant parameter included in this data set and
that void width, while showing apparently high significance in correlation and
linear regression has no actual functional significance in its own right.

Attempts to determine the best functional parameters of the structural
Gestalt from the larger choice of parameters offered in Data C have encountered
difficulties which apparently trace to the high cross-correlations in this data
set and which may result from omission of the commercial alloys as strangers to
the spherical particle system. Generally no more than five dimensions could be
included in one regression equation and only one parameter of the voids could be
evaluated at one time. Regression equations in several 3 to 5-dimensional
selections have resulted in the general map of apparent relationships shown in
Figure 2. Relative weights are shown as the product of t times the fraction of
range accounted for.

Failure of any of the simple void measurements to assume a dominant status
led to the development of "relative surface coverage," the ratio of the measured
$S_V$ of the amalgam, as generated by voids, to the $S_{V_O}$ value characteristic of
dry alloy particles of the nominal particle size. This constructed parameter
is of moderate functional significance when used in combination with total void
content (which contributes to $S_V$) but becomes the dominant functional parameter
in a three-dimensional analysis including mercury content. Mercury content has
an adverse effect as logically expected but in contrast to the small positive
influence indicated by the correlation and linear regression computations.

In this example of analysis of less than adequate data, it is found
that representation of alternate Gestalt models by hyperplanes in data spaces
leads to logically acceptable representation of functional effects. In contrast, two-dimensional analyses were usually ambiguous and frequently directly
contrary to the conclusions from controlled multidimensional analysis. Use of
single viewpoints when analyzing complex data arrays thus is shown to
involve hazards in interpretation which are analogous to those encountered when
viewing real three-dimensional systems from a single viewpoint.

REFERENCES

1. Moore, G. A., "Gestalt Properties of Aggregate
   Materials" - Practical Metallography, Vol. 9,
   No. 2 (2/72) p. 76-97.

2. Eden, G. T. and Waterstrat, R. M., "Effects of
   Packing Pressure on the Properties of
   Spherical Alloy Amalgams" - JADA - Vol. 74
   (4/67) p. 1024-9.

NOTE - Discussions of multiple regression analysis
will be found in:

3. Draper, N. and Smith, H., "Applied Regression
   Analysis," New York, Wiley Interscience, 1966.

4. Daniel, C. and Wood, F., "Fitting Equations to
   Data: Computer Analysis of Multifactor Data
   for Scientists and Engineers," New York, Wiley
   Interscience, 1971.

*National Bureau of Standards Special Publication 431*
*Proceedings of the* FOURTH INTERNATIONAL CONGRESS FOR STEREOLOGY
*held at* NBS, Gaithersburg, Md., September 4-9, 1975 (Issued January 1976)

## THREE-DIMENSIONAL SHAPE PARAMETERS FROM PLANAR SECTIONS

by Ervin E. Underwood, *President of the International Society for Stereology*
*Visiting Scientist, Max-Planck-Institut für Metallforschung, Stuttgart, Germany*
*On leave from Georgia Institute of Technology, Atlanta, Georgia 30328, U.S.A.*

Shape parameters for features of the microstructure may be expressed in many ways, depending on the type of system being studied and the purpose in mind (1). The preferred shape parameters are those based on the quantitative, assumption-free, statistically exact equations of stereology (2), obtained from simple microstructural measurements.

In this paper we utilize general relationships that equate measurements made on the random 2-d microsection to the spatial quantities belonging to the 3-d features of the microstructure. Five 2-d and five 3-d quantities have been selected for study.

| 2-d Quantities | | 3-d Quantities | |
|---|---|---|---|
| $A$, | intercept area | $V$, | volume |
| $L_p$, | perimeter length | $S$, | surface area |
| $d$, | tangent diameter | $D$, | tangent diameter |
| $L_2$, | intercept length | $L_3$, | intercept length |
| $k$, | curvature of lines | $K_m$, | curvature of surfaces |

These quantities may be expressed simply for convex bodies, or aggregates of bodies, in space or in the section plane. (Note that primes denote projected quantities.)

$$\bar{A} = V/\bar{D} = A_A/N_A \qquad\qquad \bar{V} = \bar{L}_2\bar{A}' = V_V/N_V$$

$$\bar{L}_p = \pi S/4\bar{D} = L_A/N_A \qquad\qquad \bar{S} = 4\bar{A}' = S_V/N_V$$

$$\bar{d} = S/4\bar{D} = N_L/N_A \qquad\qquad \bar{D} = \bar{A}'/\bar{d} = N_A/N_V$$

$$\bar{L}_2 = 4V/S = L_L/N_L \qquad\qquad \bar{L}_3 = \bar{A}/\bar{d} = L_L/N_L$$

$$\bar{k} = 8\bar{D}/S = 2N_A/N_L \qquad\qquad \bar{K}_m = \pi/2\bar{d} = \pi N_A/2N_L$$

For systems of convex particles of the same shape, we can express the 3-d quantities in terms of an arbitrary linear dimension or "size", X (such as the mean intercept length $\bar{L}_3$, or mean tangent diameter, $\bar{D}$), and a dimensionless Form Factor, $k_n$.

| Monodispersed | Polydispersed | Log-normal Size Distribution |
|---|---|---|
| $V = k_3 X^3$ | $\bar{V} = k_3 M_3$ | $\bar{V} = k_3 \exp(3\ln X_g + 4.5\ln^2\sigma_g)$ |
| $S = k_2 X^2$ | $\bar{S} = k_2 M_2$ | $\bar{S} = k_2 \exp(2\ln X_g + 2\ln^2\sigma_g)$ |
| $D = k_1 X$ | $\bar{D} = k_1 M_1$ | $\bar{D} = k_1 \exp(\ln X_g + 0.5\ln\sigma_g)$ |
| $K_m = k_{-1} X^{-1}$ | $\bar{K}_m = k_{-1} M_{-1}$ | $\bar{K}_m = k_{-1} \exp(-\ln X_g + 0.5\ln^2\sigma_g)$ |

The moments of the size distributions are defined by $M_n(X) = \Sigma f_i x_i^n / \Sigma f_i = \overline{x^n}$. For log-normal distributions, $X_g$ is the geometric mean, $\sigma_g$ is the geometric standard deviation, and the moments about the origin $\mu_i'(X) = \exp(i\ln X_g + (1/2)i^2\ln^2\sigma_g)$. (2)

Various dimensionless groups of the above four Form Factors were examined in order to determine their usefulness as Shape Parameters. Several basic classes are distinguishable. Class I Shape Parameters are functions of $\delta/\bar{K}_m$, where $\delta$, the specific surface, equals $S/V$. Class II

Shape Parameters are functions of $\delta D$, while Class III parameters are functions of $\overline{D} \, \overline{K}_m$. Class IV parameters consist of miscellaneous categories, including ratios that are not dimensionless (3, 6).

Examples of the four classes of Shape Parameters are given below for monodispersed systems of convex particles, for polydispersed systems, and for log-normal size distributions.

Class I Shape Parameters (functions of $\delta/\overline{K}_m$)

| Monodispersed | Polydispersed | Log-normal |
|---|---|---|

$$\frac{k_2^2}{k_1 k_3} = \frac{S^2}{DV} = \frac{16d}{L_2} = \frac{4P_L^2}{N_A P}. \qquad = \left(\frac{4P_L^2}{N_A P}\right)\left[\frac{M_1 M_3}{M_2^2}\right]. \qquad = \left(\frac{4P_L^2}{N_A P}\right)\exp(\ell n^2 \sigma_g).$$

Class II Shape Parameters (functions of $\delta D$)

$$\frac{k_{-1} k_2^2}{k_3} = \frac{2\pi DS}{V} = 8\pi\frac{A'}{A} = \frac{4\pi N_A P_L}{N_V P}. \qquad = \left(\frac{4\pi N_A P_L}{N_V P}\right)\left[\frac{M_3}{M_2^2 M_{-1}}\right]. \qquad = \left(\frac{4\pi N_A P_L}{N_V P}\right).$$

Class III Shape Parameters (functions of $\overline{D} \, \overline{K}_m$)

$$\frac{k_1^2}{k_2} = \frac{D^2}{S} = \frac{D}{4d} = \frac{N_A^2}{2P_L N_V}. \qquad = \left(\frac{N_A^2}{2P_L N_V}\right)\left[\frac{M_2}{M_1^2}\right]. \qquad = \left(\frac{N_A^2}{2P_L N_V}\right)\exp(\ell n^2 \sigma_g).$$

Class IV Shape Parameters (miscellaneous)

$$\frac{k_2^2 k_1}{k_3} \cdot \frac{S^2 D}{V} = \frac{16A'^2}{A} = \frac{4N_A P_L^2}{N_V^2 P}. \qquad = \left(\frac{4N_A P_L^2}{N_V^2 P}\right)\left[\frac{M_3}{M_2^2 M_1}\right]. \qquad = \left(\frac{4N_A P_L^2}{N_V^2 P}\right)D_g^{-2}.$$

It can be seen that the Class I Shape Parameters need only counting measurements for their evaluation, as noted by DeHoff (4) and Fischmeister (5). Consequently, several Class I parameters were formed from the five 2-d quantities for possible application to different types of microstructures.

Examples of Shape Parameters deemed suitable for expressing particle shape are $\overline{L}_p/L_2$ and $\overline{L}_p/\overline{Ak}$; for interlocking microstructures, $\overline{L}_p^2/\overline{A}$ and $\overline{kL}_p^2/L_2$; and for single-phase materials, $\overline{d}/\overline{L}_2$, $\overline{d}/\overline{Ak}$ or $\overline{L}_p/L_2$. Of course, other parameters (6) may perform as well or better, depending on the particular application desired. In any case, the above treatment provides a generalized framework for devising new or modified parameters that relate features on the microsection to the 3-d spatial quantities of the microstructure.

References:
(1) Underwood, E. E., "Stereology in Automatic Image Analysis", The Microscope, 22, First Qtr. (1974) p. 69.
(2) Underwood, E. E., Quantitative Stereology, Addison-Wesley Publ. Co., Reading, Mass. (1970) p. 24; 140.
(3) Underwood, E. E., in Quantitative Analysis of Microstructures in Medicine, Biology and Materials Development, Edited by H. E. Exner, Dr. Riederer-Verlag GmbH, Stuttgart (1975) p. 234.
(4) Quantitative Microscopy, Edited by R. T. DeHoff and F. N. Rhines, McGraw-Hill Book Co. (1968) p. 133.
(5) H. F. Fischmeister, Zt. Metallk. 65 (1974) p. 558.
(6) Underwood, E. E. "Quantitative Shape Parameters for Microstructural Features" Inter/Micro-75, Cambridge, England, June 23-26, 1975. To be published in The Microscope, 24, First Qtr. (1976).

*National Bureau of Standards Special Publication 431*
*Proceedings of the* FOURTH INTERNATIONAL CONGRESS FOR STEREOLOGY
*held at* NBS, Gaithersburg, Md., September 4-9, 1975 (Issued January 1976)

INTERPRETATION OF SOME OF THE BASIC FEATURES OF FIELD-ION IMAGE PROJECTIONS
FROM A HEMISPHERICAL TO A PLANAR SURFACE USING MOIRE PATTERNS

by P. Darrell Ownby*, Robert M. Doerr,** Walter Bollmann,***
*Ceramic Engineering Dept., University of Missouri, Rolla, Missouri 65401, U.S.A.
**U.S. Bureau of Mines, Rolla, Missouri 65401, U.S.A.
***Battelle Institute, Geneva, Switzerland

A common problem in projection geometry is that of analyzing the pattern
formed in a planar observation surface by the projection of point, lineal, or
areal features on a hemispherical surface. In this study, the most prominent
ellipses representing successive ledges on off-axis oriented planes, are
simulated by the Moiré patterns produced by the intersection of a simple grid
representing the crystal lattice parallel to the projection plane and a
spherical projection of concentric circles representing successive ledges of
atomic planes proceeding normal to the projection plane. The Moiré patterns
are analyzed mathematically by considering the coincidence of the equivalent
points in both the circle set and the parallel line (grid) set. Computer
plotted patterns using this analysis are shown to coincide with the Moiré
patterns. Examples of orthographic and stereographic projections are shown
for comparison with the actual field-ion image.

The surface geometry of a single crystal which produces a field-ion microscope image is to a first approximation, a hemispherical surface which is
intersected by sets of lattice planes. For cubic crystals, three sets of orthogonally oriented planes intersect the surface. These intersections determine
the ledge-step geometry of the surface which is seen projected onto a plane in
atomic detail in field-ion microscopy.

In the present work the intersection of each plane of the three sets with
the surface of the sphere is projected to a plane, forming three families
of lines. Each of these sets of curves interacts with the others to form a
Moiré pattern which faithfully represents many of the main features of the
spherical surface geometry and therefore resembles, to a limited degree, the
field-ion micrograph.

Curves representing the Moiré interference bands can be produced by an
analytical calculation without actually drawing the families of lines. The
guiding principle for calculating the Moiré pattern between two parametrized
sets of curves is the "coincidence of equivalent positions" (1).

For simplicity consider first the orthographic projection. The (001)
oriented, face-centered cubic crystal lattice geometry will be used for illustration. Orthographically projected Moirés of other orientations and Bravais
lattices are shown elsewhere (2). The first set of projected curves becomes a
spherical projection of concentric circles. The circles tend to overlap as the
great circle†† is approached, making a full hemisphere projection impractical.
The second and third sets of curves are simply two orthogonal sets of parallel
straight lines. Note that in the FCC lattice those lines are 45° to the cube
axes. The resulting indexed Moiré pattern is shown in figure 1. Each set of
parallel lines produces Moiré ellipses along a circle diameter perpendicular to
the lines. The three sets of lines combine to produce the ellipses which lie
along the cube axes (at 45° to the straight line sets).

For the analytical calculation of the Moiré ellipses, consider the interaction of a set of parallel equidistant straight lines numbered from $-\infty$ to $\infty$
and a set of concentric circles numbered from 1 to N, from the smallest to the

---

†† The radius of the reference sphere and the maximum radius of the projected
circles in both projections shown herein was 4 inches (101.6 mm) before
reduction for publication.

largest. The "Coincidence of equivalent positions" are the intersections of circle No. 1 with line No. 1, of circle No. 2 with line No. 2, etc. Then we consider a continuous distribution of lines and circles so that, i.e., line No. 1.376 intersects circle 1.376, etc. This gives a continuous curve of intersections which represents a curve of the Moiré pattern. A second set of curves (ellipses in the present example) is obtained by the intersection of circle No. 1 with line No. 2, circle No. 2 with line No. 3, circle No. 1.376 with line 2.376, etc.

For the set of straight lines parallel to the y-axis, $x = na$ where a is the spacing of the lines (for FCC (001), a is the unit cell dimension/ $\sqrt{2}$ ) and n is an integer. The other set of straight lines are given by $y = na$. The set of circles is given by N equidistant cuts of thickness d through a hemisphere, (for FCC (001) this thickness is $\frac{1}{2}$ of the unit cell dimension) so that $N \cdot d = R$, the radius of the sphere. The radius, $\rho$, of each circle is given by

$$\rho^2 = x^2 + y^2 = (Nd)^2 - [(N - n)d]^2$$
$$= d^2 (2Nn - n^2) \tag{1}$$

Now we change the integer n into a continuous z and introduce in x the integer b, which is the difference in the numbering of the two sets.

$$x = a(z + b) \tag{2}$$
$$\rho^2 = x^2 + y^2 = d^2 (2Nz - z^2) \tag{3}$$

Now we calculate z from these equations as the solution of a quadratic equation.

$$z = \frac{Nd^2 - ba^2}{a^2 + d^2} \pm \left[ \left( \frac{Nd^2 - ba^2}{a^2 + d^2} \right)^2 - \frac{b^2a^2 + y^2}{a^2 + d^2} \right]^{\frac{1}{2}} \tag{4}$$

Next we eliminate z by introducing equation (4) into (2); this yields the equation of an ellipse.

$$\frac{(x - x_o)}{a_o^2} + \frac{y^2}{b_o^2} = 1 \tag{5}$$

with
$$x_o = \frac{ad^2 (N + b)}{a^2 + d^2} \tag{6}$$

$$a_o^2 = \frac{a^2[N^2d^4 - (2Nb - b^2)a^2d^2]}{(a^2 + d^2)^2} \tag{7}$$

$$b_o^2 = a_o^2 \left( 1 + \frac{d^2}{a^2} \right) \tag{8}$$

where $x_o$ is the center and $a_o$ and $b_o$ are the two semi-axes of the ellipse. By varying b, the whole set of ellipses is obtained. However, analytically, the values of b cannot be chosen fully arbitrarily. There is an upper and lower limit of b which depends upon the relative placement of the origin of the intersecting sets of curves, and is determined by the condition $a_o^2 > 0$. For $a_o^2 = 0$, we obtain the equation

$$[N^2d^2 - (2Nb + b^2) a^2d^2] = 0 \tag{9}$$

which is a quadratic equation for b with the solutions:

$$b_{\substack{max \\ min}} = \dot{N} \left[ \pm \left( \frac{d^2}{a^2} + 1 \right)^{\frac{1}{2}} - 1 \right] \tag{10}$$

The next integer smaller than $b_{max}$ corresponds to the smallest ellipse of the set and the smallest integer larger than $b_{min}$ to the largest ellipse.

Other sets of Moiré ellipses are obtained·if e.g. the first circle intersects the second line, the second circle the fourth line, etc. This means that the Moiré appears as though the spacing, a, were doubled or tripled, etc., i.e. a is replaced by 2a, by 3a, and so on. The smaller the ratio d/a, the more circle-like become the ellipses which is seen from equation (8). As can be seen on Figure 1, ellipses perpendicular to the line sets, i.e. those along the cube diagonals, have the spacing a = nd/√2. For ellipses inclined by 45° to the straight line sets, i.e. those along the cube axes, the spacing a becomes nd.

If a Moiré pattern such as on figure 1 is already given, the data can be analyzed in the following way. From the ratio of the square of the radius of the inner circle with number n, to that of the limiting circle,

$$\frac{\rho^2}{R^2} = \Phi = \frac{2Nn - n^2}{N^2} \tag{11}$$

we obtain

$$N = \frac{n}{\Phi} \left[ 1 + (1 - \Phi)^{\frac{1}{2}} \right] \tag{12}$$

Since $\Phi$ depends on n, N should prove to be independent of n. In the case of Figure 1, for n > 10, (where the errors are small) N = 100, and since Nd = R, d = 1.016 mm and the unit cell dimension in the projection = 2.032 mm (before reduction).

In the second quadrant of Fig. 1 some analytical results are shown. Only a few of the ellipses from each set were drawn and they show very good agreement with the Moiré. For example, for the set where a = d, 83 ellipses are possible according to equation (10), whereas only 5 ellipses were drawn. Also, not all poles predicted by the analysis were drawn – only those corresponding to the most obvious ones in the Moiré pattern. The "b" value for each ellipse is shown as well as the "a" value for each set. The poles are indexed in the fourth quadrant.

Having given the explanation of the approach, we will proceed to the stereographic projection, which has the following advantages::

(a) The complete hemisphere can be visibly projected.

(b) It more closely resembles the projection obtained in the field-ion microscope.

(c) The angular relationships between poles is preserved so that indexing with a Wulff net is facilitated. This characteristic of stereographic projections renders the pole steps circular instead of elliptical, preserving their spherical surface character.

The computer program by which the Moiré is plotted is a general one which considers three (or four in the hexagonal case) sets of planes intersecting the surface of a sphere. The interplanar spacings for·each set and the angles between sets are variables. The point of projection can also be varied.

In stereographic projection, all three sets of circles on the reference sphere produced by the planar-sphere intersections are projected as circles

(as opposed i.e. to the orthographic, where two sets project as straight lines)
as shown on Figure 2. The Moiré pattern is seen to include many more poles
and more closely resembles the FIM. The analytical Moiré circles are shown in
the second quadrant. They were produced by projecting the analytical ellipses
from the orthographic projection, back to the reference sphere, and then repro-
jecting them stereographically. As in the orthographic case only a few of the
possible circles are drawn for each pole (not necessarily the same ones),
however, in this case, all possible poles from the orthographic analysis are
represented by at least one circle. The general position of these circles
coincides very closely with the Moiré pattern as before. The exact size of the
small circles in the Moirés varies considerably and is not important here.
Small changes in the relative position of each of the original sets of planes
make large differences in the size of the smallest Moiré fringes. This should
not surprise anyone who has observed the field evaporation process in a field-
ion microscope where all of the circles representing the atomic plane edges are
continually collapsing to zero radius as the last atom from each plane evapor-
ates. Consequently, in field-ion images the size of the smallest ledge circle
from the same type of plane will vary from quadrant to quadrant in the same
image as is also true and can be seen in the Moirés.

The most notable discrepancy between the stereographic Moiré pattern and
the analytical circles produced by reprojection from the orthographic is that
many prominent outer poles are not predicted. This is because they require that
the circles from the 2nd and 3rd sets be projected as non-straight lines. This
requires a modification of the details of the described procedure, but the same
approach is applicable and should yield equally good agreement with all of the
Moiré poles. Differences between the FIM image and the Moirés arise with
deviations from sphericity, changes in local radii of curvature, and physical
properties beyond the scope of purely geometrical considerations such as finite
sizes of atoms, ionization potentials, field sublimation energies, directional
bonding, distribution of surface charges, etc.

The use of Moiré models for the geometrical description of crystalline
interfaces has been well established (1). The present work has demonstrated
that the same general geometry which produces the successive rings of planar
edge atoms around low index poles in field-ion images can be used to produce
Moiré patterns of rings around "low index poles" which greatly resemble those
of the basic field-ion image.

The assistance of Dr. H.L. Lukas with the electronic calculations and graphics
is gratefully acknowledged.

REFERENCES

1. Bollmann, W., (1970), Crystal Defects and Crystalline Interfaces, Springer-
   Verlag, Berlin.

2. Doerr, R.M., and Ownby, P.D. (1975), Moiré Simulation of Field-Ion
   Micrographs, Pract. Metallog. 12, 78.

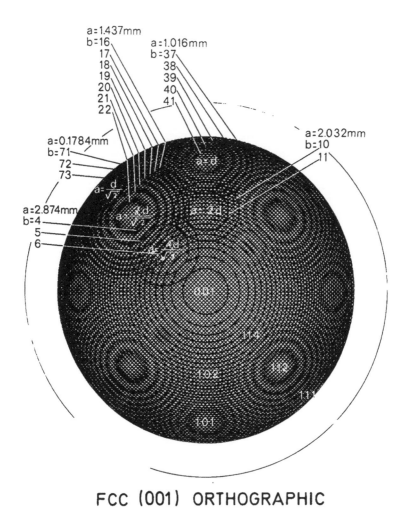

## FCC (001) ORTHOGRAPHIC

Figure 1:   Indexed Orthographic Moiré with corresponding analytical ellipses labeled.

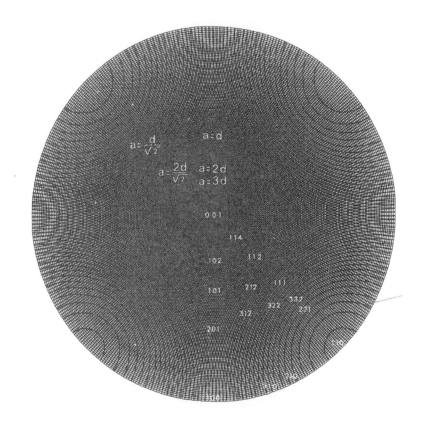

# FCC (001) STEREOGRAPHIC

Figure 2: Indexed Stereographic Moiré with re-projected orthographic
analytical ellipses labeled.

*National Bureau of Standards Special Publication 431*
*Proceedings of the* FOURTH INTERNATIONAL CONGRESS FOR STEREOLOGY
*held at* NBS, Gaithersburg, Md., September 4-9, 1975 (Issued January 1976)

## NETWORK SIMULATION OF CONNECTIVITY AND AGGREGATE SIZE IN TWO PHASE MIXTURES

by W. D. Leahy* and J. H. Steele, Jr.**
*Division of Minerals Engineering, Virginia Polytechnic Institute and State University, Blacksburg, Virginia 24061, U.S.A.*

INTRODUCTION

The theory of random clumping(1) or clustering encompasses some of the most difficult unsolved problems in stereology. One such problem involves predicting the distribution of aggregate sizes and connectivity in a random mixture of polyhedral grains of two phases. Cahn(2) has described a model which predicts aggregate connectivity in an average sense, that is connectivity per grain. His model does not, however, provide statistical estimates of aggregate size. Bishop et al.(3) have also used Cahn's model for describing connectivity in a graded mixture of two phases. In this note initial results of a Monte Carlo computer simulation using a network model for a polycrystalline structure will be described.

DESCRIPTION OF COMPUTER SIMULATION

A network model(4) for a polycrystalline microstructure can be defined if each vertex or node represents a polyhedral-like grain and each edge represents a face shared by an adjoining grain. This type of model allows aggregate properties to be calculated after the nodes are randomly associated either with one or the other two phases forming the mixture. For illustrative purposes the phases will be called $\alpha$ and $\beta$. A random process such as this is similar to a coloring(5) where two colors, e.g. black and white, are associated with the nodes of the network.

The network information was stored within a digital computer using its vertex matrix, $(V_{ij})$ where $V_{ij} = 1$ if the ith and jth grains form a face, and $V_{ij} = 0$ otherwise. In order to store the required information for a 1000 node network, involving $(1000)^2$ matrix values, a sparse matrix technique was utilized. Briefly this involved two lists, the first indicating the number of edges emanating from each node and the second indicating the number labels for each of the connecting nodes.

In order to define a network which would be a representative model for a polycrystalline aggregate, the following procedure was used. First, 1000 points were positioned randomly within a unit cube using a random number generator to select $(x,y,z)$ coordinates. Then all of the points within a fixed radial distance of each point were found and used as neighbors to define the faces for the grain at each point. This procedure is similar to that used by Meijering(6) and Gilbert(7) to define random tesselation models for cellular aggregates such as the polycrystal. Varying the radial distance changes the distribution and average number of neighbors. A radial distance which produced an average of 14 faces (or edges) per grain (or node) was used to define the network. Periodic boundary conditions were set up so that external surface effects on the unit cube would be minimized.

Random labeling of the grains or nodes as either $\alpha$ or $\beta$ was carried out to produce a mixture within the network structure. The initial results presented are based upon a single network model with different number fractions of grains applied randomly upon it.

Aggregate size and distribution are determined from the vertex matrix by a sorting algorithm which partitions the matrix according to the separate parts. The procedure involves a search and a relabeling of the nodes in a connected part so that they occur sequentially within the vertex matrix. In this manner the number of separate parts and their size can be obtained from the partitioned vertex matrix. An example network with its partitioned vertex matrix is shown in Figure 1.

*Presently with Corning Glass, Corning, New York
**Presently with Armco Steel Corporation, Middletown, Ohio

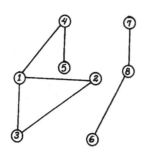

| | 1 | 2 | 3 | 4 | 5 | 6 | 7 | 8 |
|---|---|---|---|---|---|---|---|---|
| 1 | 0 | 1 | 1 | 1 | 0 | | | |
| 2 | 1 | 0 | 1 | 0 | 0 | | | |
| 3 | 1 | 1 | 0 | 0 | 0 | | | |
| 4 | 1 | 0 | 0 | 0 | 1 | | | |
| 5 | 0 | 0 | 0 | 1 | 0 | | | |
| 6 | | | | | | 0 | 0 | 1 |
| 7 | | | | | | 0 | 0 | 1 |
| 8 | | | | | | 1 | 1 | 0 |

Network                               Partitioned Vertex Matrix

Figure 1. Example of network with its partitioned vertex matrix.

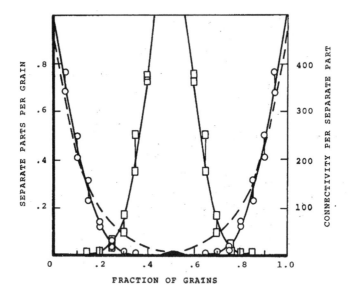

FRACTION OF GRAINS

Figure 2. Variation in connectivity per separate part (data points shown as squares) and separate parts per grain (shown as circles) with the fraction of grains associated with α phase. Note that the plot is symmetric about 0.5. The dashed curve is a plot of the function exp. $(-8V_V)$ for the minor phase.

RESULTS AND DISCUSSION

The effect of increasing number fraction (and also volume fraction) upon the number of aggregates per grain and upon the connectivity per aggregate is shown in Figure 2. An aggregate is defined as a connected clump or cluster of grains of the same phase. Connectivity in this model represents the number of independent circuits in a network sense in each aggregate. Number of separate parts per grain indicates by its reciprocal the number of grains per separate part.

An extremely good fit for this function (separate parts per grain) is given by $e^{-8V_V}$ where $V_V$ is the volume fraction of the phase. This is interesting since this exponential represents the probability for an isolated sphere among randomly positioned spheres in three dimensions. The dashed line in Figure 1 shows how well this random isolation probability estimates the aggregate size function for volume fractions up to 15%.

Distribution data for the various aggregates are presented in Table I for the initial simulation run. Although the single simulation does not provide quantitative statistical estimates of the probabilities of grain pairs, triplets, etc., it does give a qualitative indication of the tendency for these configurations. It should be noted from Figure 2 and the tabulated data that the random labeling process is symmetric with regard to the minor constituent. Hence, one simulation run provides two statistical data points. Approximate probabilities for a randomly selected grain belonging to a given size configuration are presented in Figure 3. These probabilities are somewhat below the functions given by Roach[1] as approximations for the pair, triplet, and higher order configurations.

Connectivity (denoted by $\beta_1$), which represents the number of linearly independent circuits, in a connected network, is given by,

$$\beta_1 = N_1 - N_0 + 1 \tag{1}$$

where $N_0$ = number of nodes, and $N_1$ = number of edges. If this equation is summed over the separate parts, then it becomes,

$$\beta_{1_{TOT}} = N_{1_{TOT}} - N_{0_{TOT}} + N_{parts} \tag{2}$$

Hence, when the vertex matrix is partitioned to allow the number of nodes and edges in each separate part to be counted, the connectivity of each separate part and the total connectivity may be calculated rather easily.

Cahn's statistical equation[2] for connectivity in a similar type aggregate involves the Euler characteristic, $Q'$, of the phase boundary. This is not the same as the connectivity plotted in Figure 1 since internal circuits among $\alpha$-grains within a single aggregate are involved in the network model, but not in Cahn's model. Effectively, Cahn's model treats each aggregate according to the properties of its phase boundaries. A comparison between the network simulation and Cahn's $Q'/G'$ (Euler characteristic per grain) can be made by noting that for each ith separate part,

$$\frac{Q'_i}{G'_i} = \frac{2 - 2\beta_{1_i}}{N_{grains}} \tag{3}$$

Hence, summing over the separate parts yields,

$$\frac{Q'}{G'} = \frac{2\left[N_{parts} - \beta_{1_{TOT}}\right]}{N_{grains}} \tag{4}$$

TABLE I - SIMULATION DATA

| Amount Of Phases | | Total No. Separate Parts | | Number of Separate Parts With N Grains | | | | | | | | | Separate Parts With N ≥ 10 |
|---|---|---|---|---|---|---|---|---|---|---|---|---|---|
| %α | %β | α | β | N=1 | 2 | 3 | 4 | 5 | 6 | 7 | 8 | 9 | α |
| 5% | 95% | 34 & 1 | | 24 | 6 | 3 | 0 | 1 | 0 | | | | |
| 10 | - 90 | 41 & 1 | | 19 | 7 | 7 | 2 | 2 | 3 | | | | 10 |
| 15 | - 35 | 34 & 1 | | 18 | 1 | 2 | 5 | 2 | 2 | 1 | | | 10-10-55 |
| 20 | - 80 | 24 & 1 | | 12 | 2 | 2 | 1 | 2 | 2 | 0 | | | 15-23-114 |
| 25 | - 75 | 13 & 1 | | 5 | 2 | 0 | 1 | 1 | 1 | 1 | 1 | | 211 |
| 30 | - 70 | 4 & 1 | | 2 | 0 | 0 | 0 | 1 | | | | | 293 |
| 35 | - 65 | 2 & 1 | | 1 | | | | | | | | | 349 |
| 40 | - 60 | 2 & 1 | | 1 | | | | | | | | | 399 |
| 45 | - 55 | 1 & 1 | | | | | | | | | | | |
| 50 | - 50 | 1 & 1 | | | | | | | | | | | |
| 55 | - 45 | 1 & 1 | | β | | | | | | | | | β |
| 60 | - 40 | 1 & 2 | | 1 | | | | | | | | | 399 |
| 65 | - 35 | 1 & 3 | | 1 | 0 | 1 | | | | | | | 346 |
| 70 | - 30 | 1 & 7 | | 6 | | | | | | | | | 294 |
| 75 | - 25 | 1 & 9 | | 8 | | | | | | | | | 242 |
| 80 | - 20 | 1 & 27 | | 10 | 4 | 6 | 2 | 2 | 0 | 1 | 0 | 1 | 130 |
| 85 | - 15 | 1 & 47 | | 24 | 5 | 6 | 4 | 4 | 1 | 1 | | | 24 & 25 |
| 90 | - 10 | 1 & 50 | | 30 | 10 | 5 | 2 | 2 | 0 | 0 | | | 17 |
| 95 | - 5 | 1 & 38 | | 31 | 5 | 0 | 1 | 1 | | | | | |

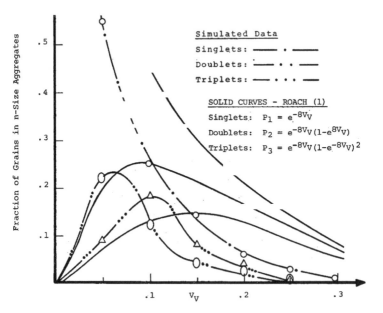

Figure 3. Distribution of aggregates which consist of separate grains, (singlets) pairs of grains (doublets), and triplets. Simulated data points are shown with dashed curves drawn through them. Solid curves are plots of approximations given by Roach (1).

Simulated Data
Singlets: —— • ——
Doublets: —— • • ——
Triplets: —— • • • ——

SOLID CURVES - ROACH (1)
Singlets: $P_1 = e^{-8V_V}$
Doublets: $P_2 = e^{-8V_V}(1-e^{8V_V})$
Triplets: $P_3 = e^{-8V_V}(1-e^{-8V_V})^2$

A plot of this characteristic per grain function using the computer simulation data is presented in Figure 4. Negative values are produced by larger connectivity, and the effect of internal circuits within $\alpha$-aggregates causes the network analysis to be significantly lower than Cahn's equation between $V_V \cong .2$ and $V_V \cong .8$.

Another property which is of interest in clustering problems is the critical percolation probability[8] (denoted $P_c$). This value represents the fraction of nodes (or grains) required in a phase to produce infinite size clusters and thus provide continuous paths through the network. Applying the techniques used by Dean and Bird[9] who plot the fraction of nodes in the largest part within the network, and a modified second moment of the cluster distribution versus fraction of nodes in the phase as shown in Figure 5. The modified second moment used is given by,

$$\mu = \sum_i n(i)\ i^2 / (\sum_i n(i)i)^2$$

where $\sum_i$ represents the sum over all size clusters, and n(i) is the number of nodes in the ith size cluster. The critical percolation probability is estimated as the fraction of grains where the maximum slope occurs for .
Thus, the estimated value is approximately, $P_c \cong .20$. This value is likely to be an over estimate because of the small size of the network utilized in this simulation. However, it is quite close to the value of $P_c$ for the face-centered cubic lattice arrangement ($P_c = .195$) which has twelve edges per node.

CONCLUSIONS

Polycrystalline aggregates may be simulated by network models of various types to study clustering, connectivity, and percolation phenomena. This study has indicated qualitatively the type of results that can be obtained using a network model and Monte Carlo simulation techniques.

Qualitatively, the results show that in a two phase polycrystalline aggregate where both phases have similar size and shape distributions, that between volume fractions of 0.2 and 0.8, both phases would be expected to have large highly interconnected clusters permeating the structure.

REFERENCES

1. Roach, S. A., *The Theory of Random Clumping*, Methuen & Co. Ltd., London, 1968.
2. Cahn, J. W., "A Model for Connectivity in Multiphase Structures," Acta Met., 14 (1966), 477.
3. Bishop, G. H., Quinn, G. D., and Katz, R. N., "Connectivity in Graded Two-Phase Structures," Scripta Met., 5 (1971), 623.
4. Steele, J. H., Jr., "Application of Topological Concepts in Stereology," *Quantitative Metallography and Stereology*, ASTM STP 504, Philadelphia, Pennsylvania, 1972, Page 39.
5. Busacher, R. G., and Saaty, T. L., *Finite Graphs and Networks*, McGraw-Hill Company, New York, 1965.
6. Meijering, J. L., "Interface Area, Edge Length, and Number of Vertices in Crystal Aggregates with Random Nucleation," Philips Research Reports, 8 (1953), 270.
7. Gilbert, E. N., "Random Subdivisions of Space into Crystals," Ann. Math. Stat., 33 (1962), 958.
8. Shante, V. K. S. and Kirkpatrick, S., "An Introduction to Percolation Theory," Adv. Phys., 20 (1971), 325.
9. Dean, P. and Bird, N.F., "Monte Carlo Studies of the Percolation Properties of Two and Three Dimensional Lattices," National Physical Laboratory Report Ma 61, Mathematics Division, November 1966.

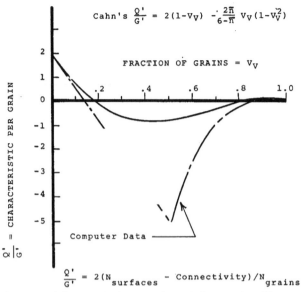

Figure 4. Connectivity per grain comparing Cahn's equation (2) to computer simulation results (dashed curve).

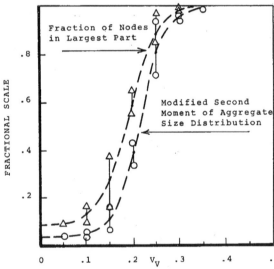

Figure 5. Results showing the variation in the fraction of nodes associated with the longest part and the modified second moment with fraction of grains associated with minor phase. The percolation probability $P_c$ is estimated from the fraction of grains at which the slope in the modified second moment is a maximum.

# 2. PATTERN RECOGNITION

*Chairpersons:*

G. Bernroider
G. C. Cheng
C.-H. Chen

*National Bureau of Standards Special Publication 431*
*Proceedings of the* FOURTH INTERNATIONAL CONGRESS FOR STEREOLOGY
*held at* NBS, Gaithersburg, Md., September 4-9, 1975 (Issued January 1976)

## WHAT CAN PATTERN RECOGNITION DO FOR STEREOLOGY

by George C. Cheng
*University of Georgia, Athens, Georgia 30601, U.S.A.*

ABSTRACT*

What is Pattern Recognition? Hundreds of scientists have devoted research
to Pattern Recognition for more than twenty years.  More than a hundred
books have been written on this subject.  Thousands of articles and reports
have been published.  The Pattern Recognition Society and its official journal
were organized eight years ago.  However, no one has seriously attempted to
define the term 'Pattern Recognition'.  Because of the concepts, methodologies,
and applications of Pattern Recognition are very diverse in research, no two
researchers would agree upon a precise definition.  There are also diverse
opinions about whether Pattern Recognition should be considered as an art or
a science.  Among the research in the past twenty years, the main interest
has been in the development of automatic systems and machines to perform
pattern recognition.

From the methodological viewpoint, Pattern recognition research has been
generally approached from one of the following directions:

1. Heuristic: cognitive process, Gestalt, artificial intelligence, machine
   intelligence, theorem proving, problem solving, robot, automata, and
   learning machine are frequently associated terms.
2. Structural or linguistic: syntax-directed methods, picture gramma, state
   diagram, sequential decision, and tree searching are related terms.
3. Mathematical/statistical: multivariate analysis, factor, clustering,
   classification, and discriminant analyses are the terms associated with
   the statistical approach.  Set theory, category theory, and formulism are
   those related with other mathematical approaches.

In addition, Perceptron, Neuron-models, and Cybernetics are also related to
Pattern Recognition.

The heuristic and structural approaches are discussed in this talk through
several simple illustrative examples.  The statistical approach is discussed
by Dr. C. H. Chen in this session.  The Category Theory oriented approach is
discussed by Dr. M. Pavel in the session on Mathematics and Pattern Recognition
of this Congress.

---

*Due to copy rights, the pictures and data presented at the Congress are
not included in this printed version.

A major research area in Pattern Recognition has been the pictorial (or visual) Pattern Recognition. Stereologists may find that this class of Pattern Recognition is most useful for their work.

From the operational viewpoint, Pattern Recognition, particularly pictorial Pattern Recognition can be considered in the following four phases:

1. Data acquisition: instrumentation is a major concern. Representative works include that of Dr. G.A. Moore (NBS), Dr. L.E. Lipkins (NIH), Dr. R.S. Ledley (Georgetown University), Quantimet, Leitz Texture Analysing System, Illiac III, etc.

2. Data preprocessing: picture or image digitization, edge/line/boundary detection, feature extraction, scale transformation, spatial transformations, etc. are associated terms.

3. Pattern Recognition (of preprocessed data): factoring, clustering, classifying, discriminant-analysing, syntax-directed recognition, tree-searching, etc. are the employed operations.

4. Theory forming: hypothese forming and testing, speculating/forecasting, decision making, etc. are the consequences of Pattern Recognition.

What has Pattern Recognition helped to accomplish in solving practical problems? The most successful application of Pattern Recognition research has probably been Optical Character Recognition (OCR). For examples, started in 1967, the Japanese automatic mail sorting machine could recognize the handwritten postal code (zip code) and sort 600 pieces of mail in a minute. The OCR machines made by IBM for the U.S. Social Security Office could recognize typed multifont characters at a speed of more than a thousand characters per second. The state of the art in OCR is about 2000 characters per second with a less than 0.02% error rate.

How does a Pattern Recognition machine recognize a handwritten character? A simplified example of OCR is presented for illustration of the structural approach of Pattern Recognition. One of the purposes of showing this example is to illustrate the picture processing operations, including the formation of the picture matrix, feature extraction from the picture matrix, and syntax-directed recognition. Another purpose is to point out that in OCR, the quality and/or form of the raw data can be controlled. For examples, the Japanese postal code is written in red colored boxes on the envelope, and for the typed and printed OCR, two new fonts of English letters were designed specially for OCR machines, namely the OCR-A font in America and the OCR-B font in Europe.

Examples of raw data that have been subjected to Pattern Recognition research. In stereology as well as in many other disciplines, the raw data cannot be controlled as well as does in OCR. Some thirty selected examples of this kind raw data that have been subjected to Pattern Recognition research are shown herewith. (Many more examples are shown by other colleagues in this Congress.) A purpose of showing these pictures is to provide comparison facility to stereologists. If there is a close similarity between the stereologist's own data and some of the tried examples, the degree of assistance that Pattern Recognition can offer to his problem and the cost of using this technique can be estimated from the previous experience. Otherwise, more effort would be needed to study his problem from the view-point of Pattern Recognition.

The purpose of showing the last set of three pictures is to point out that in the computer processing of pictorial data, the data is usually preprocessed and converted into line drawings before Pattern Recognition is applied. For this reason, the following examples used to illustrate the heuristic approach of Pattern Recognition are all line drawings.

A purpose of showing the example of a 'bear behind a tree' is to point out that in the real world, we use information from the environment, the context of the situation, the purpose of our investigation, and our build-in cognitive system as aids in doing pattern recognition. When we see four paws on the tree trunk, it is not difficult for us to recognize that there is a bear behind it. However, in an automatic Pattern Recognition system or machine, the raw data is converted into a line drawing like that shown in the slide. The information from the environment or context is lost. In most Pattern Recognition research, the contextual information is neither presented to the researchers nor requested by the researchers to be provided. In other cases, the mathematical models are so rigidly designed, there is no room for the exogenous information to be inserted.

The purpose of showing the 'man or girl' example is to point out the existence of the overlapping situation. Statistics provides many sophisticated techniques for handling the population-overlapping situations. However, a situation like the 'man or girl' would create a gap between the raw data and the kinds of measurement required by the statistical models.

What can Pattern Recognition do for Stereology? There is no Pattern Recognition cook-book available for stereologists to use. The immediately useful contribution of automatic Pattern Recognition to Stereology would be in the instrumentation of data acquisition and data preprocessing. The transfer of Pattern Recognition technology from one discipline to another is not straightforward. This is mainly because the effectiveness of a particular Pattern Recognition technique depends on the quality and quantity of the data. Either a stereologist needs to devote a lot of effort to learn Pattern Recognition or the Pattern Recognition researchers need to devote effort to learn about the data and problems of the stereologist. The cost

109

of these trainings is high, but it may be worthwhile if the case is like
that of OCR. From the viewpoint of the Pattern Recognition discipline, any
research effort in this area by stereologists would be beneficial to its
advance. We hope that it would also be beneficial to Stereology. In
conclusion,

- If your data is or can be controlled in quality and in·
  quantity as in OCR, the structural and/or statistical
  approaches of Pattern Recognition can be of great
  assistance to Stereology
- If you fancy the modern algebraic way of thinking, the
  mathematical Pattern Recognition concept may provide
  you with mental enjoyment and may lead to some break
  through.
- In one degree or another, we all use heuristic approaches
  in solving problems. The concept of heuristic Pattern
  Recognition may reinforce your intuitive attitude in
  solving Stereological problems.

Selected Pattern Recognition bibliography for Stereologists

I.  Review articles:

Cormack, R.M., "A Review of Classification," the Journal of Royal Statistical
Society, V.134, (1971) pp.321-367. (227 references were included.)

Duran, B.S. & Odell, P.L., Cluster Analysis - A Survey, Springer-Verlag, 1974.
(409 references were included.)

Das Gupta, S., "Theories & Methods in Classification - A Review," in ·
Discriminant Analysis and Applications, edited by T. Cacoullos, Academic
Press, 1973. (260 references were included.)

Nagy, G., "State of the Art in Pattern Recognition," IEEE Proceedings, V.56,
No.5, (May, 1968) pp.836-862. (148 references were included.)

Toussaint, G.T., "Subjective Clustering and Bibliography of Books on Pattern
Recognition," Information Sciences, V.8, (1975) pp.251-257. (117
references were included.)

II.  Pictorial and 3-dimensional objects Pattern Recognition:

Barnhill, R.E. & Reisenfeld, R.F., (ed.) Computer Aided Geometric Design,
Academic Press, 1974.

Cheng, G.C., Ledley, R.S., Pollock, D.K., Rosenfeld, A., (ed.) Pictorial
Pattern Recognition, Tompson Book Co., 1968.

Gips, J., "A syntax-directed program that performs a three-dimensional
perceptual task," Pattern Recognition, V.6., ·(1974) pp.189-199.

Grasselli, A. (ed.) Automatic Interpretation & Classification of Images,
Academic Press, 1969.

Ivancevic, N.S., "Stereometric Pattern Recognition by artificial touch," Pattern Recognition, V.6. (1974) pp.77-83.

Maruyama, K., A study of visual shape perception, University of Illinois, University of Computer Science. UIUCDS-R-72-533, Oct., 1972.

McKee, J.W., & Aggarwal, J.K., "Finding the edges of 3-D objects," Pattern Recognition, V.7, (1975) pp.25-52.

Muller, W., "The Leitz Texture Analysing System," Leitz Scientific & Technical Information, (April, 1974) pp.101-116.

Rosenfeld, A., Picture Processing by Computer, Academic Press, 1969.

Serra, J., "Theoretical Bases of the Leitz Texture Analysing System," Leitz Scientific & Technical Information, (April, 1974) pp. 125-136.

Sobel, I., "Towards Automated 3D Micro-Anatomy Via Reconstruction from Serial section micrographs," 1975 Engineering Foundation Conference on Automatic Cytology.

Swann, P.R., Humphreys, & Goringe, (ed.) High Voltage Electron Microscopy, Academic Press, 1974.

Tenenbaum, J.M., Garvey, T.D., Weyl, S., Wolf, H.C., An Interactive Facility For Scene Analysis Research, Stanford Research Institute, Artificial Intelligence Center, Tech. Note 87, Jan., 1974.


III.  Books:

Andrews, H.C., Introduction to Mathematical Techniques in Pattern Recognition, John Wiley & Sons, 1972.

Arbib, M.A., The Metaphorical Brain, John Wiley & Sons, 1972.

Banerji, R.B., Theory of Problem Solving - An approach to artificial intelligence, American Elsevier Publishing Co., 1969.

Bongard, M. (ed. by Hawkins, J.K) Pattern Recognition, Spartan Books, 1970.

Braffort, P., L'intelligence artificielle, Press Universitaires  De France, 1968.

Carne, E.B., Artificial Intelligence Techniques, Spartan Books, 1965.

Chen, C.H., Statistical Pattern Recognition, Spartan Books, 1973.

Duda, R.O., Hart, P.E., Pattern Classification and Scene Analysis, John Wiley & Sons, 1973.

Findler, N.V., Meltzer, B., (ed.) Artificial Intelligence & Heuristic Programming, American Elsevier Pub., 1971.

Fukunaga, K., Introduction to Statistical Pattern Recognition, Academic Press, 1972.

Grenander, U., A Unified Approach to Pattern Analysis, Brown University, Center for Computer and Information Sciences, 1969.

Kanal, L.N., (ed.) Pattern Recognition, Thompson Book, 1968.

Kolers, P.A., Eden, M. (ed.) <u>Recognizing Patterns, Studies in Living &</u>
<u>Automatic Systems</u>, the MIT Press, 1968.

Meltzer, B. & Michie, D. (ed.) <u>Machine Intelligence</u>, Am. Elsevier, 1971

Mendel, J.M., Fu, K.S. (ed.) <u>Adaptive, Learning, & Pattern Recognition</u>
<u>Systems</u>, Academic Press, 1970.

Minsky, M., Papert, S., <u>Perceptrons, An introduction to computational</u>
<u>Geometry</u>, the MIT Press, 1969.

Moore, G.A., <u>Application of Computers to Quantitative Analysis of Micro-</u>
<u>structures</u>, National Bureau of Standards, Report No. 9428, Oct.,1966.

Nilsson, N.J., <u>Learning Machines</u>, McGraw-Hill, 1965.

Nilsson, N.J., <u>Problem Solving Methods in Artificial Intelligence</u>, McGraw-
Hill, 1971.

Patrick, E.A., <u>Fundamentals of Pattern Recognition</u>, Prentice-Hall, 1972.

Pavel, M., <u>Fondements mathematiques de la Reconnaissance des Structures</u>,
Actualistes Scientifiques, No.1342, Hermann, Paris, 1969.

Sklansky, J. (ed.) <u>Pattern Recognition - Introduction & Foundation</u>, Dowden,
Hutchinson & Ross, 1973.

Tsypkin, Y.Z., <u>Foundations of the Theory of Learning Systems</u>,·Academic Press,
·1973.

Watanabe, S., <u>Methodologies of Pattern Recognition</u>, Academic Press, 1969.

Watanabe, S., <u>Frontiers of Pattern Recognition</u>, Academic Press, 1972.

Young, T.Y., Calvert, T.W., <u>Classification, Estimation, and Pattern Recognition</u>,
American Elsevier Publishing Co., 1974.


IV. Additions:

Bellman, R., Kalaba, R., Zadeh, L., "Abstraction and Pattern Classification,"
J. of Mathematical Analysis and Applications, V.13, (1966) pp.1-7.

Kirsch, R., Cahn, L., Ray, L.C., Urban, G.H., "Experiments in Processing Pictorial Information with a Digital Computer," <u>Proceedings of Eastern Joint</u>
<u>Computer Conference</u>, (1957) pp.221-229.

Moore, G.A., "Gestalt Properties of Aggregate Materials," <u>Practical Metallo-</u>
<u>graphy</u>, V.IX, No.2, pp.76-97.

Zobrist, A., & Thompson, W.B., "Building a Distance Function for Gestalt
Group," IEEE Trans. on Comp., C-4. #7, (July, 1975) pp. 718-728.

*National Bureau of Standards Special Publication 431*
*Proceedings of the* FOURTH INTERNATIONAL CONGRESS FOR STEREOLOGY
*held at* NBS, Gaithersburg, Md., September 4-9, 1975 (Issued January 1976)

THEORY AND APPLICATIONS OF IMAGERY PATTERN RECOGNITION

by C. H. Chen
*Southeastern Massachusetts University, North Dartmouth, Massachusetts 02747, U.S.A.*

ABSTRACT

In recent years, there has been an increasingly great demand for automatic recognition of imagery patterns which arise in biomedical, space and
stereology and a number of other applications. As the computer hardware cost
decreases rapidly, automatic imagery recognition will soon become a reality
in many of these applications. In this paper, a unified theory of imagery
pattern recognition is developed which includes: (1) image representation and
compression, (2) preprocessing, (3) feature extraction, and (4) pattern classification. Status of image recognition applications is reported. Practical
considerations such as the computational complexity and operational expenses
are discussed.

I. Introduction

Since the development of pattern recognition area in early fifties, there
have been much efforts made to develop methods of processing pictorial information by computer, optical devices or hybrid systems. Most·patterns in real
life are imagery in nature. The demand for automation is great due to the
rapidly increasingly volume of pictorial patterns. Extensive research on
various image recognition application has been reported in the literature.
For example, Refs. 1-3 provide excellent discussions of automatic image analysis in stereology or vice versa. Extensive list of references on this subject
is also available (4) (5). However the automatic recognition has not been
cost effective for desired performance and the gap between theory and practice
remains wide open. Much effort is needed to make effective use of pattern
recognition theory, which offers a number of mathematical tools, on applications. With this objective in mind, we discuss in this paper both the theoretical and practical considerations of imagery pattern recognition.

II. Image Representation and Compressions

To start with, the image data are in analog forms such as continuous gray
levels or intensity values. For computer processing it is necessary to digi-
tize the images and quantize the discrete set of data points. The result may
be called a picture function which represents the original images completely
with the exception of quantization errors. The picture function consisting
of two-dimensional array of data points usually contains too much information
to be handled effectively by the computer. To reduce the amount of data without any significant loss of useful information, image representation and data
compression must be considered. If the images are considered as stochastic
processes, the Karhunen-Loeve transform is optimum in the sense of minimizing
the mean square error between the original and transformed pictures. Although
the symmetric property of covariance matrices for imagery data can be used to
simplify this optimum transform, the required computation is still excessive.
Two-dimensional orthogonal transforms such as Fourier, Walsh-Hadamard, Haar,
and discrete cosine transforms can be computed much more efficiently because
fast algorithms are available. The area of data compression for picture band-
width reduction has been well studied in recent years. Its objective is to
reconstruct the transmitted image at the receiving site as accurately as
possible. For the purpose of pattern recognition, data compression is to
remove the irrelevant details.and to retain only the discrimination information.
Thus the data compression objective is to reduce the amount of computation for
classification with minimal loss of recognition accuracy. A simple data compression technique is to reduce the original picture to a binary (two-level)
picture which can further be compressed with almost no loss of discrimination

information (6). Picture encoding methods may also be used for image representation and compression.

## III. Image Preprocessing Techniques

An important class of preprocessing techniques is the position-invariant operations. This includes linear operations such as convolution and correlation and point operations such as requantization and gray level normalization which may be linear or nonlinear operations. Image enhancement for smoothing to reduce the noise effect or sharpening to improve the boundaries or edges are often necessary before any features can be derived. The enhanced image which is better for human visualization is also good for computer recognition. Statistical information of a picture or a subpicture may be derived from gray level histograms. An optimum threshold can be determined from the histogram. The threshold is useful to generate a binary picture for picture segmentation, picture data compression, or even boundary detection (6), (7).

Spatial smoothing of a region can be performed by convolution or frequency filtering. If noise spectrum is considerably different from the signal, then bandpass filtering can suppress noise by deleting a selected band of spatial frequencies while leaving enough at high frequencies to keep edges unblurred. Optimum (Wiener) filtering may be performed but the boundaries will be blurred. There are many ways to sharpen or deblur a picture. For example, this can be done by high-pass spatial filtering or high-frequency emphasis operation. We know that integration (i.e., averaging) blurs a picture, it is natural to use some sort of differentiation operation to deblur a picture. Gradient operation serves this objective; but the resulting isolated bursts have to be removed by further processing. Gray level normalization and distribution equalization also can sharpen the picture. A combination of sharpening and smoothing is usually needed to arrive at the best result. Gray level slicing is particularly effective in improving certain characteristics of the picture (7). Other preprocessing problems include image matching, restoration, etc. Digital signal processing techniques such as the use of hormomorphic systems (8) are also very useful for image preprocessing.

## IV. Feature Extraction

Feature extraction, consisting of deriving certain characteristic measurements from the input picture function, is considered by far the most difficult step in image recognition. Our approach to the feature extraction problem, applicable to a restricted, although large class of picture functions, is to derive measures of a statistical nature. Such measures are most appropriate with textural patterns. The second-order gray level distribution can be used to derive a set of texture features or average local property measures. Features may also be derived from two-dimensional transforms which exploit global property. At present features are extracted fairly heuristically and the effectiveness of such features is evaluated by the probability of error as well as the information and distance measures (9). It is usually more desirable having a small set of effective features followed by a simple classifier than using a large number of features in conjunction with a complicated classifier. To select good features really requires a good understanding of the basic (physical) properties of the pattern. Strictly relying on mathematical features usually cannot lead to the best possible performance. Thus in automatic image analysis, the stereologists can be very helpful to provide good features.

## V. Pattern Classification

As a final portion of the recognition system design, classification is a decision making process which assigns the input pattern to one of several possible classes (categories). The design objective in most recognition problems is the minimum number of errors incurred in decision making. Statistical decision theory provides an excellent formal approach to the decision making. Typical classification rule are (10): 1) Bayes decision rule, 2) maximum likelihood decision rule, 3) nearest neighbor decision rule, 4) sequential decision procedure. To reduce the computational effort, a table look-up approach has been proposed.

## VI. Practical Considerations

The following are some important practical problems:

1. The sample size, i.e., the number of samples available for learning and classification is always limited. Pattern recognition theory is based on infinite sample size. This may cause discrepancy between practical result and theoretical prediction. Large number of samples, however, may require excessive computation.

2. Computational complexity including the number of computation steps and the memory requirement is directly related to the cost for the recognition task. There is almost no theoretical work on the prediction of computation cost. Programming complexity is another problem. At present it all depends on experience and experimentation. A few simple rules may be useful. If the probability density cannot be conveniently defined, the nearest neighbor decision rule is always useful though the computation of Euclidean distance is time consuming for large sample size. If Gaussian assumption is reasonable, then the maximum likelihood decision rule is useful.

3. The cost difference between human recognition and machine recognition is still considerable. For example human interpretation of a photo may cost only a few cents but it may require over a hundred dollars for machine interpretation. Although the basic problems here are not just computer hardware and software, good computational algorithm is always important in reducing the cost. Larger and faster computers alone cannot solve the problem.

To resolve some of the practical problems, it is important to note that the effective preprocessing and feature extraction can reduce considerably the overall computation cost. And the finite sample size effect must always be taken into account.

## VII. Applications

The great demands for automatic imagery recognition in various application areas have prompted much theoretical studies on imagery pattern recognition. New image processing techniques, on the other hand, have also been used in various applications. An excellent discussion on various applications is in the book by Cheng, et.al. (11). There is more success in image processing than in feature extraction and classification. At present the imagery recognition has much more success in space, biomedical areas than in military applications. It appears, however, to be a long way to a fully automated recognition system. Thus much more work is needed in both theory and applications.

C. H. Chen

References

1.  G. A. Moore, "Recent progress in automatic image analysis", Journal of Microscopy, vol. 95, Pt. 1, pp. 105-118, Feb. 1972.

2.  E. E. Underwood, "Stereology in Automatic image analysis", Microscope, vol. 22, pp. 69-80, 1974.

3.  I. A. Cruttwell, "Pattern recognition by automatic image analysis", Microscope, vol. 22, pp. 27-37, 1974.

4.  A. Jesse, "Information sources in automatic image analysis", Microscope, vol. 22, pp. 81-88, 1974.

5.  A. Jesse, "Bibliography on automatic image analysis", Microscope, vol. 22, pp. 89-109, 1974.

6.  C. H. Chen, "Theory and applications of image pattern recognition", TR EE-75-3, April 1975.

7.  C. H. Chen,et.al., "Image analysis, enhancement, display, and classification study with the aerial photographic data", TR EE-75-8, August 1975.

8.  A. V. Oppenheim and R. W. Schafer, "Digital Signal Processing", Prentice-Hall 1975.

9.  C. H. Chen, "On information and distance measures, error bounds, and feature selection", Information Sciences Journal, to appear.

10. C. H. Chen, "Statistical Pattern Recognition", Hayden Book Co., 1973.

11. G. C. Cheng, et.al., "Pictorial Pattern Recognition", Thompson Book Co. 1968.

National Bureau of Standards Special Publication 431
Proceedings of the FOURTH INTERNATIONAL CONGRESS FOR STEREOLOGY
held at NBS, Gaithersburg, Md., September 4-9, 1975 (Issued January 1976)

A NEW, FAST AND STORAGE-SAVING IMAGE ANALYSIS PROCEDURE FOR INVESTIGATING INDIVIDUALS BY A DIGITAL COMPUTER

by M. Rink
Institut für Geophysik der TU Clausthal, D-3392 Clausthal-Zellerfeld, Germany

INTRODUCTION

The problem of automatic counting and sizing of particle projections or sections, or any other maybe complex-shaped, but individual parts of a distinct image component - here simply called figures - is a very old one, and many different principles have been applied to its solution. At present the extension of the problem towards recognizing special individuals of a given collective by a morphometric characterization by means of a versatile detailed analysis of individuals has become of interest. The analysing capabilities of modern automatic image analysers are not yet satisfying in this respect. Some of the limitations of these instruments arise from the restrictions of an economically reasonable perfection of the logical units and of their storage capacities.

The rapid development in computer technology suggests the use of a general purpose computer for the solution of this problem. It is the merit of G.A.Moore and co-workers of having given for the first time, about in 1957, a computer method resulting in an exact analysis of such individual figures (Kirsch et al., 1957; Moore and Wyman, 1963). But his way of solution, based on the so-called arrested-scan-principle, is somewhat tedious. For getting information on the interesting figures, the whole digitalized image must be stored. During the automatic analysis always just one single figure is copied out in a second blank image field, reserved in the computer, and is quantitatively investigated there. Then this auxiliary image field is cleared again, the next figure in the original image is searched for, copied out, numerically analysed, a.s.o. An advantage of this method of time serial figure by figure analysis is the easy understandability and straight forwardness of the program concept, whereas it is relatively storage- and time consuming in operation. The latter fact, among others, has apparently led Moore (1972) himself to a somewhat pessimistic view on the future chances of the analysis of individuals.

The author has developed a universal image analysis procedure (IAP) for digital computers permitting the speedy and comprehensive analysis of individual figures without any restrictions to shape and connectivity (Rink, 1970).

THE IMAGE ANALYSIS PROCEDURE

The digitalized - and with respect to the light intensity dualized - image values serve as input into the computer. The image digitalizing points define the nodes of a square mesh net. They are arranged in lines and columns of distance x, and can be exclusively either a so-called figure point or a complementary point.

The main feature of the IAP is the continuous line scan principle, marked by proceeding step by step in line direction and continuing line by line. In this characteristic point the IAP differs from earlier image analysis methods for computers. During operation, information of two adjacent lines only is necessary at one particular time. In the following these two lines will be named comparison- and investigation line.

The basic problem of such a two-line comparison consists in recognizing which individual figure intercept belongs to which particular figure. The principle of solution involves the quasi parallel resp. simultaneous recognition, unravelling, and proper assemblage of all figure parts hit by the scan. This is achieved by using the well-known criterion of overlapping figure intercepts supplemented by a continuous monitoring and reordering of the intermediate figure parameters belonging to the figure parts already under treatment.

By shifting the bit pattern of an image line within the computer this information is converted into the equivalent form of the socalled intercept data, which represent the series of pairs of column coordinates of the left and right figure intercept ends of this line. The intercept data of the actual comparison- and investigation line are stored in separately reserved arrays. The association by overlapping of intercepts then is found by arithmetic relations between these intercept data.

Directly gainable figure parameters are built up successively with each increase in the next line. Some of them are figure integral values, as e.g. figure area, and other ones are figure-specific comparison values, as e.g. the column coordinate of the leftest figure point. To each figure just being investigated, a set of intermediate parameter values belongs, with strictly observed arrangement. All these sets of intermediate parameter values are stored until the figures are recognized to be complete. The assignment of such a set of intermediate parameter values to the corresponding figure is realized by an address calculation resulting from a systematic figure numbering. Reasons for changing these numbers are the beginning of a new figure, the merging of several figure parts, and the completeness of a figure.

If an intercept of the investigation line is overlapped by intercepts of the comparison line more than once, the remaining number of the so far recognized figure part will be the smallest number of the overlapping figure parts. In this case, the updating of the intermediate figure parameter values consists - besides considering the usual intercept increase - in taking over the intermediate parameter values of the multiply overlapping figure parts into the now valid intermediate parameter set of that so far recognized figure part.

During the scanning of the investigation line, newly recovered figure parts are characterized as such and their so-called starting parameter values are stored intermediately as a reduced set of intermediate parameter values.

A criterion for the completeness of a figure is derived from the increment of a monotonously increasing intermediate figure parameter (e.g. figure area) being checked at the end of each line.

The inserting of the starting parameter values of new figures, and the sorting-out of the intermediate parameter values of complete figures as well as those no longer needed ones of multiply overlapping figure parts, at the end of a line, in conjunction with the two-line principle, keeps the working storage at a minimum. Parallel to this rearrangement of the intermediate parameter values also the numbers assigned to the intercepts of the investigation line have to be actualized. Because of figure parts possibly branching in downward direction those numbers then represent a sequence beginning with one, but being not necessarily a monotonous sequence.

In the computing process the two-line cycle is reflected by overwriting the content of the storage region of the comparison line by that of the storage region belonging to the present investigation line. This can be done at the end of each line since then the former intercept data of the comparison line are not needed any longer. At the same time the storage region of the investigation line is available again. After this redefinition of the former investigation line as new comparison line, the next line can be investigated in the same manner.

When a figure has proven to be complete, all their directly

gainable parameters are available in the computer. By mathematical operations a series of meaningful derived individual figure parameters can be calculated from the direct ones. So after just one analysing computer run, the wanted direct and derived parameters of all figures of the image are determined. Of course, the recognition of individual figures also permits their exact counting.

To demonstrate the capability of the IAP a compilation of figure parameters will be given.

Direct figure parameters are:

the area a, proportional to the number of figure points;

the smallest (le resp. lo) and largest (ri resp. hi) column resp. line coordinate, which characterize the figure-circumscribed rectangle;

the multiple projection $p$ resp. q on the column resp. line direction;

the maximum chord length $s_{max, 1}$ in line direction, also known as so-called Krumbein diameter. [1] It should be noticed that the line-by-line analysis permits determining this parameter – often used as an orientation dependent and therefore statistical one-dimensional size parameter – with reference to the line direction only, whereas it is possible to measure the multiple and single projection in the perpendicular direction, too.

the number v of rectangular inward corners at the boundary of the digitalized figure. This information is very important for the determination of the length of the true figure circumference.

the connectivity, informing about the number of holes within the outer figure limit;

the image frame contact of a figure, which is of interest for excluding errors caused by frame-cut figures.

Derived figure parameters are:

the height h = (hi-lo+1)x resp. the width w = (ri-le+1)x as the single projection on the column resp. line direction or so-called statistical Feret diameters;

the mean chord length $s_{m,1}$ = a/p resp. $s_{m,c}$ = a/q in line resp. column direction. These linear figure measures are still orientation dependent, but they are weighted by the figure area.

the nominal diameter after Wadell $d_w = 2\sqrt{a/\pi}$ as diameter of the circle being area-equivalent with the figure.

an approximation to the longest figure dimension 1 of convex figures as a non-statistical, unique measure for figure length $1 = (s_{max}/w)^2 max(h,w) + (1-(s_{max}/w)^2 \sqrt{h^2+w^2})$;

the mean dimension $d_m$ = a/l perpendicular to 1;

the minor axis $e_3$ = 4a/l$\pi$ of the ellipse of area a and major axis 1;

aspect ratio factors, derived from the ratio $1/d_m$ resp. $1/e_3$;

the compactness factor $k_{ct}$ = a/hw as the area ratio of figure and circumscribed rectangle (Moore, 1968);

the so-called concavity factor $k_{cv}$ = (p+q)/(h+w), exactly giving information about the re-entrance as well in line as in column direction;

the length $c_p$ = 2(p+q) of the rectangular polygon limiting the digitalized figure;

a good approximation to the true figure circumference $c = c_p - v(2-\sqrt{2})x$;

The combination of the two orientation independent parameters area and circumference yields further essential information about the figure, e.g.:

the so-called hydraulic radius m = a/c as a size and shape dependent parameter, and

mere shape factors expressing the circularity or roughness of the

circumference as the so-called Heywood factor $k_{01} = c/2\sqrt{a\pi}$ or its
squared inverse value $k_{02} = 4\pi a/c^2$ ;
further possibilities of combining area and circumference are
the transformations of the figure into the equivalent ellipse and
rectangle after Moore (1968).

At present the program will be extended by the evaluation for
the areal center of mass and the inertial moment - parameters
surely of interest to morphological and stereological problems.

With the great number of those morphometric parameters a characterization of individual figures is possible, and geometrical
filters can be set up to perform image cleaning and shape selection.

The data storage in the computer also permits the derivation
of image related integral values as well as the immediate statistical analysis of the figure parameters. So the program is provided for computing the mean, the variance, and the standard deviation, and higher statistical moments if desired, of all geometrically meaningful figure parameters, further their frequency
distributions in logarithmic or linear scale. This advanced statistical treatment of the figure morphometry is the necessary prerequisite for any attempt of a three-dimensional particle number-,
size-, and shape analysis.

The essential program part responsible for the figure recognition and evaluation of the direct figure parameters is written in
assembler language, all other parts are written in FORTRAN. The
assembler part including working storage requires about 4200 computer words, the FORTRAN part needs about 3500 words.

With the described IAP, the computing time for analysing an
image of 107 200 image points containing about 800 - 900 figures,
and for listing their statistical results, runs up to about 150 sec,
with the TR4 computer having a word length of 48 bits and a cycle
time of 9 $\mu$sec.

The difficile task of developing such a program, once done,
pays by the speed and enormous saving in storage space. Therefore
this software method seems well suited for the combination of a
computer with a fast television input device.


References

Kirsch,R.A., Cahn,L., Ray,C. & Urban,G.H.(1957) Experiments in
   processing pictoral information with a digital computer. Proc.
   Eastern Joint Computer Conf., 221-229, Washington D.C.
Moore,G.A.(1968) Automatic scanning and computer processes for
   the quantitative analysis of micrographs and equivalent sub-
   jects. In: Pictorial pattern recognition, 275-325, Thompson
   Book Co. Washington
Moore,G.A.(1972) Recent progress in automatic image analysis. In:
   Stereology 3 (ed.Weibel et al.) 105-118, Blackwell Sci.Publ.
   Oxford London Edinburgh Melbourne
Moore,G.A. & Wyman,L.L.(1963) Quantitative metallography with a
   digital computer: Application to a Nb-Sn superconducting wire.
   J.Res.NBS 67A, 127-147
Rink,M.(1970) Automatische morphometrische Bildanalyse mit Hilfe
   eines elektronischen Digitalrechners. Diss. Techn. Univ. Claus-
   thal, BRD

*National Bureau of Standards Special Publication 431*
*Proceedings of the* FOURTH INTERNATIONAL CONGRESS FOR STEREOLOGY
*held at* NBS, Gaithersburg, Md., September 4-9, 1975 (Issued January 1976)

## SYSTEM FOR COMPUTER INPUT AND PROCESSING OF TWO-DIMENSIONAL PICTURES

by Ivan Krekule and Miroslav Indra
*Institute of Physiology, Czechoslovak Academy of Sciences, Prague, Czechoslovakia*

A hard and software system, based on the laboratory computer LINC 8 /DEC/ is described, which makes it possible i/ to input contours of two-dimensional structures from a micrograph into the computer with the aid of an operator-manipulated x-y plotter /resolution better than 0.5 mm/, ii/ to display the plot, respectively correct and store these contours on magnetic tape, iii/ to evaluate their parameters /as area, perimeter, length of axes/ and statistics of these parameters.
Application of the system is illustrated by examples taken from a morphological study on the development of the rat brain.

An important prerequisite for practical application of stereological methods to research in morphology of the nervous system /2/ is an input of two-dimensional data /micrographs prints or slides/ into a digital computer. When comparing the information content of a micrograph and the memory capacity of the computer, it is of use to carry out some kind of preprocessing of the picture /i.e. recognition of elements of structures and their classification/ concurrently with its input into the computer. However, the complexity of the preprocessing task may increase owing to some imperfections of the picture processed. Therefore , if a digital laboratory computer is used for this purpose, a human operator providing the classification and interpolation of the structure during the input of the picture is very often employed /1/.

This communication deals with a system for a/ semiautomatic input of pictures into a laboratory computer LINC 8 /DEC/; its b/ editing and c/ elementary processing features.

a/ The input is based on application of a x-y plotter, the pen holder of which is manually controlled to follow contours of the chosen element /structure/ of the picture. Two types of plotters were used: the analog /BAK 4T Aritma Czechoslovakia/ and the incremental /CALCOMP 565 U.S.A./. The former approach is simpler as far as the control is concerned, since the holder is moved directly by the operator and the computer samples analog signals fed from precision potentiometers of the plotter. Sampling cycles are running continuously and values obtained are compared with those in a previous cycle. Whenever a difference is detected, the sample is stored as a new element of the contour.

121

The application of the incremental plotter is more convenient as far as speed and processing of the input signal are
concerned, since the operator controls the movement of the
plotter carriage via the computer by using set of push button switches. In both cases the resolution is better than
0.5 mm and the matrix is 512 x 512 points in the former
or 1024 x 1024 in the latter case. The sampled contour
/structure/ of the picture is displayed concurrently on
a CRT. Each structure is stored in a separate file composed
of blocks of 0.25 K words 12 bits each, on DECTAPE.
    Individual structures can be divided into 26 classes
labelled by letters. The library of all files /i.e. table
of labels, starting blocks on the DECTAPE and coordinates
of initial points/ is stored in two initial blocks of the
DECTAPE as soon as the input of the whole picture is terminated.
    b/ Editing consists of i/ checking and manual correct-
ing of erroneous segments of the selected individual structure and its renaming by using CRT display /featuring cursor
and scaling/ and a teletype console. ii/ Simultaneous display of initial points /read from the library/ of all structures belonging to the chosen classes or sequential display
of full contours of these structures. Instead of the display
the x-y plotter can be used as an output device producing
a hard copy of sampled structures, which can thus be compared with the original picture. iii/ Generation of a new set
of files by selecting files of the same structure sampled
from different micrographs can be done. This will be used
for three-dimensional reconstruction of that structure.
    c/ Processing of each structure enables evaluation of
its area, perimeter, length and orientation of axes. Arithmetic mean and standard deviation of these values for different classes of structures can also be computed.
    Programming was accomplished in the LAP 6 assembler,
the processing benefits from double-precission arithmetic.
The system has modular structure and some modules are also
used for input and processing of time functions or data in
autoradiography.
    The described system is now applied to the quantitati-
ve study of development of nervous tissue in the hippocampus
of rats. It may, however, be used also for linear problems
like the study of the lengths of different kinds of neuronal
axons.

REFERENCES

1/    Cowan W.M. and Wann D.F.: A computer system for the
         measurement of cell and nuclear sizes. J. of
         Microscopy 99:331-348, 1973
2/    Haug H.: Stereological methods in the analysis of
         neuronal parameters in the central nervous
         system. J. of Microscopy 95:165-180, 1971

National Bureau of Standards Special Publication 431
Proceedings of the FOURTH INTERNATIONAL CONGRESS FOR STEREOLOGY
held at NBS, Gaithersburg, Md., September 4-9, 1975 (Issued January 1976)

## PATTERN RECOGNITION ON THE QUANTIMET 720 IMAGE ANALYZING COMPUTER

by Katheryn E. Lawson
Sandia Laboratories, Albuquerque, New Mexico 87115 U.S.A.

The Sandia Laboratories Image Analyzing Facility consists of an IMANCO Quantimet 720 Image Analyzing Computer with an assortment of modules selected to tackle a wide diversity of problems in materials characterization. The measurements most frequently required are particle count, particle area, and particle size distribution. The term "particle" refers to any feature of a material which contributes to or can be related to its physical or mechanical behavior. Thus it may be, for example, a void in a material or an inclusion in a matrix.

On the 720 an image is produced from a microscope or epidiascope, then projected onto a high resolution scanner. The video signal corresponding to the image is then detected; i.e., interesting parts of the field of view are isolated on the basis of grey level criteria. The detector output signal is then fed into a computer module which makes the required measurement, which may range in complexity from a simple area measurement to an elaborate pattern recognition. Numerical values for the measured parameter are shown on a display monitor screen which can also show exactly the features measured. The 720 is interfaced to a Texas Instrument Silent Data Terminal which tabulates all measurements made on each field and if desired, records on tapes data for input to a PDP-10 time-sharing computer to which the system is acoustically coupled.

The modules available for operating in the pattern recognition mode include (1) the MS3 standard computer which accepts the detected or amended video, performs chord sizing and controls the remaining modules; (2) two function computers which measure area, perimeter, horizontal and vertical Feret diameters, and horizontal and vertical projections; (3) a form separator which computes a shape function involving measured parameters combined in a way suitable for the separation of mixed populations of features; and (4) a classifier-collector from which is released only those data meeting the pre-selected criteria. The data flow is shown in Figure 1.

A number of metal and metal oxide powders, including Fe, W, and oxides of U and Pu have been subjected to classification and particle sizing. Most often, the classification has been based on area sizing primarily because the particles have been near spherical and $\pi r^2$ varies faster than the linear parameters.

As an example of these classifications, Figure 2 shows the monitor display for the analysis of a Pu oxide aerosol. The aerosol was produced and dispersed with a Lovelace aerosol particle separator (LAPS) designed at the Lovelace foundation for Medical Education and Research in New Mexico. LAPS is essentially a spiral centrifuge which continuously separates micron and submicron particles according to their aerodynamic equivalent diameters, i.e. "the diameter of a spherical particle of unit density having the same settling speed in air as the particle in question."[1] The particles are collected on electron micrograph grids strategically placed along the spinning spiral duct of the separator. Individual area and perimeter measurements from the 5000X electron micrographs were made as the particles were moved successively into the live frame. The variable frame and the translational-rotational capability of the epidiascope stage were particularly useful for this analysis. As shown, the particles are generally non-spherical and non-uniform. Even so, area was considered to be a suitable criterion for classification and was used as the basis for the distribution shown by the histogram in Figure 3. The mean area of the particles was $0.68 \pm .07 \mu^2$. Manual measurements of some of the larger particles agreed with those obtained by the Quantimet within $\approx 5\%$. The software for this histogram was developed at Sandia. It is a family of subroutines directed by a driver program to input statistical data from the keyboard or to process data already in a file on the PDP-10 by means of a Tektronics 4010 terminal. The program is interactive and has many options for output presentation.

The measurements made on the aerosol utilized the MS3 standard computer, one function computer and the collector of the pattern recognition system. They could very well have been made using only the MS3, which can be programmed to provide area-perimeter data. Ideally, this analysis would have utilized the MS3, one or two function computers, a sizing distributor, and the classifier-collector. The inability to fully utilize the pattern recognition mode is related to detection problems and the quality of the micrographs. This situation will again be referred to under limitations of the system.

A more extensive application of the pattern recognition mode is illustrated for an airborne dust sample used in the abrasion of solar reflector materials. The sample which had passed a Tyler standard screen sieve #35 (417μ opening) was dispersed dry onto microscope slides. Some 4000 particles were available for the size distribution. As shown in Figure 4, the particles of interest cover a wide range of sizes. Pattern recognition was used to accept and provide data for particles falling within a certain area range and rejecting all others, as shown in Figure 5. A size distribution into eight classes using automatic staging and focusing resulting in a curve shown in Figure 6. An alternate method of sizing would use the form separator set to calculate the dimensionless area/perimeter$^2$ factor for each feature. The smaller particles have a value closer to that of a sphere for this function and can be sized separately.

Finally, the suitability of this or any pattern recognition system for stereological analysis is not without its drawbacks. Inherent sources of error should be fully appreciated when attempting quantification of a microstructure through the image analysis technique. Few specimens have the high surface perfection and contrast needed for full utilization of all the functions of an automated image analyzing computer. The choice of modules for use in any one analysis is dictated not only by the information needed, but also by the quality of the input image. Even when elaborate specialized preparation techniques are resorted to, as was the case for the Pu oxide aerosol previously discussed, the quality of the input to the system is often inadequate. In this particular case it was virtually impossible to focus on a large area of the electron micrograph. This resulted in an increase in analysis time, since it was necessary to employ semi-automatic and even manual operation.

The most frustrating source of error originates in the detection. Even with the most sophisticated automatic detectors, such as the 2D of our system, the wanted feature is often improperly detected. The operator must often manually adjust the threshold of detection to insure that all feature boundaries are correctly recorded. The recently available image editor module which permits the operator to modify the image for more accurate detection of wanted features will be added to our system shortly and hopefully will alleviate this problem.

Reference

1. P. Kotrappa and M. E. Light, <u>Rev. Sci. Inst.</u>, <u>43</u>, 1106-1112, 1972.

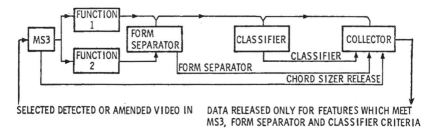

Fig. 1:  Modules for Pattern Recognition

Fig. 2:  Framing of a Pu Oxide Aerosol Particle for
Image Analysis (5000X Electron Micrograph)

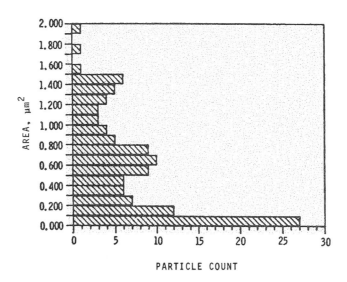

Fig. 3:  Distribution of Particles iñ Pu Oxide Aerosol

125

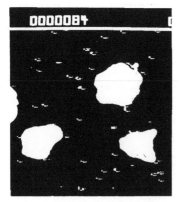

Fig. 4:  Full field count of
airborne dust particles

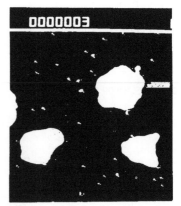

Fig. 5:  Field count of airborne
dust particles having an
area > 500pp$^2$

## DUNE DUST DISTRIBUTION

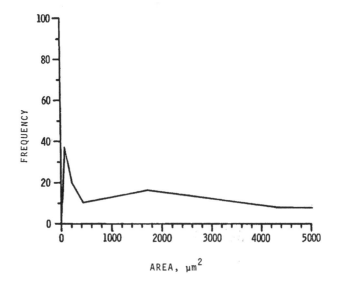

Fig. 6:  Size distribution of airborne dust particles

*National Bureau of Standards Special Publication 431*
*Proceedings of the* FOURTH INTERNATIONAL CONGRESS FOR STEREOLOGY
*held at* NBS, Gaithersburg, Md., September 4-9, 1975 (Issued January 1976)

RECOGNITION AND CLASSIFICATION OF STRUCTURE BY MEANS OF STEREOLOGICAL METHODS
IN NEUROBIOLOGY

by Gustav Bernroider
*Zoologisches Institut der Universität, A-5020 Salzburg, Austria*

SUMMARY

The paper presents methods which connect pattern recognition procedures
with the probabilistic formalism of stereology. By taking individual measurements and transforming them to a d-dimensional feature space, the most informative datas can be extracted. Now the question: what are the most informative
variables describing the particular structure?, can be answered. By this supervised data reduction much more information can be obtained than by using mean
values as estimates. An important result of these methods is the possible combination of position invariant and dependent characteristics which is strongly
desirable for many neurobiological questions. The realisation of these problems
is based on two software packages named ACOFS and SWIFA (Automatic Comparative
Feature Selection) and (Structure Within Function Analysis). Both are demonstrated by flow charts. Some results, which are obtained by analysing hypothal-
amic nuclei of the developing postnatal ratbrain, are shown. The general concept
itself is not versus stereological principles but can be a well integrated extension of them.

INTRODUCTION

Among others, there is still one great problem in Stereology which is worth
while to be considered and the solution of this may lead to a new
concept in today's Stereology. Especially in biology many questions deal with
arrangements of cells into certain groups, where the properties of a single
cell are of as much interest as the specific position of this cell and possible
relation to near or far neighbours. The best example for this is the brain. In
neurobiology we are concerned with two main 'neuronal specificities', the first
being the position invariant properties of a single neuron, the second position
dependent locus specificities (Jacobson M. 1973). Now the question arises: how
can both types of information be obtained by using stereological principles, e.g.
taking random sections and extrapolate from 2-dim to 3-dim structural information. In this paper an approximate solution is discussed, which is based on
'supervised data reduction' after having transformed single particle variables
from 3-dim observation space to a d-dim feature space.

It should be noted that a rigorous mathematical notation of this problem
is not subject to this paper and after completion will be discussed at another
place. The result of such data extraction methods is numerous. Firstly we can
combine position coordinates and their properties. Secondly we can select those
variables which contribute most information to our problem. This 'information'
content of each variable is tested by its 'discrimination power' e.g. by its
ability to discriminate between a class of neurons $A^*$ and a second class $B^*$.
The concept of SWIFA (Structure Within Function Analytical System) mainly
lies in the selection of 'high density boxes' by prior knowledge. This means
that domains $(D(1), D(2), ..., D(k))$ are selected where the probability of
occurrence of single cell characteristics is considerably high.

Rather than program details, the concept of a possible connection between
individual and mean valued measurements will be discussed, since this concept in
general enables various practically most useful procedures. Some problems like
comparative feature selection can only be mentioned briefly and the reader who
is not very familiar with pattern recognition terminology may find some topics
rather difficult to understand. He may refer to one of the many excellent books
written about this subject. We shall first consider some information -theoretical
relations and apply them afterwards to the principles of feature selection methods. Classification algorithms are not the decisive part in pattern recognition
and their efficiency is strongly dependent on proper feature selection.

INFORMATION CONTENT AND CONSISTENT MEAN VALUES
Stereological analysis of real structure is based on expected values which result from various statistical experiments, as random division of space, random interaction of an element with its structural component etc. As a result we obtain estimates for the fundamental stereological relations

$$( P_P, L_L, A_A, V_V, N_V) \tag{1}$$

If such an experiment (e.g. the result of an interaction between a structuring element and a component $\alpha$ on the section plane) is denoted by Z, writing

$$Z = \left( \begin{matrix} z_1, z_2, \ldots, z_n \\ p_1, p_2, \ldots, p_n \end{matrix} \right) \tag{2}$$

where $(z_1, z_2, \ldots, z_n)$ are its possible results and $(p_1, p_2, \ldots, p_n)$ the corresponding probabilities, a measure of Z is H(Z)

$$H(Z) = E(p_1, p_2, \ldots, p_n) = - \sum_j p(j) \lg p(j), \tag{3}$$

H(Z) is the well known information content of experiment Z.
Using 'mean values' on the test plane , possible results of the experiment (Z) are $[\ 1/m \cdot \sum C(i)\ ]j$ where $j = 1, 2, 3, \ldots, n$ and where C stands for any measur - able characteristic (e.g. Area).
Of course, the probability that a certain characteristic C(j) appears within the mask of a test area depends on the relative size of this area (e.g. the random variable 'occurrence of C(J) within a sufficient small test plane will follow a Poisson distribution with considerably small p(j).) On the other hand the above mean value $[\ 1/n \cdot \sum C(i)\ ]j$ is supposed to be consistent, symbolically,

$$\lim_{n \to \infty} P\ [|1/n. \sum_i C(i) - E(C)| < \epsilon] \to 1, \tag{4}$$

While, concerning the total experiment $Z(z_1, z_2, \ldots, z_n)$, the average information content H(Z) merely depends on the n-tupel $P(p_1, p_2, \ldots, p_n)$ with $0 \leqq p(J) \leqq 1$ and $\sum_j p(j) = 1$.
We also know that H(Z) has its maximum where P is equally distributed with $p(j) = 1/n$. In other words, if there are messages (sections and test planes) with large p(j), the average information contents per message is poor. Dealing with consistent mean values only brings a considerable large loss of information per message about our domain D, since by replication for every message, it is most probable that the case

$$[1/m \sum_i C(i)]j = E(C) \tag{5}$$

occurs. The information concerning each single C(i) is almost lost. This fact might seem to be obvious now, but nevertheless is hardly ever accounted for in theory nor in practice. It can be simply demonstrated by comparing the computer storage used before and after calculation of consistent statistics. This loss of information becomes a decisive problem, if 'individual properties' of particles are important and even more is their position within the domain D is also of interest.

SINGLE CELL CHARACTERISTICS
As a single cell characteristic we simply define an r-dim column vector

$$C = \begin{bmatrix} c_1 \\ c_2 \\ \cdot \\ \dot{c}_r \end{bmatrix} \tag{6}$$

Such an r-tuple is presumed to be the result of the feature extraction operation Since feature extraction is normally a reduction of a large amount of data to a

smaller amount of data with the same useful information content, $r \leqslant k$ if the number k denotes the dimension of the column vector in 'observation space'. The amount of objects in observation space (e.g.cells,nuclei etc) leads to a sequence of t such vectors, denoted by $C(*)_t = [C_1,\ldots\ldots,C_t]$.
FIG I demonstrates a single cell characteristic of a neuron (a typical neuron of Nucleus Supraopticus). In this example r=5 and the variables measured are Area (normal), Area(paralysed),Vertical Projection,Horizontal Projection and Density of the cell nucleus.

## A Problem in Measurement Space

Each individual cell characteristic C is obtained from measurements on a section with finite thickness. Because we deal with individuals and not with 'aggregate mean values', each characteristic C(j) of the jth objects must be a good representative. In other words, a consistent estimate for each vector component should be optained.
Now, if $c(1)^*$, $c(2)^*,\ldots,c(r)^*$ are the true parameters, we have to find an estimate $c(1), c(2),\ldots\ldots,c(r)$ for which

$$\lim P \ [ \left| c(i)_n - c(i)^* \right| < \varepsilon ] \rightarrow 1, \text{ as } n \rightarrow \infty \qquad (7)$$

for $i = 1,2,\ldots,r$ and $\varepsilon$ being a real number however small. How to obtain such estimations within the slice sections has been discussed previously.
(Bernroider,1974)

## Transformations to Feature Space

The above written sequence of vectors are characterized by their components and their dimensions. The selection of those is most decisive for the following processes. The underlying philosophy of a selection mechanism is the retention of class discriminatory information and reduction of class common information. There are too many different methods of feature selection that they can be discussed at this place. As we have seen above, the average amount of information can be maximized if p(j) = 1/n for all n. This implies that we learn most from every message ( = object ) if the probability of occurrence of a single cell characteristic is uniformly distributed. We also know, that, if there are two experiments Z and Q, the amount of information which is contained in Z about Q is just

$$I(Z/Q) = H(Q) - H(Q/Z), \qquad (8)$$

where H(Q/Z)

$$H(Q/Z) = - \sum_i p(i) \ (\sum_k P(k/i).\lg P(k/i), \qquad (9)$$

H(Q/Z) is the mean conditional uncertainty of experiment Q under the assumption that the result of Z is known. If we now apply this to a conditional probability $P(W(k)/c(i)\ell)$ which says that P is the probability that an object is allotted to the kth class W(k) under condition that the ith component c(i) takes a value of $\ell$, the amount of information for a given dimension becomes

$$I(W/c(i)) = H(W) - H(W/c(i)), \qquad (10)$$

Note: If this amount of information is high, $c(i)\ell$ also carries high information, thus ordering the coordinate dimensions $(c_1,\ldots\ldots,c_r)$ such that

$$I(W/c_1) \geqslant I(W/c_2) \geqslant \ldots\ldots \geqslant I(W/C_r), \qquad (11)$$

we can reduce our $r$ - dim observation space by retaining the largest such dimensions only. If we select the first m components we have reduced this space by r/m. (FIG II represents a selected 2 - dim feature space for n vectors)
Note: Datas can now be reduced with only small loss of information!

FIG I

A single neuron and its image
display,demonstrating a 5 - dim
single cell characteristic C(5).
Also from the detected image
itself.characteristics can be
extracted. If in such a case the
most informative m features
are extracted, the result of it
will be the well known
'skeletonization' of picture
processing.

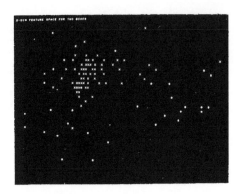

FIG II

2 - dim feature space for
a sequence of n vectors.
The symbol 'X' stands for a
density box where it is highly
probable that a neuron of
'Nucleus Suprachiasmaticus'
appears. 'O' is the box of
Nucleus Paraventricularis
The components are Area of
the cell nuclei and their
optical density.
The cluster of 'X' is clearly
visible.

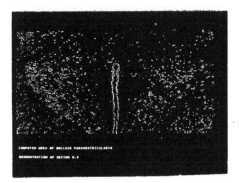

FIG III

The computed area of box 'O'
show that the neurons tend to
form a physical cluster.
Do they also form a cluster
on the basis of their single
cell characteristic ?

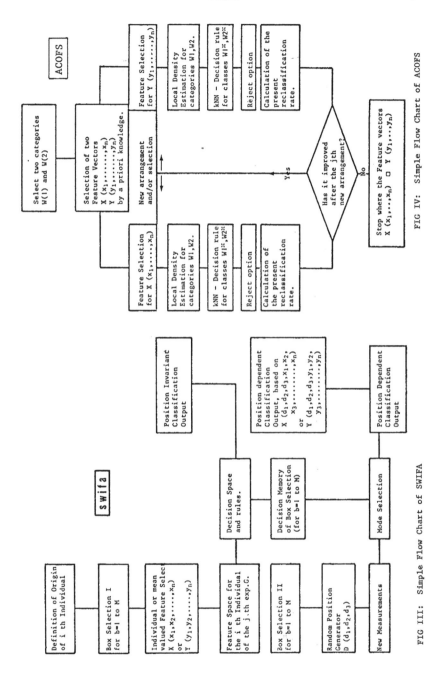

FIG IV: Simple Flow Chart of ACOFS

FIG III: Simple Flow Chart of SWIFA

## WHAT IS WHERE? RANDOM POSITION DEPENDENCE

When we are interested in the position of a single cell characteristic we have to select the 3 physical dimensions and transform them to feature space, obtaining

$$C_r = [d_1, d_2, d_3, c_1, \ldots c_n] \tag{12}$$

with $r = n + 3$ dimensions.

The problem itself lies in the fact that each 'location' should have the same chance to be selected. This can be obtained by a 'uniform random number generator' and a selection of 'high density boxes' where the probability of occurrence of a single cell characteristic can be adjusted with the number of locations necessary.

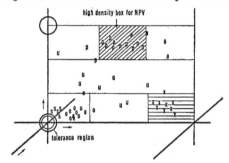

**FIG IV**

Demonstration of 'high density boxes' and the origin of a random coordinate frame. The origin of the coordinates lay within a certain tolerance region which is due to the normal spread of biological individuals.

## AUTOMATIC COMPARATIVE FEATURE SELECTION (ACOFS)

If we can set up two extremely different feature vectors X and Y and select clearly different categories W(1) and W(2), under replication feature selection for both vectors will lead to a point where both vectors dicriminate both categories W(1), W(2) best. As a criterion of "goodness" serves the reclassification rate, i.e. the percentage of objects of category W(1) and W(2) respectively which have been allotted to the corresponding class $W(1)^*$, $W(2)^*$.

Flow chart 2 shows a simplified diagram of ACOFS. The procedure stops where both kind of information have the same effect and allow 'correlative Interpretation'. (e.g. a feature vector obtained from electron microscopical investigations and another one from light microscopy)

## CONCLUSIONS

Individual data extraction allows selection of the most informative variables and measurements describing a physical structure. Further , using Pattern Recognition and Cluster techniques a number of operations can be performed which enable the linkage of position dependent and invariant cell characteristics and comparative feature selection for "interpretation problems". Compared with conventional estimates the amount of information gained can be decisive for many problems, especially those of Neurobiology.

## ACKNOWLEDGMENT

I would like to thank Prof. Dr. Adam for his encouragement and support, further Dr. A. Goldschmied for his anatomical advice and Mrs. H. Langer for excellent assistance in preparing the biological material. The work was kindly supported by the Fond zur Förderung der wissenschaftlichen Forschung, Austria, (Progr. 1838/n 39)

## REFERENCES

Andrews,H.C.(1972) Introd.to Mathem. Techniques in Pattern Recognition, Wiley
Bernroider,G.(1974) In:Spec.Issues of Pract.Metallogr.,5,pp.177-184,Dr.Riederer V.
Haug,H.(1972) J.Microsc.,95,pp.165 - 180.
Miles,R.E.,(1972)J.Microsc.,95,pp.181 - 195.
Miles,R.E.,(1969)Adv.appl.Prob.1, pp.211.

*National Bureau of Standards Special Publication 431*
*Proceedings of the* FOURTH INTERNATIONAL CONGRESS FOR STEREOLOGY
*held at* NBS, Gaithersburg, Md., September 4-9, 1975 (Issued January 1976)

## PROJECTORS IN PATTERN RECOGNITION CATEGORIES

by Monique Pavel
*Université René Descartes, Paris, France*

ABSTRACT

In Pattern Recognition, as well as in particular in Stereology, one aims to identify patterns starting from incomplete information; we are interested in this paper in mathematical methods of Pattern Recognition which can, in particular, be useful in Stereology. We proceed in two steps.

(1) We present for the first time a general formalism of recognition problems, using a categorical setting. If the objects $I \epsilon I$ are classes of identified configurations into a same image, and if the family $\Delta$ of deformations $\delta$ is a semigroupoid, then we prove that $(I, \Delta)$ is a category. If $\Delta$ contains the group $\Gamma$ of similarities, then $(I, \Gamma)$ is proved to be the associated recognition category. We characterize probes and recognition functions as being invariant functors of categories; probes may in particular be projectors.

(2) Patterns can be rebuilt and studied starting from their projections and the lifting properties of the latter. We present for the first time a characterization of projectors by means of retracts of images and of categories of images. In the category $(I, \Delta)$, every image $I \epsilon I$ is a retract of its similar images. If $\Gamma$ is a normal subgroup of $\Delta$, and if $\delta_{\mu\nu} \epsilon \text{Mor}_\Gamma \mathcal{D}(I_\mu, I_\nu)$ is a retraction in $(I, \Delta)$, then its class of equivalence is a retraction in the quotient category $I^\Delta/\Gamma$. $I$ being a recognition category associated with $I^\Delta$, if the category $P$ of patterns is $P = I/\Gamma$, then $P$ is a retract of $I$.

INTRODUCTION

We present in this paper the first general mathematical setting for recognition problems and for projections. Objects or images can be described by their primitive components and their composition, and (or) be defined by their global and (or) local topologically invariant properties; recognizing a pattern then means detecting an equivalence, or isomorphism, with regard to a specified set of transformations, between the given object and an element of a set of "canonical patterns," or templates. This identification process has to be accomplished starting from incomplete information about objects and about possible transformations between them.

In order to systematize the concept of recognition problem, it will be helpful to introduce and adapt some elementary concepts from category theory, and then to define recognition categories and invariants: the interest of recognition categories; in which all transformations are isomorphisms, lies in the fact that they provide the adequate setting for the identification problem associated with the original category, in which transformations are deformations.

The equivalence mentioned above is generally established by computing and comparing a "sufficient" number of probes, i.e., functions which are invariant with regard to certain transformations. Probes often used are projectors. The interest of retracts is that they provide the general mathematical setting for projectors in recognition categories and that they focus our attention on their particularly useful invariance lifting properties.

Only a slight acquaintance with the elementary definitions of category theory (see B. Mitchell [1]) is needed to understand the discussion in this paper.

1. IMAGES, PATTERNS, RECOGNITION

We shall refer in this section to M. Pavel [2].

1.1 Let $S$ be a set of elements $s_i$, called <u>signs</u>, possibly belonging to different types $\sigma_1, \ldots, \sigma_p$. <u>Configurations</u> $c_i = (s^1_i, \ldots, s^1_m) \in C$ are built by defining on $S$ a composition $+$; a group $G$ of transformations $g$, called <u>similarities</u>, is also defined on $S$, $G: S \to S$. An identification relation $R$, or a simplification operator $\rho$, both of which are equivalence relations and $G$-invariant, define $I = C/R$, respectively $S = C/\rho$; the elements $I \in \mathcal{I}$ are classes, called <u>images</u>: $I = \{c; cR_{c_I}\}$, respectively, $I = \{c'; \rho(c') = \rho(c_I')\}$. When defined, the mappings $(I_1, I_2) \to I_1 + I_2$ and $(g, I) \to gI$ give rise on $\mathcal{I}$ to an additive semigroup structure such that $g(I_1 + I_2) = gI_1 + gI_2$ and $g_1(g_2 I) = (g_1 g_2) I$: $\mathcal{I}$ is called a "grammar" of images.

The <u>patterns</u> $P_1, \ldots, P_n$ retain only the same "meaning" of possibly different images (e.g. behaviour with regard to transformations, or types which have been used, etc.): the patterns $P_\nu$ are defined by means of a family $P$ of disjoint and $G$-invariant sets $P_1, \ldots, P_n$.

1.2 The recognition problem for $\mathcal{I}$ consists in finding a mapping, or recognition function, $R: \mathcal{I} \to P$, which assigns an image $I \in \mathcal{I}$ to a pattern $P_i \in P$ (if it exists). If in particular $P$ is the family of equivalence classes induced by $G$, we shall write $P = \mathcal{I}/G$ and then $R(I_\mu) = R(I_\nu)$ iff $I_\nu = gI_\mu$.

1.3 If, more generally, a set $T$ of transformations $t$, $T: \mathcal{I} \to \mathcal{I}$, containing the group $G$ of similarities but not necessarily being a group is given on $\mathcal{I}$, and if there exists a set of <u>templates</u> (i.e., images $I_\nu \in \mathcal{I}$ such that $R(TI_\nu) = P_\nu$) we shall say that $I_\mu$ and $I_\nu$ are <u>synonymous</u> iff $R(I_\mu) = R(I_\nu)$, i.e., iff $I_\nu = tI_\mu$, $t \in T$. Synonymy is generally established by means of <u>probes</u> $P$, which map $\mathcal{I}$ on a set $V$ (generally different from $\mathcal{I}$) and which are constant on synonymy classes: $I_\nu \equiv I_\nu(T) \Rightarrow P(I_\mu) = P(I_\nu)$; probes can in particular be projectors.

1.4 Let $D$ be a set of transformations $d$, called <u>deformations</u>, defined on $\mathcal{I}; \mathcal{I}^D$ will stand for the set of elements $dI$, with $I \in \mathcal{I}$ and $d \in D$. We shall suppose that $\mathcal{I}^D \supset \mathcal{I}$, and that similarity transformations as well as the composition $+$ on $\mathcal{I}^D$ coincide (on $\mathcal{I}$) with those originally introduced on $\mathcal{I}$. A <u>recognition function</u> $R: \mathcal{I}^D \to P$ then assigns a deformed image $dI \in \mathcal{I}^D$ to a pattern $P_\tau \in P$ (if it exists): we shall write $R(dI) = P_\tau$. If deformations $d$ are homomorphisms ($d(I_1 + I_2) = dI_1 + dI_2$, $d(gI) = g(dI)$), then $\mathcal{I}^D = \{dI; I \in \mathcal{I}, d \in D, \cdot \mathcal{I}$ a grammar and $d$ homomorphic$\}$ is a "grammar" of deformed images.

2. RECOGNITION CATEGORIES

2.1 *Definitions*.

Categories will be designated by capital script letters, objects of categories by capital Roman letters A through I, and morphisms of categories by lower case Greek letters. If A and B are objects of the category $A$, the expression $\text{Mor}_A(A,B)$ denotes the set of morphisms with domain A and range B. The composition of the morphism $\alpha$ with the morphism $\beta$ is written $\alpha\beta$, i.e., $\beta$ is followed by $\alpha$. A morphism $\alpha \in \text{Mor}_A(A,B)$ is called an <u>isomorphism</u> if it has a two-sided inverse, i.e., there exists $\beta \in \text{Mor}_A(A,B)$ such that $\alpha\beta = 1_B$, $\beta\alpha = 1_A$. We write $A \simeq B$ if there is an isomorphism $\alpha \in \text{Mor}_A(A,B)$. It is easy to see that $\simeq$ is an equivalence relation. The capital Roman letters J, P, R, S, and T represent functors from one category to another; only covariant functors will be encountered in this paper. If S and T are naturally equivalent functors from a category $A$ to a category $B$, we shall write $S \simeq T$.

Definition 2.1. A <u>recognition category</u> $A$ is a category in which all morphisms are isomorphisms.

In the rest of this section, $A$ will denote a recognition category.

Definition 2.2. If $A, B \in \mathrm{Ob}\ A$, the <u>recognition problem</u> associated to $A$ consists in deciding whether or not $A$ is isomorphic to $\bar{B}$, i.e., whether or not $\mathrm{Mor}_A(A,B)$ is non-empty.

Definition 2.3. An <u>invariant</u> $J$ of $A$ is a functor $J:A \to E$, where $E$ is a discrete category (i.e., the only morphisms of $E$ are the identity morphisms $1_E$, $E \in \mathrm{Ob}\ E$).

The last definition means that $J$ is an invariant of $A$ iff $A \simeq B$ in $A$ implies $J(A) = J(B)$ in $E$; $J$ is <u>full</u> (or <u>complete</u>) if the converse is also true.

## 2.2 *Recognition categories*.

Let the objects of $\mathrm{Ob}\ I^\Delta$ be the sets of classes $I_i \in I$ of $n_i$-uples of $R$-identified or $\rho$-simplified configurations defined in section 1; taking for $\mathrm{Mor} I^\Delta$ the family of mutually disjoint sets $\{\mathrm{Mor}_{I\Delta}(I_\mu, I_\nu)\}$ for all objects $I_\mu$, $I_\nu \in \mathrm{Ob}\ I^\Delta$, whose elements $\delta_{\mu\nu} \in \mathrm{Mor}_{I\Delta}(I_\mu, I_\nu)$ are defined by the triples $\delta_{\mu\nu} = (I_\mu, d_{\mu\nu}, I_\nu)$, where $I_\nu \ne d_{\mu\nu} I_\mu$, $d_{\mu\nu} \in D$, and "forgetting" for the moment the algebraic structure of $I^D$, we prove:

Theorem 2.1. If $\Delta$ is a semigroupoid, then the family $(I, \Delta)$ is a category.

We have thus organized images and deformations in a category, and shall designate this category by $I^\Delta$.

Corollary 2.1. The category $I^\Delta$ in which the objects are the sets $I$ and the morphisms are the deformations $\delta_{\mu\nu}$, is the image by a forgetful functor of the category $\bar{I}^\Delta$ in which the objects are grammars and the morphisms are homomorphisms.

Similarities being elements of a group $G$, and defining $\Gamma = \{\gamma_{\mu\nu};\ \gamma_{\mu\nu} = (I_\mu, g_{\mu\nu}, I_\nu),\ g_{\mu\nu} \in G\}$ we prove:

Theorem 2.2. If $\Delta$ is a semigroupoid of deformations containing the group $\Gamma$ of similarities, then $(I, \Gamma)$ is a recognition category associated to the category $I^\Delta$.

The only morphisms of $(I, \Gamma)$ are similarities, therefore isomorphisms, and hence $\mathrm{Mor}_{(I,\Gamma)}(I_\mu, I_\nu) \ne \emptyset$ means that $I_\mu$ and $I_\nu$ are similar. We shall designate this category by $I$.

Corollary 2.2.

(i)  $I$ is a (non-full) subcategory of $I^\Delta$.

(ii)  $P$ is a discrete category.

(iii)  If $\Delta$ is a group containing $\Gamma$ as a normal subgroup, then the category $P = I/\Gamma$ is mapped by the deformations belonging to the quotient group $\Delta/\Gamma$ on the category $I^{\Delta/\Gamma}$.

## 2.3 *Recognition problem*.

Given the category $I^\Delta$ and an equivalence or synonymy law $T$, the recognition problem for $I^\Delta$ reads as follows: find a way to decide if $I_\mu \equiv I_\nu(T)$. By associating to $I^\Delta$ its recognition category $I$, the recognition problem for the latter reads: given $I_\mu, I_\nu \in \mathrm{Ob}\ I$, decide whether $I_\mu \equiv I_\nu(\Gamma)$, i.e., whether $\mathrm{Mor}_I(I_\mu, I_\nu) \ne \emptyset$. The solution of the problem can be expressed by:

Theorem 2.3.

(i) Probes P of the category $I$ are (forgetful) invariant functors from $I$ to a discrete category $E$:

$$I_\mu \equiv I_\nu(\Gamma) \text{ in } I \Rightarrow P(I_\mu) = P(I_\nu) \text{ in } E.$$

(ii) Recognition functions R defined on $I$ are invariant functors from $I$ to the discrete category $P$ of patterns:

$$I_\mu = gI_\nu \Leftrightarrow I_\mu \simeq I_\nu \text{ in } I \Rightarrow R(I_\mu) = R(I_\nu) \text{ in } P.$$

(iii) If $P = I/\Gamma$ then R is the canonical mapping of I into its equivalence class [1].

(iv) If P is a full probe, then

$$P(I_\mu) = P(I_\nu) \Leftrightarrow I_\mu \equiv I_\nu(T) \Leftrightarrow R(I_\mu) = R(I_\nu).$$

See for the proofs of the theorems in this section, M. Pavel [3].

Let us note that: (1) the equation in (i) is the reason for which probes, and in particular projectors, are used even if they do not solve the problem entirely: if $P(I_\mu) \neq P(I_\nu)$, then certainly $I_\mu \not\simeq I_\nu$. (2) G-invariance is an elementary component of synonymy T; it shows off one of the numerous characteristics of the images in $I$, G-invariant probes forgetting all but invariance with regard to the similarity $g \in G$. (3) Total discrimination requires a full (or complete) invariant, which is often built on a basis whose components are "simpler" but each of them being incomplete: one actually operates on "elementary" characteristics of images. (4) The recognition problem for $I^\Delta$ cannot be reduced to that of constructing a full invariant of $I$; in order to obtain significant results, one has to take $P_\mu = R(TI_\mu) \not\supseteq I_\mu$: $P_\mu$ has to contain also other images than those obtained from I by similarity.

## 3. PROJECTORS

Probes that are often used are projectors. Let P be a projector, i.e., a mapping from a space $B$ of objects to the subspace $A$ of $B$, $P : B \to A$, such that $P(B) = A$ and that $P^2(b) = P(b)$ (which means that $P(a) = a$ for all $a \in A$). In other words, if P is a projector to $A$, then its restriction on $A$ is the identity mapping of $A$: $P/A = Id_A$; this comes to the existence of a mapping $\bar{P}$ from $A$ to $B$ such that $P\bar{P} = Id_A$. If instead of taking for $A$ a subspace of $B$, we more generally take for $A$ an arbitrary space, if we consider the mapping $S : B \to A$, and replace $\bar{P}$ by an arbitrary mapping $T : A \to B$, then, in order to preserve the projector's property we need to ask that ST be, if not equal to the identity of $A$, at least "equivalent," in some sense to be precised, to the identity of the space $A$: this can be expressed by the general concept of retract.

The utility of retracts in recognition problems proceeds from their lifting properties: suppose, more generally, we are given a functor $T : A \to B$, $A$ and $B$ categories. Then T can be used to lift every invariant of $B$ to an invariant of $A$; in fact, it is obvious that if J is an invariant of $B$, then the composite functor JT is an invariant of $A$. Many familiar invariants are obtained in this way.

### 3.1 *Definitions*.

Let us define first retracts of objects in a category and then retracts of categories.

**Definition 3.1.** A morphism $\alpha:A \to B$, $A,B \in Ob\ A$, is a <u>coretraction</u> (or a <u>section</u>) if there exists a morphism $\beta:B \to A$ such that $\beta\alpha = 1_A$. A is called a <u>retract</u> of B.

**Definition 3.2.** A morphism $\alpha:A \to B$ is called a <u>retraction</u> if there exists a morphism $\gamma:B \to A$ such that $\alpha\gamma = 1_B$. If $\alpha$ is a retraction and a coretraction, then it is an isomorphism (see Section 2.1).

**Definition 3.3.** Let $A$ and $B$ be recognition categories; $A$ is a <u>retract</u> of $B$ if there exist functors $S:B \to A$ and $T:A \to B$ such that the composition ST be naturally equivalent to the identity functor $Id_A$. We shall in this case write $A \dashv_S {}_T B$ or simply $A \dashv B$ if there is no possible confusion. The condition $A \dashv B$ agrees very well with the intuitive idea that the recognition problem for $A$ is no more difficult than the recognition problem for $B$.

**Definition 3.4.** Two categories $A$ and $B$ are <u>equivalent</u> if there exist functors $T:A \to B$ and $S:B \to A$ such that $ST \simeq Id_A$ and $TS \simeq Id_B$. Obviously, if $A$ is equivalent to $B$, then $A$ is a retract of $B$ and $B$ is a retract of $A$, but the converse fails.

**Definition 3.5.** Let $A$ be a recognition category. For $I \in Ob A$, define $Aut(I) = Mor_A(I,I)$. For $\alpha \in Mor_A(I_\mu, I_\nu)$ and $\psi \in Aut(I_\mu)$, define $(Aut(\alpha))(\psi) = \alpha\psi\alpha^{-1}$.

### 3.2 *Projectors in pattern recognition categories*.

**Theorem 3.1.** In the category $I^\Delta$, every image $I$ is a retract of all similar images $gI$, $g \in G$.

**Theorem 3.2.** If $\delta_{\mu\nu} \in Mor_{I^\Delta}(I_\mu, I_\nu)$ is a retraction in $I^\Delta$, and if $\Delta$ is a group containing $\Gamma$ as a normal subgroup, then the equivalence class $[\delta_{\mu\nu}]$ of $\delta_{\mu\nu}$ is a retraction of the quotient category $I^{\Delta/\Gamma}$.

**Theorem 3.3.** If $I$ is a recognition category associated with the category $I^\Delta$, and if the category $P$ of patterns is $P = I/G$, then the category $P$ is a retract of the category $I$.

Probes $P:I \to Y$, $P(I) = Y$, $Y$ a discrete category, and in particular projectors, can then be used to lift every invariant of $I^\Delta$ to an invariant of $\widetilde{I}^\Delta$; and every invariant of $I$ to an invariant of $I^\Delta$; and every invariant of $P$ to an invariant of $I$.

**Theorem 3.4.** A recognition category $I$ is equivalent to the discrete category $P$ of patterns iff $Aut(I)$ is a one element group for every image $I \in I$.

See for proofs of the theorems in this section, M. Pavel [4].

Ordering recognition categories following the concept of projector given in Definition 3.3, the last results can be interpreted as follows: (1) discrete image categories, such as the category $P$ of patterns, are minimal; (2) every recognition category majorizes some discrete category; and (3) a recognition category which is equivalent to a discrete category need not be discrete but is very particular: each set of similarities $Mor_I(I_\mu, I_\nu)$ is either empty, or else contains exactly one isomorphism; in particular, $Mor_I(I,I)$ is a one element group for every image $I \in I$.

BIBLIOGRAPHY

[1]  B. Mitchell.  Theory of categories.  Academic Press, 1965.

[2]  M. Pavel.  Fondements mathématiques de la reconnaissance des structures.
     Actualités Scientifiques no. 1342, Hermann, Paris, 1969.

[3]  M. Pavel.  Reconnaissance des structures dans les catégories.  Comptes
     Rendus de l'Académie des Sciences de Paris, série A, v. 280, 1975,
     pp. 295-298.

[4]  M. Pavel.  Rétractes dans les catégories de reconnaissance.  Comptes
     Rendus de l'Académie des Sciences de Paris, série A, v. 280, 1975,
     pp. 285-287.

# 3. INSTRUMENTATION

*Chairpersons:*

H. P. Hougardy
R. Ralph
H. Haug
R. Kirsch
G. A. Moore

*National Bureau of Standards Special Publication 431*
*Proceedings of the* FOURTH INTERNATIONAL CONGRESS FOR STEREOLOGY
*held at* NBS, Gaithersburg, Md., September 4-9, 1975 (Issued January 1976)

AUTOMATIC IMAGE ANALYSING INSTRUMENTS TODAY

by Hans P. Hougardy
*Max-Planck-Institut für Eisenforschung GmbH, Düsseldorf, Germany*

SUMMARY

At this time, there are 12 different instruments for automatic image analysis on the market. Most of them are modular
giving an infinite number of sub-systems. Advices are given for
a technical description of the performance of such instruments.
A survey on the modern pattern recognition systems is given with
some remarks, how to choose the 'best' suited instrument.

Since the last ISS congress in 1971 there has been an
explosion in the development of instruments for automatic image
analysis. The variety of only one instrument makes it impossible
to describe a single system in detail. Therefore the basic rules
for the design of such instruments are given as well as some
advice to the user how to form an opinion of such a system and to
test them. The available commercial instruments are listed with a
short characterisation.

PRINCIPALS OF AUTOMATIC IMAGE ANALYSIS

The reason for using automatic image analysis is to make the
measurement faster and more accurate. The image to be analysed may
be a print, a slide or immediately the output of a light- or
electron microscope. In general, special features must be selected
for measurement from all the other parts of the image. That means
that all the information of the image must be transferred to the
analyser, but only a small part is really needed for the measurement. Therefore there must be a rapid decision about each information bit, if it is needed or not, to get a sufficient working
speed.
At this time the fastest processing and data handling systems
are analog or digital computers. Consequently the first step in an
automatic image analysing instrument is a conversion of the
optical information (the image) picture point by picture point
into an electric signal suited for the particular computer used.
The converter - see figure 1 - may be a TV camera, a photo diode,
or a photomultiplier, as well as an x-ray tube. If the image
already has the form of an electric signal - as in scanning
electron microscopy - only an adaption of the voltages and a
synchronisation is necessary.
All instruments 'detect' the features to be measured first
according to the signal modulation - the grey level in an image -
using a signal discriminator (1). From these detected features, in
a second recognition process, those having the desired shape,
orientation or other criteria are next separated and analysed to
get parameters such as area or perimeter. The accuracy of such a
measurement of course depends mainly on the correct image conversion and detection (1,2,9). Therefore the main efforts of the
manufacturers in the last few years were to improve the converter
accuracy and to develop automatic detectors, in order to be
independent from subjective threshold setting by the operator
(1,2). This was the basis for the realisation of extended pattern
recognition systems (3-7).

DESIGN OF MODERN IMAGE ANALYSERS

Low cost instruments are built according to figure 1 without a 'feature recognition' unit. In this case the 'feature analyser' measures only 'field specific data'. That means, for example, that the value of 'projected length' is the mean value of all the features in a field of view. It is not possible to measure area or perimeter of each feature in a field of view separately, except setting the measuring frame to each feature separately by hand. This is a limitation, but for many applications these instruments give all the desired results with sufficient accuracy.

If there are several parts of the image having the same grey level as the features to be measured, the grey level detection must be extended by a feature recognition according to their shape, which is realised in modern instruments using three different principles: The parameter method, the feature modification and the interactive method. According to figure 1 the first step is in every case a grey level detection to reduce the amount of information to be handled. To give a survey, all principal systems available at the present time are combined in figure 2. Any particular system can be redrawn from figure 2 by deleting one or several units or connections. Of course, depending on the instruments, the 'units' of figure 2 consist of one or several hardware or software modules with different capabilities. In some cases, the arrangement of the units as given by figure 2 can be changed to get another sequence of operations.

The parameter system. Using this method, the shape of features is defined by the parameters determined by the 'feature analyser' (3,4). These parameters are interconnected in a 'shape recognition' unit to get such values as 'area divided by the perimeter squared' which is used as description of a circular shape, for example (6-8,16). Only the parameters of features satisfying such a preset shape condition are fed into the analysing and output computer by gate 4. This method is quite powerful, provided the features have sufficient contrast and a sufficient size on the monitor screen (2,9). Otherwise the scatter in values of the shape descriptor is too great, giving an over-lapping in identification of the desired shape with other shapes. If, for example, the coefficient of variation for repeated measurements of one field of view for feature parameters is 3 % - due to insufficient contrast of the features - it is not possible to distinguish exactly between circles and squares using the shape descriptor mentioned above (4).

The feature modification. The two-dimensional analysis (1,4, 10-12) allows a modification of features which can be used for identification. By 'erosion' (12), for example, features which appear as two overlapping circles can be separated. Additionally small particles or scratches can be eliminated (4). On the other hand, particles can be identified according to their distance by the opposite operation, the 'dilatation' (4,12). Therefore this method is well suited to define features by their neighbourhood, in addition to their shape, especially if a sequence of erosion and dilatation is used. The separated features may be analysed to get the standard parameters or, for example, they may be sized by a subsequent erosion (12-14).

In those machines that provide manual interaction the features are recognized by marking them with a light pen (4,7,5). This, of course, cannot be done with the speed of a TV system, for example. Therefore at first glance manual interaction does not seem to be part of an automatic but of a semiautomatic system. But in fact it is much faster than common semiautomatic instruments. If a feature cannot be recognized automatically, it is only necessary to redraw its outlines with a light pen. Also the most complex parameters are then measured automatically by the instrument. According to figure 2 this interaction is stored in the appropriate unit and gated into the measuring line, in general by gate 2. If an instrument allows a connection to gate 3, the interactive system can be used to correct the automatic feature recognition only, if necessary, which of course speeds up the operation. Additionally, interactive systems can be used to draw any form of measuring frame or to compare a measured feature with another one which may be a standard shape (4).

According to figure 2 these systems can be combined to any degree of complexity, which of course will be very expensive. This is only recommended if complex measurements are to be done frequently.

PROGRAMMING OF INSTRUMENTS

To program a system as given by figure 2 takes a lot of time if it must be done manually by interconnection of each module or by the use of patchboards. To simplify this, for most the instruments a computer connection is available which enables the use of software systems which ask just for the kind of detection, the shape of the features to be recognized and the parameters to be measured. All other processing is done automatically (4). This is very useful, if the parameters to be measured and the detection criteria must be changed often. If only one kind of measurement is necessary, specialized instruments are available with a hard wired program, operated just by pushing the button 'start'. They can be used for control purposes by unskilled operators (1,4).

METHODS FOR EVALUATION OF INSTRUMENTS

In general, it is not possible to decide - even after an extended test - that this is the 'best' instrument. The best system is that which provides only the measurements that are necessary with the desired accuracy. This instrument has, in general, the lowest price compared with the capability used. The first step to find the most suitable instrument is therefore to define the parameters to be measured and the method for feature detection. If the features can be separated by their grey level only, no feature recognition units are needed. Otherwise the shape descriptor which is valid only for the desired features must be defined, whereupon, on the other hand, this specifies the feature recognition system.
The second step is to decide if field specific data are sufficient, or whether feature specific data must be measured.
The third step is to decide if hardware programming is sufficient or if software programming is to be preferred. The latter requirement frequently occurs in a complex system where the kind of measurement changes often. With these steps the type of instrument, and its analysing capability is determined according to figure 2.
The fourth step is to define the accuracy necessary and to select an adequate instrument. The accuracy is determined mainly by the first two units in figure 1, the image converter, and the detector. All the other modules handle electrical information by arithmetic or logic rules using, in general, digital, or in some instruments, analog methods. The latter give a well known accuracy

which is generally better than that of the converter and detector. Therefore it is important, to get values for the accuracy of these modules. The capability of the converter and the detector can be described by four values, which can be measured by each user (2): the resolution, the separation of grey levels, the reproducibility of threshold settings, and the statistical accuracy of a system. There are values not derived from the common technical description of such systems but from the needs of a user of automatic image analysers.

From this point of view, following a proposal (2), <u>resolution</u> is defined as the diameter of the smallest circle whose area can be measured with an error less than ± 6 %. Of course other values can be taken. For 6 %, a diameter of 1.5 % of the frame width is a reasonable value for this resolution. This value is defined for a contrast of 1 ! At lower contrast the error increases for a circle of the same diameter. The resolution limit is very important, for the error arising from the measurement of particles smaller than the given limit is not detectable in the result by any statistic calculation of accuracy !

<u>The separation of grey levels</u>. Two phases which should be separated by the detector must have a minimum difference in their grey level. This difference can be measured quantitatively using additional equipment: The camera is illuminated without any lens system by parallel monochromatic light, starting with the maximum light intensity which can be measured by the camera system used. For this value, the threshold is set to get 99 % area of the live frame detected. Then the light source is dimmed until the area reading is only 1 %. This is one grey step. After changing the threshold until again 99 % area is measured, the light is dimmed again and so on. 30 grey level steps determined by this method show acceptable performance (2). Using neutral density filters the absolute sensitivity of the system can be measured. Comparing these results with photometric measurements of specimens shows whether a specific feature is detectable due to its grey level or not. The number of grey levels which can be separated depends on the wave length of the light used for illumination, the sensitivity of the camera system, and the detector used. It is only valid for features greater than the resolution limit defined above and depends on the supposition that the <u>image</u> has no shading. In general, this is not given if a microscope is used. Otherwise the values measured as described can be realised only with an image shading corrector (2).

<u>The reproducibility of threshold setting</u> is given by the area of a feature as a function of the threshold setting. Standard detectors give no advice about the correct setting. Therefore special methods must be used to avoid over- or under-detection (9). With automatic detection systems there is a plateau at the correct setting (1) which means that the value of the area does not change with a change of the threshold. The difference in the area measurement for a ± 1 % change of the threshold setting gives a value for the reproducibility of the threshold setting. It changes with size and contrast of the features (2).

<u>The statistical accuracy</u> of a system is given by the standard deviation of a set of measurements on the same field of view. For small particles or low contrast the coefficient of variation can exceed 1 % and go up to 20 % (2). This statistical error must be added to the error, calculated from different fields of view.

These four values, if known for an instrument, enable the user to decide if his problem can be solved by a particular system and he can calculate the accuracy of the parameter measurements. From this, the scatter of the shape descriptors - as far as used - can be estimated which gives a decision if two shapes can be separated or not, according to step 1 mentioned above.

INSTRUMENTS AVAILABLE AT THIS TIME

There are 12 instruments manufactured by 8 different companies on the market at this time. Most of them are modular, giving an infinite number of possible combinations and special systems. Therefore, here we characterize only briefly the basic instruments. If not mentioned, the input is a TV system. Details are given elsewhere (1,4).

The Bausch and Lomb (20) instrument, called 'Omnicon', is a modular system which can be extended to include automatic detection and complex parameter feature recognition, supplemented by an interactive system, which is then called 'P.A.S.' (Pattern Analysis System). The sequence of operations, including feature recognition, is programmed just by pushing buttons on a keyboard.

Ealing-Beck (21) has developed recently an instrument called 'Histotrak' (19,22) which is combined with a microscope to provide an automatic change of magnification. It is designed for easy handling, measuring the most important parameters by means of an interactive system.

The Hamamatsu Corporation (23) manufactures a variety of systems for special applications such as the 'Iriscorder' for ophtalmology. General purpose instruments are the 'Multianalyser', an analog instrument for simple measurements, and the digital 'Polyprocessor' which is hardware programmed using read only memories. The main part of the processing and data evaluation are done by software.

The 'Quantimet 720', manufactured by Imanco (24), is a system which can be extended from a simple instrument to a computer processed and programmable system containing all the capabilities described in figure 2. Special systems are available for special purpose as the 'Quantimet 360', a push button instrument for the measurement of nonmetallic inclusions in steel.

The Kontron (26) instrument, built in cooperation with Wild (27), is called 'Digiscan' and uses a photo diode as image converter which also gives a high grey-level resolution. The specimen is moved for scanning by a mechanical stage. The 'Epiquant' and the Kontron instruments are powerful systems for measuring features with low contrast.

Leitz (28) has developed a system which measures the linear parameters of features, the 'Classimat'. They also build the 'Texture Analysing System' (T.A.S.). This instrument uses the principles of the random set theory for a two-dimensional feature analysis. Feature modification, as well as two-dimensional sizing, can be realised (1,12,13).

VEB Carl Zeiss Jena (25) has developed a system called 'Epiquant' specially designed for investigations of opaque specimens. The image converter is a photomultiplier which has in principle a higher grey level resolution than TV systems. All measurements of a lineal analysis can be done.

The 'Micro Videomat', manufactured by Zeiss (29), is a compact system for making the main important measurements. The application is aided by a set of software for processing and data handling, and for statistical calculations to get the accuracy of a measurement.

FURTHER DEVELOPMENTS

Looking at the instruments today, they represent different compromises between the optimal system and the costs. For example, to increase the speed of a measurement it is necessary to have a simultaneous data processing which can be realised by an increasing amount of hardware and - increasing costs. On the other hand, for many applications a digital storage of the image is desired to have all the advantages of a software analysis. Until now this is realised only for special applications (17,18) due to the high costs of core memories necessary for image storage and the relatively long time necessary for analysis. But the costs of computer memories decrease, the speed of modern computers increases. Therefore a digital image storage and a software analysis may be used more and more in the future. But in every case, to store the total information of an image needs too much memory. Therefore it may be useful to reduce the information as far as possible. In principle, figure 2 includes highly sophisticated instruments which scan the image in a first step very quickly with low resolution to get a 'survey'. In a second step the image converter is positioned - directed by the computer (see figure 2) - two points of special interest to get more detailed information. This is the most economical way of image analysis which is used by human operators if they have to describe an image (15). It depends on the development of computer costs and speed, if such a system will be a good solution for some applications.

New developments of instruments will go in three directions. First, a series of systems for special purposes will be designed as push button instruments for quality control and routine measurements. Only such an instrument guaranties reproducible measurements working with unskilled operators which again is a question of the costs of quality control. Of course, if a special system is not yet available, a research instrument may be used making all switches and knobs for adjustment and calibration unavailable and starting the measurement by a remote button. Second, very complex instruments, combined with large computers according to figure 2, will be built for very difficult, especially research problems. In between, instruments of medium capacity will be available which are self contained or which utilize only some modules of a large system. They are - due to their low costs - very useful, if an image analyser is not used full time but it is too time consuming to work manually or to use semiautomatic instruments. For all systems the accuracy of the image converter and the detector will be improved further, at least as an option, using new conversion principles or additional modules like the shading corrector. Therefore the number of instruments, especially of different versions of one system, available at the present time will increase further. But if the ability to do the measurement is determined as described above, defining the desired parameters and their accuracy and comparing these values with that of an instrument, it should not be difficult to select from an increasing number of instruments the most suitable one for a given application.

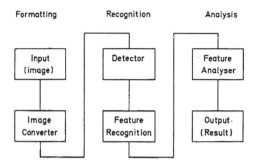

Fig. 1   Schematic drawing of automatic image analysing systems.
The input information must be formatted for the
recognition system which extracts all the features to be
measured.

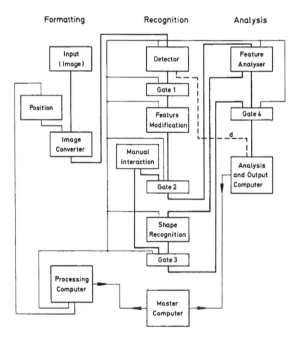

Fig. 2   Schematic drawing of a complex image analysing system
with different pattern recognition modes, automatic
processing and digital image storage (connection d).
Heavy lines: Information flow.
Thin lines: Processing connections.

The author thanks Mr. E.E. Underwood for his help in writing this paper.

REFERENCES

1) Hougardy, H.P. (1974) Instrumentation in automatic image analysis. The Microscope 22, 5/26.
2) Hougardy, H.P. Determination of automatic image analyzing instruments accuracy. Praktische Metallographie (in press).
3) Cruttwell, I.A. (1974) Pattern recognition by automatic image analysis. The Microscope 22, 27/37.
4) Hougardy, H.P. (1976) Recent progress in automatic image analysis instrumentation. The Microscope 24, January.
5) Fraser, P.V. (1975) New operator interactive light pen system. INTER/MICRO-75 (Cambridge, GB, 23-26 June 1975).
6) Lawson, K.E. (1975) Pattern recognition on the Quantimet 720 image analyzing computer. 4th International Congress for Stereology (Gaithersburg, MD, 4-9 Sept. 1975).
7) Morton, R.R.A. (1975) An image analysis system based on feature descriptors. 4th International Congress for Stereology (Gaithersburg, MD, 4-9 Sept. 1975).
8) Kelly, T. & Heil, R.H. (1975) Automatic data collection for the analysis of porous materials. 4th International Congress for Stereology (Gaithersburg, MD, 4-9 Sept. 1975).
9) Hougardy, H.P. (1975) Fehler bei Messungen mit automatischen Bildanalysatoren. In: Quantitative Analyse von Mikrostrukturen in Medizin, Biologie und Materialentwicklung. Praktische Metallographie, Sonderband 5, 73/80.
10) Griggs, R. (1975) Application of two-dimensional image amendment in automatic image analysis. INTER/MICRO-75 (Cambridge, GB, 23-26 June 1975).
11) Hunn, W. (1975) Texture analysis of stereology specimens. 4th International Congress for Stereology (Gaithersburg, MD, 4-9 Sept. 1975).
12) Müller, W. (1974) The Leitz-Texture-Analyzing-System. Scientific and Technical Information, Suppl. I, 4, 1974, p. 1Q1/116.
13) Delfiner, P. (1972) A generalisation of the concept of size. J.Microscopy 95, 203/16.
14) Serra, I. (1974) Theoretical bases of the Leitz-Texture-Analyzing-System. Scientific and Technical Information, Suppl. I, 4, 1974, p. 125/136.
15) a) Hougardy, H.P. (1972) Gefügebildanalyse im Prinzip und ihre technische Verwirklichung. Atti della giornata di studio sui metodi quantitative di analisi delle imagini in metallographia. Nel quadro di met' 72-3a mostra europea della metallurgia. Torino, 23 settembre - 2 ottobre 1972.
b) Hougardy, H.P. (1974) Il principio dell' analisi strutturale e sua realizzazione tecnica. La metallurgia italiana. Atti notizie. Vol. XXIX, n.3-1974, p. 183/85.

16) Terrell, A.C. (1975) Characterization of particles by size, shape, etc. INTER/MICRO-75 (Cambridge, GB, 23-26 June 1975).
17) Rink, M. (1975) A New Fst, storage-saving image analysis procedure for investigation individual features. 4th International Congress for Stereology (Gaithersburg, MD, 4-9 Sept. 1975).
18) Rink, M. (1975) A computerized image process of isolating individual features in a netted pattern. 4th International Congress for Stereology (Gaithersburg, MD, 4-9 Sept. 1975).
19) Hopkins, B. & Healy, P. (1975) The use of a commercially available image analyser in obtaining stereological information on corpora lutea in the mouse ovary. 4th International Congress for Stereology (Gaithersburg, MD, 4-9 Sept. 1975).
20) Bausch & Lomb, 41834 Bausch St., Rochester, NY 14602, USA.
W.A. Howe Co. Ltd., 88 Petersborough Road, London SW6, England.
Fica Route de Levis St. nom, F-78 Le Mesnil St. Denis, France.
D-8043 Unterföhring, Postfach 11 28, Germany.
21) Ealing-Beck Ltd. Graycaine Road, Watford WD2 4PW, England.
22) Healy, P. (1975) The design and applications of a new multi-disciplinary automatic image analyser (Histotrak). INTER/MICRO-75 (Cambridge, GB, 23-26 June 1975).
23) Hamamatsu TV Co., Ltd., 1126, Ichinocho, Hamamatsu, Japan.
3000 Marcus Ave., Lake Success, N.Y. 11040, USA.
D-8031 Hechendorf/Pilsensee, Hauptstr. 2, Germany.
24) Image Analysing Computers Ltd. Melbourn, Royston, Herts SG8 6EJ, England.
21, bd de Port Royal, F-75 Paris 13, France.
D-605 Offenbach, Schreberstr. 18, Germany.
40 Robert Pitt Drive, Monsey, NY 10952, USA.
25) Jenoptik Jena GmbH, DDR-69 Jena, Postfach 190.
26) Kontron, D-8 München 50, Lerchenstr. 8-10, Germany.
27) Wild Heerbrugg Ltd., CH-9435 Heerbrugg, Switzerland.
28) Ernst Leitz, D-633 Wetzlar, Postfach 210, Germany.
30 Mortimer Street, London W1N 8BB, England.
17-19 Rue Danton, F-94 Le Kremlin-Bicetre, France.
Rockleigh, NJ 07647, USA.
29) Carl Zeiss, D-7082 Oberkochen, Postfach 35, Germany.
31-36 Foley Street, London W1, England.
Les Bureaux de la Collin de Saint Cloud, F-92 213 St. Cloud, France.
444 Fifth Avenue, New York, NY 10018, USA.

*National Bureau of Standards Special Publication 431*
*Proceedings of the* FOURTH INTERNATIONAL CONGRESS FOR STEREOLOGY
*held at* NBS, Gaithersburg, Md., September 4-9, 1975 (Issued January 1976)

TEXTURE ANALYSIS OF STEREOLOGICAL SPECIMENS

by Werner Hunn
*E. Leitz, Inc., Rockleigh, New Jersey 07647, U.S.A.*

The major development effort for the commercially available
image analyzers has been directed toward automation of the
measurement process. The classical measuring procedures are
essentially applied and implemented in television speed logics.
Even the individual feature analysis is no exception. What
was lacking were new concepts for the quantization of a measuring problem so that data can be extracted from the image
which correlate to the physical characterization of a sample.

New possibilities in the framework of mathematical morphology
were developed by Matheron and Serra (1,2) and found their technical realization in the Leitz Texture Analyzing System (3).
Its comprehensive concept of automatic image analysis with
structuring elements contains the classical procedures as
specialized cases where it generally reaches deep into the
analysis of textures.

This is achieved by a generalization of the concept of size (4).
the measurement of image correlation functions to probe for
spacial interrelationships, the determination of the rose of
direction for structural anisotropy description and the two
dimensional image transformations. Some of these novel concepts should be briefly discussed.

Covariogram (5,6): This distribution measures image correlation.
It is defined as the frequency of occurrence of point pairs in
direction within the image whereby the distance between the
two points increases incrementally and is not "cared" for.
Only the two end points are tested to be a "hit" in any one
structure of the interesting component.

If we consider the covariogram more closely, it becomes obvious that this function reveals very powerful information
about the morphology of a complex body.

1.  Periodicities of sample structures in the micro
    or macro range which can even be disturbed or
    hidden by regional phenomena.

2.  Characterization of object groups of various
    sizes or spacing.

3.  The interrelationship of different sample phases
    which are probed in pairs in the so-called cross-
    covariogram.

4.  The statistical estimation (7). How many sample
    fields and polished section have to be analyzed,
    for example, to extract a volume fraction of in-
    clusions in steel with a certain accuracy (8).
    The morphological answer is uniquely given by the
    covariogram.

149

The _rose of direction_ is a powerful tool to describe orientations or anisotropies within a plane sample. Schmidt (9) proposed in 1917 a polar plot to show the frequency of a certain
parameter in any direction (the number of intersections with
surfaces per unit length in function of the variable direction
α). . The result allows a quantitative evaluation of the degree of orientation. Serra found from mathematical morphologic consideration . . .

$$\rho(\alpha) = P_L(\alpha) + \left[P_L(\alpha)\right]''_\alpha$$

. . . which is extremely powerful because it yields the perimeter length degree by degree. It basically adds one term
(the second derivative in α) to Schmidt's definition which
increases the sensitivity of this function to directional
changes considerably, and anisotropy directions of various
orders of magnitude become obvious (see Figure). The machine
is capable of measuring this polar plot by means of a programmed
optical image rotation in one degree increments in a matter of
minutes.

Two Dimensional Transformation: If a structuring element of
two dimensional extension and isotropic shape, such as a circle, is used then the original image is transformed by various
yes-no operations:

Erosion: All points which can be occupied by the center
point of the circle so that the circle itself is fully
contained within the original image structure define a
new image which is outlined by a minimum distance from
the original structure periphery. This image transform
yields some interesting results. Surface extremities are
smoothened, structural bridges smaller than the circle
are broken open and detail smaller than the circle completely suppressed. When this operation is repeatedly
applied with increasing size of the circles, the original
image structures can, for example, be size classified on
the basis of maximum circle to be inscribed. Whereby, we
do not request a priori that the image consist of individual features but rather can be a complex pattern of,
for example, porosity.

Opening: A different transform is obtained if we consider all the points which are covered by the circle as
long as the circle stays within the original structure.

Here no marginal zone is deleted and testing with increasing size circles leads to a structural sieve-
analysis or porosity analysis by mercury pressure
method implemented in morphometric tv operations. This
distribution lends itself particularly to the analysis
of complex pattern in form of a size-area fraction
distribution as is often required in an intersection
plane of a body. For example, to analyze a lung tissue
network or a continuous, interspersed phase in a metal
section.

Dilation: If we ask for the image transform which is
given by all the points covered by the circle center
point so that the original structure and the circle
have at least one point in common, then we see that the
image components are expanded in all directions by the
same amount and we probe for neighborhood relationships,
such as, orientation independent distance distributions.

150

Closing: If we take this latest transform and erode it again by the same size circle, then we reconstitute the original image very closely except for bridges which remain between structures closer spaced then the circle diameter and holes are filled in. This transform finds application in, for example, the spacial analysis of stringer inclusions in steels.

Serra shows in a beautiful example the power of these new techniques. Mineral sections of volcanic rock of various geographical locations over ten square kilometers have been studied as to their genesis history. The porous medium has been measured in an opening-closing distribution, whereby the size-area fractions are plotted starting from openings at large circle sizes decreasing to zero circle dimension to closings with step-wise increasing circles. The results shown are interesting. We can see (Fig. 3) that the curves of the porous samples are grouped. The bottom group of nearly parallel lines with each other and the x-axis reflects a progressive invasion of the space by the porous medium. Two other groups of curves have various amounts of incline and come to an intersection to the left of the y-axis. The first fact reveals that the pores and the distance between them increase simultaneously; the second one that a small size porosity seems superimposed on the larger one as if it were coming from different genesis.

1. G. Matheron: Les variables regionalisees et leurs estimation, Masson, Paris, 1963

2. J. Serra: Introduction a la Morphologie Mathematique. Les cahiers du Centre de Morphologie Mathematique, Fasc. 3, 1969 – Ecole des Mines de Paris

3. W. Mueller: The Leitz - Texture - Analyzing - System, Leitz - Mitt. Wiss. u. Techn. Suppl. 1, Nr. 4, June 1973

4. P. Delfiner: A generalization of the concept of size, Journal of Microscopy, Vol. 95, Pt. 2, April 1972, pp. 203-216

5. J. Serra: Un critere nouveau de decouvertes des structures, le variogramme, Sci. Terre (sous presse)

6. R. Alberny, J. Serra and M. Turpin: Use of Covariograms for Dendrite Arm Spacing. Measurements, Trans. of the Met. Soc. of AIME, Volume 245, Jan. 1969-55

7. J. Serra: Echantillonnage et estimation locale des phenomenes de transition minieres, These, Nancy, 660 pp

8. J. Serra: L'echantillonnage des inclusions dans les aciers, Journies d'etudes de la metallurgie, Turin, 1972

9. W. Schmidt: "Statistische Methoden beim Gefugestudium kristalliner Schiefer," Sitz. Kaiserl. Akad. Wiss. Wien, Math.-nat, Kl., Abt. 1, 126 (1918) 515

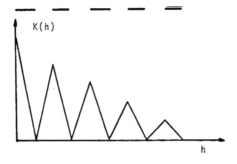

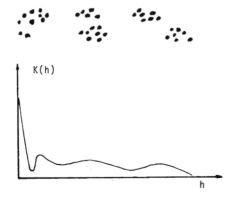

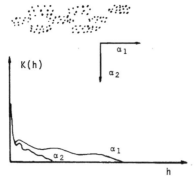

Figure 1: Various schematic structural configurations and the corresponding covariograms

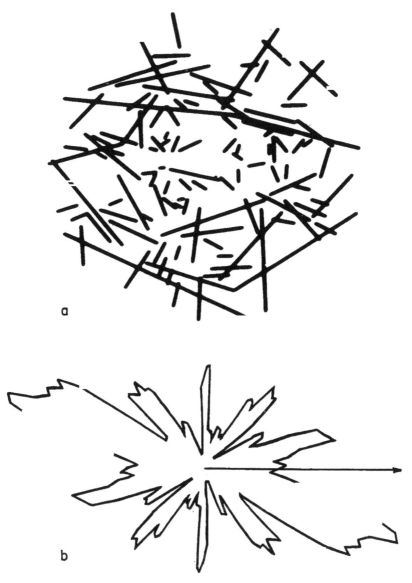

Figure 2a.b:  Pattern of fault lines and the resulting
"rose of direction" (b)

153

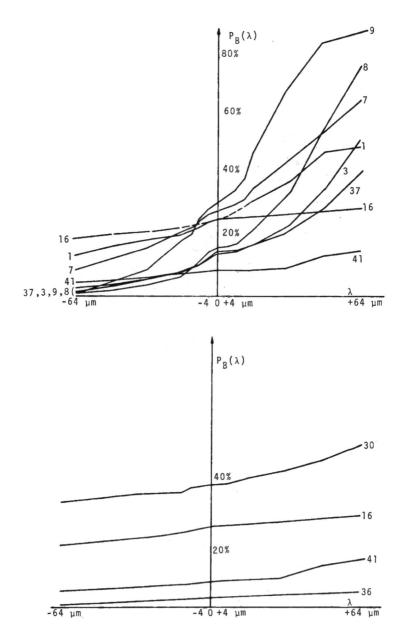

Figure 3: Opening-closing percentage area distribution of various samples (see text)

154

National Bureau of Standards Special Publication 431
Proceedings of the FOURTH INTERNATIONAL CONGRESS FOR STEREOLOGY
held at NBS, Gaithersburg, Md., September 4-9, 1975 (Issued January 1976)

A COMPUTERIZED IMAGE PROCESS FOR ISOLATING INDIVIDUALS IN AN ORIGINALLY NETTED PATTERN

by M. Rink

Institut für Geophysik der TU Clausthal, D-3392 Clausthal-Zellerfeld, Germany

INTRODUCTION

In nature, objects mostly are much more complicated than is accessible to theoretical model treatments. In the theories describing microstructures, the structure-forming elements are approximated by relatively simple geometric bodies. Such an example in the field of sediment petrophysics is given by the attempt of estimating hydraulic permeability of a porous rock, as an important three-dimensional material constant, from the microstructure of the pores, measurable in plane sections only. To accomplish this, most mathematical considerations start from much simplified capillary bundle models. However, the sections through a porous rock show very clearly the netting of the pore channels. Especially for a weakly cemented matrix, larger compact pore cross sections are multiple-connected by narrow bridges, and often extend over wide parts of a section. In such an original section image, it is not possible without restriction to talk of single pore cross sections as individual figures. Therefore, the section image of the pore pattern must be simplified so that individual pore cross sections only remain, which contribute considerably to the permeability, and which are simultaneously simplified in shape so far that their so-called hydraulic radii ( = area/circumference) are good enough for the description not only of hydrostatic but also of hydrodynamic effects. In this case, the problem of figure individualization can be solved in principle by artificially separating the narrow interconnecting figure bridges.

Since the author had developed a computerized image analysis procedure (Rink,1970,1975), which yields statistical values of a lot of morphometric figure parameters of all figures of the investigated image in just one computer run, it became desirable to apply this image analysis procedure even to images, not containing classical individuals a priori. So the above mentioned pattern simplification had to be realized. (It may be noticed that this demand does not result from insufficiencies in the image analysing program, but from the necessity of gaining meaningful statistical values of the figure parameters.)

KNOWN IMAGE PROCESSES

Useful auxiliary image processes for figure individualization are the areal erosion and dilatation of figures or combinations thereof, known by the notions opening and closing (Matheron,1967; Serra,1969).

The author has written computer programs for performing the said image processes. Of course, the theoretical background is the same for a hardware or software solution. For practical realization it is an important fact that the opening figures may be constructed from the original figures eroded at first and dilatated afterwards, both with the same structuring element. Furthermore just the programming of an erosion and dilatation subroutine for the smallest structuring element only is necessary. By repeated subroutine calling, the proceeding of the image processes is achieved, corresponding to the use of increasing structuring elements.

The effect of the opening process consists in eliminating all figures resp. figure parts which are smaller than the pre-set structuring element, being a circle area in the ideal theoretical case. This leads to a separation of larger more compact figure

parts, originally interconnected by narrow bridges. In most cases
the bridges have different width, however. To complete the figure
separation, the opening process therefore must be continued with
increasing structuring elements. But thereby also small isolated
figures are lost. Finally the largest and most compact figures only are left. Also, during the opening process a certain shape
variation towards the shape of the structuring element occurs with
all figures involved.

The usual evaluation of such opening series consists in acquir-
ing so-called granulometries. But this shall not be discussed here.
It is intended rather - starting from a complicated netted original
pattern - to create one image containing all those - eventually
artificially individualized - figures that shall be statistically
evaluated in a common manner.

## THE CUT PROCESS

For this problem, the mere opening process does not lead to the
wanted aim. Indeed, that process removes narrow bridges between
compact figure parts, and thus separates originally complicated
figures into smaller and simpler ones, as mentioned above. But its
disadvantage is that the already isolated figures, which are smaller than the pre-set structuring element, disappear.

For avoiding this loss, a first idea could be collecting in an
auxiliary image in advance all those figures that would be eliminated by the next opening step. Thus, before the n-th opening step,
this auxiliary image contains figures - shape simplified by opening - of all size classes smaller than the remaining opening figures. Then the superposition of these last opening figures and
the figures of the auxiliary image yields an individualized image
pattern.

This method requires a criterion for recognizing in advance the
figures smaller than the pre-set structuring element and a procedure for transferring them to the auxiliary image field. Such a
criterion may be derived from the information, whether a digitalized figure consists of boundary points only or not. A continuous
three-line association is necessary for achieving this information.

This way of solution was tried, but it turned out to be unsatisfactory for its time-consuming complexity of the copying process.
Furthermore, the figure separation was of limited success only
because of overlapping and contacting of figures after the necessary image superpositions.

A better effect of individualization is obtained in the following even simpler and faster way. At first, in the original pattern,
all figure points on the outline of the circumference of the figures, resulting from the first opening step, will be erased. By
that means the narrow interconnecting bridges and small figure
parts will not be eliminated totally - as by mere opening - but
only separated at the contact points to the larger figures. That
is why this procedure may shortly be called 'cutting'. For advancing the process of figure individualization, successive opening is
continued as an auxiliary process, always erasing the outline
points of the corresponding opening figures in the just preceding
cut image. This process can be continued until no further figure
separation occurs.

By the continued cutting, many very small relics of the original bridges result that must not be taken into account. To remove
them, a final opening has to be applied to the so far derived individualized image. However, in this so-called clear cut image the
smallest original figures are missing now. Often this clear cut
image is already welcome for the quantitative analysis. But when
the loss of those smallest original figures cannot be tolerated,
they have to be filtered out of the original, and added to the
clear cut image, after the above mentioned method. Such a so-

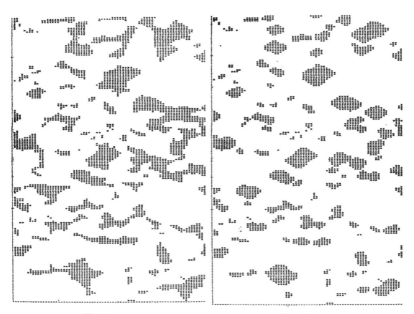

Fig.1                                    Fig.2

Parts of computer outputs of a cross section of the porous sandstone specimen B50. Fig.1:Original image; Fig.2:Clear cut 5+ image

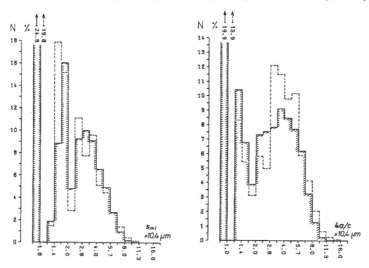

Fig.3                                    Fig.4

Number-normalized frequency distributions of the mean chords in
line direction (Fig.3) resp. hydraulic radii, times 4 (Fig.4) of
pore cross sections of specimen B45. 777 Original; - - - Clear cut 4+

157

called clear cut 5+ image is partially shown in Fig.2, 5 being the number of cut steps.

The elementary processes erosion and dilatation in this software solution are based on a successive processing of three adjacent image lines. For an advanced image processing however, it is favourable to reserve storage space for four image fields. The cutting then is only a logical word by word combination of the three images involved.

EXAMPLE OF APPLICATION

The pore space of specially prepared sandstone specimen was investigated by the described image analytic computer methods (Rink, 1973a,b,1974). The cutting effect can be seen by a comparison of Fig.1 and 2. A comparison of data of original and clear cut images clearly shows the size reduction, shape simplification and number increase of the individualized figures. The frequency distributions of figure parameters reflect details of the same facts (Fig.3 and 4).

For the stereological determination of usual parameters as porosity or specific surface, integral values of the original images must be used. But for estimating the hydraulic permeability after a refined Kozeny-Carman theory, three-dimensional statistical pore values have to be incorporated. A first calculation has shown that the original images are not suited for this purpose, whereas images, individualized by the cut process, furnish much better results.

The presented cut process - relying on the individualizing property of the opening process, but granting at the same time the coexistence of isolated figures of various sizes in one image - offers extended possibilities for the analysis of complicated image patterns.

REFERENCES

Matheron,G.(1967) Eléments pour une théorie des milieux poreux. Masson, Paris

Rink,M.(1970) Automatische morphometrische Bildanalyse mit Hilfe eines elektronischen Digitalrechners. Diss. Techn. Univ. Clausthal, BRD

Rink,M.(1973a) Porengeometrische Untersuchungen an Sedimentgesteinen mit einem Bildanalyseverfahren für Digitalrechner. Z.Geophys.,39,989-1005

Rink,M.(1973b) An image analysis procedure for digital computers and its application to investigations on pore structures. Proc. RILEM/IUPAC Int.Symp. 'Pore Structures and Properties of Materials', Prague, Final Report II, C497-C506

Rink,M.(1974) Untersuchungen am Porenraum von Sedimentgestein mit bildanalytischen Methoden im Digitalrechner. Z.Prakt.Metallogr. Sonderband 5,308-322

Rink,M.(1975) A new fast and storage saving image analysis procedure for investigating individuals by a digital computer. Proc. 4th Int.Congr.Stereology, Stereology 4, Washington D.C.

Serra,J.(1969) Introduction à la morphologie mathématique. Cahier du Centre de Morphologie mathématique, Fasc.III, Ecole des Mines de Paris

*National Bureau of Standards Special Publication 431*
*Proceedings of the* FOURTH INTERNATIONAL CONGRESS FOR STEREOLOGY
*held at* NBS, Gaithersburg, Md., September 4-9, 1975 (Issued January 1976)

## AN IMAGE ANALYSIS SYSTEM BASED ON FEATURE DESCRIPTORS

by Roger R. A. Morton
*Bausch & Lomb, Rochester, New York 14625, U.S.A.*

Image analysis is finding increased acceptance in a diversity of fields. As a direct result of this acceptance, new applications are continually being found which require image analysis equipment with more flexible and powerful capabilities.

To meet these requirements, an image analysis system must be able to quickly solve advanced analysis problems. Furthermore, the control of the instrument by the operator as well as the instrument's communication to the operator, must be simple and rapid. Clearly, there is little point in having an instrument which can analyze an image rapidly, but which requires a much greater time for the operator to tell it what to do.

These considerations were foremost in designing the Omnicon[TM] Pattern Analysis System. As a result, the two modes of operation were chosen. One employs a special purpose keyboard which operates in conjunction with an alphanumeric presentation on the display. The other uses the BASIC language to sequence and control the system operation. When using the BASIC language, a second general purpose keyboard is employed which operates interactively with both the alphanumeric display and a printer. There are many advantages to this new type of image analysis control. For example, expandability of the system is not limited by the control panel; incorrect or ambiguous commands may be flagged directly and the control status of the system is continuously summarized on the display.

In the keyboard mode, five function phases are defined. They are: Selection, Measurement, Process, Output, and Mode.

The Selection phase is used to select those features within the sample which it is desired to measure. The operator selects, using the keyboard, the ranges of specific measurements, shape factors, orientation factors and so on, that a feature must have if it is to be selected. The instrument then performs the desired measurements and indicates with a tag on the display the features which meet these criteria. The parameters upon which a selection may be based are termed "descriptors", and they may be classified as follows:

### Gray Level

The gray level of a feature may correspond to its density, its absorbance, reflectance or transmittance, depending on how the object is imaged. Selection may be based on maximum gray value per feature, the average value or the requirement that the entire feature exceed a preset value.

### Size

Size of a feature may be defined in terms of large number of different measurements. Examples include area, area of holes, longest dimension, breadth, perimeter, convex perimeter, or derived measurements. (Derived measurements are simply expressions involving a number of measurements of the type just mentioned.)

## Shape

Shape factors may be constructed from measurements of the type
already discussed, or from additional measurements such as the
number of vertices of a feature or the number of holes within a
feature.  Shape factors are dimensionless expressions and are in-
varient with both size and orientation of the feature.

## Orientation Factors

Orientation factors define the orientation of a noncircular feature.  A variety of orientation factors can be constructed using
measurements like the ones discussed  or measurements such as the
direction of the longest dimension or direction of bréadth which
may be used directly for such factors.

## Position

Features may be selected based on their position within the field
of view or position within the sample.  The position of the feature may be defined in terms of either the position of the topologically defined point corresponding, for example, to one of the
extremities of the feature or the position of the centroid of a
feature.

## Adjacency

Adjacency is a position measure which is based on the distance to
other features, or, for example, determines whether a feature
is surrounded by another feature.

## Focus

The measurement of average sharpness of the edge of a feature or
some other focus measure is a useful descriptor when the features
of interest do not lie in the same plane.

Notice that these descriptors are in a sense independent of each
other.  One may even say orthogonal in the mathematical sense.
The value of any descriptor may be changed for a particular feature
without affecting the value of other descriptors of the feature.
Alternatively, they may be considered as dimensions in a decision
space which is used to define the properties of desired features.

## Measurement

Selected features may be measured using any of the measurements or
descriptor types already described.  Measurement expressions are
scanned to determine their dimensionality and calibration factors
are appropriately applied.

The Measurement operation produces a single result for every feature in the field of view in a fraction of a second.  This large
amount of data is usually of little use to the operator.  Consequently, the Process phase occurs concurrently with the measurement operation to reduce this data into the format desired. Typical formats include statistical summary indicating sum of measurements, mean, standard deviation, number of features measured,
maximum or minimum values, or size distribution based on linear
or logarithmic intervals.  Alternatively, a list of the feature
measurements may be presented on the display.

Also, occurring during Measurement and Processing operations is the Output operation, which involves passing the processed data to a printer, onto a display or into the BASIC language section of the software for further processing using operator selected programs.

BASIC Language
The BASIC compiler implements a full repertoire of BASIC commands. Additional commands have the same effects as keyboard operations while others allow the transfer of data from the processing section of the software into BASIC variables. In addition, BASIC commands permit control of the microscope stage and sense the number of the microscope objective currently in use.

Consequently, a BASIC programs may be written to perform complex analysis and recognition operations. Examples include: agglomerating detected areas into single feature for sample where a single feature may appear as a number of "white blobs", to measure Feret's diameter at right angles to the longest dimension, and a program to measure agglomerates, and separate the agglomerates into individual features based on prior knowledge of the shape of the individual features.

Block Diagram
Figure 1 illustrates the major functional blocks of the Pattern Analysis System. The scanner receives the image from either microscope, macro unit, or transmission electron microscope and generates a corresponding video signal. This signal is passed to the detection or thresholding electronics where the boundaries of the features to be measured are defined using one of a number of detection criteria. The detection circuits identify those transitions which are to form the boundaries of features to be analyzed using either the fixed threshold level or using a threshold level defined with respect to the background of the image. Once a transition has been identified as forming part of a boundary of a feature to be analyzed, the position of the boundary with respect to the transition may be defined by determining the maximum and minimum video amplitudes in the vicinity of the transition and computing the boundary as occurring at a weighted average, 50% of the maximum and minimum. The exact weighting of this average is determined by the gamma of the scanner tube.

One option which may be placed between the scanner and the detection electronics is the "Shading Corrector". This module corrects shading which may arise in both scanner and image formation optics. This shading is classified into white or multiplitive shading, and black or offset shading. The white shading corresponds to variations in illumination of the sample vignetting in the optics variations of sensitivity across the faceplate of the scanner, while the black shading will correspond to dark current offsets in the scanner and nonuniform glare in the optics, especially when using incident light illumination. The Shading Corrector independently assesses these two sources of shading by means of two reference images. Firstly, a black reference image which is formed from a uniform sample is viewed through the microscope or other image source to produce the darkest gray levels or density levels anticipated in the specimen. The video profiles corresponding to this resulting black image is stored in the Shading Corrector. Secondly, a white reference image is produced from a uniform sample which produces an image corresponding to the lightest or least dense areas of the specimen. By storing the white reference and comparing it with the black, the Shading Corrector is able

to determine the multiplicative or white shading components and
treat them separately from the offset or black level components of
shading determined on the black reference.

Using the Shading Corrector, the number of shades of gray which
can be separated (measured using the 10%/90% method) is typically
50 and can be as high as 66. In other words, the shading plus
noise taken between the ten percentile extremes is typically
1/46 or 2.2% to 1/66 or 1.5% of the black to white video range.

The detected signal is passed to the Pattern Analysis Module (PAM)
comprising the Pattern Analysis System electronics and associated
processes.

In this module, the video signal is processed on an intercept
basis in accordance with the commands entered through the keyboard
of the system. A keyboard compiler in the processor processes
the keyboard instructions, accesses the appropriate subroutines,
and generates the appropriate code to cause the system to execute
the required operations.

The processor has a number of peripheral devices connected to it
through an input/output buss. These include the display processor,
which generates graphics, characters and other numerics on the
display, a printer to print the results of the operation; the
special purpose Pattern Analysis System keyboard and its associated
status matrix; a general purpose keyboard for communication to the
BASIC compiler; and the stage control X,Y and Z motions of the
microscope stage. Other peripherals may be added including: automatic focus, and system sequencer to control the functions of other
hardware modules.

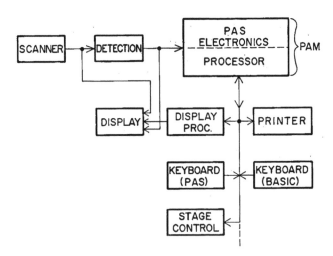

Figure 1:  Block diagram of Omnicon Pattern Analysis System.

*National Bureau of Standards Special Publication 431*
*Proceedings of the* FOURTH INTERNATIONAL CONGRESS FOR STEREOLOGY
*held at* NBS, Gaithersburg, Md., September 4-9, 1975 (Issued January 1976)

IMPLICATIONS OF RECENT ADVANCES IN IMAGE ANALYSIS FOR STEREOLOGICAL THEORY

by Holger Schmeisser
*IMANCO Bild-Analysen-Computer GmbH, D-605 Offenbach, Schreberstrasse 18, Germany*

INTRODUCTION

One of the most recent technical advances in electronic image analysis is the facility to modify image features in two dimensions in a well defined manner. Unfortunately this innovation has caused considerable confusion. While having been introduced to the system QUANTIMET 720 as an additional option which increases the structure descriptive power of the system considerably (Schmeisser 1974), the same technique has independently been featured as the heart of a completely new concept of describing structures and interpreting morphometrical properties. From discussions in the recent past, it is felt that there is a considerable lack of understanding of the relations between "traditional" stereology and the new approach. This paper attempts to clarify some of the aspects involved. Because of their relevance to practical problems, the concept of two dimensional amendment and the measurement of spatial distributions have been chosen as examples.

THE CONCEPT OF TWO DIMENSIONAL AMENDMENT

While structures displaying well defined features in a background-matrix allow very detailed characterization by means of feature specific data, one finds oneself drawn back to fairly general descriptors as soon as penetration structures are under study. Overall measurements e.g. area fraction or mean intercept length are very useful in many instances, particularly because they are part of a well evaluated framework of stereologic relations which allow direct interpretation of the data (De Hoff & Rhines 1968, Underwood 1970). There are, however, a number of situations in which more detailed information is required e.g. where distributional measurements are to be performed. Due to the technical layout of most image analyzers, a chord size distribution is measured.

Before discussing its problems, the actual method of measurement should be recalled to fig. 1: In order to determine whether or not a given chord exceeds a preset length l a proportion of the chord corresponding to l is suppressed electronically as shown in fig. 1a. Only chords which "survive" this procedure are then taken into consideration. So effectively, the first step involves a modification of the image and the result of this amendment is then subject to the actual measurement.

Detailed study shows that the results are independent of the particular way in which the amendment is performed. If one suppresses half the chord length at the leading edge of a feature and half at the trailing edge (fig. 1b), this is exactly the same as suppressing the full chord length at either the leading or trailing edges. The system illustrated in fig. 1b, however, would be referred to as "erosion of the feature with respect to l". The reverse operation, amendment by extending the chords, results in the corresponding "dilation".

The obvious problem with chord sizing – or in general with linear amendment – is the directional dependence of the results. The importance of this fact is demonstrated in fig. 3.

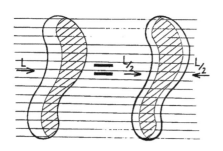

Fig. 1:  Equivalence of chord sizing and linear
         erosion

As this difficulty originates from the presence of a preferred direction in the
evaluation process, the natural solution is to treat all directions equivalently.
This results in a concept of two dimensional amendment as shown in fig. 2: Instead of shifting the boun-
daries in one direction only,
"elementary shifts" are applied which result in a new
boundary. This concept is
similar to the well known
principle of wave optics.
Fig. 2 shows that also in
this case there are two obvious derived structures -
an eroded and a dilated one.
Again it turns out that this
simple generalization of the
concept of chord sizing
gives exactly the same results as a strict mathematical treatment.

Fig. 2: Relationship between original structure
(bold) and the eroded (dashed) and dilated structures respectively

The practical useful-
ness is demonstrated in

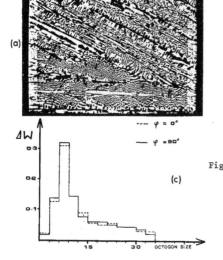

(a)

(b)

(c)

Fig. 3: Comparison of directional dependence of chord size distributions (b) and distributions obtained by combined application
of erosion and dilation (c) -
both diagrams show two distributions obtained in perpendicular orientations of the sample
shown in (a)

fig. 3c, which shows the result of successively performed erosions/dilations in
two perpendicular orientations of the sample. The elimination of the ambiguity of
the chord size distributions of fig. 3b is clearly noticeable.

So in the same sense in which the step from linear analysis to true two dimensional image analysis enabled the description of features by real two dimensional feature specific parameters (Gibbard, et. al. 1972) rather than individual
chord information, the concept of two dimensional amendment allows a more comprehensive and less ambiguous description of complex structures than simple chord
sizing.

It is noteworthy that two dimensional amendment in combination with feature
specific measurements is particularly useful for some applications (Schmeisser
1974). A direct application is e.g. feature sizing with respect to the "biggest

inscribable octagon" as explained in fig. 4. This is a valuable supplement to
the normally available maximum chord length distribution, although it must be
emphasised that it is completely different from a true area size distribution
based on feature specific data (fig.4). The results obtained will, however, only
be meaningful if 2D amendment is applied in combination with a full feature count
logic as "keying" signal - otherwise the well known effect of feature break up
(Cole 1971) will destroy the measuring accuracy (fig. 4).

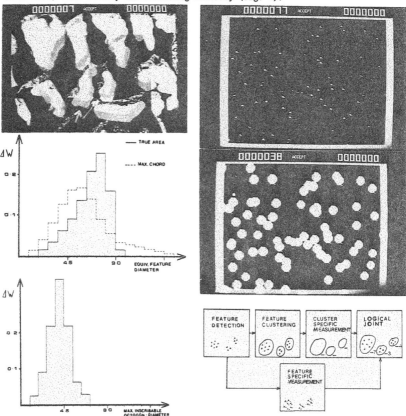

Fig. 4:   Sizing with respect to the
biggest inscribable octagon
problem of feature break up
(a) and comparison of ob-
tained size distribution (b)
with maximum chord length
and true area distributions
(c).

Fig. 5:   Two dimensional dilation clusters
particles with certain distances
(a), (b). Appropriate logical set
up of the instrument enables particle group analysis (c).

Two dimensional dilation is useful for determining neighbourhood relations between features as shown in fig. 5. Besides the measurement of shortest distance-
distributions a more sophisticated combination of 2D amendment and feature specific
measurements enables particle group analysis (fig. 5c).

## MEASUREMENTS OF SPATIAL DISTRIBUTIONS

A well known way of determining the spatial relationship of components in a
structure is the application of a stepping measuring frame. Fig. 6a shows a
slightly different arrangement: the total frame is chopped into many little sub-
frames which are assessed individually. Fig.6b shows the result of an area fraction measurement in each of the individual frames plotted against the respective
position - the banded structure is obvious. As the wave length of such a banding
pattern is an important parameter of the materials properties the distribution

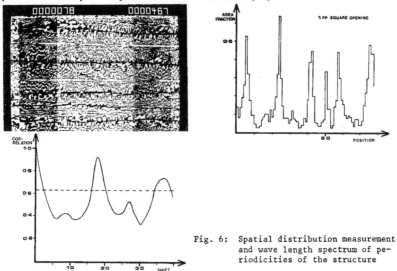

Fig. 6: Spatial distribution measurement
and wave length spectrum of periodicities of the structure

curve in fig. 6b is converted into a wave length spectrum by means of a simple
algorithm (fig.6c). Applying the algorithm to the distribution curve of only one
phase gives the respective wave length spectrum, cross correlation of two phases
shows the spatial relationship of those. Being comparable to a "covariance"-mea-
surement this algorithm is also easily applicable to e.g. concentric patterns.

## CONCLUSION

Considering the results presented here the new concepts represent a very natural extension of the existing approach to morphological measurements. Although
there is certainly no necessity for a complete reconsideration of the existing
formulation, the theory which accommodates the new measuring feasibilities to the
framework of stereology still has to be evaluated. Looking on the structure descriptive power of these measuring aids, however, it should be a real challenge
to the theoretically working stereologists to fill this gap very soon.

## REFERENCES

Cole, M. (1971) "Instrument Errors in Quantitative Image Analysis", Microscope
19, 1

De Hoff, R.T. & Rhines, F.N. (1968) "Quantitative Microscopy", Mc. Graw-Hill

Gibbard, D.W., D.J. Smith, A. Wells (1972) "Area Sizing and Pattern Recognition
on the Quantimet 720" Microscope 20, 37

Schmeisser, H. (1974) "Zweidimensionale Bildveränderung - eine Erweiterung bild-
analytischer Meßmöglichkeiten" in Special Issues of Practical
Metallography No. 5, 60

Underwood, E.E. (1970) "Quantitative Stereology", Addison Wesley

*National Bureau of Standards Special Publication 431*
*Proceedings of the* FOURTH INTERNATIONAL CONGRESS FOR STEREOLOGY
*held at* NBS, Gaithersburg, Md., September 4-9, 1975 (Issued January 1976)

EXPERIENCES WITH OPTOMANUAL AUTOMATED EVALUATION-SYSTEMS IN BIOLOGICAL RESEARCH, ESPECIALLY IN NEUROMORPHOLOGY.

by H. Haug
*Medical School Lübeck - Department of Anatomy, D-2400 Lübeck, Ratzeburger Allee 160, Germany*

The principles of stereology have a mathematical basis and this science is independent of a practical use (15, 17, 21). However, stereology is only significant in the case of its application in morphological sciences (2, 5, 6, 9, 21). The practice needs on the other hand some technical helps for an expedient evaluation. In the last years this can be observed by the development of more and more sophisticated instruments (3, 12, 17). Many papers of this congress deal with the construction and use of such instruments. The most modern instruments have a full-automated device where the scientist operates mainly with electronic switches of the computer equipment (4, 11, 13, 14, 15).

Scan-instruments are guided by their computer and in this way the instrument can distinguish between distinct gray-intensities of different structural details in a tissue and can calculate the results from this evaluation by means of stereological formulas. However in many cases this distinction is difficult, due to:

1) the gray-steps are too low, 2) more than one structure has an equal gray-intensity, 3) one structure has different kinds of gray-steps dependent on some factors in the section (7, 1o, 12, 22). In such cases the scan-computer system is unable to differentiate the structures and we are forced to evaluate with older and simpler but time-consuming procedures. Another possibility is the extension of the scan-instruments for operating with the human eye-brain system. This eye-brain system is very well suited for the discrimination of the above mentioned difficulties.

This paper deals with the advanced practice for operating by the eye-brain system. The output of the eye-brain system is the hand and therefore such devices are called opto-manual instruments. All of the newer developments of these instruments work with computers. These enables one to restrict the function of the eye-brain system to the important point of the diagnosis.

It is also possible to improve the full-automated scan-devices for optomanual treatments, but, such devices need besides the relatively expensive scan-instruments further expensive hard and soft-ware. Therefore it is important that in the last years some instruments have been developed which can operate directly with the eye-brain system and are on this basis less expensive than the full-automated scan-devices.

Before a description of the possibilities of these optomanual systems will be given, it is necessary to describe two biological examples which will make clear that a scan-instrument is unable to work in tissue with minute differencies. The first one deals with macroscopy of the brain and the second one with electron-microscopy of the nervous tissue.

A macroscopic section through the human brain shows white and gray matter. The white matter consists of connections between the different parts of the gray matter by nervous fibres. The gray matter contains many nerve cells and the complicated network of their processes. However, every gray matter represents another functional part of the brain. Therefore it must be differentiated by an evaluation between the distinct gray matters which have, in many cases, but not all, nearly the same gray value. Such a differentiation can only be done by the eye-brain system.

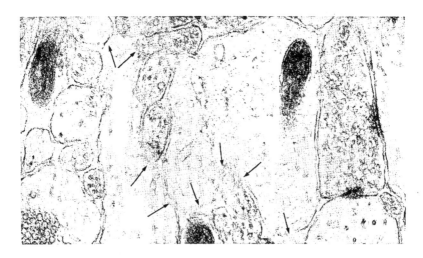

Fig.1: Neuropil of the cortex cerebri with entangled cell processes. The arrows indicate the possible locations of oblique sectioned cytomembranes (4o,ooo : 1).

The second example shows an electronmicrograph of the, so called, neuropil (Fig. 1).

This neuropil is lying between the perikarya of the nerve cells in the gray matter and contains the interconnections between the network of nerve cell processes. The network is composed of three kinds of processes with different functions: one is for input, one is for output and one is for different auxiliary functions. The three kinds of processes have a similar image and the human eye needs a certain period of experience to recognize these subtle differencies. The functional interconnections between the in- and output processes are the synapses with a specific, but, not too different structure.

Furthermore the diagnosis of these processes will be ·implicated by the fact, that the surface of it, which is represented by cytomembranes, have different angles to the surface of the section resulting in different gray-values. In this case the structural composition can only be differentiated by the eye-brain system and not by the scaning-instruments.

The optomanual systems.
The main problem in the optomanual system is the manual transfer of the optical diagnosis into an advanced system of registration. Today various solutions of this problem exist.

1) A simple instrument was and is a counter with a keyboard for different structures. Such instruments have in the meantime been developed as a receiver for a computersystem. The first construction was introduced by

WEIBEL. Such instruments are able to register values for more than one structure (2o).

2) An older instrument, the TGZ 3 of Zeiss, has recently been automated and can now be used for registration of more than one diameter of a structure. At the same time it registers the number of measured structures (6, 7).

The following instruments are based on the automatic registration of a
manually tracing of structural details with an instrument. The technical
solution of such devices are very different. However, three of these devices
need a flat plane for evaluation. The extension of this plane should not be
too small, because our brain-hand-instrument should have a certain space for
an exact manipulation of the specific instrument to diminish inexact tracing
errors. The following three devices have solved this problem in a similar
way. It is possible to examine slides within a microscope with the help of
a drawing mirror. All other structures may be evaluated by their photomicrographs or any other kind of projection. In this way these systems are
independent of the instrument of observation or of the magnification of the
image.

3) One instrument has been developed by COWAN and WANN. It registers the
topographical coordinates of the manual traced instrument by an ultrasonic
device in a fixed board of evaluation. This registration is transfered to
a computer (1, 19).

4) The MOP II from Contron comp. registers with the help of a light stylus
the crossing of a line in a lattice as a point. Dr. Rohr suggested the
development of such instrument. The optical registered points are counted in
different registers of the instrument and can, but must not be transfered on
or off line to a computer-system. This instrument can also be used as a point-
counter. There are different kinds of lattices (e.g. after WEIBEL, MERTZ etc.)
and these can be used on every flat surface (7).

5) The tracing-instruments of some computer-systems which have been developed
for the automated cartography can also be used in stereology. Such a device
has been developed by BLACKSTADT in Aarhous (Denmark) for investigations in
the nervous system. The different kinds of technical solutions make it
impossible to gather experience with all the mentioned devices. Therefore
the following part of the report is mainly restricted on the application of
the MOP II in point counting and tracing, and of the TGZ 3 Zeiss in measuring
diameters. But it seems that the other both devices can be similarly handled
like the MOP II-instrument.

In this past year the MOP II and the TGZ 3 of Zeiss have been installed in our
institute and they are combined with a smaller computer (Wang 22oo B). Now
it is possible to report on the first experience with these instruments. The
dimension of this equipment has been directed by financial possibilities. The
supporting instruments of the computer, especially, the Teleprint, are
relatively slow. But, I think it is not worthwhile to discuss on a two, three
or four times faster system because the speed of evaluation is surely slower
than one of the computer equipment.

The following report shows that an advanced investigation needs as well
sophisticated optomanual instruments as a good computer program. One important
problem in biological research especially in brain-research is the
inhomogeneous structural distribution of details in the nervous tissue. For
example the cortex cerebri, in which the consciousness is located, is
composed of different layers with different functions and structures.

The nerve cells and all other structures of the cortex cerebri are arranged
in layers. For example, the nerve cells place themselves in different
densities and sizes. Thus it is ne cessary to evaluate the different densities
because each layer has another functional significance. Our way to solve this
problem is the following: The measuring of all desired stereological
parameters will be done from one evaluated microscopical field to the next
field (8). The fields are arranged in rows perpendicular to the surface of
the cortex. However, the distance of every field to the surface as well as
the measured values are stored automatically by a tape-punch of a teletyp.

169

Twelve different values for each field are possible.

The evaluation of one cortical area needs about 15 to 3o rows, each 2o to 5o fields long. The number of fields in one row depends on the thickness of the cortex. The content of the punched tape will be transferred to our computer system and for all other calculations stored on a magnetic tape. With the help of a program developed in our institute we can compute our results under different stereological assumptions, as numbers per volume, or surface areas or length of a structure or the volume-part of one or more details. An important further advantage of our program is the possibility to divide our samples in the perpendicular direction as well as in the horizontal one. The calculation in horizontal direction can be done by a rotation of the matrix of values. Such a new calculation needs the stored original values. In this way it is possible to study distinct details in our material without a new measurement step. The fig. 2 shows a result of such an investigation, in which the average values and the confidence limits are plotted by the computer. To avoid irregularities in our samples the computer program makes it possible to smooth the values by averaging over one, two or three fields.

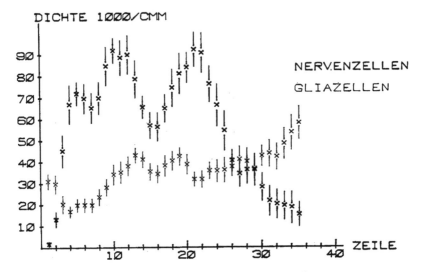

Fig.2: Computer diagram: Ordinate: Celldensity looo/mm$^3$ (p = o.o5). Abscisse: Distance from surface in counting fields.

This short example of the evaluation in the cortex cerebri shows the importance that every single value of one field should be stored. The reason is that in the beginning of a scientific evaluation it is not always known in which way the individual results should be arranged. The storing of the original values makes it possible to compute results based on other and newer arrangements of the values without a new measuring. This is, in my opinion, a further advantage of our system. In this way it compensates its slower evaluation speed with respect to the scanning devices.

Summary

Some factors of the image, especially low degrees of gray-intensities, lead to problems for full-automated scan-devices. It is therefore important to evaluate with the aid of an opto-manual intervention. Newly developed instruments do not work with scan-procedures, but are based on the visual diagnosis of structure and a manual transfer of the information to a full-automated system. By this way the human eye-brain-system can work like a highly-programmed scan-computer. Three of five optomanual systems are briefly described. Some more details are given for the TGZ 3 and the MOP II. Furthermore it will be pointed out that it is necessary to store the original values in order to calculate the different structural arrangements. Therefore it is possible to calculate certain topographical details without a new reevaluation. A computer-program has been developed for such investigations in tissues which are arranged in layers. As an example, investigations in the cortex cerebri of the man are reported upon.

References

1) Cowan, W. M. and Wann, D. F., J. Microsc. 99 (1973) 331 - 348.

2) Elias, H., Hennig, A. and Schwartz, D. E., Physiol. Rev. 51 (1971) 158 - 2oo.

3) Exner, H. E., in Newsletter 73 in Stereology, Kernforschungszentrum Karlsruhe KFK Ext. 6/37-2 (1973) 6 - 18.

4) Gahm, J., in Quantitative Analyse von Gefügen in Medizin, Biologie und Materialentwicklung, Ed. H. E. Exner, Dr. Riederer GmbH, Stuttgart (1975) 29 - 46.

5) Haug, H., Z. Anat. Entwickl.-Gesch. 118 (1955) 3o2 - 312.

. 6) Haug, H., J. Microsc. 95 (1972) 165 - 18o.

7) Haug, H., in Quantitative Analyse von Gefügen in Medizin, Biologie und Materialentwicklung , Ed. H. E. Exner, Dr. Riederer GmbH, Stuttgart (1975) 91 - 1o2.

8) Haug, H., Kebbel, J. und Wiedemeyer, G.-L., Microsc. Acta 71 (1971) 121 - 128.

9) Hennig, A., Bestimmung der Oberfläche beliebig geformter Körper mit besonderer Anwendung auf Körperhaufen im mikroskopischen Bereich. Mikroskopie Wien 11 (1956) 1 - 2o.

1o) Hougardy, H. P., in Quantitative Analyse von Gefügen in Medizin, Biologie und Materialentwicklung, Ed. H. E. Exner, Dr. Riederer GmbH, Stuttgart (1975) 73 - 8o.

11) Kluge, N., in Newsletter 73 in Stereology, Kernforschungszentrum Karlsruhe KFK Ext. 6/73-2 (1973) 29 - 39.

12) Moore, G.A., J. Microsc. 95 (1972) 1o5 - 118.

13) Müller, W., in Newsletter 73 in Stereology, Kernforschungszentrum Karlsruhe KFK Ext. 6/37-2 (1973) 19 - 28.

14) Ondracek, G., in Newsletter 71 in Stereology, Kernforschungszentrum Karlsruhe (1971), 1 - 1o9.

15) Schmeißer, H., in Quantitative Analyse von Gefügen in Medizin, Biologie und Materialentwicklung, Ed. H. E. Exner, Dr. Riederer GmbH., Stuttgart (1975) 6o - 72.

16) Underwood, E.E., Stereology and quantitative metallography, ASTM STP 5o4, American Society for Testing and Materials (1972) 3 - 38.

17) Underwood, E. E., Microscope 22 (1974) 69 - 1o9.

18) Underwood, E. E., in Quantitative Analyse von Gefügen in Medizin, Biologie und Materialentwicklung, Ed. H. E. Exner, Dr. Riederer. GmbH. Stuttgart (1975) 8 - 14.

19) Wann, D. F., Woolsey, T. A., Dierker, M. L. and Cowan, W. M., IEEE Trans. Biomed. Eng. BME 2o (1973) 233.

2o) Weibel, E. R., Stereology Proceedings of the Second International Congress for Stereology Chicago, Springer, New York (1967) 275 - 276.

21) Weibel, E. R. and Elias, H., in Quantitative Methods in Morphology, Ed. E. R. Weibel and H. Elias, Springer, Berlin, Heidelberg, New York (1967) a 3 - 19.

22) Weibel, E. R. and Elias, H., in Quantitative Methods in Morphology, Ed. E. R. Weibel and H. Elias, Springer, Berlin, Heidelberg, New York (1967) b 89 - 98.

*National Bureau of Standards Special Publication 431*
*Proceedings of the* FOURTH INTERNATIONAL CONGRESS FOR STEREOLOGY
*held at* NBS, Gaithersburg, Md., September 4-9, 1975 (Issued January 1976)

A USER-ORIENTED AUTOMATIC SCANNING OPTICAL MICROSCOPE

by J. E. Hilliard
*Department of Materials Science and Engineering, Northwestern University, Evanston, Illinois 60201, U.S.A.*

ABSTRACT

A "Digiscan" optical stage scanning microscope has been interfaced to a DEC PDP8/e computer. .Two important factors in the design of the instrument were flexibility and ease of operation. The latter has been achieved by making the instrument self-instructional and by the use of dialog with the user in order to eliminate the large number of controls that are normally necessary with an instrument of this complexity.

INTRODUCTION

The following were the design objectives for an automatic optical scanning microscope that has been assembled for use in the metallography facility at Northwestern University: (1) Maximum flexibility - it must be capable of being easily updated as new types of analyses and computational procedures are developed. (2) Ease of operation - since there will be several different and changing users it is important to avoid the need for lengthy instructions in the use of the instrument. (3) Must allow for visual discrimination in those cases where there is insufficient difference in contrast between constituents to permit automatic discrimination. (4) State of the art performance with respect to resolution and linearity.

It was decided that these requirements would best be met by a computer-controlled stage scanning microscope, SSM, rather than a TV scanning instrument. The use of a computer provides the required flexibility, since modifications in the functioning of the instrument require only changes in the programming and not in the hardware. Also, a computer permits the employment of certain self-instructional features that will be discussed later.

In an SSM, scanning is accomplished, as the name implies, by a motor driven stage; a photodetector is used to determine the light intensity at a small spot at the center of the field and, provided there is sufficient difference in contrast between the constituents, the output from the detector can be used for automatic discrimination. Alternatively, the user can signal an event (the crossing of a boundary, for example) by pressing one of five keys on the base of the microscope. Although it is theoretically possible to make closely spaced scans, this would be very time consuming. Thus, in practice the SSM is limited, when automatic discrimination is used, to measurements that can be made with a linear probe. This limitation is not too serious if, as in our case, one is primarily concerned with measurements on a section through a three-dimensional structure. Most of the properties of interest (volume fractions, boundary area per unit volume, size distribution, shape factors, etc.) can be obtained from linear probe data. [The situation is different when analyses are being performed on a structure that is essentially two-dimensional (dust particles on a slide, for example). In this case there are certain properties, such as the shape of individual particles, which may be of importance and which cannot be determined with a linear probe.] In other respects, an SSM compares favorably with image or TV scanning instruments: (1) All measurements are made on the optical axis of the microscope and therefore maximum resolution is achieved. Also, it is possible to operate with the field diaphragm almost completely closed, thereby reducing glare and enhancing the contrast. (2) There is no "shading" problem. For this reason, and because of the slower scanning speed, the gray level resolution is limited only by the stability of the photodetector and light source. (3) The scanning is inherently linear. (4) Because long traverses can be made, end effects are negligible or, at most, require a minor correction. (5) An SSM is easily operated in a visual discrimination mode when automatic discrimination is not possible.

DETAILS OF THE INSTRUMENT

The instrument comprises a Wild-Kontron "Digiscan" optical microscope which has been interfaced to a DEC PDP8/e computer. The stage is driven in the x and y directions by pulsed DC motors; the stage velocity can be varied over a range of 150:1 with a maximum of 1 mm per second. Even at maximum velocity under an oil immersion objective there is no stage vibration; thus measurements can be (and are) made with the stage in motion. (This is in contrast with some other SSM's in which the stage is stepped through a finite interval in between measurements.) A variable joystick is available for manual control of the stage. The response to this joystick is exponential so that precise control can be obtained at low velocities while still allowing for high-velocity slewing of the stage. Motion of the stage is monitored by optical encoders having a resolution of 0.5 μm. A solid-state photodetector is used and the stability of the system is good enough to resolve over 500 gray levels. The spatial resolution varies inversely as the power of the objective; for a 100X objective it is 0.9 μm for a 5 to 95 pct change in intensity across a sharp boundary.

Components of the computer include an 8-channel A/D converter (used for interfacing the joystick and photodetector), a programmable time clock, and a 12-channel digital input and output interface (the output is used for pulsing the stage motors and other purposes; the input side is interfaced to the optical encoders and the keys on the microscope). A dual magnetic "DECtape" unit permits rapid loading of programs and the storage of data. A video terminal is used for high-speed text display (900 characters per second) and a teletype is available when a hard copy of the output is required. The computer also has a hard-wired, high-speed (150,000 band) connection to a network maintained by the Department of Computer Sciences. This allows use of the network's magnetic disc and line printers and also permits direct transmission of data to the University's CDC 6400 computer whose processing is beyond the capabilities of the PDP8/e.

The DEC OS/8 operating system is used for the loading of programs and also for their editing and assembly. Software developments have included the development of a high level language FOCUS. This is an extensively modified version of FOCL/F written by D. E. Wrege which, in turn, is a variant of DEC FOCAL-8. (The latter is an interpretive language similar in form to BASIC. Unlike FORTRAN, the user's program does not have to be assembled; as a result, debugging is very much faster.) FOCUS contains the machine-language routines required to drive the stage, to monitor the optical encoders and photodetector, and to handle interrupts from the five microscope keys. Since FOCUS eliminates most of the need for machine-language programming, new programs for specific analyses can be prepared very rapidly even by persons who have had little previous experience in programming. It should also be added that the interfacing to the microscope does not interfere with the regular use of the computer. Thus, when the scanning microscope is not being used, a reasonably powerful computing system is available.

OPERATION OF THE INSTRUMENT

As noted in the introduction, one of the design objectives was simplicity of operation. This has been accomplished in two ways. First, insofar as possible the unit has been made self-instructional. Secondly, most of the information required by the instrument is elicited by dialog rather than having the user operate switches. (In fact, excluding the microscope keys and the switches on the video terminal, there are only two switches and one knob on the whole instrument that have to be set by the user. The teletype, for example, is turned on and off under program control.)

In order to start operation, the user enters a two word command at the keyboard. From then on all programs are chained in automatically as required. After the initial program has been loaded, it checks that the right magnetic tapes have been loaded and then runs a 3-second diagnostic which tests all the principal components and reports any malfunctions detected. The user is then asked "Do you need instructions for aligning the microscope (Y/N)". If the response is "Y" the user is queried with respect to the mode in which the microscope will be used (reflected or transmitted light, bright- or dark field). Step-by-step instructions are then given on the video terminal. The user advances to the next frame by giving a carriage return; the option is also available for backing up to any previous frame. At the completion of the alignment instructions (which occupy about 20 frames) a directory is displayed of the analyses currently available. The user enters his choice at the keyboard and the program is automatically loaded. As an illustration of these programs we will consider one that involves visual discrimination and another using automatic discrimination.

### Point Counting

This is for a volume fraction determination using visual discrimination. The user is first given the option of receiving instructions for performing the analysis. A request is then made for (a) Specimen identification (this is printed out with the results); (b) Power of the objective (this determines the spacing between counting positions); (c) Approximate area of specimen (used for the estimation of the minimum coefficient of variation, CV, that can be achieved; where CV is the ratio of the standard deviation to the average). All responses are checked and any which are apparently in error are queried (for example, an objective power outside the range of 10 - 100X). The user is then instructed to proceed. A 9-point grid inscribed on a reticule in a focusing eyepiece is used for counting. One of the microscope keys is used to signal an occupied point, other keys are used for advancing the stage to the next counting position, reversing direction in the x-direction and incrementing the stage in the y-direction; recording of 1/2 points and the deletion of erroneous entries. After 20 fields have been counted a buzzer sounds and the user is informed on the video terminal of: (a) The minimum CV that can be attained in the estimate of the volume fraction, assuming that the whole area available for analysis is sampled, and; (b) The counting time required to achieve this CV. These estimates are based on a theoretical estimate of the CV [Hilliard (1975)] and the time taken by the user to count the first 20 fields. The user can then enter any higher value for the CV and will be given the corresponding counting time. (The time estimates turn out to be surprisingly reliable; usually they are within 5 pct) If a higher CV than the minimum is selected, the option is provided of increasing the distance between counting positions so as to provide a uniform sampling of the specimen.

During the analysis, a current estimate of the CV is displayed on the terminal and this is updated after every 10 frames have been counted. At any time the user can obtain a summary of the results on the video terminal and, if desired, a full print-out on the teletype. The user can then continue the analysis or return to the directory of programs, one of which is the shutdown procedure.

### Intercept-Length Distribution

This program permits the determination of the distribution of intercept lengths in two different constituents. As before, the user has the option of receiving instructions for the analysis; these also include the steps required to adjust the balance control on the photodetector. The next step is to position the stage at various positions within the two constituents, the identity of which is signaled by one of the microscope keys. As each entry is made, the intensity is plotted with an identifying letter on the video terminal. After a sufficient number of entries, the user decides whether automatic discrimination is possible and, if so, enters lower and upper discrimination levels. (If, because of overlap on the gray scale, automatic discrimination is not possible, the option is provided of branching to a version of the program that allows visual discrimination.)

During the analysis the stage can be driven with the joystick or at a preset velocity by pressing one of the microscope keys. The intercept lengths in the two constituents are classified into 128 class intervals. (If necessary this number can easily be increased.) Initially, the width of the intervals is set at 0.5 μm giving an upper limit of 64 μm for an intercept length. If an intercept length longer than this is encountered, permission is requested to regroup the data. If this is given, adjacent class intervals are combined so as to produce 1 μm intervals with an upper limit of 128 μm. This regrouping procedure is repeated each time an intercept is encountered that exceeds the current upper limit. (The grouping of the intercepts is treated separately for the two constituents.)

At any time, a histogram of the results can be displayed either on the video terminal or on the teletype. For display purposes the distribution can be grouped into any number of intervals that can be expressed as a power of two. Provision is also made for deleting any entry that appears erroneous. In addition to the distribution of intercept lengths, the volume fractions and the boundary area between the two constituents are reported, together with the experimentally estimated standard deviations of these quantities. After the measurements have been completed, various programs can be called in for the analysis of the data. These include routines for unfolding the intercept distribution to determine sphere-size distributions, testing for a random spatial dispersion of particles, and the determination of shape factors. Also, as previously mentioned, the data can be transmitted directly to the CDC 6400 computer.

ACKNOWLEDGEMENTS

This instrument would not have been possible without the programming aid of several undergraduate students; I am particularly indebted to L. E. Suk, who developed FOCUS and was responsible for several of the operating programs.

REFERENCES

Hilliard, J. E. (1975) Assessment of sampling errors in stereological analyses. [This volume, Eq. (13).]

*National Bureau of Standards Special Publication 431*
*Proceedings of the* FOURTH INTERNATIONAL CONGRESS FOR STEREOLOGY
*held at* NBS, Gaithersburg, Md., September 4-9, 1975 (Issued January 1976)

THE DETERMINATION OF PARTICLE SHAPE AND SIZE DISTRIBUTIONS USING
AUTOMATIC IMAGE ANALYSIS TECHNIQUES

by Jeff Slater and Brian Ralph
*Department of Metallurgy and Materials Science, University of Cambridge, England*

1)   INTRODUCTION

One of the commonly occurring structures in materials science comprises a population of particles dispersed in a matrix. In general the particles within the population are of the same origin and chemical composition, consequently having similar three dimensional shape throughout, though there is a variation of particle size due to the kinetics controlling their formation.

Various workers (e.g. Hennig and Elias, 1963) have developed methods for the determination of particle shape and size distributions from manual measurements where particles belong to a simple analytical shape class. The development of this approach has been retarded by the amount of data required for the analysis.

The larger image analysing computers, currently available on the market, provide the means for the acquisition of the large amounts of feature specific data (i.e. parameters that may be related to each feature in a field of view) quickly and efficiently. The aim of this paper is to outline a general method for the classification of a population of particles, by size and shape, using data from automatic instruments.

2)   THE MATHEMATICAL APPROACH

Before a solution to the particle sizing problem can be achieved, the following assumptions must be made:

i)   The volume size distribution of the particles contains particles of discrete sizes;

ii)   The observed distribution of the particles on the section may be classified as the number of particles in fixed size intervals.

The relationship between true and observed size distributions then becomes:

$$N_{Vj}\cdot(P_{ij}) = N_{Ai} \qquad (1)$$

where the probability matrix $P_{ij}$ relates the number of particles of class j to the observed number of features in class i. For three dimensional shapes that may be expressed analytically, the probability matrix can be calculated exactly and the equation may then be inverted to give the relationship between observed and true sizes. This is the method used by Saltykov (1958), Hyam and Nutting (1956) and others.

This method has been extended to cover the case of particles of non-analytical shapes (Schwartz and Ralph (1969), Schwartz (1972)). The particles in space are described by a faceted model which is then sectioned randomly in a computer to examine the probability of attaining a section of size i from a particle of size j and an equation of type (1) is assembled. However, due to the fact that it is only possible to make a finite number of random sections, there are inherent errors in the calculation of the probability matrix. In this case the final errors are smaller if the equation is not inverted but an iterative procedure is used.

Traditionally the model shape has been determined by serial sectioning or by making an intuitive estimate from observations on the section; or has been assumed to be spherical so that the simplest sizing procedures could be followed. Elias (1963) and Myers (1962) independently developed a particle shape determination method in parallel with the sizing procedures where by a three dimensional particle of known shape was related to a distribution of observed shapes on a section by a probability matrix. Both of these methods were developed before the advent of large image analysing computers and were therefore tailored to manual measuring methods for specific model particle shapes; both excluded the relationship between feature size and the measured shape.

177

When a non-simple particle shape is being evaluated the form factor (e.g. Fischmeister, 1974) being used must define the shape of the feature on the section as well as possible. With an automatic image analysing computer, like the Quantimet 720, the feature specific parameters of interest are: Area, Perimeter, Projections and Feret Diameters at angles of $0^{\circ}$, $45^{\circ}$, $90^{\circ}$. and $135^{\circ}$ to the television scan direction. To involve all of these parameters in a single form factor would result in an unwieldy relationship. It is instructive to use the available parameters to develop two independent form factors for each feature such that the meaning of each form factor can be interpreted in terms of an equation:-

$$S^3 . (Q_{nm}) = (S^2_{nm}) \qquad (2)$$

The measured distribution $(S^2_{nm})$ is then represented not as a curve but a surface defined by the number of features having form factors in the limits $S^2_n$ and $S^2_m$ (this is simply the three dimensional analogue of a histogram); for ease of presentation the surface is best shown as a contour map. In principle the shape of the particles in the population can then be calculated by setting up a model of $S^3$ in a computer program then random sectioning to compute a value for $S^2_{nm}$ for comparison with the measured $S_{nm}$. The best fit value of $S^3$ can then be obtained by following an iterative procedure.

## 3) EXPERIMENTAL AND COMPUTING PROCEDURES

a) *Experimental*

The required feature specific experimental data was collected using a Quantimet 720 image analysing computer with pattern recognition, a Calculator Field/Feature Interface and a 2D Automatic Detector. A test specimen was constructed by accurately cutting several thousand pieces of copper wire such that they all had the same length to diameter ratio. These were then embedded in cold mounting medium as randomly as possible. The amount containing the rods was then cut into arbitrary pieces which were then remounted as randomly as conditions would permit.

The parameters measured for each feature were Area, Perimeter and the four Feret Diameters (at $0^{\circ}$, $45^{\circ}$, $90^{\circ}$ and $135^{\circ}$) to the T.V. scan direction). This data was accumulated in a Hewlett Packard 9830 programmable calculator and the values of $S^2_{nm}$ were calculated directly.

b) *Computing*

The computed values of the $S^2_{nm}$ array were determined using an IBM 370/165 computer and the IBM PL/1 compilers. A particle of a given three dimensional shape $(S^3)$ was set up in the program using a many faceted polyhedral approximation, the particle being described by the Cartesian co-ordinates of its vertices and a look-up table to define which vertices were connected by edges. The particle was randomly sectioned a large number of times and the co-ordinates of the vertices of each generated feature were stored. The computed data at this point represents the expected distribution of features produced by a section through a dispersion of particles all of the same size and shape. The parameters required for the generation of the form factors were then computed and the computed $S^2_{nm}$ array assembled.

## 4) RESULTS AND DISCUSSION

In taking the faceted computer model approach to particle shape and size determination, one of the first points to be considered is the number of facets on the model required to represent the model shape adequately. It was found that in the general case a 48 sided model was sufficient; though in special cases much simpler models could be used. In simple situations where the actual particle shape is faceted and contains a low number ($\sim$12) of facets the actual shape determination procedure reverts to a simpler situation. In such cases

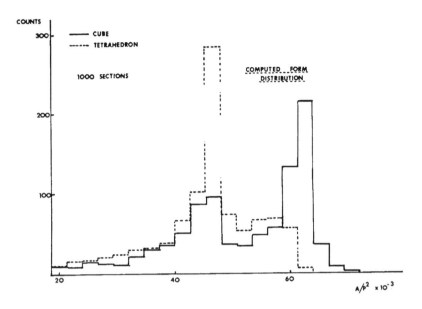

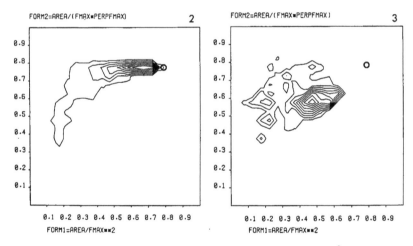

FIGURE 1.   Frequency of occurrence of form factor (Area/Perimeter$^2$) for random
            sections of cubes and tetrahedra.
FIGURE 2.   Computed form factor map for random sections of 6:1 aspect ratio
            rods.
FIGURE 3.   Measured form factor map for random sections of 6:1 aspect ratio
            rods.

179

the distribution of only a single form factor is required. Figure 1 shows
unique computed distributions of $A/P^2$ (Area/Perimeter$^2$) for cubes and tetrahedra
where the principle of identification of a particle shape from a form factor
distribution curve can be seen readily.

Figure 2 shows the computed form factor map for a distribution of rods of
aspect ratio 6:1. As is expected the majority of the generated features have
form factors close to the limiting case of a circle, (the point marked 'O' on
the figure corresponds to a circular section). In the corresponding
experimental data, presented in figure 3, the expected circular sections have
been displaced towards the origin. This is thought to be the compound effect
of resolution errors in the input peripheral and errors in the measurement of
the constituent parameters of the form factors.

Any particle of finite size when sectioned will produce sections varying
in size from an upper limit to zero. As the size of the feature defined on the
section approaches zero it becomes much harder to define the shape of the
feature as the relative errors in the measuring process increase. This effect
is of particular importance when automatic instrumentation is used to make the
measurements where analogue to digital conversion of the image takes place. The
computer program that generates $S_{nm}$ must therefore contain the same digital
logic that is used by the image analysing instrument for the detection and
measurement of the computed 'image' of the features. Once this digital logic
has been included comparisons can be made between the measured and computed
values of $S_{nm}^2$.

It must be emphasised that once the instrument digital logic has been
included in the shape determination procedure the apparent shape of the features
on the section is inter-related with the size as well as the shape of the
particles in the population. It is not possible therefore, to measure the size
distribution without knowledge of the shape distribution; the converse is
also true but for different reasons:

i) Particle shape influences the values of the elements in the probability
matrix $P_{ij}$ in equation (1), therefore the particle shape must be known
before particle sizing can be accomplished.

ii) The particle sizes present determine the image magnification used in a
Quantimet type experiment thus influencing the relative spacing of the
'picture point grid' and hence the shape measurements with respect
to the particle size.

It is not currently feasible to write a computer program to solve the
interrelated problems of size and shape distributions simultaneously, largely
because of the enormous amount of computer time that would be required by such
a program. The problem of particle shape determination must therefore be solved
by scaling the 'picture point grid' in the computer model so that it has the
same correspondence with the mean feature size (computed) as does the actual
picture point grid with the mean feature size (measured), the mean feature size
being a readily measurable parameter. The iterative sizing procedures may then
be utilised.

REFERENCES
Fischmeister, H.F. (1974) Shape Factors in Quantitative Microscopy; Z.
Metallkde, 65, 558.
Hennig, A. and Elias, H. (1963) Theoretical and Experimental Investigations on
Sections of Rotary Ellipsoids; Z. wiss Mikroskopie, 64, 133.
Hyam, E.D. and Nutting, J. (1956) The tempering of Plain Carbon Steel; J. Iron
and Steel Inst., 184, 148.
Myers, E.J. (1962) Quantitative Metallography of Cylinders, Cubes and Other
Polyhedrons. Ph.D. Thesis, University of Michigan.
Saltykov, S.A. (1958) Stereometric Metallography, Moscow.
Schwartz, D.M. (1972) A General Method for Calculating Particle Size
Distributions; J. Microsc., 96, 25.
Schwartz, D.M. and Ralph, B. (1969) The Analysis of Particle Size Distributions
from Field Ion Microscope Data, Phil. Mag., 19, 1061.

*National Bureau of Standards Special Publication 431*
*Proceedings of the* FOURTH INTERNATIONAL CONGRESS FOR STEREOLOGY
*held at* NBS, Gaithersburg, Md., September 4-9, 1975 (Issued January 1976)

SOME SHAPE FUNCTIONS FOR USE IN AUTOMATIC IMAGE ANALYSERS FOR THE CLASSIFICATION OF
PARTICULATE MATERIALS

by Adrian C. Terrell* and R. J. Willes**
*Image Analysing Computers Ltd., Melbourn, England
**Image Analysing Computers Inc., New York, U.S.A.

INTRODUCTION

Television based Image Analysers are now able to measure 2 Dimensional image
features *separately* at very high speed. The various available measurements can
be combined to form the most appropriate *size* parameter for a particular
application. Dimensionless form functions may also be derived from the measurements so that a *shape* distribution can be used to characterise a material or,
alternatively, a single shape class can be chosen to pick out a particular
population of features so that its number/density, location and/or size
distribution can be obtained separately. The choice of *form function* depends
not only on its theoretical ability to recognise the geometric, densitometric or
texture characteristics of the required features but also on the precision with
which it can be measured; given the available algorithms of the hardware and
software employed and the signal to noise ratio of the optics and electronics
used to form, scan and digitise the image.

AVAILABLE MEASUREMENTS

Some typical *feature* measurements are shown in fig 1. Many other feature
measurements are available from specially equipped machines; for example the
features may be classified by integrated brightness or by optical density. The
number of *sub features* within the main feature may be used as a classification
parameter. The sub features as well as the main feature, may be subject to
size and shape tests. 2 Dimensional Amendment can be used to generate new
features representing groups of image features by agglomeration in 2 dimensions.
The lobes of interconnected features may be separated so that they can be
treated individually.

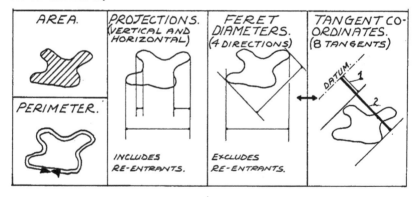

Figure 1. Quantimet 720, some feature measurements.

SPEED

The Authors are applying two distinct image analysis systems to a wide
variety of applications problems. The first system uses a Quantimet 720 interfaced to a Hewlett Packard 9830 calculator so that *either* the results of
size/shape distributions made in the Quantimet can be transferred to the
calculator *or* the individual feature measurements can be transferred for more

181

sophisticated size and shape tests. Using the former operating mode, particles might be classified into 14 size classes with selection by a simple shape test at a speed of about 4 seconds per microscope field. In a typical powder problem this would represent some 40,000 particles per hour. The speed is slightly slower at the highest optical microscope magnifications because of the time required for the autofocus mechanism to operate. In the feature data mode the system transfers all the required measurements from one feature at a time. Much more sophisticated shape tests may be used and the size distribution can be in any number of classes but the speed is reduced to some 5,000 particles per hour.

The second system uses a fast *Direct Memory Access* method to transfer the data to the core of a PDP 11 computer. Four separate measurements plus co-ordinates of all the features in a field are transferred in one scan (about 1/10th second). Overall speeds of some 400,000 features per hour are attainable.

The hardwired classification, by size and/or shape, is frequently used together with the calculator or computer methods. In the case of the calculator this is to save time by skipping the 'out of range' features from the feature sequencing. In the case of the computer method it is to save memory by allowing smaller arrays to be dimensioned for the feature data.

CLASSIFICATION PARAMETERS

For specialised applications the Quantimet can be programmed to measure only the required parameters. For general feature size and shape distributions it is useful to derive a number of non directional classification parameters from the raw feature data. It is convenient to express all of these parameters in the same dimensions and with a constant of proportionality for each one chosen to make them mutually comparable; for example if the parameters are calculated as *diameters* then we might use the following selection.

1. area diameter - diameter of circle having same area as feature
2. perimeter diameter - diameter of circle having same perimeter as feature.
3. longest dimension - max feret diameter
4. mean feret diameter - a good approximation to the diameter of the circle having the same convex perimeter as the feature
5. convex area diameter - the diameter of the circle having area equal to the convex area of the feature, derivable from the 8 tangent co-ordinates
6. mean width - equal to twice the area divided by the perimeter, not an independent function as it can be derived from 1 and 2 but useful for many classifications

Minimum feret diameter would be a useful indication of the width of the envelope for elongated features but the feret diameter would have to be sampled at, say, 16 directions to give a reliable value.

FORM FUNCTIONS

The advantage of arranging the basic size measurements as a set of comparable *diameters* is that any pair of them can easily be ratioed to give a dimensionless form function which has a value of 1 for a circular feature. The six chosen parameters give rise to 13 separate form functions (excluding those which are equivalent to, or reciprocals of, each other). It is not easy to visualise the geometric meaning of each one of them in a general sense but their meanings are quite clear in respect of some limited ranges of allowable shapes. The selection of the best form functions for any particular image analysis problem depends not only on the geometric interpretation of the function in terms of the expected feature shapes but also on the precision with which the function can be determined for the range of shapes, sizes and orientations of features which will be found.

For our general purposes feature data programs, used to characterise a wide range of particulate materials; we have found it useful to allow for two classifications, each of which applies a series of limits to *either* one

of the *classification parameters,* as a *size* test, or one of the *form functions*
as a shape test. If we choose a single bin (two limits) for one of the
classifications then we have a selection, usually by shape, and a
classification, usually by size but the same program will give a matrix
classification when both classifications have more than one bin.

SHAPE DISCRIMINATION

We can predict the variations in the measured values of form functions,
for any particular size and shape of feature from a knowledge of the relevant
system properties: Detection and Digitisation errors have greatest effect on
small features. Large features are affected by detection errors if they have
narrow elements as occur in images of thin fibres. Detection errors are
largely eliminated by the latest auto delineating detectors which establish
local feature boundary positions by using 2 Dimensional *look around* logic
to establish the half amplitude point of the video signal across the feature
boundary, independently of the angle which the scan direction makes with the
local boundary. (Wadlow et al, 1972) Algorithm errors can give rise to small
variations in some parameters with feature orientation. These errors occur
with perimeter and diagonal feret diameter measurements for a few feature
shapes which happen to lie in unfavourable orientations.

As 'the proof of the pudding is the eating thereof' we decided to compare
some of our predictions with measurements made on a few feature shapes. A
single feature was measured many times, typically 100 times, and the mean
and standard deviation of values for the 6 basic classification parameters and
some Form Functions considered useful for the shapes concerned were printed
out. The measurements were repeated with the features being moved small
distances to check digitisation errors, and larger distances to check
distortion errors and finally with randomly changed orientation to check
algorithm errors. We also repeated the measurements for variously sized
features having the same shape. (Note that the variations in a form function
value cannot be predicted for a knowledge of the standard deviations of the
parameters from which it is derived because there is often an unknown
correlation between them) A lot of information is still to be gleaned from
the data but some conclusions can be drawn at once.

PRELIMINARY CONCLUSIONS

In one experiment we picked a family of elliptical features having the
smooth sharp edges and covering the eccentricity range from one (circle) to
10. From the results we estimated the number of shapes which could be discriminated within the eccentricity range; a) with constant orientation b) with
random orientation. (The discrimination criterion used was that the widths of
the shape classes should accept the full range of form function values found in
a set of 100 measurements.) The rapid change in discrimination with feature
size reinforced our view that any image analysis involving shape as well as
size tests on features should be confined to features larger than a lower
limit which is considerably higher than that usable for ordinary size
distributions. The printout can be made to show the overlapping parts of the
distribution separately or to merge the curves by taking the high magnification
figure only when a statistically significant count was achieved.

The second tentative conclusion is that the best form function for separating the family of shapes concerned will often not be the same one that is best
for eliminating other features not belonging to the same family of shapes. For
this reason we now sometimes prefer to store the basic parameters for all the
features on cassette or disc data files so that different shape tests can be
applied to the original data in subsequent experiments without having to
repeat the microscopy. With the high speed computer interfaced Quantimet we
can apply a series of different tests to the features from one field,
displaying the results of the tests on the Quantimet screen against each feature.
The sophisticated display facilities with this instrumentation greatly aid
the choice of form function and the selection of separation limits to be
applied to it. Once a classification regime has been established it might be
preferable to set up a working system using the less expensive calculator
system which does not have the computer generated display facilities. In

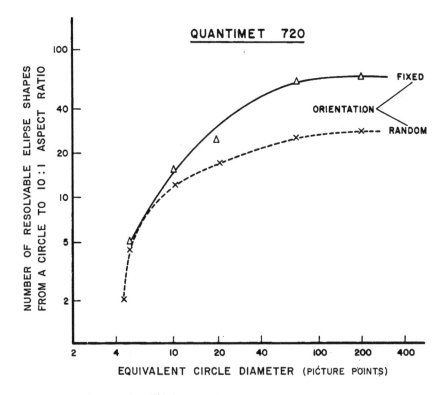

Figure 2. Quantimet 720 image analysis of a family of elliptical features.

other cases the computer system might be required for use with the developed application because of its inherently faster operating speed.

STEREOLOGICAL IMPLICATIONS

The shape tests we have described are limited to 2 Dimensions. We have reached the stage that we can now generate fast reasonably accurate feature specific data on 2 dimensional image features. It remains to see if this data can be used to derive reliable 3 dimensional information in each of a variety of applications.

REFERENCE

WADLOW, D.E., HOPKINS, B.M., GARDNER, G.M., and FISHER, C., "New Detection Systems for the Quantimet 720", The Microscope, 20, 183-202, (1972)

National Bureau of Standards Special Publication 431
Proceedings of the FOURTH INTERNATIONAL CONGRESS FOR STEREOLOGY
held at NBS, Gaithersburg, Md., September 4-9, 1975 (Issued January 1976)

DETERMINATION OF PARTICLE SIZE AND SHAPE DISTRIBUTION BY AUTOMATIC FEATURE ANALYSIS

by J. R. Schopper

Institut für Geophysik der TU Clausthal, D-3392 Clausthal-Zellerfeld, Germany

SUMMARY

Particle type, shape and size analysis of mixtures of multi-shape multi-size particles of various geometrical types, from section images is possible in principle, since a sufficiently fast computer method is available for a thorough evaluation of size and shape of section features. Such features can be classified according to their origin from certain type particles, and a statistical analysis of distinct geometrical feature parameters can be used for a stereological transformation into space, resulting in the number per volume of the different type particles and the mean and variance of their size and shape parameters. Even some topological information on cell structures or networks might be gained from such a section image analysis. - As an example, ways for a morphometric and granulometric analysis of a mixture of approximate balls, prolate spheroids, cylinders, cubes and tetrahedrons of non-uniform shape, size and orientation are indicated, final solutions yet pending.

1. INTRODUCTION

There is more information about spatial structures contained in a section image than can be extracted by point and lineal analysis. There has been considerable theoretical work already in the past, about shape and size relations between structures in space and in sections, but the first one to recognize the practical need for an extensive individual section feature analysis, to recommend computer processing, and to develop himself a working - though yet slow - computer program, has been George Moore (Moore & Wyman, 1963; Moore, 1964, 1966).

The development of a sufficiently fast computer procedure for a most complete statistical analysis of the geometry of individual section features (Rink, 1970), enables stereology now to draw more efficiently on the complex statistical signal from space, a section image represents. Thus this paper deals with the problem of particle size and shape analysis, of agglomerates of different type particles, of different sizes, and of different shapes (consistent with each type), using geometrical information from section images that can be automatically obtained by such an image analysis program.

The here considered collection of model particles of rather regular geometry is very well suited for approximating practically occurring real particles, so far as they are singly connected and do not have extreme concavities. However, in the present preliminary stage of work, it really is the intention of this paper to discuss ways and means rather than to present ready solutions.

2. QUALITATIVE EVALUATION OF FEATURE SHAPES IN SECTIONS

Already a common sense approach can clearly show the vast amount of morphometric information available from a section image. Each particle shape in space results in some characteristic feature shapes in sections, and by the selection of most significant feature shapes, much can be said about the presence or absence of particles of various shapes. Such 'geometrical filtering' of the section image can also be done automatically by computer.

Of course, as the section features become more complex and diversified, more uncertainty is involved in their interpretation. How-

ever, little outside information can greatly reduce this uncertainty in many practical cases.

## 3. QUANTITATIVE EVALUATION OF FEATURE SHAPES IN SECTIONS

With the aid of computer processing, the previously discussed
approach can also be quantified. The frequency of appearence of various characteristic shapes or shape groups of intersects tells
about the frequency of certain type particles in space. Eventually
this information can be used for a complete particle type, shape
and size analysis, as will be shown below.

Myers (1964) lists the probabilities for the intersects of various polyhedrons being polygons of certain numbers of corners. Thus
an analysis of an agglomerate of mixed one-size polyhedrons, resulting in the frequencies $N_V$ of the different polyhedrons, is possible by counting the frequencies $N_A$ of the different polygons in the
section plane. Note that this method is size independent in principle, and can be adapted to mixtures of multi-type multi-size
polyhedrons. Even certain type-consistent shape variations can be
permitted.

In a similar way other types of particles can be treated. Quite
a number of authors investigated the geometrical probabilities of
sections through ellipsoids and cylinders, and there is no fundamental problem in expanding those findings to cones etc.

## 4. PARTICLE TYPE, SHAPE AND SIZE ANALYSIS

From the well-known fact that a particle size determination from
sections requires the knowledge of the particle shape, it follows
that shape analysis must always precede the size analysis. But
while the measure of size is a single number, shape requires a - generally large - set of numbers, and a really general mathematical
expression for shape can be very complex and impractical. To simplify the shape description, the occurring particles should be pre-
classified - exactly or approximately - into particle 'types', represented by basic geometrical bodies as cylinders, spheroids, etc.
Shape - within each class - then can be expressed by one or a few
numbers, e.g. an axial ratio.

A rigorous definition of size should be independent of type and
shape and thus based on the particle volume. The cubic root of the
volume or the radius of a volume-equivalent sphere would be such a
proper measure of size. However, it is often handier to use other
size parameters, differently for each type; always some characteristic length like e.g. a main axis. This latter way is preferred
here for practical reasons.

The determination of the complete size distribution, with many
particle types of variable shape present, is a very complex problem, and a mathematically rigorous attack - as that of Wicksell
(1925) - becomes quite involved and requires the solution of integral equations. However, knowledge of the complete distribution is
seldom desired. By means of a binomial or Taylor development, any
distribution can be fully expressed by its arithmetic mean and an
infinite series of moments about the mean. But there is hardly ever
a need in practice for extending it further than to the fourth order. In most cases the mean and the variance will suffice, and this
is presumed here in this paper.

The problem of particle type, shape and size analysis thus is
reduced to the determination of the types, and the mean and variance of proper shape and size parameters. This results in a much
easier solution. But there still remains the problem of untangling
the intertwined effects of size and shape variation. That is why
most publications dealing with particle size analysis assume con-
stant-shape particles, or at least such of the same type. Here,
ways for an analysis without such limitations will be shown by an
example of a mixture of balls, prolate spheroids, prolate cylinders

of varying size and shape and of regular tetrahedrons and cubes of
varying size; with type, shape and size assumed statistically independent.

## 5. EXAMPLE FOR QUANTITATIVE ANALYSIS OF AN AGGLOMERATE

The aforesaid model mixture results in triangles, quadrangles,
pentagons, hexagons, circles, ellipses and truñkated ellipses in
the section image. For its analysis, Fullman (1953) could be followed. However, instead of intersect and intercept, other section
feature parameters·are often used with more advantage, and from
their mean and mean square, mean and variance of particle size and
shape quantities can be derived.

### 5.1. ANALYSIS FOR CYLINDERS

The minor axes of elliptical sections of prolate cylinders always equal their diameter. Symmetrically truncated ellipses, necessarily belonging to cylinders, provide those minor axes too, and
also tell about the cylinder length. A proper mean and mean square
analysis then can yield mean and variance of cylinder radius·and
length. Finally, the total number of cylinders derives from a count
of some particular or all of the truncated ellipses.

### 5.2. ANALYSIS FOR POLYHEDRONS

Hexagons and pentagons in the section result from cubes only.
Quadrangles might result from cubes or tetrahedrons or - less probable, and valid for rectangles only - from cylinders. Triangles
might result from cubes or tetrahedrons. If necessary, the rectangles resulting from cylinders can be calculated from the information of the previous subsection (5.1.) and excluded prior to, or
during analyzing the set of polygons.

With Myers's (1964) probabilities for an i-sided polygon resulting from a k-faced polyhedron, the relative numbers of cubes and
tetrahedrons can be obtained. Subsequently the mean and variance of
the size of either type polyhedron and their number per volume can
be determined in various ways.

### 5.3. ANALYSIS FOR SPHEROIDS, INCLUDING BALLS

Ellipses in the section result from cylinders and spheroids. The
part from cylinders can be calculated according to the data of subsection 5.1. and eliminated from the analysis for spheroids.

For prolate spheroids, the minor axes $e_1$ of section ellipses represent intercepts in planes normal to the rotational axis, while
the major axes $e_2$ are always intercepts of the generating ellipse.
The circular intersects are oriented normal to the rotational axis.

Thus the mean of the elliptical axes and the circular areas
yield mean and variance of the radial axis and the mean of the rotational axis of the spheroids, rendering the over-all mean intercept and intersect redundant. A variance analysis of $e_2$ could provide an estimate of the variance of the rotational axis too. With
mean shape known, the number per volume can be derived.- If desired,
balls can be treated separately after elimination of the circular
intersects pertaining to the non-spherical objects.

### 5.4. GENERAL BALANCE AND CHECK FOR MODEL FIDELITY

There are more criteria for double-checking the results obtained
so far, and possibly improving them iteratively. Namely the volume
fraction, specific surface and number per volume must balance.

Such a check will also inform about the applicability of the
assumed model agglomerate. On the other hand, larger deviations
from the model will already show up in the section image by particularly shaped features and might be semi-quantitatively assessed
therefrom.

Nevertheless, often not so great an accuracy·at all is required

in practice, so that the application of the theoretical concept discussed here to a complete morphometric image analysis promises considerable merits for practical particle size and shape analysis.

## 6. AIDS FOR DETERMINING TOPOLOGICAL PROPERTIES

It has been repeatedly stressed, especially in various papers of
DeHoff and Rhines (cf.1968) that complete topological information
about completely irregular spatial structures cannot be gained from
single sections, but by serial sectioning only. However, a complete
knowledge of the topology of a structure is not always required.
Into the hydraulic permeability of a porous material e.g., just a
single topological constant enters, namely the ratio of the number
of nodes to the number of branches of the pore channel network
(Schopper, 1972).

Furthermore, often simple shapes can be assigned (truly or approximately) to the structural elements, thus permitting the application of a morphometric principle to the problem of finding topological numbers per unit volume from similar topological numbers
per unit area.

Consider a network of channels, topologically to be viewed as a
network of branches; or, consider a granular pack, that can be described topologically as a system of space-filling polyhedral cells.
Both those structures can be thought topologically identical by
thinking of the branches as being edges of polyhedrons.

In the section plane, a plane network of polygonal meshes results, of which the corners (nodes) represent the piercing points
of the spatial branches and the sides (branches) are the traces of
the polyhedron faces.

Thus Myers's (1964) probability coefficients can be applied for
calculating the frequency of different type polyhedrons occurring
in the cellular structure and for finally arriving at a 'mean cell
polyhedron', the number of cells per unit volume, and the mean
cell size.

If the common assumption is made, that two cells meet in a face,
three cells in an edge and four cells in a corner, immediately the
number of faces, edges and corners per unit volume is known too,
and e.g. the above hydraulically interesting topological constant
can be calculated.

However, such an assumption reduces generality unnecessarily,
since just the meeting of two cells in a face is obviously a general truth for the considered structure. Even without such an assumption, the number $N_V$ of branches (edges) per unit volume can be determined independently by counting the number of piercing points
$N_A$ per unit area on the section according to $N_V = 2N_A/\bar{l}$ and deriving the mean edge length $\bar{l}$ from the mean polyhedron already calculated. Then the numbers per unit volume of cells, faces and edges
(branches) are known and that of corners (nodes) follows from
Euler's polyhedron theorem, as does the wanted constant.

References

DeHoff,R.T. & Rhines,F.N.(1968) Quantita-
tive Microscopy. McGraw-Hill, New York.
Fullman,R.L.(1953) Measurement of Particle
Size in Opaque Bodies. Trans.AIME 197,447
Moore,G.A. & Wyman,L.L.(1963) Quantitative
Metallography with a Digital Computer.
J.Res.NBS 67A,127
Moore,G.A.(1964) Direct Quantitative Ana-
lysis of Photomicrographs by a Digital
Computer. Photogr.Sci.Engng. 8,152
Moore,G.A.(1966) Application of Computers
to Quantitative Analysis of Microstruc-
tures. In: Ceramic Microstructures,

Wiley & Son, New York.
Myers,E.J.(1964) Sectioning of Polyhedrons.
1st Congr. Stereology, Vienna, 15
Rink,M.(1970) Automatische morphometrische
Bildanalyse mit Hilfe eines elektronischen
Digitalrechners. Doct. Diss., Clausthal.
Schopper,J.R.(1972) Theoretische Untersu-
chung elektrischer, hydraulischer und an-
derer physikalischer Eigenschaften poröser
Gesteine mit Hilfe statistischer Netzwerk-
modelle. Habilitationsschrift, Clausthal.
Wicksell,S,D.(1925) The Corpuscle Problem.
Biometrica 17,84

*National Bureau of Standards Special Publication 431*
*Proceedings of the* FOURTH INTERNATIONAL CONGRESS FOR STEREOLOGY
*held at* NBS, Gaithersburg, Md., September 4-9, 1975 (Issued January 1976)

STEREOLOGY AND THE AUTOMATIC LINEAR ANALYSIS OF MINERALOGICAL MATERIALS

by M. P. Jones and G. Barbery
*Mineral Technology Department, Imperial College, London, England*

Abstract.

The authors have designed an automatic method of linear analysis based on an electron probe X-ray microanalyser. The linear data that are produced are stereologically transformed by a method of moments and can be used to determine volumetric grain size, grain shape and other parameters.

The numerical equivalence of point-, length-, area-, and volume-proportions in modal analyses have been known and used by mineralogists for a long time. The application of these and other stereological relationships to mineralogical measurements has been limited by the tedium of collecting the necessary data and by the inconsistencies of the results that are obtained by manual measuring methods. Consequently, a number of attempts have been made to develop automatic methods of measurement and these include a linear-measuring system that has been devised by the authors. The measuring equipment is based on a Geoscan electron probe X-ray microanalyser and this device is controlled by a small dedicated computer. The Geoscan was selected because of its excellent ability to discriminate between mineral phases; an ability that is achieved by simultaneously using some of the many characteristic X-ray signals that are produced when a specimen is bombarded by a narrow beam of high energy electrons. This phase discriminating ability of the Geoscan (Jones and Shaw, 1973) is at least as good (and is often much better) than that achieved by the optically-based, automatic, linear, image analysers that are commercially available.

In the Geoscan a suitably prepared specimen is moved from point to point under the electron beam by computer-controlled stepper motors. The 'points' are about 2 μm in diameter and overlap and thus provide a means of carrying out a linear analysis. The X-rays (and, if necessary, other signals) from the specimen are fed into the control computer. The X-ray signals are characteristic of the elements and of the proportions of those elements that occur in the small volume of specimen (about $10^{-18}$ m$^3$) being irradiated by the electron beam. These signals are used to identify the mineral phases that pass under the stationary electron beam. During such a traverse the instrument measures only a single mineral phase (although programmes are being prepared that may enable 5 or 6 phases to be measured during a single traverse) and the following details are collected:

> T = total traverse length; N = number of grains of the selected mineral; $\Sigma(L)$ = total intercept length on the selected mineral; $\Sigma(L^n)$ = the sum of the n$^{th}$ power of each individual intercept, where n is an integer between 1 and 4; A/B = the number of contacts between minerals A and B.

In addition, each intercept is categorised, either in arithmetic or geometric progression, and listed in the form of histogram data. The results form the basis of a series of stereological transformations that provide information on grain size distributions, grain shape coefficients, specific surface areas, etc.

Grain Size.

The size of a simple geometrical form such as a sphere can be clearly defined by a single value 'd', its diameter. However, since it is difficult

to define the shape of an irregular particle it also becomes difficult to
define its size. The most commonly used method for determining the size
distribution of 'free' mineral grains is to pass the grains through a series
of apertures of defined shape and size (i.e. by sieving). This procedure
involves an undetermined grain shape factor and, therefore, the 'size' is only
poorly defined. The size of 'locked' grains cannot, in any case, be determined
by this means.

The method of random intercept length measurement allows the shapes of all
convex grains to be mathematically defined. It is then possible to transform
the linear intercepts derived from grains of unknown size (whether 'free' or
'locked') into the true three-dimensional size distribution (or equivalent
screen size) of those grains. The basic equation for the transformation was
derived by Barbery (1974) and the values required can be established by linear-
measuring methods.

$$\mu_n' (D_c) = \frac{K_1}{K_{n+1}} \quad \frac{\mu'_{(n+1)}(L)}{\mu_n' (L)}$$

where $D_c$ is the three-dimensional, 'convenient' size (usually the screen size);
$\mu_n'(L)$ is the $n^{th}$ moment (from the origin) of the random chord length distribution; and $K_n$ is the $n^{th}$ order shape factor determined by linear measurement of
uniformly-sized grains of statistically uniform shape.

The first 3 moments of $(D_c)$ can be calculated from the first 4 moments of the
intercept lengths and a variety of methods can be used to establish estimates
of the actual values of the distribution of grain size. For example, the incomplete beta function can be used to determine estimates of the cumulative
fraction that forms size $(D_c)$. (Jones and Barbery, 1975).

Other Measurements

Linear measurements provide sufficient data to calculate volume proportions,
specific surface areas, particle densities, mean free distance between particles,
etc. Thus,

a) Volumetric proportion is equivalent to $\frac{\Sigma L}{T}$ (see Table 1).

b) Specific surface area, $S_v = \frac{4}{\mu_1'(L)} = \frac{4N}{\Sigma(L)}$ cm$^2$/cm$^3$

c) Interparticle distance, $\lambda = \frac{(1 - V_v)T}{N}$. $\lambda$ is a shape-independent

parameter and is the uninterrupted inter-particle distance averaged over all
possible pairs of particles in a specimen.

d) Particle density, $N_v = \frac{B(p - 3,q)}{B(p.q)} \cdot \frac{3 K_1}{\pi K_4} \cdot \frac{1}{D_{c_{max}}^3} \cdot V_v$

where p and q = are parameters of the incomplete beta function
$B(p,q)$ = beta function
$K_n$ = $n^{th}$ order shape factors
$D_{c_{max}}$ = maximum particle size - as defined by screening
$V_v$ = proportion by volume of the particles.

## TABLE 1.

__Linear Measurement of Shape Coefficients etc. for Monodisperse Beach Sand Zircon Grains__

(after Simovic, 1973)

| | | |
|---|---|---|
| T | = total traverse length across specimen | 684,000 μm |
| ΣL | = total length of intercept on zircon grains | 33,914 μm |
| $\Sigma(L^2)=$ | | 3,027,656 μm$^2$ |
| $\Sigma(L^3)=$ | | 312,817,616 μm$^3$ |
| $\Sigma(L^4)=$ | | 37,345,460,660 μm$^4$ |
| N | = total number of zircon grains | 478 |

Shape factors $= K_n = \dfrac{\mu'_n(L)}{D_c^n}$ where $D_c$ = 97.5 μm, i.e. the zircon grains all

passed a 100 μm precision screen and were held on a 95 μm screen.

$K_1 = 0.7277$    $K_2 = 0.6663$    $K_3 = 0.7061$    $K_4 = 0.8646$

The above data can also be used to determine:

   a.   volumetric proportion of zircon $= \dfrac{\Sigma L}{T}$ $\qquad = 4.96\%$

   b.   specific surface area of zircon $= \dfrac{4}{\mu'_1(L)}$ $\quad = 564$ cm$^2$/cm$^3$

   c.   particle density $\qquad\qquad\qquad\qquad = 60.7/$mm$^3$

   d.   mean free inter-particle distance ($\lambda$) $\quad = 1360$ μm

   e.   average nearest neighbour distance ($\Delta_3$) $\quad = 141$ μm

### Grain Shape.

The shape of a convex grain can be described uniquely by the length distribution of random intercepts through that grain. The length distribution of random intercepts through a simple geometric form can be determined by calculation or by Monte Carlo simulation. Unfortunately, the length distribution of random intercepts through an irregular-shaped grain can only be established by experimental measurements. One method of doing this is to prepare a large number of grains of uniform size and statistically uniform shape and then disperse these grains at random in space (Simovic, 1973). The moments of the length distribution of random intercepts through these monodisperse grains provide mathematical expressions of the average shape of those grains, i.e.

$$K_n = \frac{\mu'_n(L)}{D_c^n}$$

where $K_n$ = dimensionless nth order shape factor; $\mu'_n(L)$ = is the nth moment from the origin of the intercept length distribution; $D_c^n$ = 'convenient' size – usually the screen size.

Shape coefficients determined by this method are given in Table 1.

It is also possible to derive shape coefficients by comparing the intercept length distribution obtained from polydisperse grains of a mineral embedded in a rock matrix with the three-dimensional size distribution obtained by sieving the unbroken grains after they have been removed from the rock by dissolving the matrix (Foo,1974). The parameters that relate the two distributions provide the average shape coefficients of that mineral over its complete size range.

e) By a small modification of the control programme it is possible to calculate a <u>locking index</u>. $P_{A/B} = \dfrac{S_{A/B}}{S_A}$ where $P_{A/B}$ = proportion of mineral A in contact with mineral B

$S_{A/B}$ = surface area of A in contact with B; $S_A$ = total surface area of A.

## References

Barbery, G. (1974). Determination of particle size distribution from measurements on sections. <u>Powder Technology.</u> 9, 231.

Foo, K. (1974). Process design study for the beneficiation of a tin ore from Queen Hill, Tasmania. <u>M.Sc. Thesis,</u> London University.

Jones, M.P. and Barbery, G. (1975). The size distributions and shapes of minerals in multi-phase materials: practical determination and use in mineral process design and control. Paper 36, <u>XIth Inter. Miner. Proc. Cong. Cagliari.</u>

Jones, M.P. and Shaw, J.L. (1973). Automatic measurement and stereological assessment of mineral data, for use in mineral technology. <u>Proc. Xth Inter. Miner. Proc. Cong. London,</u> Ed. M.J. Jones, p.737.

Simovic, M. (1973). Stereological determination of the size distributions of minerals. <u>Min. Tech. Res. Prog. Rept.,</u> Imperial College, London.

*National Bureau of Standards Special Publication 431*
*Proceedings of the* FOURTH INTERNATIONAL CONGRESS FOR STEREOLOGY
*held at* NBS, Gaithersburg, Md., September 4-9, 1975 (Issued January 1976)

## A NEW OPTO-MANUAL SEMI-AUTOMATIC EVALUATION SYSTEM

by H. P. Rohr
*Department of Pathology, University of Basel, Basel, Switzerland*

The introduction of stereological principles to determine volumes, surfaces or the number of particles offers new aspects and possibilities to assign interdisciplinary importance to morphological results. As mostly a greater number of such test samples must be evaluated for statistic security of the results, the broadest possible rationalisation and economy of image analysis have become extremely desirable to-day.

Basically there are two approaches for image analysis: 1. The image analysis through the so-called scanning methods which are usually performed mostly without the human eye and 2. the opto-manual semi-automatic evaluation systems. In spite of the rapid and very promising development in the field of scanning, some limitations of this method have become more and more obvious: As with this method image analysis is achieved mostly by a differentiation of various shades of grey, a clear identification of biological structures is quite often not possible. Therefore, we believe that the human eye will for long remain a determining factor in image analysis, at least in biology. This assumption is finally supported by the fact that at present opto-manual systems develop rather rapidly.

To-day the determinations of volume can be realized rationally according to the principles of Delesse and Glagoleff. Surface determinations are based mainly on the Rosival-principle. In volume and surface determinations the measuring procedures are reduced to simple counting procedures. In both cases test plates are used. The simplest way to transmit the counting results is a transmission of the data by hand to a key board which is joined with an adequate memory unit.

At the present time, no suitable appliance which would render the human eye superflous for automatic picture analysis exists. Aware of this fact, an instrument (MOP, Manual Optical Picture Analyser) was developed, which allows a simpler and more efficient way for conventional picture analyses (Fig.1). The principle of this semi-automatic picture analysing instrument consists therein, that elec-

tric current is carried through the test lines which are placed
upon a plexiglass plate. By every contact of a special pencil with
these lines, an impulse is released which is registered by a storage unit. How are volume and surface densities of certain structures determined with this opto-manual system ?

To determine a distinct point density for the calculation of the
corresponding volume density, the intercepts lying within the structure are scanned with the contact pen. At each crossing with a
testline (= test point in conventional counting) an impulse is released (Fig.2).

For the calculation of the surface density the number of intersections ($I_i$) per test plate can be determined by tracing the contours
of the structure profiles to be analysed with the contact pen (Fig.3)
The number of structure profiles per test area in order to calculate the numerical density is realized by another pen. The setting
on of this mechanical pen to each structure profile releases an
impulse.
Finally this MOP-system can be used as simple manual counter.
The counting data, or basic values, of up to 24 different structures can be accumulated and stored for each sample. An outprint
of the data is possible. With adequate small computers these un-
processed data can be converted into the according volume, surface or numerical densities with regard to the magnification and
the characteristics of the test plate. Furthermore, particle size
distributions can also be performed with this system. If the opto-
manual system is used in combination with a calculator, it is e.g.
possible to accumulate the results of a preliminary evaluation on
magnetic cards or another data carrier.

Every structure definable by the human eye is accessible by an
image analysis with consideration of the premises given by the
stereological principles. In image analyses with optomanual systems depending on the principles of stereology, the contact test
plates can be adjusted to the structures to be analysed. The lines
of the contact test plates can be arranged very closely. Therefore,
the average values of the samples offer a greater statistical security. Marking of the analysed structures is possible.

Summarizing we have with this instrument a very simple picture analysing system at hand, which allows a very fast morphometric evaluation of structures at light and electron microscopic level.

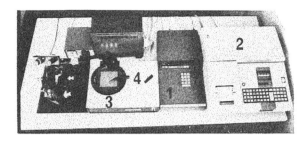

Fig. 1 : Manual optical sample analyser (Kontron Messgeräte GmbH, Munich, Germany)
   1: Electronic count, memory and display unit
   2: Desk computer
   3: Illumination box with incorporated test plate
   4: Light and mechanical pen

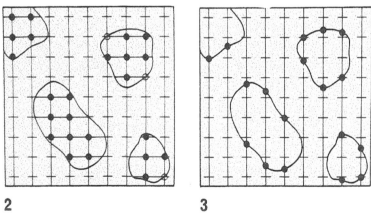

Fig. 2 : Principle of determination of volume density by the manual optical sample analyser (MOP KM 2)

Fig. 3 : Principle of determination of the surface density by the manual optical sample analyser (MOP KM 2)

*National Bureau of Standards Special Publication 431*
*Proceedings of the* FOURTH INTERNATIONAL CONGRESS FOR STEREOLOGY
*held at* NBS, Gaithersburg, Md., September 4-9, 1975 (Issued January 1976)

## AUTOMATED MEASUREMENTS FOR DETERMINATION OF THE STATISTICAL NATURE OF THE DISTRIBUTION OF INCLUSIONS IN STEEL

by George A. Moore
*National Bureau of Standards, Washington, D. C. 20234, U.S.A.*

An ASTM committee [1] is composing a recommended practice for quality control inspection of metals with respect to inclusion content. This must include minimum inspection requirements together with methods of determining when further inspection is necessary. Knowledge of the statistical nature of the distribution of field measurements is therefore necessary. Data were published by Allmand and Coleman [2] for concentration $V_V$ and size classified island counts for 1800 fields on each of three steel specimens. These data exhibit an unexpectedly high coefficient of variation and extreme abnormality as compared with either a Gaussian distribution or a positive half Gaussian. High readings are several times as abundant as expected in a normal distribution. In the presence of this abnormality, it cannot be stated that their three specimens are statistically different. It was thus estimated that $10^4$ to $10^5$ fields must be measured on specimens of one steel to determine the true nature of the distribution. High level statistical analysis is required to estimate the inflation ratio to be applied to the confidence limits.

A task of this magnitude requires automation at the computer-assisted-by-man level. Measurement acquisition and data processing are required to proceed automatically without continuous supervision. We are operating a quantitative television microscope assembly under full control of a mini-computer. A software program of over 25,000 computer words in BASIC operates in several sections. The system first instructs the operator in validating magnification, setting up the specimen, and setting necessary machine controls. A test exercise determines that all parts of the system are operational and that the area threshold has in fact been set at that level where addition or removal of edge delineation does not significantly alter the area measurement. Necessary identification and control information is then demanded and supplied by the operator. The system then proceeds to automatic measurements of $10^3$ fields on the specimen. Protection from off-specimen positioning and damaged or dirty fields is provided by an area magnitude check, with a decision required from the operator if a new area maximum is higher than reasonably anticipated.

Analysis is made in sets of 30 to 100 fields forming a coherent band along the longitudinal direction of the specimen. The first four powers of each measured parameter are accumulated in both set and specimen stores. The class size of each measurement is tallied. Two statistical analyses are made at the end of each set. The first examines the data for agreement with a mode zero distribution of positive half Gaussian shape. Normal, RMS, and 4th root of 4th moment averages are shown. The 2nd, 3rd, and 4th moments are listed and the kurtosis computed. The ratio of this kurtosis to the value, 3, for a half Gaussian is a measure of the lack of conformity to this form of distribution.

For the second analysis the four moments are recomputed about the mean. The statistical parameters usually determined at this level of analysis are computed. The confidence ranges about the mean are known to require inflation whenever the coefficient of variation (CV) is excessive or the distribution abnormal. The confidence limit (CL) value is arbitrarily flagged as unreliable if the CV exceeds 0.22 and a suggested maximum CL computed from the kurtosis.

At the end of the final set similar statistical analyses are made for the observations over the whole specimen. The standard deviations of the RMS average and of the standard error of the mean are determined, permitting a new estimate of the inflation factor for the confidence limit. An analysis is also made for the distribution of set means about the grand mean. The distribution of set means can be expected to be more nearly normal than the distribution of observations, and therefore to yield a more reliable estimate of the true CL.

The final data for each specimen are systematically stored on magnetic tape files. These can be recalled and merged to yield a combined analysis for several surfaces of the same specimen or for all the specimens from one heat.

A logical decision has been made that total inclusion concentration, $A_A = V_V$, and the transverse projection value of surface to volume ratio, $S_V'$, should be functionally related to the mechanical behavior of steel. $S_V'$ is the stereological equivalent of totaled feature length in a standard field area but is computed directly from the intercept count. This count is made at a threshold higher than the proper setting for area determination, but below the snow level. This threshold is normally set at a level where addition of edge delineation does not significantly increase the intercept count.

As no functional relationships are presently known between measurements of individual inclusion islands and mechanical behavior of the heat, the additional machine time to measure individual features does not appear justified for quality control applications. The research system does, however, include total feature counts at each threshold level and a count of large inclusions more than 20 μm long. Computations to reveal possible dependence of measurements on a Poissonian distribution of inclusion particles are included in the final specimen analysis. Maximum values of each field measurement within each set are retained and an attempt is being made to find a relation of these to the probable maxima for the material.

The material being measured, a heat of steel intended for wire drawing material, contains small dark isotropic particles presumably of indigenous origin, together with lighter stringers presumed exogenous. These stringers are of duplex structure, frequently incorporating the small particles at their surface. As the two types presumably cooperate in determining mechanical properties, no attempt is currently being made to measure them separately. No evidence of a bimodal distribution has been seen in the histogram tables.

The present research system is memory bound in the data acquisition program section and computation bound in the rate of processing incoming data. Hence, advantage is taken of the averaged scan mode for higher precision. About 225 fields are measured per hour, requiring over five hours per specimen. While acceptable for research purposes, this rate is not acceptable in quality control applications. Scanners exist which operate at about 500 fields per minute. These could be fitted with parallel high-speed computation systems to accomplish whatever portion of the present computations may prove necessary or useful.

Presently available data are only preliminary. Analysis of the first 12,000 fields (120 sets) of the test steel confirms a high coefficient of variation between fields for all five measurements. The distribution apparently is much less uniform than Poissonian. High positive skew values are found. The kurtosis is an order of magnitude higher than normal. Thus confidence limits computed on the assumption of normality could be even an order of magnitude too small! Fortunately, the distribution of averages for sets of 100 fields does appear to approach normality. Analysis by sets apparently will be acceptable in control practice. Confidence ranges computed from set averages are 2.5 to 6 times wider than those computed from the observations as a whole. Thus results based on traditional examination of 100 to 300 fields do not characterize the material. Data on ten sets taken from different specimens may be sufficient to determine if a material is close to an imposed control limit. Measurement of 30 to 100 sets may be necessary for critical decisions.

REFERENCES
1. Am. Soc. for Testing and Materials, Committee E4 on Metallography, Task Group 14.04 (W. D. Forgeng, Jr., Chm.).
2. T. R. Allmand and D. S. Coleman, The Microscope 20, 1 (Jan. 1972) 57-81.

# 4. THREE-DIMENSIONAL RECONSTRUCTION

*Chairpersons:*

R. Gordon
L. D. Peachy
R. Moore

*National Bureau of Standards Special Publication 431*
*Proceedings of the* FOURTH INTERNATIONAL CONGRESS FOR STEREOLOGY
*held at* NBS, Gaithersburg, Md., September 4-9, 1975 (Issued January 1976)

RECONSTRUCTION FROM PROJECTIONS: A SURVEY OF APPLICATIONS

by Richard Gordon
*Image Processing Unit, National Cancer Institute, National Institutes of Health, Building 36, Room 4D28
Bethesda, Maryland 20014, U.S.A.*

The problem of reconstructing a function in n dimensions
from its projections to a lower dimensional space arises
in many fields, from astronomy to electron microscopy. The
most active application is in medicine, in which the new
x-ray scanners are revolutionizing the practice of radiology
by providing clear cross sections of the brain and body.
Nuclear medicine is also making advances in the reconstruction
of the distribution of emitting radioisotopes. In astronomy
there is a need to obtain good two dimensional images from
the whole spectrum of waves and particles using instruments
which record strip integrals. Within our own solar system
three dimensional features may be reconstructed, such as
the solar corona, the planetary atmospheres, and the earth's
ionosphere. Reconstruction from projections in electron
microscopy is providing low resolution structures of biological
macromolecules, promising to become as important as x-ray
crystallography in molecular biology. There is a whole unex-
plored range of applications to light microscopy.

REFERENCES

A. The following is a good current sampling of the range
of interdisciplinary activity in reconstruction from projections
and also includes a bibliography of the whole field:

    1. Digest and Supplement of Technical Papers, Topical
       Meeting on Image Processing for 2-D and 3-D Recon-
       struction from Projections: Theory and Practice in
       Medicine and the Physical Sciences, 4-7 August,
       Stanford University, chairman R. Gordon. Optical
       Society of America, Washington, D.C.

B. The following collections of papers are also valuable:

    1. Tomographic Imaging in Nuclear Medicine, Soc. Nuclear
       Medicine, New York, 1973.

    2. IEEE Trans. Nucl. Sci., volume NS-21, Number 3, 1974.

    3. Techniques of Three-Dimensional Reconstruction,
       Brookhaven National Laboratory, July 15-19, 1974,
       BNL 20425.

    4. Workshop on Reconstruction Tomograph in Diagnostic
       Radiology and Nuclear Medicine, San Juan, Puerto Rico,
       17-19 April, 1974, New York University Park Press,
       in press.

C. The following are reviews of the algorithms and the whole field:

1.  R. Gordon & G. T. Herman (1974). Three dimensional reconstruction from projections: A review of algorithms. Int. Rev. Cytol. 38, 111-151.

2.  R. Gordon (1976) A Treatise on Reconstruction from Projections. Plenum Press, New York, in preparation.

National Bureau of Standards Special Publication 431
Proceedings of the FOURTH INTERNATIONAL CONGRESS FOR STEREOLOGY
held at NBS, Gaithersburg, Md., September 4-9, 1975 (Issued January 1976)

MORPHOMETRIC ANALYSIS OF NEURONS IN DIFFERENT DEPTHS OF THE CAT'S BRAIN CORTEX AFTER
HYPOXIA

by O. Hunziker,* U. Schulz,* Ch. Walliser* and J. Serra**
*Basic Medical Research Department, Sandoz Ltd., CH-4002 Basel, Switzerland
**Centre de morphologie mathématique, École des Mines de Paris, F-77303 Fontainebleau, France

A morphometrical method is described, which allows to determine size and
form of neurons (opening-procedure) in the cat's sensori-motor cortex with the
Texture Analyzer of Leitz. The data obtained from control animals are compared
with those of cats having resided a 3 weeks period at high altitude. For this
reason the opening values were statistically treated with the multivariate
analysis of Benzecri, which yields the classification of neurons in different
depths of the cerebral cortex.

INTRODUCTION
    The aim of this study is to investigate quantitatively neuronal cells of
the cat's cerebral cortex using light microscopy. During morphometric investigations of this kind many problems arise from the great variety of shape (Gihr,
1963) and size (Colon & Smit, 1970; Haug, 1967, 1972; Ramon – Moliner, 1961;
Sholl, 1967) of cortical neurons. Thus, the Texture Analyzer (TAS) of Leitz
(Serra & Müller, 1973) as an instrument of optical-electronic image-analysis
facilitates considerably quantitative histological studies on the neuronal
system. In order to demonstrate another object of the study and the technique
of the TAS, neurons under normal conditions were compared with experimentally
altered cells.

EXPERIMENTAL AND MORPHOMETRIC PROCEDURE
    The experimental part of our investigation was performed as follows:  3
male cats were kept during 3 weeks in a conditioned low-pressure chamber and
exposed to a simulated high altitude of 5025 m as previously described (Hunziker
et al., 1975). During the same time 3 other cats remained as controls at 250 m
altitude in a standard laboratory. At the end of the experiment all animals
were deeply anaesthetised with Pentobarbital and the brains perfused with 2.5%
glutaraldehyde in phosphate buffer (pH 7.4). The neurons of the temporo-parie-
tal Suprasylvian Gyrus were demonstrated by a Nissl-staining with Gallocyanin-
Chromalum (Fig. 1) and subsequently measured with the TAS (magnif. factor of
the objective: 40x). For the investigation 9 serial frozen sections (thickness
= 14µ) per brain region and animal were randomly selected. In each section 3
measuring directions were localised, the directions being identical for all
sections. For each direction the scanning stage of the TAS was preprogrammed
to start from the pial surface and to reach the white matter by steps of 50µ.
The total number of fields per measuring direction was 30, covering a total
thickness of 1500µ, i.e., approximately the effective thickness of the temporo-
parietal cortex. Using an electronically set measuring field (mask) of the TAS,
1 neuron, of which the nucleolus was visible in the section (Haug, 1972), was
selected from each measuring field. The optical input system of the TAS is a
microscope combined with a black-and-white television camera. The theoretical
background involved in the TAS is mathematical morphology and the hit or miss
transformation (Matheron, 1975). In the following investigation we particularly
used the concept of isotropic opening, defined as follows. Let us consider a
circle with a given radius r (structuring element) moving everywhere inside the
neuron and satisfying the condition, not to cut the pericaryon border. The set
covered by the circle when sweeping over the neuron is called its morphological
opening. The smaller the radius r is, the clearer are the details of the
shape, which the circle can reach. When r increases, the opening area decreases
to zero for $r > r_{max}$ ($r_{max}$ = radius of the circle within the neuron). For
$r < r_{max}$ the neuron shrinks progressively (continuous part of the opening), and
the remaining part vanishes at one for $r = r_{max}$ (discontinuous part). Experimentally, a hexagon is sufficient for our purpose. A digital hexagonal frame
can produce very small regular hexagons. Here the elementary step (side of the

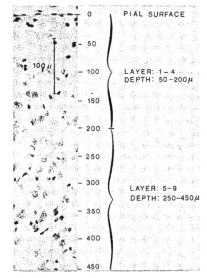

Fig. 1: Neurons in the upper layers of the cat's cerebral cortex. Controls. Gallocyanin-Chromalum (magnif. 230x).

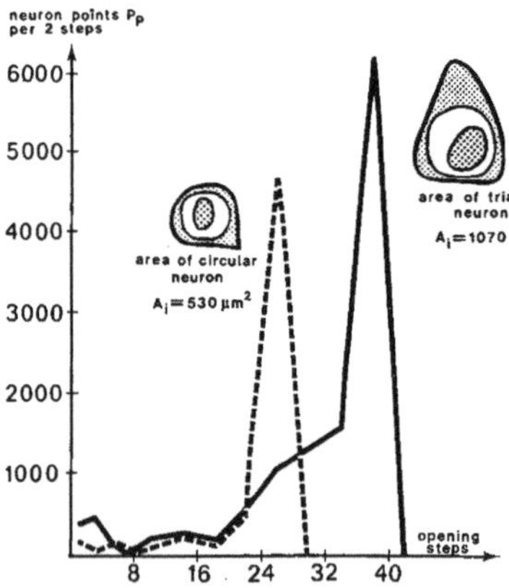

Fig. 2: Opening curves of a triangular and a circular neuron.

$$A_i = \frac{a^2 \sqrt{3}}{2} P_P$$

hexagon) values a = 0.33μ. It was enlarged, step by step, to 6a and by increasments of 2 to 32 a, which represent 18 classes of opening increments. Thus the data are a matrix of 72 × 30 elements (18 opening increments × 4 cats = lines, 30 depths = columns). As an example we plotted 2 curves of opening increment area (Fig. 2) for 2 different neurons. They differ: i) by their size or integral of the curve, ii) by the ratio of the discontinuous part over the total area, or circulatory factor , which is for the triangular neuron = 0.51 and for the circular one = 0.77, iii) by the distribution of the continuous part (opening steps 16–32 resp. 16–24). All these properties can be quantified more precisely (Serra, 1975). Here they are evaluated by a multivariate analysis treatment. The key point is, that a shape is characterized not by a finite number of parameters, but by a function.

STATISTICAL TREATMENT

There are several methods of multivariate analysis. The method of choice was the correspondence analysis theory of J. P. Benzecri, of which the advantage is to provide results independent from the selected opening classes: If 2 points representing 2 steps of openings are close together in the main factor subspaces, the 2 corresponding classes may be regrouped in a single one (they are proportional) without disturbing the remaining constellation of points. Fig. 3, 4 and 5 show the first factorial plane (axis Nr. 1 and 2). Both depth-points (stars) and openings (circled points) are simultaneously represented. The first 2 axes explain 45% of the total inertia of the point cloud (resp. 28 and 17%). Their meaning is provided by the projection of the points upon them. The closer they are to the origin, the less significant they are. Considering the opening points up to Nr. 9, they may be assumed non-significant on the first 2 axes. After Nr. 9 their first axis coordinate increases systematically, showing that this axis classifies the neurons according to their size. The second axis opposes 2 phenomena: the medium sized openings (10, 11, 12) from the larger ones (14, 15), and the large openings (17) from the largest one (18). The first opposition can be explained by the form factor. This interpretation of both axes

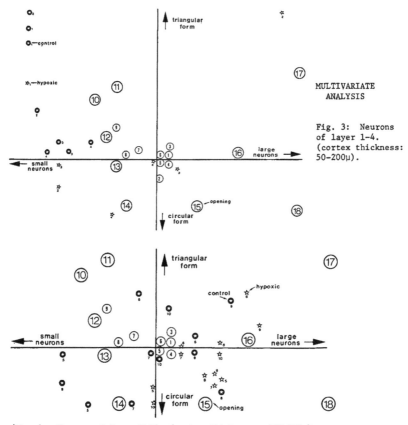

MULTIVARIATE
ANALYSIS

Fig. 3: Neurons
of layer 1-4.
(cortex thickness:
50-200µ).

Fig. 4: Neurons of layer 5-10. (cortex thickness: 250-500µ).

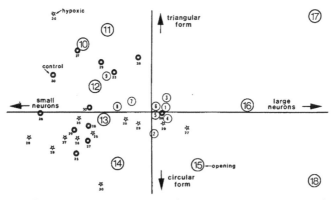

Fig. 5: Neurons of layer 25-30. (cortex thickness: 1250-1500µ).

allows the differentiation between certain groups of measuring fields, depending on their depths 1 to 30 from the pial surface. Hence it helps to distinguish certain layers of the cerebral cortex. With reference to the depth of the measuring fields, we decomposed the diagram into 4 parts. Fig. 3 contains 2 parts: In the superficial layers 1 and 2 the neurons of the hypoxic cats have a more spherical shape, but do not seem to be larger. The layers 3 and 4 beyond exhibit a predominance of larger neurons, all having a similar shape. In the depth of 250-500μ (Fig. 4) the neurons of the hypoxic animals are enlarged, without changing their shape. It is interesting to notice that the stressed neurons seem to resemble each other more than the normal neurons do. In the layers of 550-1200μ no apparent differences are observed. From 1250-1500μ (Fig. 5) the neurons of all animals are smaller than in the layers of 250-500μ. Between the normal and the hypoxic animals there is no difference in size, whereas there is one in their shape. These neurons appear to be more circular. The presented morphometrical method allows to determine quantitatively neuronal changes after longer residence of animals at high altitude.

REFERENCES

Colon,E.J.&Smit,G.T.(1970) The organization of the cerebral cortex.Quantitative aspects.Acta Morph.Neerl.Scand.7,309-319.

Gihr,M.(1963) Rekonstruktion von Nervenzellen.Int. Congr.F.Stereologie 18,1-37.

Haug,H.(1967) Probleme und Methoden der Struktur- zählung im Schnittpräparat, in:Quantitative Methods in Morphology (Ed.by E.Weibel),pp. 57-78, Springer-Verlag,Berlin-Heidelberg- New York.

Haug,H.(1972) Stereological methods in the analy- sis of neuronal parameters in the central nervous system.J.Microsc.95,165-180.

Hunziker,O.,Schulz,U.,Veteau,M.J.,Weihe,W.H. and Meier-Ruge,W.(1975) Morphometric capillary parameters in the cat's brain cortex after chronic hypoxia. In:Quantitative Analysis of Microstructures in Medicine,Biology and Materials Development (Ed.by H.E.Exner), pp.168-176, Dr.Riederer-Verlag,Stuttgart.

Matheron,G.(1975) Random sets and integral geome- try.Wiley-Intersciences,New York.

Ramon-Moliner,E.(1961) The histology of the post- cruciate gyrus in the cat. I.Quantitative studies. J.Comp.Neurol.117,43-62.

Serra,J.(1975) Mathematical morphology and image analysis. SIAM,Washington D.C.

Serra,J.&Müller,W.(1973) The Leitz TAS. Leitz- Mitt.Wiss.Tech.Suppl.1,4.

Sholl,D.A.(1967) The organization of the cerebral cortex. Hafner Publ. Company, New York- London.

National Bureau of Standards Special Publication 431
Proceedings of the FOURTH INTERNATIONAL CONGRESS FOR STEREOLOGY
held at NBS, Gaithersburg, Md., September 4-9, 1975 (Issued January 1976)

THREE-DIMENSIONAL RECONSTRUCTION FROM SERIAL HIGH VOLTAGE ELECTRON MICROGRAPHS

by Lee D. Peachey, Caroline H. Damsky and Arthur Veen
Department of Biology, University of Pennsylvania, Philadelphia, Pennsylvania 19174 and Department of
Molecular, Cellular and Developmental Biology, University of Colorado, Boulder, Colorado 80302, U.S.A.

INTRODUCTION

The process of three-dimensional reconstruction from a series of two-dimensional profiles, as in the case of serial slices of biological tissue examined microscopically, is simple and straightforward in principle. In practice, however, it can be tedious and, if not done accurately, the results can be misleading. Because of the great potential usefulness of three-dimensional reconstruction of fine details of cellular structures from relatively thick serial slices of biological tissue using high voltage electron microscopy (HVEM) with accelerating voltages of about 1 MV, we decided to reconsider the methods that could be used in such a reconstruction process. Our hope was to improve the reliability and accuracy of the process, to increase its versatility in terms of the kinds of structures that could be reconstructed, and to improve the rate of reconstruction while reducing its tediousness. Our approach has been to consider each step in the process as separate from the others and subject to its own improvements, and to consider the use of digital computers to replace human operations whenever the replacement really made sense. Our goal was not a fully automated system. We felt that such a system, even if it could be produced, probably would not have reasonable accessibility by many workers because of its complexity and cost. We also felt that a highly automated system would be likely to have less versatility than a system with a greater degree of human involvement at certain critical steps in the process of generating the reconstruction. Therefore, the system we have evolved can be considered as a result of a series of compromises between human and machine operations.

CHARACTERISTICS OF HVEM IMAGES

HVEM provides images that have a high degree of lateral resolution compared to the thickness of the specimens. Lateral resolution of the order of 10 nm. or less can be obtained with specimens of the order of 1 μm. in thickness, with the entire thickness of the specimen in focus in the image. Thus the positions of structures or boundaries of structures, when oriented favorably, can be determined very accurately in two dimensions, but information on the third dimension is lost in single images. Pairs of images, taken with the specimen tilted an appropriate amount between exposures, can be viewed stereoscopically, and this often provides considerably improved visualization of the third dimension. However, the greatest extent of three-dimensional information can be obtained with multiple images of serial slices. These can be used even when the structure of interest is too large to be contained within a single slice. In this case, reconstruction of the original structure from a series of micrographs becomes a necessity.

RECONSTRUCTION

Reconstruction from serial micrographs is, in essence, a form of reassembly of a microscopic structure that previously was disassembled when it was cut into a series of slices. If the assembly is to be valid, then information of how the structure originally was arranged in three dimensions must be retained or regained as the reconstruction is done. A common form for such reconstructions is the real physical model. These can be built from a series of slabs of plastic or other material, with each slab cut out to represent the profiles of the structures in one slice or micrograph. If the slab thickness is chosen correctly in relation to both the slice thickness and the magnification of the micrograph, and if the slabs are aligned properly one upon the next, then the model can be a rather accurate and useful representation of the original structure.

## STEPS IN RECONSTRUCTION

Assuming that one has an adequate set of micrographs of known magnification from a set of consecutive serial slices of known thickness, the process of reconstruction can be divided into a sequence of steps, as follows:

1. Recognizing and selecting in each micrograph profiles of the structures to be included in the reconstruction.
2. Obtaining data on the relative locations of points along the boundaries of the chosen profiles.
3. Aligning the profile data for each slice in relation to adjacent slices to restore the original spatial relationships, which were distorted when the slices were cut and separated.
4. Generating a representation of the complete set of profiles in three dimensions as a physical model, artist's drawing, or computer graphic display.

## RECOGNITION, SELECTION, AND OBTAINING PROFILE DATA

We have chosen to do steps 1 and 2 visually and manually, rather than to attempt to develop an automated optical scanning and pattern recognition system. Using a tracing stylus linked through a backlash-free cable system to a pair of high-resolution potentiometers, the operator traces around the selected profiles. Voltages from the potentiometers, proportional to polar position of the stylus, are converted automatically to digital values and fed to a PDP-8/L computer. Significant movements of the stylus, usually greater than about 0.3 mm., are coded and punched on paper tape. Information on magnification, locations of calibration points, number of slices, a title for the reconstruction, etc. are input to the computer from a teletype and transferred to the paper tape along with the profile data. Step 2 is completed when this paper tape is read into a PDP-6 computer and the data are converted into a series of points in rectangular coordinates and are packed efficiently into 36-bit words on a magnetic storage device. The data also are scaled to fit within a square of unit size, so that all X and Y coordinates are in the range 0.0 to 1.0.

## ALIGNMENT OF DATA FROM SUCCESSIVE SLICES

An important part of the reconstruction process, and one apparently not often considered fully, is editing of the data to achieve valid alignment of each slice with respect to the next slice. This must be done objectively and with information not biased incorrectly by any assumption of what the structure originally was like. For this purpose, we select certain features of the microscopic images that are external to the features being reconstructed and that can be expected to maintain positional and/or rotational alignment from one micrograph to the next. These include spherical objects cut near their centers and thin filamentous or tubular structures cut exactly transversely. Lines drawn between such structures lying at some distance from and on opposite sides of the area being reconstructed are traced, and the data on their positions and orientations are carried along with information on the reconstruction profiles and fiduciary profiles in the data set.

A program called EDIT is used for visual presentation on an oscilloscope terminal of the data for individual slices or pairs of slices before, during, and after editing for alignment. The operator uses a set of teletype commands for specifying translation and/or rotation of the data set of one slice with respect to that of the next slice. Alternately, the criteria for alignment can be specified, e.g. two lines should be parallel, two closed profiles should be superimposed, etc., and the computer calculates and carries out the rotation and translation required to meet the specified criteria as well as possible. Sizes of individual profiles, or of the whole slice, are changed either isotropically or anisotropically, when such adjustment is specified and can be justified. The number of data points can be reduced, for efficiency of storage, and profiles used for alignment but not wanted in the final reconstruction can be eliminated from the data set. When the same operations are to be done on a series of more than two slices, global commands can be used. The result of the EDIT operations is a new data set in which the data for successive slices have been recomputed for best alignment from each slice to the next, and step 3 above has been completed.

## THREE-DIMENSIONAL DISPLAY

Step 4 starts with a conversion of the data for each slice into a tabulation of intersections of profiles with a set of about 200 evenly spaced, parallel lines in the plane of the slice, by a program called DIGIT. The output of this program then is displayed in three dimensions by a program called FIG3D. The display, on an oscilloscope terminal or hard copy device, is presented in proper perspective, from any viewing angle in the forward hemisphere, and from any viewing distance.

The three-dimensional appearance of the final display is enhanced by a number of features, some optional. Hidden lines are not displayed, so the reconstruction appears solid rather than transparent. Several lines can be drawn for each slice, giving it a slab-like appearance. Each slab has an apparent thickness in proportion to the individual slice thickness and to the magnification and perspective of the display. A calibrated box can be displayed enclosing the reconstruction. Two displays, presented at slightly different viewing angles, can be viewed stereoscopically in three dimensions.

## SUMMARY

We have described a method for generating three-dimensional representations of structures reassembled from a set of micrographs of serial slices of the original structures. A combination of human and machine operations has been selected for each of the sequential steps in making the reconstruction, and particular care has been taken to achieve valid and accurate realignment of profiles in successive slices. This computer based procedure is more rapid than construction of a real model, and is likely to be more precise in the key step of aligning successive slices in the reconstruction to produce a valid representation of the original structure.

## ACKNOWLEDGMENT

This work was supported by grants from the National Institutes of Health, Division of Research Resources (No. RR-15 and RR-592), the National Heart and Lung Institute (Pennsylvania Muscle Institute, HL-15835), and the Muscular Dystrophy Associations of America, Inc. (Henry M. Watts Neuromuscular Disease Center, Philadelphia).

## FIGURES

The figures show various stages in the reconstruction of yeast cell mitochondria in two consecutive slices about 0.4 μm. thick from a series of HVEM micrographs. The top four panels show HVEM micrographs of the two slices, with fiduciary lines from distant objects drawn on the micrograph, and the two corresponding data sets as displayed by the EDIT program (EDIT supplies the square enclosing the profile data, and the lettering was added by an artist). Each display shows three mitochondrial profiles as well as a tracing of the large vacuole in the cell, and one or two fiduciary lines.

The bottom four panels show stages in the editing and display process. Above at the left is the unedited display of both slices superimposed. To the right is the display after editing using one line and the vacuole for alignment. Below on the left is the display from EDIT after all fiduciary lines and profiles have been removed. At the bottom right is a three-dimensional display of the two slices after editing. In this display, only the slab effect and the hidden lines routine have been used.

209

L. D. Peachey, C. H. Damsky and A. Veen

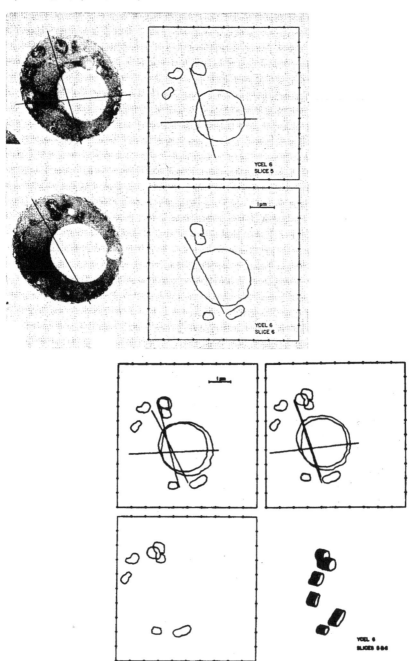

210

National Bureau of Standards Special Publication 431
Proceedings of the FOURTH INTERNATIONAL CONGRESS FOR STEREOLOGY
held at NBS, Gaithersburg, Md., September 4-9, 1975 (Issued January 1976)

DETERMINING THREE-DIMENSIONAL INTRACELLULAR STRUCTURE DATA FROM SCANNING PROTON MICRO-
SCOPY LUMINESCENCE DATA

by Daniel G. Oldfield
Department of Biological Sciences, De Paul University, Chicago, Illinois 60614, U.S.A.

INTRODUCTION

The spatial resolution presently attainable in cellular electron microscopy has improved visualization of specimen detail one or two orders of
magnitude beyond the limits allowed by optical microscopy. However, the chemical resolution attained in electron microscopy--the ability to distinguish
various chemical species that may be present in the spatially resolved region--
falls distinctly short of that permitted by optical microscopy (e.g., utilizing
cytochemical techniques).

Advances in scanning electron microscopy and, more particularly, the
advent of scanning proton microscopy (Levi-Setti, 1974) prompt one to consider
the possibility of using small-diameter, minimally scattered charged-particle
beams to excite luminescence in small, beam-delimited regions of cells because
luminescence spectra so produced may provide at least a partial chemical
characterization of the cell region penetrated by the beam.

The purpose of this paper is (1) to formulate a basic theoretical model
for the use of heavy charged particles such as protons to excite luminescence
in localized regions of intact cells or organelles; and (2) to examine factors
which affect the spatial resolution attainable using particle-induced luminescence.

ANALYTIC BASIS OF THE METHOD

*Beam and specimen specifications*

Consider an initially monoenergetic, monodirectional charged-particle beam
propagating in the direction of a positive $z$-axis fixed in space. If we
position the specimen in the beam, the intersection of the beam with the specimen surface defines a point in space. The location of this point relative to a
coordinate system fixed in the specimen is assumed to be determinable from the
(known) shape of the specimen via transformation equations such as Hilliard
(1972) has investigated for the analysis of SEM images. More generally, we are
assuming that, given the coordinates x,y,z of any point in the specimen expressed in the space coordinate system, we can determine via transformation
equations the coordinates of the same point expressed in the coordinate system
fixed in the specimen.

We limit consideration to dried cellular or organellar specimens having an
average linear dimension of less than 100 microns. Assuming unit density for
the specimen, its average thickness would be 10 mg/cm$^2$.

We assume that the specimen contains a luminor which is either an exogen-
ously added substance known to specifically bind to, or be sequestered by, the
specimen or an endogenous substance which is chemically extractable from the
cell. For example, the luminor might be one of the oxdiazoles used in fluoro-
metric Feulgen techniques for the determination of chromosomal desoxyribonu-
cleoprotein (Ruch, 1966); or the luminor might be a naturally occurring
nucleotide or protein or pigment such as riboflavin, globulin, or carotene
(Udenfriend, 1969).

Using the transformation equations above, the luminor concentration can
be expressed in the space variables x,y,z as $C(x,y,z)$.

*D. G. Oldfield*

*Interaction specifications*

Following the development employed by Fano (1963), we define: $p(E)dE$, mean pathlength (cm) of particles with energy in $dE$ at $E$; $X_{Aj}(E)$, cross-section (cm$^2$) for interaction (by inelastic collision) exciting an atom of type A to energy level $E_j$ above its ground state per particle of energy E.

The function $p(E)$ is a probability density function for mean pathlength. Its integral over the energy interval 0 to $F_0$ is the range of a particle with initial energy $E_o$. If straggling is neglected, $p_A(E)$ is simply the reciprocal of the specific energy loss (stopping power) $-dE_A/ds$. Therefore, assuming additivity for the various types of atoms present and summing, we write for the overall interaction in the specimen $-dE/ds = 1/p(E)$.

Thus, $p(E)$ can be estimated experimentally from specific energy loss measurements provided that data for low energy particles (0 to 10 mev per nucleon) and thin specimens (0.001 to 10 mg/cm$^2$) are obtainable. Some progress in this area has been reported (Jung, 1967; Dennis, 1972; Watt, 1972; Johnson, et al., 1973). Also, theoretical efforts are continuing (Inokuti, 1971; Massey & Gilbody, 1974).

ANALYSIS OF LUMINESCENCE PRODUCTION

*General*

Energy is transferred from a beam particle to all atoms along the track of the particle; but luminescence will occur only for energy transfers to atoms from which absorbed energy may migrate to the luminor molecule. Therefore, luminescence from a specimen will depend on: luminor concentration $C(x,y,z)$, the cross-section $X(E)$ (cm$^2$) for luminor excitation by a particle of energy E, and the quantum yield $q(k,l,E)dkdw$ of luminescent photons with wave number k emitted into the solid angle w in direction l per luminor molecule. Noting that $X(E)p(E)dE$ is the average differential volume within which excitation occurs, the luminescence produced by a particle which stops in the specimen is determined by the (spectral and spatial) density function

$$G(k,l,E_o) = \int_0^{E_o} C(x,y,z)X(E)q(k,l,E)p(E)dE. \qquad (1)$$

Photon counting systems for luminescence detection are available in the wavelength range 400-700 nm.

*Test and reference specimens*

It is convenient to refer the analysis of the biological test specimen to luminescence parameters measured in an atomically similar but molecularly simpler non-biological reference specimen containing a known uniform concentration of the luminor. For example, thin sheets of polystyrene or polyvinyltoluene containing the luminor might be suitable reference specimens. Thus, we define a structure factor $S = Xq/X_eq_e$, where the subscript e refers to the reference specimen. If the differential luminescence from reference specimens of known thickness and known luminor concentration are measured for various initial beam energies and with only small energy losses per particle, the quantity $X_e(E)q_e(k,l,E)$ can be determined from the density equation (1). Also, since the dependence of this quantity on reference specimen thickness is known, the quantity $X_e(z)q_e(k,l,z)$ can be found (the independent variable now being the penetration depth z into the specimen rather than the particle energy E).

212

Finally, defining $C_b(z)$ as the average concentration of luminor over the cross-section $b^2$ of the beam, the density function determining luminescence in the test specimen is

$$G(k,1,R) = \int_0^R C_b(z) X_e(z) q_e(k,1,z) S(k,1,z) dz, \qquad (2)$$

where R is the projected range.

## SPATIAL RESOLUTION

Consider two beam particles with initial energies $E_o'$ and $E_o''$ which are successively incident on a test specimen and penetrate to projected ranges R' and R''. The difference in density functions is $g = G(k,1,R'') - G(k,1,R')$. If $R'' - R' = r$ is very small, the functions within the integrals will change only negligibly over this range of z. We then obtain an equation for the net change in density function over the distance r in the specimen:

$$g(k,1,\bar{R}) = rC_b(\bar{R}) X_e(\bar{R}) q_e(k,1,\bar{R}) S(k,1,\bar{R}) \quad ; \quad \bar{R} = 1/2(R'+R''). \qquad (3)$$

We see from equation (3) that $g(k,1,\bar{R})$ will refer to luminescence photons produced <u>only</u> in the volume $b^2r$ at $z=\bar{R}$ provided that the time-averaged photon spectrum produced from $z=0$ to $z=R'$ by beam particles with initial energy $E_o'$ is exactly the same as that produced over the same distance by beam particles with initial energy $E_o''$. We expect this would be true (Curran, 1953).

We conclude therefore that in the absence of range straggling and multiple scattering, the spatial resolution limits for determining the basic specimen characteristics of luminor concentration $C_b(\bar{R})$ and structure factor $S(k,1,\bar{R}_z)$ would be given by the range difference r along the beam axis and the beam area $b^2$ perpendicular to it. Neglecting practical considerations of instrumental sensitivity and stability, specimen heating and damage, luminescence lifetimes and attenuation in the specimen, r could in theory be made arbitrarily small, and the beam area reduced to the cross-section of a single track (as determined by impact parameter and lowest excited state considerations). The irreducible stochastic limitations on resolution would then be determined (1) along the z-axis, by range straggling; (2) in the x,y plane, by multiple elastic scattering.

For protons having an initial energy of a few mev, the increase in penetration due to range straggling would be estimated to be a few percent of the range; but the increase in beam cross-sectional width would be estimated to be several tens of percent of the range (Fano, 1963). Therefore, for test specimens (such as organelles or flattened cells) in which moderate penetration depths are sufficient for analysis, the resolution obtainable along the z-axis would exceed that of optical microscopy for penetrations of 5 microns or less; for the equivalent resolution laterally, penetration depths would have to be restricted to less than 1 micron. It is possible, however, that these limitations could be eased by the use of stereographic image enhancement, pattern recognition, or other techniques.

## STEREOLOGIC ASPECTS

It is evident that the use of charged-particle beams at several discrete energies in the manner described here constitutes a statistical sampling of specimen structure along the z-axis to which standard stereologic methods could be applied. When heavy-charged-particle microscopy has been further developed, stereologic analysis will undoubtedly be called upon to more rationally interpret the data.

REFERENCES

Curran, S. C. (1953) *Luminescence & the Scintillation Counter.*
Academic Press, New York, p. 104.

Dennis, J. A. (1972) Interaction of low energy protons with ribonuclease.
*Physics in Med. & Biol. 17,* 304.

Fano, U. (1963) Penetration of protons, alpha particles, and mesons.
*Ann. Rev. Nuclear Science, 13,* 1.

Hilliard, J. E. (1972) Quantitative analysis of scanning electron
micrographs. *J. Microscopy 95,* 45.

Inokuti, M. (1971) Inelastic collisions of fast charged particles with
atoms & molecules--the Bethe theory revisited. *Rev. Mod. Phys., 43,* 297.

Johnson, R. E., Trevisani, E. T. & Harberger, J. H. (1973) Charge transfer
in bio-molecular damage by low energy protons. *Physics in Med. & Biol.,
18,* 287.

Jung, H. (1967) Inactivation of ribonuclease by elastic nuclear collisons.
*Rad. Res. Suppl., 7,* 64.

Levi-Setti, R. (1974) Proton scanning microscopy: feasibility and promise.
*Proceedings of the Seventh Annual Scanning Electron Microscopy Symposium,*
pp. 125-134, IIT Research Institute, Chicago.

Massey, H. S. W. & Gilbody, H. B. (1974) *Electronic & Ionic Impact Phenomena,*
V. 4, Oxford Univ. Pr., New York.

Ruch, F. (1966) Determination of DNA content by microfluorometry. In :
*Introduction to Quantitative Cytochemistry* (Ed. by G. L. Wied), p. 281.
Academic Press, New York.

Udenfriend, S. (1969) *Fluorescence Assay in Biology and Medicine,* Academic
Press, New York.

Watt, D. E. & Hughes, S. (1972) Bio-molecular damage by low energy heavy
particles. *Physics in Med. & Biol., 17,* 306.

ACKNOWLEDGMENTS

The author wishes to acknowledge support of the National Science Foundation
via a grant to De Paul University, and of the Faculty Research Participation
Program at Argonne National Laboratory of the United States Energy Research and
Development Administration.

*National Bureau of Standards Special Publication 431*
*Proceedings of the* FOURTH INTERNATIONAL CONGRESS FOR STEREOLOGY
*held at* NBS, Gaithersburg, Md., September 4-9, 1975 (Issued January 1976)

ARTIFACTS IN THREE DIMENSIONAL RECONSTRUCTION FROM MEDICAL RADIOGRAPHIC DATA

by E. L. Hall*, G. C. Huth**, R. A. Gans**, and I. S. Reed*

*Dept. of Electrical Engineering, University of Southern California, Los Angeles, California 90007, U.S.A.*
**Laboratory of Nuclear Medicine & Radiation Biology, University of California-Los Angeles, Los Angeles, California 90024, U.S.A.*

ABSTRACT

Artificial objects (artifacts) produced in reconstructed images[1,2] from medical radiographic projection data are a serious problem. In the extreme case, an artifact in a diagnostic image could influence the medical treatment. More commonly, artifacts produce a background noise which obscures the desired image and creates confusion in interpretation of the images. The purpose of this paper is to systematically describe the major sources of artifacts for the diagnostic radiology and nuclear medicine use of reconstruction techniques. An awareness of the source and occurrence of artifacts can often lead to complete elimination of the artifacts through the use of correction techniques and at least provides prior information to the viewer. For definiteness this study will be restricted to the use of radiographic projection data and images reconstructed with the simple convolution algorithm, although many of the conclusions may be generalized to other situations.

INTRODUCTION

Artificial objects in reconstructed images may arise from four major sources: 1) physical assumptions, 2) engineering implementation in device design, 3) computation algorithms, and 4) image display. Each of these sources will now be considered in detail.

PHYSICAL ASSUMPTIONS

The physical assumption which forms the basis of the reconstruction technique is that the attenuation of photons passing through an object is described by exponential decay:

$$I = I_0 \exp\left\{-\iint \mu(P, Z, E)dEdS\right\}$$

where $I_0$ is the number of photons entering the object, $\mu(P, Z, E)$ is the linear attenuation coefficient at position $P$ and $dS$ is the elemental path length of the ray. The attenuation coefficient, $\mu$, is a function of the composition of the object as indicated by its atomic number, $Z$. It is also a function of the energy of the energy spectra, $E$, of the photon beam. The basic linear equation is obtained by simply taking the log of the previous equation:

$$D = -\log \frac{I}{I_0} = \iint \mu(P, Z, E)dEdS.$$

The simplification of this linear equation for use in a reconstruction algorithm requires several assumptions. For geometrical simplicity, one may assume that the ray is positioned along the x axis and the position is x. The first major assumption is that the continuous position integral can be replaced by a summation:

$$D = \sum_i \int \mu(x_i, Z_i, E_i)dE_i \Delta x_i.$$

Note that this assumption is equivalent to assuming that the physical composition of the matter along the length, $\Delta x_i$, is homogeneous. This fact is denoted by replacing $Z_i$ with an average atomic number, $\overline{Z}_i$. The consequence of this assumption in a reconstructed image is illustrated in Fig. 1 which shows that an element composed of a mixture of bone and air can appear identical to an element of tissue.

215

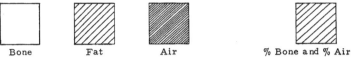

Figure 1.  An element composed of a mixture of bone and
air appears identical to an element of tissue.

The continuous dependence of the attenuation coefficient upon the energy
spectra must also be simplified by replacing the integral by

$$\mu(x_t, \overline{Z}_t, \overline{E}_t) = \int \mu(x_t, Z_t, E_t) dE_t$$

where $\overline{E}_t$ is the effective energy along the elemental area.  The averaging effect of this assumption is similar to the atomic number dependence, in that two
photons of low and high energy can produce the same result as a single photon
of medium energy.

A much more serious assumption is made in the reconstructed process.
This is that the attenuation coefficient is only a function of the physical composition, $\overline{Z}$ and not of the energy, $\overline{E}$.  This assumption is implicitly made when
the algorithm estimates the contribution of the projection data resulting from
an elemental area independent of the path length and object composition.  The
variation of energy spectra passing through a material is a well known phenomena.. Furthermore, the amount of attenuation coefficient variation can be significant.

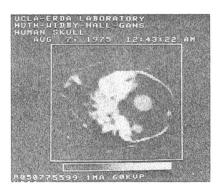

Figure 2.  Motion artifact resulting from subject motion in
the reconstruction of a human skull.  The two
circular objects are simulated tumors.  The linear object is the result of motion.

ENGINEERING IMPLEMENTATION

Several limitations in the accuracy of computer tomography are the result
of engineering implementation in the design of the mechanical motion portion of
a scanner, the x-ray source and detector, the timing signals used for control
and the sampling or pulse counting of the project data.  A serious limitation
is the resolution element size of the projection data.  This resolution is typically

on the order of 1.0 to 3.0 mm and thus 140-160 elements are used to span approximately 50 cm. Another limitation is the signal to noise ratio or number of counts collected at each data point. Typically, 50,000 to 100,000 counts are collected for each projection point although this time is controllable so that any desired signal to noise ratio could be achieved. Subject motion, one of the most serious artifacts, is also a function of the device design. An example of the linear artifact produced by subject motion is shown in Figure 2. The seriousness of this artifact is indicated by the fact that most new scanners use a fan beam geometry rather than rectangular to minimize data collection time and thus the possibility of motion artifacts.

## COMPUTATION ALGORITHMS

The computational algorithm also seriously influences the final reconstruction image. For definiteness, only the popular convolution algorithm will be considered. This method involves a log conversion of the projection data, a convolution for deblurring, and an interpolation to back project the data into the correct image element. The accuracy of log conversion and interpolation are considered well known. However, it is not well known that the convolution operation can be computed with zero roundoff error using finite field methods. This method will be briefly described.

Recently[3] it has been shown that the convolution of two finite sequences of integers $(a_k)$ and $(b_k)$ for $k = 1, 2, \ldots, d$ ($d = 160, 256, 320$, etc.) can be obtained as the inverse transform of the product of two transforms which were similar to but different from the usual discrete Fourier transform.

The transforms were of the form

$$A_k = \sum_{n=0}^{d-1} a_n 2^{nk} \bmod b$$

where b is either a Mersenne number

$$b = 2^p - 1, \qquad p \text{ a prime}$$

or b was the Fermat number

$$b = 2^{2^m} + 1, \qquad m \text{ an integer.}$$

The usual DFT may be written as

$$F_k = \sum_{n=0}^{d-1} a_n w^{nk}, \text{ where } w \text{ is the } d^{th} \text{ root of unity.}$$

The primary advantage of this type transform over the DFT is the fact that multiplication by powers of w are replaced by multiplication by powers of 2 which can be implemented by a simple binary shift. Thus, by proper scaling a zero roundoff error computation may be obtained. Also, the method is potentially faster than the fast Fourier transform which is faster than direct convolution for more than 128 points.

This advantage must be weighed against the difficulty of computing the result modulo b, numeric constraints relating word length, length of sequence, d, and compositeness of d imposed by the two choices of b.

## IMAGE DISPLAY

The final source of artifacts is perhaps the most important since a diagnosis is made by visually observing a reconstructed image.

Two important aspects of computed tomography reconstructed images should be mentioned. These are the large number of distinguishable gray levels and the use of pseudocolor. Computed tomography has the capability of producing images with about 1000 distinct shades of gray. The storage tube displays used on the first computed tomography device[4] had a capability of fewer than 30 gray

levels. Also, the image was displayed at a temporal rate which produced noticeable flicker. More recently, digital refresh TV displays have been announ-
ced which present a flicker free display. These monochrome display devices
can produce more than 30, but still fewer than 100 levels. Thus, the gray
level windowing technique must be used. This technique permits viewing of at
most 10% of the information available. Therefore, the relation of small objects to the overall structure may be obscured by the limited viewing range.
The limitation to fewer than 100 gray levels for monochrome image presentation has not prevented the widespread use of computed tomography. A method
for increasing the number of levels which may be perceived to the order of
1000 is by the correct use of pseudocolor. The human is capable of distinguishing several thousand colors. For example, a popular paint manufacturer
sells more than 700 colors of paint. Furthermore, even though the number of
colors, reproducible on a TV display is significantly less than perceivable
colors, it is still on the order of 1000. The difficulty with the use of pseudocolor is the fact that a perceptual distance function in color space which could
be correlated with the numerical distance in the reconstructed images has not
yet been discovered. Thus, confusion in relationships and especially artifact
edges are easily produced. This effect may be minimized by the use of a familiar color mapping such as the rainbow spectra.

## SUMMARY

In this paper we have attempted to point out the importance of an awareness to artifacts in medical reconstructed images. The application of finite
field transformations for the convolution can produce a zero roundoff error.
The other sources of artifacts can at least be controlled by careful device design and usage. The revolutionary new area of computed tomography has been
a significant breakthrough in medical imaging; however, like all scientific instruments the results must be carefully scrutinized.

## REFERENCES

1.  R. Gordon, G. T. Herman, and S. A. Johnson, "Image Reconstruction
    from Projections," Scientific American, October 1975.

2.  G. C. Huth and E. L. Hall, "Computed Tomography and Its Application
    to Nuclear Medical Imaging," Computers in Medicine, (to be published).

3.  I. S. Reed and T. K. Troung, "The Use of Finite Fields to Compute Con-
    volutions," IEEE Transactions on Information Theory, vol. IT-21, no. 2,
    March 1975.

4.  G. N. Hounsfield, "A Method of and Apparatus for Examination of a Body
    by Radiation such as X or Gamma Radiation," Patent Specifications
    1283915, Patent Office, London.

*National Bureau of Standards Special Publication 431*
*Proceedings of the* FOURTH INTERNATIONAL CONGRESS FOR STEREOLOGY
*held at* NBS, Gaithersburg, Md., September 4-9, 1975 (Issued January 1976)

ARTIFICIAL STEREO: A GENERALIZED COMPUTER ALGORITHM FOR COMBINING MULTI-CHANNEL IMAGE DATA

by William G. Pichel, R. L. Brower, D. R. Brandman and R. J. Moy

*National Environmental Satellite Service, National Oceanic and Atmospheric Administration (NOAA)*
*Washington, D. C. 20233, U.S.A.*

MULTI-CHANNEL INTERPRETATION PROBLEM

Recent advances in satellite scanning radiometer technology have produced an increase in the number of channels employed to sense radiation simultaneously in different spectral regions. However, advanced techniques for rapidly interpreting the vast number of resultant images have not been developed. An artificial stereo technique shows promise of being helpful in solving this problem.

The NOAA-4 polar orbiting satellite carries a two-channel imaging scanner having a .52- to .73-$\mu$m visible channel and a second channel sensitive to thermal radiation in the 10.5- to 12.5-$\mu$m region of the infrared (IR) spectrum. Figures 1a and 1b show, respectively, a visible and an IR image constructed from NOAA-4 data. Orbital data are merged into mosaic image arrays on a polar stereographic map base with each image element representing an area approximately 10 km in width at 40° North Latitude. A 512 X 640 element subset of each mosaic is depicted. An analyst interpreting the adjacent images will alternately refer to one image, then the other, to develop a mental picture of the vertical structure of the clouds. The time required to perform this analysis could be decreased considerably if a stereogram of the visible channel were available.

Satellite images have occasionally been used to form stereograms. For instance, stereograms have been assembled from images in the overlap region of two geostationary satellites (Bristor and Pichel 1974). Unfortunately, conventional stereograms have not been produced using scanner data from polar orbiting satellites, since these scanners have not been designed to view the same geographic area twice from different vantage points. However, it was found (Pichel et al. 1973) that with a simple computer algorithm, one can create an artificial stereogram of the polar scanner visible image in which the height of each cloud element is deduced from its corresponding IR channel temperature.

ARTIFICIAL STEREO ALGORITHM

The perception of depth in a fused stereo pair requires only the presence of parallax. Parallax occurs if distances between corresponding image elements measured horizontally between the two halves of a stereo pair are different for different pairs of elements. Those elements having the smallest relative distance will appear the highest in the fused image. The results of a computer algorithm which introduces parallax into the visible channel image using height information obtained from the IR channel is shown in Figure 2. This stereogram is created a row at a time from the visible image array. Each element in a row of the IR image array is converted to a height "h" where "h" may vary from 0 (the surface level) to 20. The stereogram is then constructed by moving elements within a row of the visible image. The position of each visible element is moved "h" elements to the left in the right half and "h" elements to the right in the left half of the stereogram. Thus, the greater the value of "h" the smaller will be the distance between corresponding image elements of each half of the stereo pair and the higher the fused element will appear. During the displacement process, if two visible elements compete for the same location in the stereogram, then the higher element is retained. After all image elements for a row have been processed, gaps in the stereogram are filled by the lower of the two image elements at the extremities of each gap.

APPLICATION TO QUANTITATIVE IMAGE INTERPRETATION

If the artificial stereogram is to be used quantitatively, IR image data must first be normalized so that IR elements at the earth's surface are always assigned to level zero and cloud elements are converted to an altitude above this surface level. Ideally, vertical temperature profiles would be used for

the normalization.  However, as a first approximation, it is assumed that the
decrease in temperature with height is the lapse rate of the U.S. Standard
Atmosphere, -6.5°C per kilometer of altitude.  Each IR temperature is converted
to a height in kilometers by finding its deviation from the sea surface temperature and dividing by 6.5°C/km.  Heights from 0.0 to 0.5 km are assigned to
level 0.  Level 1 is 0.5 to 1.0 km, level 2 is 1.0 to 1.5 km, and so forth, in
0.5 km increments up to level 20 (>10 km).  Figure 2 is the stereogram resulting
from the application of the above normalization scheme.  Accuracy is a function
of the deviation of: 1) the actual vertical temperature profile from the U.S.
Standard Atmosphere, and 2) the actual emissivity of the cloud element from the
assumed value of 1.0.  However, if all errors were eliminated, the appearance of
the corrected stereogram would probably not differ significantly from Figure 2.

REFERENCES
Bristor, Charles L., and Pichel, W. G., "3-D cloud viewing using overlapped
    pictures from two geostationary satellites," Bulletin of the American
    Meteorological Society, Vol. 55, No. 11, November 1974, pp. 1353-1355.
Pichel, William G., Bristor, C. L., and Brower, R. L., "Artificial Stereo: A
    technique for combining multi-channel satellite image data," Bulletin of the
    American Meteorological Society, Vol. 54, No. 7, July 1973, pp. 688-691.

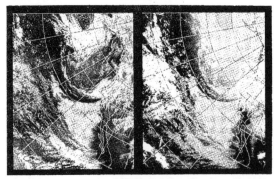

Figure 1a                   Figure 1b
NOAA-4 Visible 3/15/75      NOAA-4 Infrared 3/15/75
Polar Mosaic               Polar Mosaic

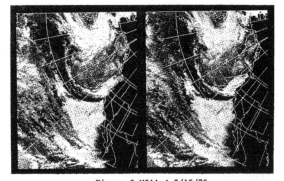

Figure 2 NOAA-4 3/15/75
Artificial Stereogram - normalized display showing clouds
at their correct altitude with grids at the surface level

*National Bureau of Standards Special Publication 431*
*Proceedings of the* FOURTH INTERNATIONAL CONGRESS FOR STEREOLOGY
*held at* NBS, Gaithersburg, Md., September 4-9, 1975 (Issued January 1976)

## STEREOSCOPIC PHOTOMACROGRAPHY AND STEREOSCOPIC PROJECTION

by Yasumichi Fujimoto and Yoshikuni Ohta
*Department of Anatomy, Osaka Dental University, Osaka, 540, Japan*

Stereoscopic photomacrography can be easily taken by placing a specimen either on a sliding or tilting stage (Figs. 3 & 2), and using a single lens reflex camera (specimen sliding method and specimen tilting method, respectively ((Fig. 1))). We have determined the optimum condition to obtain a correct stereoscopic image which is applicable to various magnifications (x1/6 to 60) and subject-lens distances. In the case of the specimen sliding method, the correct value of the sliding distances (i.e. the stereo base) is one of $D/10$, $D/20$ or $D/40$, when the subject-lens distance is D. The value of $D/10$ or $D/20$ with a viewer, $D/20$ or $D/40$ with a screen projection is recommended. In the case of the specimen tilting method, the convergence angle of $12°$ with a viewer or $4-6°$ with a screen projection is considered proper. The relationship between the specimen tilting method and the specimen sliding method is expressed as follows:

$$B = 2 \sin\theta \cdot D \qquad \text{where, } B = \text{stereo base and } 2\theta = \text{convergence angle.}$$

A stereoscopic projection is facilitated by a polarization projection method. "Stereo Black Screen" (Sun Screen Mfg. Co., Ltd.) has been found to be the far better screen than the aluminum powder-coated screen or the aluminum plate screen. The Stereo Black Screen is coated with scale-leaves of the scabbard fish. The projected images of a stereo pair must be superimposed on the screen precisely. Since this procedure is difficult to align by hand, two different kinds of equipments have been designed. First, ordinary projectors are placed on two different types of tables, one being inclined and the other is able to be rotated (Fig. 4). Secondly, ordinary projectors are placed back to back and the images are projected onto two mirrors, one inclined and the other is capable of being rotated (Fig. 5). Both of these set-ups are driven finely by reversible motors and can be remotely controlled by a speaker. Therefore, in both cases, the projected images can be adjusted precisely to give an excellent stereo image.

### EXPLANATION OF FIGURES

Fig. 1  A schema of the specimen sliding method and the specimen tilting method.

Fig. 2  The specimen tilting stage.

Fig. 3  The specimen sliding stage.

Fig. 4  Set-up of stereoscopic projection (1). Inclining stage IS, rotating stage RS, remote control devise C, polarizing filters P and polarized eyeglasses G.

Fig. 5  Set-up of stereoscopic projection (2). Inclining mirror IM, rotating mirror RM and power supply box S.

Figs. 6 & 7. Stereo-photographs.

Fig. 6  A resin cast of the vascular pattern in the pig spleen. Specimen sliding method (sliding 0.6mm, i.e. $D/20$). x43.

Fig. 7  A cleared specimen of the intestinal villi of the rat. The vascular system is injected with India ink and red resin. Specimen tilting method (convergence angle $12°$). x 20. (Specimen courtesy of Assistant Professor T. Tokioka.)

### BIBLIOGRAPHY

1) Fujimoto, Y.: Stereoscopic photomacrography and stereoscopic projection, J. Med. Photogr., 11:43-59 (1975)

2) Martinsen, W.L.M.: Gross specimen photography, Med. biol. Ill., 2:179-190, (1952)

3) Morgan, W.D. and Lester, H.M.: Stereo realist manual, 283-287, 335-359, (1954), The Fountain Press, London.

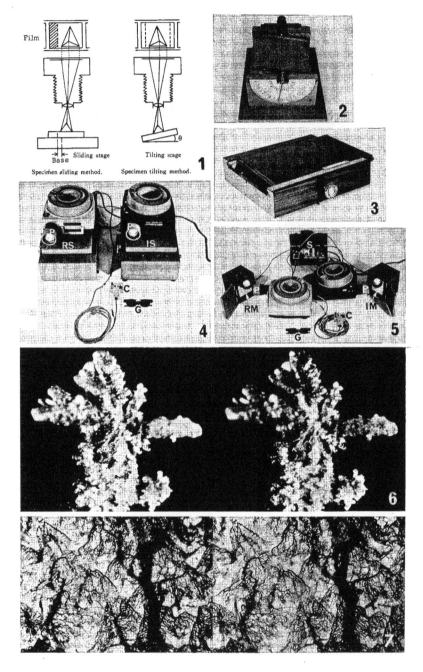

National Bureau of Standards Special Publication 431
Proceedings of the FOURTH INTERNATIONAL CONGRESS FOR STEREOLOGY
held at NBS, Gaithersburg, Md., September 4-9, 1975 (Issued January 1976)

HOLOGRAPHIC METHODS FOR IN-DEPTH VIEWING OF MICROSCOPIC SUBJECTS

by R. J. Schaefer and J. A. Blodgett
Naval Research Laboratory, Washington, D.C. 20375, U.S.A.

ADVANTAGES OF HOLOGRAPHY

For the study of microscopic subjects, holography has two primary advantages over conventional optical methods. First, the hologram reconstructs an image of the subject in full depth, so that viewing is not confined to a single plane of focus. Second, the reconstructed image contains phase information, and it can thus be analyzed by interferometric techniques. For the study of transient phenomena occurring at unpredictable locations these characteristics are especially valuable. We have therefore used holography to study the micromorphology of growing crystals, which is frequently influenced by randomly distributed defects or erratic convection currents. Our apparatus and results have been described in detail elsewhere (1).

HOLOGRAPHIC TECHNIQUE

In a holographic microscopy system, image resolution can be traded off with volume of space recorded. Some systems have been built (2) in which the image is first magnified by a microscope and then recorded holographically. These systems can attain excellent optical resolution, and the image is recorded in much greater depth than that recorded by a conventional microscope (3), but the field of view is no larger than that of the conventional microscope. For our studies of crystal growth, we wished to observe a volume of one or more cubic centimeters, and therefore had to be satisfied with slightly diminished optical resolution. Our approach was generally to reconstruct a real image of the subject at unit magnification, and then to study this image with a high resolution microscope. In some cases, a lens was used between the object and the hologram in both recording and reconstructing stages (4), but although this method gave slightly improved resolution it was rather sensitive to the alignment of the lens and was therefore infrequently used. More often, the object was placed a few centimeters in front of the holographic film or plate with no intervening optics. Holograms of a high resolution test target indicated that 300 to 500 lines per millimeter could typically be resolved in the reconstructed image.

A high quality undistorted real image for microscopic examination was obtained by using a collimated reference beam when recording the hologram. Reconstructing with a collimated beam then produced either a real or a virtual image at unit magnification, depending on the orientation of the hologram. For recording holograms in rapid sequence, 35-mm film in a motorized camera body was used, but for the highest quality images glass plates were required.

HOLOGRAPHIC INTERFEROMETRY

Interferometric techniques were generally used for analysis of the reconstructed images. In many cases a Mach-Zender interferometer arrangement was used to combine the reconstructed image with a simple plane wave and thus obtain an interferogram showing spatial variations in the specimen's optical thickness. This technique requires good mechanical stability of the optical system during reconstruction.

223

Double-exposure holography is a unique method which was used for interferometric measurement of the changes in shape of an object. Two holographic exposures were recorded in sequence on the same plate, and the hologram then reconstructed two images of the object, with interference fringes delineating any change which took place in the interval between exposures. The fringes in this case defined contours of equal displacement of the object surface.

APPLICATIONS

We have used holography to study the growth of crystals of certain transparent organic materials which exhibit crystal growth forms similar to those found in solidifying metals, thereby revealing processes which occur, but cannot be seen, in metal ingots and castings. The phenomena which have been studied include dendritic crystallization, morphological stability, and crystalline anisotropy. The Mach–Zender interferometer technique was frequently used in the reconstruction stage, so that fringe spacings and orientations could be adjusted for convenient analysis of the transient features recorded by the holograms.

We have also studied the electropolishing of copper in phosphoric acid solutions, using double-exposure holography to measure the convective flow of copper-enriched electrolyte and the resulting inhomogeneity in the rate of removal of metal from the copper specimen. For this study we also used real-time holography, in which a holographically reconstructed image of the specimen is superimposed on the specimen itself, and moving fringes are seen directly as the specimen changes.

SUITABLE SUBJECTS FOR HOLOGRAPHIC MICROSCOPY

With any imaging system using coherent light, objects which are out of the plane of focus produce clearly visible diffraction patterns which can interfere with observation of objects in the image plane. Therefore the best holographic microscopy images are obtained when the object under study is relatively isolated from other objects in the field of view. In other words, whether the specimen chamber contains growing crystals, protozoans, or particulate pollutants, the density of such objects should be kept small.

Similarly, objects with a rough or diffuse surface are difficult to observe in detail with coherent light. In particular, double-exposure interferometry effects are eliminated if the irregular roughness features on the object surface change between exposures, as occurred for example when the metal surface of an electropolishing specimen became etched.

In general, it is well to bear in mind that holography is not the final step in the study of a microscopic subject. Analysis and documentation of an experiment generally require that the reconstructed hologram be photographed and that features in the photographs be measured.

REFERENCES
1. J. A. Blodgett, R. J. Schaefer, and M. E. Glicksman, Metallography, 7, 453 (1974).
2. R. F. vanLigten and H. Osterberg, Nature, 211, 282(1968).
3. K. C. Lawton and R. F. vanLigten, in Developments in Holography, S.P.I.E. Seminar Proceedings, Vol. 25, B. J. Thompson and J. B. DeVelis, editors, (1971).
4. L. Toth and S.A. Collins, Appl. Phys. Lett., 13, 7(1968).

*National Bureau of Standards Special Publication 431*
Proceedings of the FOURTH INTERNATIONAL CONGRESS FOR STEREOLOGY
held at NBS, Gaithersburg, Md., September 4-9, 1975 (Issued January 1976)

EVALUATION OF A NEW TECHNIQUE FOR DETERMINING THE VOLUME OF THE HUMAN LEFT VENTRICLE FROM ONE-PLANE RADIOGRAPHIC MEASUREMENTS

by Richard Moore, Ivo Beranek, Sung Kim, and Kurt Amplatz
*Department of Radiology, University of Minnesota Hospitals, Minneapolis, Minnesota 55455, U.S.A.*

In clinical medicine, especially in cardiovascular radiology, it is necessary to calculate the volume of the human left ventricle (3). The results are useful to aid in the diagnosis of cardiac disease. They can suggest the best decision about therapy, either medical or surgical; and they can help to evaluate the effects of treatment (3).

The left ventricle is a three-dimensional object. In clinical practice, a two-dimensional image is available from a single direction, or at most two directions, by means of cineangiography. This produces an X-ray movie of the contrast-filled left ventricle.

There are many methods in use to calculate the volume of the left ventricle based on radiographic measurements made using the image (1,2,3,4). Most methods are based on representing the heart by an assumed mathematical model, such as an ellipsoid of revolution and others (3,4). A cine frame is chosen, the silhouette of the ventricle is traced, and the tracing is measured or planimetered. In order to calculate the volume from such measurements, besides a factor which corrects for the nonparallelism of the X-ray beams, several assumptions are necessary (3). However, it is not necessary to assume that the volume must be approximated by a regular geometric solid (1,2,3). One can use Simpson's rule to approximate the volume of an irregular solid (1,2,3).

SCOPE OF THE STUDY

The aim of our contribution was to compare the results of calculations of left ventricular end diastolic and end systolic volume from single plane left ventriculographic measurements using two methods. The first method used, as a mathematical model, an ellipsoid of revolution. It was described by Sandler and Dodge (4) and has been most commonly used in clinical practice. The second method used was a modification of Simpson's rule. It was described by a Swedish radiologist named Ahlberg (2), who verified its applicability only using dogs. It is a new method, and its correctness has never before been verified using humans. Both methods are based on the assumption that the left ventricle has a circular cross-section perpendicular to its longest axis (3). The Ahlbert method appears intuitively to be a better one than that of Dodge, especially calculating the volumes of the left ventricle if distorted by ischemic heart disease.

METHOD

Both methods were used to calculate the end diastolic and end systolic volume in 24 patients with ischemic heart failure. Left ventriculography was performed on all patients in the RAO (right anterior oblique) position, in which the longest axis of the left ventricle is parallel to the film. Twelve of the patients had a normal ventricular shape, and twelve had abnormalities in the shape of the left ventricle. From each cine was taken one pair of silhouettes representing the left ventricle in end diastole and in end systole. From each of these tracings, the volume of the ventricle was calculated by both the Dodge and Ahlberg methods. This procedure eliminated several sources of variation due to the magnification factors, possible error in selecting the cine frame traced, and subjective error in judging the border when the tracing was

R. Moore, I. Beranek, S. Kim, and K. Amplatz

made. All the measurements and calculations for one method were done by one
person, thus eliminating inter-observer variation. Separate persons performed
each method, thus eliminating biasing the results. In effect, a double blind
study was performed.

RESULTS

Table 1 shows the values of the coefficients of correlation between results,
from the area-length method and the modified Simpson's rule method, for the
values of end systolic volume. 1a gives the coefficient of correlation, and its
level of significance, for both groups, totaling 24 shapes, 1b gives the values
for the twelve normal shapes, and 1c gives the values for the twelve abnormal
shapes. Table 2 shows the corresponding values for the end diastolic volume for
the same three groups, respectively.

DISCUSSION

It was surprising that there were no significant differences between the
results from either method. We believe that this finding can be explained by an
examination of the Dodge method. This method includes a kind of integration,
which is used for calculating the minor diameter from the planimetrically-
measured irregular area (3).

From the point of view of clinical practice, there are great differences
between the speed and ease of using these two methods. The Dodge method can
be performed manually and produce results in a short time (ten-fifteen minutes/
case). However, the Ahlberg method is very tedious for manual calculation
(requiring two or more hours/case). It is susceptible to subjective errors,
and it requires a dedicated or time-sharing computer to make its application
practicable. We have found that the Ahlberg method doesn't provide any improvement in the accuracy of the results for clinical practice. From the mathematical
point of view, both methods give almost the same results.

Davilla and San Marco (2) reported a comparative study of several roentgen
volumetric methods based on images made from two directions. Their study included the comparison of the results for ventricular volume calculated using
an ellipsoid of revolution as a model with the results calculated using other
methods based on Simpson's rule. They found a high correlation among the
results from such methods. Consequently, their results are consistent with our
results, also showing a high correlation coefficient between measurements made
from one direction and comparing results using the Dodge method with a new method
based on a modification of Simpson's rule. However, an empirical correction
factor must be applied to the results from both methods (3).

SUMMARY

The determination of left ventricular volume is an important problem in
clinical medicine. We investigated two methods for the determination of the
three-dimensional left ventricular volume on the basis of single plane silhouettes. One method used an area-length measurement, and the other method used
a modification of Simpson's rule. We found that for the left ventricular volume,
both methods gave results which were not significantly different. However, one
method is relatively easy and simple and can be done quickly by hand; and the
other method is complex and requires either a computer or a lengthy manual
calculation.

This work was supported in part by Public Health Service Grant No.
HL-16484, from the National Heart and Lung Institute.

226

REFERENCES

1. Ahlberg, N. E.; Paulin, S.; and Seeman, T.: Left Ventricular Changes During Coronary Artery Occlusion in Dogs. Acta Radiologica Diagnosis 12: 798, 1973.

2. Davilla, J. C. and San Marco, M. E.: An Analysis of the Fit of Mathematical Models Applicable to the Measurement of Left Ventricular Volume. Amer. J. Cardiol. 18: 31, 1966.

3. Mirsky, I.; Ghista, D. N.; and Sandler, H.: Cardiac Mechanics: Physiological, Clinical, and Mathematical Considerations. New York, Wiley, 1974.

4. Sandler, H. and Dodge, H. T.: The Use of Single Plane Angiocardiograms for the Calculation of Left Ventricular Volume in Man. Amer. Heart J. 75: 325, 1968.

VALUES OF THE COEFFICIENTS OF CORRELATION BETWEEN
RESULTS FROM THE AREA-LENGTH METHOD (1)
AND THE MODIFIED SIMPSON'S RULE METHOD (2)

TABLE 1 END SYSTOLIC VOLUME

|  | Number of Patients | Coefficient of Correlation | Level of Significance |
|---|---|---|---|
|  | N |  | p |
| a. Both groups | 24 | 0.98 | >0.05 |
| b. Normals | 12 | 0.99 | >0.05 |
| c. Abnormals | 12 | 0.99 | >0.05 |

TABLE 2 END DIASTOLIC VOLUME

|  | Number of Patients | Coefficient of Correlation | Level of Significance |
|---|---|---|---|
|  | N |  | p |
| a. Both groups | 24 | 0.99 | >0.05 |
| b. Normals | 12 | 0.98 | >0.05 |
| c. Abnormals | 12 | 0.98 | >0.05 |

FIGURE 1

Tracing of the left ventricular cavity. The straight line represents the longest chord, L, through the tracing. The area, A,
of the tracing is determined by planimetry. The volume, V, is
determined using

$$V = (8/3\pi) \cdot A^2/L$$

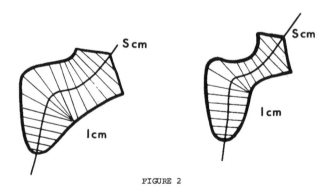

FIGURE 2

Tracings of the left ventricular cavity in end-diastole and end-
·systole with the midline (S) and perpendicular diameters (1) drawn at intervals of 0.5 cm.

228

*National Bureau of Standards Special Publication 431*
*Proceedings of the* FOURTH INTERNATIONAL CONGRESS FOR STEREOLOGY
*held at* NBS, Gaithersburg, Md., September 4-9, 1975 (Issued January 1976)

MOVIE: PRIM—9
VIEWING HIGH DIMENSIONAL POINT CLOUDS

Narrated by J. W. Tukey, *Department of Statistics, Princeton University, Princeton, NJ 08540, U.S.A.*
Produced at SLAC by Bin 9 Productions
Presented by Prof. G. S. Watson, *Princeton University*

## Review

This movie demonstrates a special visual display terminal associated with a special program for the Stanford computer. Multidimensional data are displayed to show any selected three dimensions simultaneously with the display constantly rotated about a selected axis in order that the 3-d form may be recognized. A manual control panel allows selection of the dimensions to be displayed and choice of the rotation axis, speed, and direction. Subsets of the data can be masked out in order that other sets can be more clearly seen.

The demonstration shows data for 500 nuclear events. Each event originally supplied 9 measured parameters. Conservation and symmetry rules allow transformation to 5 new orthogonal parameters which were displayed. The data grouped into several distinct and separate clouds in hyperspace. One cloud could be seen to have a triplet layered structure when viewed from one specific direction.

This reviewer believes that the original Prim-9 system could be applied directly to feature recognition or feature classification using several measured parameters of each feature of a microstructure. Such data are presently available from the quantitative television microscope systems. Reasonable expansions of the program should permit display of functional equation solutions in data systems. Three-dimensional display of reconstructed real solids should be possible when the necessary information is available from serial sections or projections.

The continuous rotation system proves to be a highly effective method of displaying 3-dimensional structure. Facilities to exploit this principle should be useful in many stereological applications.


George A. Moore

# 5. APPLICATIONS IN MATERIALS

*Chairpersons:*

F. N. Rhines
J. H. Steele, Jr.
H. E. Exner
J. Gurland
P. Stroeven
G. Ondracek
R. M. Doerr

*National Bureau of Standards Special Publication 431*
*Proceedings of the* FOURTH INTERNATIONAL CONGRESS FOR STEREOLOGY
*held at* NBS, Gaithersburg, Md., September 4-9, 1975 (Issued January 1976)

APPLICATIONS OF QUANTITATIVE MICROSCOPY IN METAL SCIENCE

by F. N. Rhines
*Distinguished Service Professor, Department of Materials Science and Engineering, University of
Florida, Gainesville, Florida 32611, U.S.A.*

Much of the daily usage of quantitative metallography is pragmatic in nature, being concerned mainly with the achievement of precision in routine measurements. My present interest is not in such applications, but rather in uses wherein quantitative microscopy provides the mathematical basis upon which new science is built. It has long been the dream of the physical metallurgist to so associate microstructure with the various properties of metals and alloys as to create an exact science. With the growth of quantitative microscopy it seemed that we had been given all that was needed to regularize the structure-property relationships, but progress in this direction has not been rapid. The geometry of microstructure has turned out to be far more subtle than was generally realized. It is my present purpose to assess the current status and to suggest where further progress may be forthcoming.

One of the first identifiable scientific uses of metal microstructure was in the development of phase diagrams. B. H. Roozeboom (1), who first assembled the scattered knowledge of phase equilibria in the form of phase diagrams, became interested in metals through the works of Sir W. Roberts-Austen (2) and of A. Sauveur (3) on equilibria in the iron-carbon system. The correspondence between microstructure and the phase transformations that had been studied thermally became apparent. Thenceforth, metallography became a standard tool of phase equilibrium investigation. In this role it furnished exact numerical data. The number of phases P, referred to in the Phase Rule (i.e. $P + F = C + 2$), is subject to exact count in the microstructure of the equilibrated alloy. No identification of the chemical nature of the phases is required. The Phase Rule asks only: "How many?" And metallography gives this information directly and precisely. This was an auspicious start for physical metallurgy and gave promise of a brilliant future.

Another early achievement of metallography came with the studies of F. Osmond (4) who, through the observation of the symmetries of micro-indents, was able to deduce the crystal structure of a phase seen with the microscope. His work was the more remarkable because he was able to apply it at high temperature as well as at room temperature. Again, Osmond's observations provided exact numerical data of fundamental scientific significance. Long before the advent of X-ray crystallography he had established the crystal structures of both alpha and gamma iron, as well as those of some non-ferrous metals. The fact that World War I intervened, after which the X-ray methods preempted the field, detracts in no way from the magnitude of Osmond's scientific accomplishment. At that point in time, it must have appeared that physical metallurgy was about to join the ranks of the exact sciences, chemistry and physics.

Early in this century, studies upon the grain structure of single phased alloys were pursued by Z. Jeffries (5) with vigor, but limited success. He proposed two kinds of measurement of grain size. One consisted in counting the number of grains, as seen in unit area of microstructure at standard magnification.

The other depended upon counting the number of grains crossed by standard length of line laid upon a microsection. Neither did the job to the satisfaction of all and ASTM, among others, complicated the situation by adding a third method based upon a comparison of the unknown microstructure with a series of "standardized photomicrographs." Each worker followed his own taste in the matter of selecting a measuring technique. The results were considerably less than satisfactory from a scientific point of view. This example has meaning for us today, because we are still enmeshed in its web of confusion and our liberation seems to be a requirement for us to resume the dream of creating an exact science of microstructure.

Meanwhile physical metallurgy has been pursuing a qualitative course of development. It is, for example, a familiar fact that the grains of a polycrystalline aggregate grow at elevated temperature, but the kinetics of the process, as they have been seen until very recently, have appeared complex and debatable. Similarly, cold worked metals are known to recrystallize upon heating, but the association between the original and the recrystallized structures remains incompletely resolved. Phase changes, like the decomposition of Austenite, have been observed since the time of Sorby. Much is known in a qualitative way of the mechanisms and their rates, but the processes, as a whole, still seem bewilderingly complex. With the advent of the transmission electron microscope, dislocations became observable and there was hope, for a time, that the integrated processes of plastic deformation would become simplified and understandable. Instead, the observations have accumulated and the ultimate synthesis of deformation remains out of reach. The field ion microscope provided the ultimate in resolution in its ability to reveal the locations of individual atoms in the metal structure and to reveal the detailed structures of such features as grain boundaries. Yet it has given us no new exact science. Most recently, the scanning electron microscope has provided the depth of focus needed to view fracture surfaces of metals. By this route we seem to have returned to the grounds of the primitive metallurgist, with greatly refined vision, but what we see in this way seems even less susceptible to expression in exact terms. And, finally, the Quantimet has taken much of the labor out of the measurements of quantitative microscopy, but has added no new metal science. We know so much about structure-property relationships in metals, but have been able to organize our knowledge so little. Surely, there is a basic element missing and I suggest that it lies in our understanding of the geometry of microstructure.

The long confusion over the measurement and expression of grain size is indicative of the presence of a very real geometric blockade, the nature of which has not generally been recognized, but which has stood in the way of our making further progress with microstructure as a science. The blockade began to yield about twenty years ago, however, when C. S. Smith (6) and others undertook the development of a system of relationships of quantitative metallography. It was a crucial aspect of this development that each of the relationships was based upon a sound mathematical derivation. The grain size measurements, and others that had proved unsatisfactory in the past, had not been so based. Seven proven relationships of quantitative metallography have thus far resulted from these efforts.

1. <u>Length of line</u> in unit volume $L_V$ is equal to twice the

number of point intercepts in unit area of microsection $N_A$ (Figure 1):

$$L_V = 2 \, N_A \qquad \text{(units cm}^{-2}\text{)}$$

2. <u>Area of surface</u> in unit volume $S_V$ is equal to twice the number of intercepts of unit length $N_L$ of test line with lineal feature in the microsection (Figure 2):

$$S_V = 2 \, N_L \qquad \text{(units cm}^{-1}\text{)}$$

3. <u>Volume fraction</u> of phase in unit volume $V_V$ is equal to the area per unit area $A_A$ in the microstructure, or the length in unit length $L_L$ of lineal traverse, or the point fraction $P_P$ of random points upon the two-dimensional microstructure (Figure 3):

$$V_V = A_A = L_L = P_P \qquad \text{(unitless)}$$

4. <u>Total curvature of surface</u> in unit volume $M_V$ summed over all of the surface in unit volume (i.e., $\int_{S_V} (1/r_1 + 1/r_2) dS_V$), is equal to pi times the number $T_A$ of tangencies $T_A$ with lineal feature in the microstructure, produced by line sweeping through unit area (Figure 4):

$$M_V = \pi \, T_A \qquad \text{(units cm}^{-2}\text{)}$$

5. <u>Total curvature of line</u> in unit volume $C_V$ is equal to pi times the number of tangencies $T_V$ occurring between lineal feature in space and unit area of surface swept through unit volume of material. This measurement can be made by the use of serial sections (Figure 5):

$$C_V = \pi \, T_V \qquad \text{(units cm}^{-3}\text{)}$$

6. <u>Number</u> in unit volume $N_V$ is a topological parameter, usually counted directly in space by serial section metallography (Figure 6). An exception exists in phase counting, which is performed upon the two-dimensional section. (units cm$^{-3}$)

7. <u>Connectivity</u> in unit volume $G_V$ is also a topological parameter and is a property of networks. It is measured directly in space, usually by counting the number of branches $B_V$ and nodes $K_V$ of the net in unit volume, by serial section metallography, (Figure 7) and by applying the relationship:

$$G_V = B_V - K_V + 1 \qquad \text{(units cm}^{-3}\text{)}$$

These geometric properties have several characteristics that distinguish them and that define a unique geometry of microstructure. It should be recognized, in the first place, that microstructures do not have repeating patterns, in the manner of crystals. It is not possible, for example, to identify a typical unit cell of microstructure and then to reproduce it repeatedly to recreate the configuration of the microstructure in space. It is true, nevertheless, that the geometric properties defined by quantitative microscopy are capable of statistical repetition throughout the structure. For this reason, the total of any geometric property in a specified volume of material can be defined exactly, can be measured with precision and can be expressed usefully in mathematical terms. All of the properties

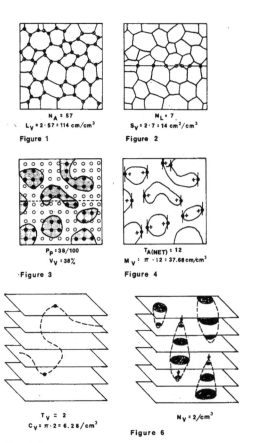

$N_A = 57$
$L_V = 2 \cdot 57 = 114\,cm/cm^3$

Figure 1

$N_L = 7$
$S_V = 2 \cdot 7 = 14\,cm^2/cm^3$

Figure 2

$P_P = 38/100$
$V_V = 38\%$

·Figure 3

$T_{A(NET)} = 12$
$M_V = \pi \cdot 12 = 37.68\,cm/cm^3$

Figure 4

$T_V = 2$
$C_V = \pi \cdot 2 = 6.28/cm^3$

Figure 5

$N_V = 2/cm^3$

Figure 6

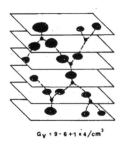

$G_V = 9 - 6 + 1 = 4/cm^3$

Figure 7

of quantitative microscopy are thus expressed as totals in unit volume.

Another characteristic of the parameters of quantitative metallography is that all are simple magnitudes, insensitive to shape in the sense that each can be measured and expressed without any knowledge whatever of the shape of the subject being measured. This means, for example, that the length of a line is independent of whether the line is straight, curved, tied in a knot or branched (Figure 8). The area of a surface is likewise independent of its outline, or whether the surface is curved or folded (Figure 9). The magnitude of a three-dimensional body is its volume and is independent of its height, thickness, width, surface area or any other measure of its shape (Figure 10). Curvature of surface is in no way affected by how the surface is deployed in space (Figure 11). The curvature of line is not changed by putting the line together in various ways or by segmenting the line (Figure 12). It is rather obvious that the number of anything is independent of the shapes of the entities being enumerated (Figure 13). And finally, the connectivity of any network, including, for example, the edges of a polyhedron, is obviously independent of form (Figure 14). Such insensitivity to shape is of great potential value in relating properties to microstructure, as will be illustrated presently.

An obvious, but often overlooked, property of microstructure is that it is continuously connected in space. Unlike a shower of snow flakes, or a heap of sand, the parts of the microstructure are fitted together so as to fill all space. This remains true even of a porous material, if its total volume is defined and the porosity is regarded as one of its phases, i.e. the gas phase. As a consequence, any change in the microstructure, that affects its volume, must bring about a compensating adjustment elsewhere, to preserve the continuity of the structure. The geometric ability to deal with this kind of requirement is provided by the topological parameters, number and connectivity. All of the other parameters of quantitative metallography are compatible with connectivity, in that none is shape sensitive.

When the fundamental parameters of quantitative metallography are combined, as ratios, they yield average properties. Thus, the reciprocal of the number of grains in unit volume $1/N_V$ is the average grain volume, an absolute measure of grain size. Also the volume of a phase divided by the area of its enclosing surface $4V_V/S_V$ is the mean free path through the phase. A special case of the latter is the mean intercept of contiguous grains $1/N_L$, which is the same as $2V_V/S_V$ when $V_V = 1$. Also the ratio of the total curvature to the total surface area $M_V/S_V$ is the average mean curvature of the surface. Essentially every ratio that can be constructed among the fundamental parameters has real geometric meaning. This greatly enlarges the scope of application of these parameters in describing metallurgical states and processes. A rather curious limitation that intrudes with the making of ratios, however, is that shape insensitivity is sometimes lost by making the ratio.

An interesting and important example of the loss of shape insensitivity is found in the case of the grain mean intercept $1/N_L$. This ratio is one of the Jeffries' measures of grain size. It has been widely assumed that it is either the average grain diameter, or is proportional to the grain diameter. C. S. Smith and L. Guttman (6) long ago pointed out that this is an incorrect assumption. The average grain diameter can be computed from the

237

Lines of Equal Length

Figure 8

Surfaces of Equal Area

Figure 9

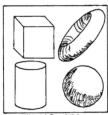

Bodies of Equal Volume

Figure 10

Surfaces of Equal Curvature

Figure 11

Lines of Equal Curvature

Figure 12

Number of Objects

Figure 13

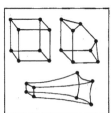

Figures with Equal Connectivity
of Edge

Figure 14

average grain volume $1/N_V$, however, which must be measured in three dimensions, as by serial section analysis, but cannot be deduced from any measurement made exclusively in the two-dimensional section. Because $1/N_L = 2/S_V$ it is apparent that the mean intercept is equal to the reciprocal of one half of the grain boundary surface area. It varies, therefore, with the area of the grain boundary, rather than with the volume of the grain. Since the total surface area of the grains can vary with the grain shape, while the grains are maintaining constant volume, it is clear that the mean intercept must be shape sensitive. It is altogether possible for two specimens to have the same grain mean intercept and yet have very different grain mean volumes and average grain diameters, or vice versa. Small differences in grain shape appear to have large effects in this respect.

The other Jeffries' (5) measure of grain size, i.e. the number of grains intercepted by unit area of microsection, is related to the measurement of the length of lineal feature, $L_V = 2N_A$. It deals with the polycrystalline aggregate as though each grain center were a node, connected with neighboring grain centers by lineal branches. The number of grains intercepted by the counting area is then equal to half the total length of the connecting branches. It can be shown that the average branch length is equal to the average grain diameter, wherefore the Jeffries' (5) grain size is seen to be half the length of chain that could be made by stringing all of the grains in unit volume beadwise upon their individual diameters. To ascertain from this the true average grain diameter it is still necessary to evaluate $N_V$, because $\bar{D} = 2N_A/N_V$. In short, none of the standard methods for measuring grain size gives any index of grain volume, or provides access to any volume property of the grain structure, such as average grain diameter. In order to get at the volume properties it is necessary to resort to three-dimensional measurement, as in serial sectioning. This is unfortunate, because of the practical difficulty of serial section analysis. There is nothing to be gained, however, by ignoring this inexorable fact.

It may seem surprising, in view of what has just been said, that some workers have found simple relationships between "grain size" and certain mechanical properties of metals. Of particular note is the supposed relation between "grain size" and the "yield point" in metals. Many workers have reported a simple relationship of this kind. When the data are carefully evaluated, however, it appears that the relation weakens, the more carefully the mechanical measurement is made. As the "yield point" is taken closer to the real elastic limit, its sensivity to grain size tends to vanish. This implies that the sensitivity to grain size is a property of the degree of plastic flow, rather than of the elastic limit. The finding of a simple relationship between "grain size" and "yield point" is evidently fortuitous. Further, since the measures of grain size that are in use are related to the grain boundary area, it seems likely that the resistance to plastic flow is, in fact, a function of grain boundary area. This example serves to illustrate the peril inherent in using unproved geometric relationships, as well as loosely defined properties of materials, in any search for scientific truth.

The implication of the foregoing example has, in fact, been borne out by other investigators who have found the resistance to plastic flow proportional directly to the grain boundary area. Of these, perhaps the first was H. T. Angus and P. F. Summers (7) who found a nearly linear relationship between the Brinell Hardness

239

Number of copper alloys and their grain boundary areas.
Subsequent studies of this relationship, (8) using proven
quantitative metallographic methods for the measurement of the
grain boundary area $S_V$, have shown repeatedly that the Brinell
Hardness Number (BHN), used as a measure of the load required to
produce a specified deformation, increases in direct and precise
proportion to the increase in the surface area (Figure 15):

$$BHN \text{ (increase)} = k\ S_V$$

where k is a constant of proportionality. This is what should be
expected if the grain boundary area were to act as a barrier to
the passage of dislocations. The resistance should, indeed, be
directly proportional to the total grain boundary area. In this
example, the shape of the grain boundary is not involved, because
$S_V$ is shape insensitive.

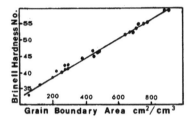

Figure 15
Brinell Hardness of annealed
70 - 30 brass versus grain
(and twin) boundary area.

A more impressive case (9) has been found in the relation of
the sintering force to surface tension, through the microstructure
of the sinter-body. The sintering force is defined as that
tensile load which, when applied to a sintering mass of metal
powder, will just prevent the shrinkage of the body in the
direction of loading. It is measured in the laboratory by spring
loading a bar composed of metal powder, suspended in a sintering
furnace at a sintering temperature. The length of the bar is
monitored continuously and the load is read when the rate of
length change is zero. At the same time, the bar is removed from
the furnace and is sectioned for metallographic analysis. The
analysis consists of reading $V_V$, $S_V$ and $M_V$ of the porosity in the
microstructure. It is then found that the sintering force per
unit area F/a is related to the surface tension $\gamma$ by the
expression (Figure 16):

$$F/a = \gamma\ V_V \cdot M_V/S_V$$

This relation is unusual in several respects. In the first place,
it is an exact relation, involving no adjustment constant.
Further, its geometric analysis involves an evaluation of the
curvature of a multiply-connected surface of indescribable
intricacy, a feat which would be unapproachable by any
conventional geometric method and which becomes possible only
through the shape insensitivity of the total curvature parameter.
It is true, however, that the ratio $M_V/S_V$ is not, of itself, shape
insensititve. The expression is applicable, therefore, only as
long as the local curvature of surface is the same as the average
curvature, a condition which is closely approximated during the
second stage of sintering. Especially, it is to be noted that the
solution has not made use of any unit cell type of model, as has
been the custom in the analysis of similar problems. In these
ways, major simplification, as well as precision, has resulted

from an application of the special geometry of quantitative microscopy.

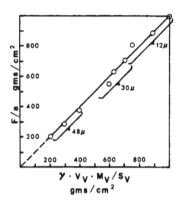

Figure 16
Directly measured sintering force F/a for copper powder in hydrogen at 1000°C versus the sintering force calculated from metallographic measurements $V_V \cdot M_V / S_V$ and the surface tension $\gamma$. Three ranges of copper powder size are included, namely: 12,30 and 48 μ.

A quite different case, one involving continuity of structure was encountered in the analysis of steady state grain growth in single phased metals and alloys. (10) Here, the total volume of the system remains fixed and grain growth occurs by the loss of grains from the system. Continuity is maintained by the grain boundaries adjusting, under the influence of surface tension, to the new topological state created by the change in the number of grains. The topological state is thus described by the number of grains $N_V$ and the average grain volume by its reciprocal $1/N_V$. The kinetics of grain growth are controlled by the rate of movement of the grain boundary, which depends upon its mobility μ, its surface tension γ and its relative curvature σ, which is equal to $M_V \cdot S_V / N_V$. During steady state grain growth the relative curvature remains constant, so that the rate of migration of the grain boundary is constant. Accordingly, grain growth is found to be proportional directly to the time t (Figure 17):

$$1/N_{V_t} = 1/N_{V_0}\{1 + 4\mu\gamma\sigma t\}$$

Once again, a very simple relationship has emerged through the use of fundamental quantitative metallographic parameters to frame the problem. There are no adjustable parameters. There is no unit cell model. The result has the characteristics of a natural law.

This brings my story to the present time. The progress that has been quoted surely indicates a trend. It seems that we should look forward to growth, perhaps accelerating growth, in the exact science of physical metallurgy, through the application of the fundamental geometry of quantitative metallography. Since my purpose in writing this paper is, in part, to attempt a glimpse into the future, I shall hazard a few guesses concerning areas that may yield to metallographic analysis. Unfortunately, the most interesting and important developments are almost certain to be ones which cannot be foreseen at this time, but I shall not allow that handicap to deter me.

241

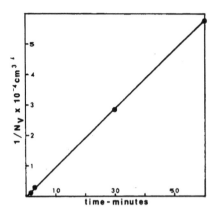

Figure 17
Grain Size of aluminum $1/N_V$
versus time at 635°C.

It is, of course, most probable that metallographic laws will arise from practical efforts to control microstructure. Such efforts are constantly in progress in various forms, such as the making of composite materials, the controlled freezing of eutectics, controlled solid state transformations, and the sintering of mixed powders. In poly-phased structures of this kind, it is usually expected that the plastic properties of the product will vary with the degree of refinement of the dispersion. There have been a number of studies of the properties of such aggregates, but none is known to this writer to have made full use of the tools of quantitative metallography. Intuitively, it seems that the plastic behavior of aggregates should be related to the area of phase interface. The application of the exact geometry of quantitative metallography should result in the development of simple and useful laws in this field.

It is to be expected, also, that exact science will arise across the broad front of metallurgical properties and processes, as old problems are recast in terms of the geometry of quantitative metallography. The large body of knowledge of structure related phenomena, that has accumulated during the past half century, is mainly couched in terms of highly complex equations, derived upon the basis of models that fail to conform to the reality of the fundamental geometry of microstructure. The reconstruction of these in suitable microstructural terms may not be easy and in some cases may prove to be impossible, but those which do emerge in simple exact form will be of greatly enhanced value. It seems, for example, that diffusion controlled processes, like homogenization and some phase transformations, where the boundary conditions have been approximated by various unit cell models, might respond to the use of a simple parameter like mean free path. In so doing, it will be important to avoid problems of shape, either by using a shape insensitive parameter, or by finding conditions under which the process occurs with shape constant. The latter state prevailed, it will be recalled, in the case of second stage sintering and also in steady state grain growth.

In dealing with the quantification of metallurgical processes, the importance of the topological properties (number and connectivity) can scarcely be exaggerated. These are probably the most fundamental of all geometric properties. It will be recalled that the application of microstructure to the Phase Rule was upon the basis of the topological property number. Again, the solution of the grain growth kinetics was made simple by the use of the topological ratio $1/N_V$. Unfortunately, few scientists are accustomed to thinking of geometrical problems in terms of their topological properties. Yet, a process such as sintering can scarcely be visualized successfully in any terms other than the connectivity of the powder stack and its subsequent topological evolution. Whenever a process is one which can be described in terms of changes in whole numbers, like the number of particles, or the number of bonds between particles, it is probable that it will be best described in terms of topological parameters.

In addition to those familiar characteristics of metals which may be amenable to scientific description in terms of an exact geometry of metallography, there may be awaiting us new discoveries in the form of processes which are new both in their physical character and in their geometry. One such seems to have emerged in the course of the recent studies upon grain growth. The new characteristic is amenability to grain growth. Some metals are much more stable structurally than others at elevated temperature. This property has often been ascribed to the presence of pinning particles. The recent grain growth studies indicate, however, that there is a much more fundamental reason for such effects. The tendency to sustain grain growth increases with the value of σ (i.e. $M_V \cdot S_V/M_V$), which is to say that it increases with the relative curvature of the grain boundary. But, it appears, the relative curvature is a function of the range of topological forms of the grains existing in the metal and this range is unaltered by grain growth. In some aluminum samples that were analyzed by serial sectioning, it was found that the topological forms of the grains ranged from three to more than fifty faces per grain in the same piece of metal. The broader the range of topological forms, the greater the tendency to sustain grain growth. The most recent addition to this discovery is that the topological distribution can be established, more or less at will, by suitable control of the conditions of recrystallization. Thus, the high temperature stability of a metal may be built in by a deliberate and exact control of its grain topology.

Speculation of this kind could be extended indefinitely. It seems, though, that enough kinds of examples have been suggested to make it clear that there exists a vast field for the application of the geometry of quantitative microscopy in the building of exact science. I believe that we are moving, slowly but surely, toward a full quantification of all of the structure-property relationships in metal systems. Progress in this direction may eventually provide us with the power to design and to construct materials to property requirements. What more could we ask of the physical metallurgy of tomorrow?

## Bibliography

1.  B. H. Roozeboom, die Heterogenen
    Gleichgewichte vom Standpunkte der
    Phasenlehre, Verlag von Friedrich Viewig und
    Sohn, Braunschweig (1901).

2.  W. C. Roberts-Austen, Fourth and Fifth Alloy
    Research Reports, Proc. Inst. Mech. Eng.
    (1897) 31-100 and (1899) 35-102.

3.  A. Sauveur, The Microstructure of Steel and
    the Current Theories of Hardening, Trans.
    Amer. Inst. Min. Eng. (1896) 26 863-906.

4.  F. Osmond, and G. Cartaud, The
    Crystallography of Iron, J. Iron-Steel Inst.
    (1906) 71 444-492.

5.  Z. Jeffries, E. A. H. Kline and E. B. Zimmer,
    The Determination of Grain Size in Metals,
    Trans. Amer. Inst. Min. Eng. (1916) 54 594-
    607.

6.  C. S. Smith and L. Guttman, Measurement of
    Internal Boundaries in Three-dimensional
    Structures by Random Sectioning, Trans.
    Amer. Inst. Min. and Met. Egrs. (1953) 197
    81-87.

7.  H. T. Angus and P. F. Summers, The Effect of
    Grain Size Upon Hardness and Annealing
    Temperature, J. Inst. Metals (1925) 33 115-
    142.

8.  W. J. Babyak and F. N. Rhines, The
    Relationship Between the Boundary Area and
    Hardness of Recrystallized Cartridge Brass,
    Trans. Amer. Inst. Min. and Met. Egrs. (1960)
    218 21-23.

9.  R. A. Gregg and F. N. Rhines, Surface Tension
    and the Sintering Force in Copper,
    Metallurgical Transactions (1973) 4 1365-
    1374.

10. F. N. Rhines, K. R. Craig and R. T. DeHoff,
    Mechanism of Steady State Grain Growth in
    Aluminum, Metallurgical Transactions (1974) 5
    413-425.

National Bureau of Standards Special Publication 431
Proceedings of the FOURTH INTERNATIONAL CONGRESS FOR STEREOLOGY
held at NBS, Gaithersburg, Md., September 4-9, 1975 (Issued January 1976)

APPLICATION OF STEREOLOGICAL TECHNIQUES TO THE QUANTITATIVE CHARACTERIZATION OF WOOD MICROSTRUCTURE

by James H. Steele, Jr.*, Geza Ifju and Jay A. Johnson
Forest Products,Virginia Polytechnic Institute and State University, Blacksburg, Virginia 24061, U.S.A.

INTRODUCTION

Wood is the product of a complex biological system, the tree, and as such it has widely variable properties which depend upon its microscopic cellular structure. The presently accepted practice of characterizing wood structure involves an ordered sequence of two-way decisions about size, distribution and shape of the anatomical elements (1). These qualitative identification keys do not provide quantitative information about the anatomical features which occur in wood microstructure. Application of stereological counting measurements upon transverse sections of several wood species for quantitative characterization is described in this paper. The results which are presented represent a preliminary and exploratory study involved in evaluating the feasibility of using stereological techniques for quantitative microstructural description and identification of wood.

Conventional wood anatomy distinguishes two types of elements, longitudinal and transverse, with respect to the direction of the tissues within the tree. Longitudinal elements may be vessels, various types of fibers, tracheids or parenchyma cells. Transverse tissues which are called rays may contain tracheids, or parenchyma elements, or both. The relative size, and size distribution of these elements within growth increments or annual rings, are used for identification and for qualitative microstructural descriptions. Since size, spatial distribution, and size distribution of the elements in highly complex structures such as wood are difficult for the human eye to recognize, minute features, or sculpturings on the elements are often used as secondary sources of identification or description.

It is often desirable to tabulate the distinguishing features and their presence or absence in the various species in order to provide and aid to the microscopist. Certain distribution patterns have to be judged subjectively according to these tables to be "gradual" or "abrupt", "diffuse-porous", or "semi-ring-porous", "up to 4 seriate", "widest more than 10", "chains", "nested", "multiples", etc.(1). All these distribution patterns are quantifiable. These subjective decisions of the presence or absence of some elements may be answered with either yes or no, or possibly not at all with any certainty. This points out the need for a statistical approach. With proper sampling and statistical treatment, the probability of the presence of a certain feature in a wood can be calculated. A quantitative approach would also provide the necessary basis for the calculation of probabilities. Although conventional descriptive anatomy can be used for distinguishing between species of wood, it does not provide any information for the prediction of properties. Description in qualitative terms is a one-way street providing basically one result, an identification.

The history of wood identification started with the advent of plant anatomy in the late 17th century (2). During the 300 years history, only a few attempts of quantifying anatomical structure of wood are documented in the literature. Most of the work was done on coniferous woods that have simple microscopic structure (3,4,5). Ladell (6) and later Kellog and Ifju (7) applied point countings to microscopic images of angiosperm wood that had complicated anatomical structures. These attempts, however, did not result in the calculation of the important parameters necessary for the complete quantitative characterization of the structures studied.

---

*Presently with Armco Steel Corporation, Middletown, Ohio

PROCEDURES

A. Material

The material for this study consisted of the wood of sassafras, aspen, yellow birch, paper birch, and sugar maple. The two birch species are diffuse-porous and are known to be anatomically very similar, with sugar maple somewhat similar to birches in its microstructure. While the birches and maple have wood of relatively high density, aspen is a low-density diffuse-porous wood of quite uniform microstructure. The wood of sassafras has a so-called ring-porous structure and therefore it is distinctly different from the diffuse-porous woods included in this study.

Transverse microtome sections of 20-40 um in thickness were prepared from small blocks of the above species. The source of wood for the blocks was a random selection from a large lumber supply. Thus, variability representative of wood could be studied. The microtome sections were stained using a fast green-safranine double-staining technique. A minimum of ten slides from each block studied were mounted on permanent glass slides for microscope projections.

B. Stereological Countings

The cross-sectional images of the transverse sections were projected through a microscope onto a 25 point square grid illustrated in Figure 1. Magnification of the image was chosen for each species such that the size of the largest longitudinal elements, namely the vessels, was approximately equal to the grid spacing. The image was rotated so that the grid lines were parallel and perpendicular to the rays. A 10-digit blood-cell counter was used to tabulate the stereological counting information. The following data were obtained for each anatomical element:

1. Point counts with each grid intersection considered a sampling point on the lumens and cell walls of fibers, vessels, and rays.
2. Number of intercepts of grid lines with vessels in a direction perpendicular to the rays. From this, the intercept density in the tangential direction $N_L(T)$ was calculated.
3. Number of intercepts of grid lines with vessels in a direction parallel to the rays. From this, the intercept density in the radial direction $N_L(R)$ was calculated.
4. Number of vessels within the grid area, $N_A$.

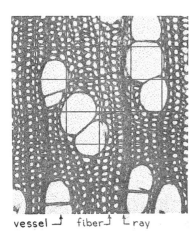

vessel ⌐ fiber⌐ ⌐ray

Figure 1. 25 point grid on typical transverse wood section.

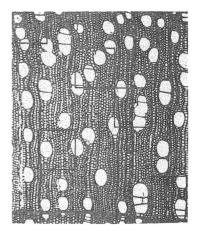

Figure 2. Five sampling fields within typical annual growth ring.

The spatial distribution of structural elements across a growth increment or
annual ring was also determined. Five zones along the radial direction within
each annual ring were selected to assess these variations. A typical sampling
field is shown in Figure 2 where it is seen that only the tangential locations
were randomly selected. Average values for $P_p$, $N_L(T)$, $N_L(R)$ and $N_A$ were determined from 20 grid placements within each of the five zones of the annual ring.
The first zone is associated with spring or earlywood, and the fifth zone with
the summer or latewood. An unfortunate aspect of using the annual ring as a
sampling unit is the lack of width uniformity within individual trees, species,
or between species. Consequently a comparison of features within the zones is
complicated by a scaling problem. However, the ring is laid down in a fixed
period of time and is a record of the tree's growing pattern which, it is presumed, remains reasonably invariant in time. Another reason for using growth
zone sampling is the possibility of correlating mechanical properties of samples
extracted from the growth zones with the stereological measurements. The techniques of testing within-growth-ring properties of wood have been developed by
Ifju (8).

RESULTS AND DISCUSSION

Although there are a number of different cell types found in hardwoods, the
three types: vessels, fibers, and wood rays predominate the anatomical organization. The characteristics of these features as found in five species of
North American hardwoods will be discussed in this section. Figure 3 shows
transverse microtome sections of the five species.

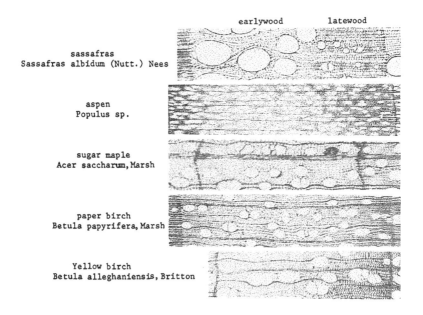

Figure 3. Examples of transverse sections of the five hardwood
species used for this study.

## A. Size and Shape Characteristics of Microstructural Elements

Average values and standard deviations (in parentheses) for the counting measurements are given in Table 1 for the five species studied. Each average is based on 100 countings, i.e. the combined value of 20 countings from each of the five growth zones. The vessel area fraction in sassafras is quite similar to that of aspen, however, this cannot be readily seen under the microscope because of their relatively small size in aspen. The last column in Table 1 shows that the reason for this similarity is that the number density of vessels in aspen is approximately 8 times that of sassafras. Note also that the birches and sugar maple take an intermediate position between the above two species. Also, it should be noted that standard deviations for the point counts on vessels of the four diffuse-porous woods appear to be independent of species, whereas sassafras, which is ring-porous, shows an appreciably larger variation in vessel area fraction across its growth rings.

Table 1. STEREOLOGICAL MEASUREMENTS FOR FIVE NORTH AMERICAN HARDWOODS

| Species | Subfeature* L = Lumen W = Wall | Point Count $\overline{P}_p$ (%) Vessel | Fiber | Ray | Total | Vessel Intercept Count $\overline{N}_L$ R T+ | Vessel Density $\overline{N}_A$ |
|---|---|---|---|---|---|---|---|
| **Ring-Porous** | | | | | | | |
| Sassafras | L--- | 27(22) | 21(11) | 7(5) | : 55(14) | R--3.2(1.1) | 17(6) |
| | W--- | 10( 6) | 27(14) | 8(6) | : 45(14) | T--3.9(1.5) | |
| | Total | 37(25) | 48(23) | 15(9) | :100(14) | | |
| **Diffuse-Porous** | | | | | | | |
| Aspen (1)# | L--- | 28( 7) | 22(12) | 3(3) | : 53(10) | R--8.6(2.9) | 113(21) |
| | W--- | 10( 6) | 33( 9) | 4(3) | : 47(10) | T-13.4(3.1) | |
| | Total | 38(13) | 55(13) | 7(5) | :100(10) | | |
| Aspen (2)# | L--- | 34( 7) | 18( 7) | 6(5) | : 58(10) | R-10.9(2.1) | 126(22) |
| | W--- | 10( 7) | 28( 9) | 4(3) | : 42(10) | T-14.1(2.2) | |
| | Total | 44(11) | 46(11) | 10(6) | :100(10) | | |
| Sugar Maple | L--- | 14( 7) | 23( 7) | 10(7) | : 47( 9) | R--5.5(1.6) | 52(12) |
| | W--- | 7( 4) | 42( 9) | 4(4) | : 53(10) | T--6.4(1.4) | |
| | Total | 21( 8) | 65( 9) | 14(8) | :100(10) | | |
| Yellow Birch | L--- | 14( 7) | 19( 7) | 9(6) | : 42( 6) | R--3.4(1.4) | 30(13) |
| | W--- | 4( 4) | 50(10) | 4(4) | : 58( 4) | T--4.2(1.3) | |
| | Total | 18( 9) | 69( 9) | 13(7) | :100( 7) | | |
| Paper Birch | L--- | 16( 7) | 16( 6) | 6(5) | : 38( 9) | R--5.7(5.1) | 49(17) |
| | W--- | 6( 4) | 53( 9) | 3(3) | : 62( 9) | T--6.1(2.0) | |
| | Total | 22( 8) | 69( 8) | 9(6) | ;100( 9) | | |

* L stands for the lumen or hollow region of the cell and W.for the wall. The numbers in parenthesis are standard deviations.

# Two different blocks of aspen were used.

+ R indicates radial and T indicates tangential direction. All units are expressed in mm.

The area occupied by fibers in the five species studied reveals several other subtle differences. For example, sassafras, maple, and aspen fibers are relatively thin-walled with less than 2/3 of the total area occupied by the wall. The birches, on the other hand, have fibers with very thick cell walls. This difference might be important from the point of view of utilization, since thin-walled slender fibers are better suited for papermaking than stiff, thick-walled ones. Also, the permeation of fluids, such as chemicals in pulping or in preservative treatments is facilitated by the larger volume fraction of cell cavity (lumen).

Ray cells occupy a relatively small percentage of the total area in the structure of all species studies. Large standard deviations in these measurements do not allow definite conclusions to be drawn in regard to their sizes. One point, however, which is indicated by the data is associated with wall thickness of the ray cells. Interestingly and rather unexpectedly, the two birches and sugar maple have ray cells with a large lumen to wall ratio when compared to aspen and sassafras. There is no apparent reason for a species like the birches with thick-walled longitudinal fibers to develop ray-cells of lower relative wall thickness than do species like the aspens that have inherently thin-walled longitudinal elements.

The total area occupied by all elements may be used as an estimate of wood density. If the density of the cell wall material is known, then the apparent density of the wood tissue may be calculated. Because of the solvent exchange procedure used in preparation of the microtome sections the accepted specific gravity of 1.5 for the cell walls cannot be used. It is estimated that a specific gravity near unity is more representative of the swollen cell wall material observed under the microscope (1). Thus the total cell wall area fraction gives a reasonable estimate of density values of published data.

This study has revealed quite clearly that there may be various reasons for differences in wood densities. Density may vary due to differences in the size or in the number of thin-walled elements such as vessels. A comparison between sassafras and aspen illustrates this point. Total cell wall area fraction of the two woods indicates that they are of similar density. Yet aspen contains some 8 times as many vessels per unit area as sassafras. Thus, the size of an average vessel in sassafras must be appreciably greater than in aspen, as shown by the data.

Intercept counts in the radial and tangential directions were used for the calculation of the size and shape parameters which are listed in Table 2. Mean chord intercept ($\overline{\lambda}_{MCI}$) and mean free path ($\overline{\lambda}_{MFP}$) are used here to describe the average length of chords within the features and between the features, respectively, which result from test lines oriented radially or tangentially within transverse sectioning planes. These are calculated from the following equations.

$$\overline{\lambda}_{MCI} = 2\overline{P}_P/\overline{N}_L$$

$$\overline{\lambda}_{MFP} = 2(1-\overline{P}_P)/\overline{N}_L$$

The mean chord intercept for sassafras is significantly greater than in the other four species studied. However, mean free path between vessels is greatest for yellow birch. Aspen again shows that its vessels are very small and that the mean free path between them is also small, suggesting that this anatomical element is quite numerous in aspen. As expected, the aspect ratios of vessels suggest an elongated shape in the radial direction, in all species studied except paper birch which has an almost circular cross-section. This subtle difference between the two birches may provide a possible diagnostic tool for identification.

249

TABLE 2. SHAPE & SIZE PARAMETERS FOR VESSELS OF FIVE NORTH AMERICAN HARDWOODS

| Species | Sampling* Direction | Mean Chord Intercept $\bar{\lambda}$ (mm) MCI | Mean Free Path $\bar{\lambda}$ (mm) MFP | Aspect Ratios $\dfrac{\bar{\lambda}\ (R)}{\bar{\lambda}\ (T)}$ MCI / MCI | Estimated First Moments $\bar{d}$ (mm) | Estimated Second Moments $\overline{d^2}$ (mm$^2$) |
|---|---|---|---|---|---|---|
| Sassafras | R | .23 | .39 | 1.23 | .075 | .028 |
| | T | .19 | .32 | | .090 | |
| Aspen (1) | R | .10 | .14 | 1.75 | .031 | .004 |
| | T | .06 | .09 | | .048 | |
| Aspen (2) | R | .08 | .10 | 1.34 | .035 | .005 |
| | T | .06 | .08 | | .045 | |
| Sugar Maple | R | .08 | .29 | 1.22 | .043 | .005 |
| | T | .07 | .25 | | .049 | |
| Yellow Birch | R | .11 | .48 | 1.36 | .055 | .009 |
| | T | .08 | .39 | | .069 | |
| Paper Birch | R | .10 | .27 | 1.08 | .048 | .006 |
| | T | .09 | .26 | | .052 | |

*R stands for "radial", and T for "tangential" on a transverse section

The first two moments of the size distribution (denoted $M_1$ and $M_2$) of vessel cross sections were calculated for each species to give quantitative indication of their size distributions. These may be obtained from the counting measurements since,

$$M_1 \equiv \int_{y=0}^{y=\infty} y\, F(y)\,dy \equiv \bar{d}\ =\ \bar{L}/\pi\ =\ 4\bar{N}_L/\pi^2\bar{N}_A$$

and,

$$M_2 \equiv \int_{y=0}^{y=\infty} y^2\, F(y)\,dy \equiv \overline{d^2}\ =\ 4\bar{A}/\pi\ =\ 4\bar{P}_p\ /\ \pi\bar{N}_A$$

Here $F(y)dy$ represents the size distribution function for spherical vessel transverse sections, $\bar{d}$ is the average diameter, $\bar{L}$ is the average perimeter, and $\bar{A}$ the average area. Although these two moments do not completely characterize the size distribution they can be used to show if significant differences occur. These moments may also be used as diagnostic quantities for distinguishing different species of wood, however, feasibility for this can only be assessed after stereological analyses of many different woods from widely different sources are made.

B.   Variation in Microstructural Features Across Growth Increments

1.   Within-Species Variation

The spatial variations (or patterns), exhibited across an annual ring, for the
vessel and fiber volume fractions of two aspen blocks are illustrated in Figure
4.   No attempt was made in this exploratory study to systematically sample all
possible forms of wood which might occur for a given species.  Although there
are some statistical differences within individual growth zones, in general the
patterns are similar.  The variation patterns across the growth increment for
the two sources are very similar; they show a decreasing vessel volume fraction
and an increasing fiber volume fraction while traversing from earlywood to
latewood.  Other parameters such as ray $\bar{P}_p$, vessel $\bar{N}_A$, $\bar{N}_L$, and $\bar{\lambda}$MFP have distinctive patterns which are similar with respect to the two sources of aspen
despite differences in individual growth zones.  Two observers were also used to
obtain stereological data for aspen with only minor discrepancies found between
their data.  Thus there appears to be consistency of the stereological analysis
within a species.

The spatial patterns of vessels and fibers are complimentary in the sense that
$\bar{P}_p$ (vessels) + $\bar{P}_p$ (fibers) $\cong$ 1.  This is based upon the observation that ray
cell volume fraction is small, and is approximately constant across the annual
ring.  Any increase in vessel volume fraction thus corresponds to a proportion-
ate decrease in fiber volume fraction.  Hence the spatial patterns may be
described using either of these two elements. Vessel distribution was utilized
in this study.

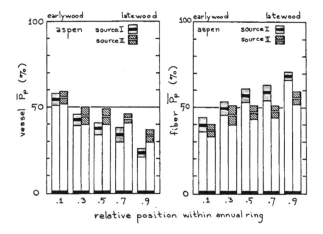

Figure 4.   Volume fraction variations for vessels and fibers within a growth
ring from two sources of aspen (in Figures 4-9) statistical scatter
is represented by + 2S.E. units, and consequently the total width
represents a 95% confidence interval.

2.  Differences Between Anatomically Widely Different Species

Having established a reasonable consistency within a species, the next area of
inquiry was the comparison between species. In Figure 5 patterns for vessel
volume fraction are shown for two species belonging to grossly different structural categories (ring-porous versus diffuse-porous). Aspen, according to
conventional classification of woods, belongs to the diffuse-porous group. The
term "diffuse-porous" is defined in wood anatomy as having vessels or pores of
uniform size evenly dispersed within growth increments. It was shown previously
that the size distribution of vessels for aspen is not uniform and also that
vessel volume fraction varies from the beginning to the end of the growth increments. An examination of the growth ring patterns for other diffuse-porous
species reveals that virtually all such woods deviate in different degrees from
the ideal uniform dispersion of vessels. These results indicate that the term
"Diffuse-porous" should either be redefined or perhaps deviations from its
classical form should be characterized preferably in quantitative terms. Sassafras, a ring-porous wood, exhibits a maximum vessel $\bar{P}_p$ in the second growth
zone, whereas aspen, which is diffuse-porous, shows monotonically decreasing
values over the entire growth increment, as illustrated in Figure 5. The contribution of wall and lumen to the total $\bar{P}_p$ indicates that the cell wall volume
fraction (which is 10% for both species) is essentially uniform over the annual
ring. Thus, sassafras and aspen allocate about the same relative amount of wood
substance per growth ring to vessels, but each species varies the vessel shape,
size, location, and number to obtain a different microarchitecture.

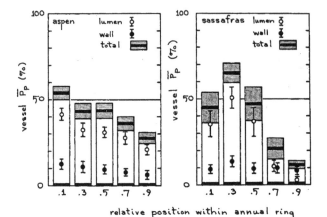

Figure 5.  Spatial variations for vessel volume fraction in aspen and sassafras.

The volume fractions of the cell wall material for all elements across growth zones for sassafras and aspen are compared in Figure 6. Since wall plus lumen equals 100%, the region above the bars represents volume fraction of lumen in the wood. The cell wall volume density for the two species is the same in the earlywood, but sassafras is considerably denser in the latewood where the wall volume fraction exceeds the lumen volume fraction.

Another comparison between these two species is shown in Figure 7. The mean chord intercepts of vessels in the tangential and radial directions for sassafras show maxima in the second growth zone. Aspect ratios for sassafras vary between 1.0 to 1.4 throughout the ring. Aspen, on the other hand, shows a decrease in size but a uniformity in shape (aspect ratio deviates only slightly from a value of 1.5) across its growth ring.

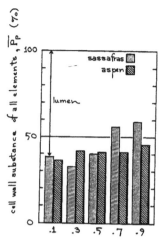

Figure 6. Volume fraction of the cell wall substance for all elements throughout the growth increment of aspen and sassafras.

The patterns shown in Figures 5 through 7 indicate inherent differences between the two species, and suggest that stereological data may be used to distinguish species. If quantification of microstructure is to be used for identification purposes, then clearly, two species which are very easy to distinguish visually must be distinguishable on the basis of stereological measurements. However, the two methods of identification need not be mutually exclusive since spatial patterns of the stereological quantities can serve to focus the observer's attention on subtle differences which exist in the complex visual images.

Figure 7. Shape and size variations of vessels throughout the annual ring for aspen and sassafras.

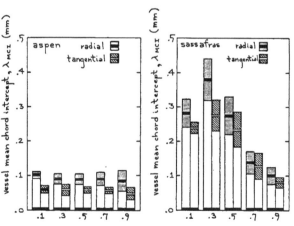

relative position within annual ring

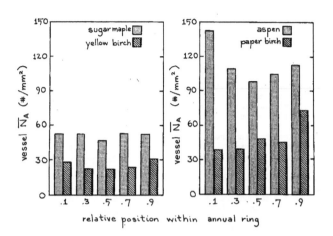

Figure 8. Variation in the density of vessels across the annual ring.

## 3. Differences Between Anatomically Closely Related Species

Within the diffuse-porous category, the structure of woods is similar, and thus it is often very difficult to visually identify individual species. Also, one should expect that stereological data would indicate the similarities as well as the differences in diffuse-porous species. To illustrate this, the number of vessels per unit area, $N_A$, for the four diffuse-porous species analyzed are plotted as a function of position in the growth increment in Figure 8. There is no problem in distinguishing aspen, however, only minor differences exist between maple and the two birches. It appears that $\overline{N}_A$ is more uniform throughout the annual ring for sugar maple with a somewhat greater value than for the birches. However, to determine whether the patterns for the maples and the birches are significantly different, more data are required. This is an important point. Extra data acquired by visually scanning more areas will undoubtedly enhance the observer's ability to distinguish a particular species, however, there is no assurance that he will decipher the information correctly. A greater statistical confidence for the assertions made can be obtained with additional stereological data. Thus, stereological characterization of wood microstructure for identification has a 'built-in' system of evaluation which is lacking in the visual method.

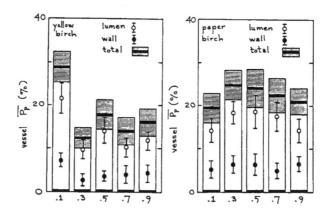

relative position within annual ring

Figure 9.    Spatial variation of vessel volume fraction within the
growth increment of two birch species.


4.    Comparison Between Two Species of the Same Genus

The final comparison involves differences between two species within the same
genus:  yellow birch vs. paper birch.  The spatial patterns of vessel volume
fraction across the growth ring are shown in Figure 9.  As with aspen and sassafras, the cell wall volume fraction appears relatively uniform across the growth
zones.  The patterns for total vessel volume fraction are quite different.  The
vessel $\overline{P}_p$ for yellow birch in the last four zones drops to 1/2 its value in the
first zone, whereas the same quantity in paper birch exhibits a variation which
is within a 10% band with a maximum at about the center of the growth ring.
These results led to visual examination of cross-sections of yellow birch and
paper birch of different sources [1].  In all cases it was found that the largest
area occupied by vessels in yellow birch was at the beginning of the earlywood.
Paper birch sections consistently exhibited a higher relative vessel area at or
near the middle of the growth rings.  This example illustrates that subtle
distribution patterns of anatomical elements within growth increments are
difficult if not impossible to recognize using conventional methods of wood
anatomy.  This is probably a result of the complexity of the microstructure of
wood which does not allow the human eye to distinguish relatively small differences without some quantification.

The distinctly different vessel area distribution patterns for the two species
of birches, sugar maple and aspen which were studied points out again the in-
adequacy of combining most of the known angiosperms or hardwoods into the category of diffuse-porous woods.  Although this study is considered the initial
step toward numerical assessment of wood anatomy, it has shown differences among
a few of the diffuse-porous species.

*J. H. Steele, Jr., G. Ifju and J. A. Johnson*

SUMMARY AND CONCLUSIONS

Microscopic transverse section of five North American hardwoods were subjected to a quantitative analysis to test the feasibility of using stereology as a basis for characterization of wood microstructure and wood identification. The growth increment or annual ring served as a microscopic unit within which three anatomical elements (vessels, fibers, and ray cells) and their features (cell wall and lumen) were sampled. The set of three volume fraction values for the elements of a given species were used to distinguish groups of species such as sassafras and aspen which were different from the birches and maple, for example. Within these groups vessel density, size, and shape were used to differentiate species.

The greatest possiblity for applying stereology as a diagnostic tool for identification of wood, as indicated by this study, would be in describing quantitatively the spatial variation of the microstructural elements through a growth ring. In this regard two sources of aspen showed consistent patterns. Since the volume fraction of ray cells is small and approximately uniform over the annual ring, the volume fraction of vessels and fibers show an inverse relationship across the growth increment. Although more work needs to be done to standardize the effect of the size and variability of the growth increment itself, the spatial patterns measured on the five species studied suggest the following conclusions:

1) None of the species classified as diffuse-porous showed a uniform distribution vessels within their annual ring.
2) Spatial distribution patterns of anatomical elements within growth increments appeared consistent and species specific.
3) Relative cell wall area measurements gave reasonable estimates for apparent density of the species studied.
4) Cell shapes, sizes and distributions in species of similar density were widely different suggesting that wood density alone is not a sufficient measure for wood properties.
5) Further studies are required to determine the effect of growth conditions on the various stereological parameters.
6) Quantitative treatment of radial and tangential planes of woods is needed to provide data for a complete three-dimensional characterization of their microstructure.

REFERENCES

1. Panshin, A. J. and de Zeeuw, C. (1970) *Textbook of Wood Technology*, Vol. 1. p. 705, McGraw-Hill, New York.
2. Kisser, J. G. (1967) History of wood anatomy. *Wood Science and Tech.* 1, 161.
3. Denyer, W. B. C., Gerrard, D. J. & Kennedy, R. W. (1966) A statistical approach to the separation of white and black spruce of the basis of xylem anatomy. *Forest Science* 12, 177.
4. Scallan, A. M. & Green, H. V. (1973) A technique for determining the transverse dimensions of the fibers in wood. *Pulp & Paper Research Inst. Can.*, Pulp & Paper Rept. No. 88.
5. Smith, D. M. (1967) Microscopic method for determining cross-sectional cell dimensions. *U. S. Forest Service*, FPL Rept. No. 79.
6. Ladell, J. L. (1969) A method of measuring the amount and distribution of cell wall material in transverse microscope sections of wood, *J. Inst. Wood Sci.* 3, 82.
7. Kellogg, R. M. and Ifju, G. (1962) Influence of specific gravity and certain other factors on the tensile properties of wood. *Forest Prod. J.* 12, 463.
8. Ifju, G. (1969) Within-growth-ring variation in some physical properties of southern pine wood. *Wood Science* 2, 11.

*National Bureau of Standards Special Publication 431*
*Proceedings of the* FOURTH INTERNATIONAL CONGRESS FOR STEREOLOGY
*held at* NBS, Gaithersburg, Md., September 4-9, 1975 (Issued January 1976)

TEXTURE ANALYZER MEASUREMENTS AND PHYSICAL ANISOTROPY OF TROPICAL WOODS

by A. Mariaux, * O. Peray** and J. Serra**
*Centre Technique Forestier Tropical, 94, Nogent-sur-Marne, France
**Centre de Morphologie Mathématique, Ecole des Mines de Paris, 77300, Fontainebleau, France

SUMMARY
    Quantitative analysis of wood structure by texture analyzer measurements, using in particular the variogram method, gives new ways of research into the description of wood anatomy. The main interest of the variogram is its ability to take in various scales of structures at the same time. We shall see in particular that it allows us to relate the punctual measurements in one field (punctual variogram) with the measurements taken between different fields (regularised variogram). Then, together with classical regression techniques, the variogram demonstrates a strong relationship between the arrangement of fibres in the wood and the anisotropy of shrinkage during drying.

INTRODUCTION
    The following study asks of Mathematical Morphology a very particular question. Consider the wood transverse section of Fig. 1 : it was cut-off perpendicularly to the trunk axis and presents a complex structure. This wood, called "Aucoumea klaineana", is one of twenty tropical woods on which we want to relate morphology to an important physical property : the anisotropy of shrinkage during drying. We obtained this property by calculating the shrinkage ratio during drying in two directions, radial and tangential (Fig. 1). Let us briefly describe the sequence of the wood structures, as we change our scale of observation. The wood anatomy is made up of a set of cells that form a skeleton of vertical tubes made of rigid walls surrounding cavities that are different in dimensions and shape. At a ten microns scale, we see the fibres (the smallest cells), lined up in strings. They have a direction parallel to the rays (the long slim cells) ; this direction is the one which radiates from the trunk centre (Fig. 2). Both these series of cells induce a strong anisotropy in the wood structure. At a hundred microns scale, the vessels appear, round and isolated, big tubes conducting the sap. Then they are more or less regrouped in small clusters, and, at a millimetric scale, we can see their density slowly change (Fig. 1). At the same scale we can also observe slow variations of darkness in the fibres ; these variations, called "tree-rings", are due to a regionalized flattening of the fibres and a thickening of their walls.  The difficulty of the study under consideration is that previous analyses have shown that neither the number, nor the shape, nor the dimension of the vessels play a role in the variations of the shrinkage anisotropy. Now it is only the vessels which appear as structures of isolated particles, the others, the rings and fibre strings, look much like a continuous medium. Thus we need an investigation method
    i) independent from the notion of particles
    ii) involving a directional character which is required by the structure
        anisotropy
    iii) able to scan the continuous succession of scales from the fibre strings
         to the rings, to be sure that no possibility of correlation with the physical parameter could escape.
    In terms of Mathematical Morphology, the only structuring element able to satisfy these three requirements is the couple of two points, the associated measurement of which is the variogram.

THE PUNCTUAL VARIOGRAM.
    By thresholding, the cavities and walls of Fig. 2 are transformed into a medium of two phases X and $X^c$. Let us consider the structuring element B, made of two points, x and x+h, ends of the vector $\vec{h}$ ( $|h|$ ,α). We shall define the punctual variogram γ(h), to be the area percentage of the set X - X ⊖ B (where ⊖ represents the erosion symbol). In other words, γ(h) is the probability that a random point x falls in the set X, and x+h falls in $X^c$. Notice that the defi-

257

nition can be extended to a medium of n phases. In this study, we have computed the variogram in two directions, for $\alpha = 0$ (tangential) and $\alpha = \frac{\pi}{2}$ (radial). For $h = 0$, $\gamma(h)$ is obviously equal to zero. The greater is h, the larger are the scales of structure that are investigated. Now, the possibility that the variogram may grow indefinitely (starting from $\gamma(0) = 0$) makes it more able to describe the superpositions of structures than the closely related notion of covariance. One can compute that the tangent of the variogram at the origin is equal to the number of intercepts of X per unit of test line, in the considered direction. For each of the twenty woods, the two variograms were calculated by taking their average value on 36 fields of Fig. 2 type, the fields being set on the points of a square grid with grid spacing of 1280µ . The maximal step h of the punctual variogram is 80µ .

STRUCTURAL INTERPRETATION OF THE VARIOGRAM.

One can observe three kinds of phenomena on the variogram of Fig. 3 :

a) a strong growth from 0 to 15µ , followed by an oscillation called "hole-effect", which corresponds to a pseudo-periodicity of the fibres. The abscissa of the minimum gives the average distance between fibres in the described direction. In the case of the radial direction, the cavities are aligned in strings and hence the abscissa is also equal to the average diameter of these fibres. The interpretation of the maximum is more delicate : one needs to simulate the given structure from a probabilistic model, and proceed from there. The difference of ordinates between minimum and maximum explains the degree of regularity in the periodicity of distances between the fibres.

b) The variograms do not reach a sill, but keep on growing slowly. This Phenomenon proves the existence of macrostructures due here to the vessels. The initial slope of this second growth gives the number of intercepts of the vessels per unit of test line.

c) When comparing the two variograms, one may notice that they are nearly identical, up to an additive constant. This loss of ordinate in the radial direction, called "zonal anisotropy", simply means that when a couple of points B falls in a ray, the two points are likely to stay in the cavities, even for high values of h.

THE REGULARISED VARIOGRAM.

Notice that the definition of the punctual variogram is equivalent to the following :

$$\gamma(h) = \frac{1}{2} D^2 \left[ f(x+h) - f(x) \right]$$

where f is the indicator function of the phase X. Now let g(y) be the proportion of the area of X in one of the 36 fields of the grid, centered on y :

$$g(y) = \frac{1}{S_y} \int_{S_y} f(s) \, ds$$

We define the *regularised variogram* (in the two directions), as :

$$\gamma_g(h) = \frac{1}{2} D^2 \left[ g(y+h) - g(y) \right]$$

where h is a multiple of the grid spacing (1280µ ). In the punctual variogram, we obtain the step by translating a point within the field ; in the regularised variogram, the step is obtained by moving the stage of the microscope. From these new definitions, we now see that the variogram quantifies, in terms of variance of the indicator function f (or of the regularised function g) the average similarity which exists between a random point (or a field) and its translation at a distance h. With the regularised variogram, we may follow the structures on a much larger scale (Fig. 4). The lack of sill in both directions again exhibits new macrostructures probably coming from the tree rings and the grouping of vessels. The parabolic growth denotes a continuity in the structure variations at the larger scale. Thus the variogram, given by the above mathematical definition, exhibits a range of important structural information. By quantifying this information, we may now try to connect the structures and the physical properties.

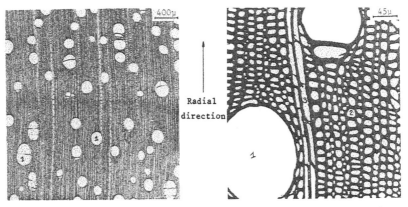

Fig. 1 : Transverse Section of
Aucoumea (x 25) with vessels (1)
structure

Fig. 2 : Same section (x 216)
after thresholding. Fibres (2) and
rays.(3)

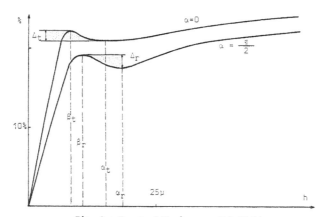

Fig. 3 : Punctual Variogram of Bridelia

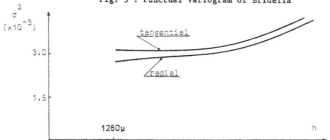

Fig. 4 : Regularised Variogram of Aucoumea

259

A. Mariaux, O. Peray and J. Serra

A STRUCTURAL EXPLANATION OF THE SHRINKAGE ANISOTROPY.
In each wood, a transverse section was cut-off in the neighbourhood of the
test-pieces. Its dimensions were 1 square centimeter by 25μ thickness. The sample was treated with Na Cl O to eliminate cell residuals, then coloured in green
to improve the contrast required by the texture analyser. The number of samples
in each species was small, because at the beginning we only wanted to find a
structural explanation for the shrinkage anisotropy, independent of the species.
The study was done with regression techniques, using, on one hand the coefficient
T/R of the shrinkage anisotropy, and on the other hand some geometrical characteristics of the variograms (see the Greek letters on Fig. 3). The first regression tried told us that only 35% of the shrinkage anisotropy variations could be
explained by the three most significant parameters of the variogram. But, after
the elimination of the four "Commiphoras" (and hence we no longer believe that
the shrinkage anisotropy is completely independent of the species), the percentage of explanation went up to 55% without a loss of statistical significance.
According to these last results, the explanation is given by the following parameters :
- the ratio $\frac{\Delta_r}{\alpha_r}$ ; where $\Delta_r$ represents the degree of periodicity present in
the distances between the cavities and $\alpha_t$ is the average value of these distances taken in the radial direction. The ratio is consistent with the fact that,
the smaller is the distance between the particles, the stronger is the regularity
in the periodicity. $\alpha_t$
- the ratio $\frac{\alpha_t}{\alpha_r}$ , where $\alpha_t$ is the average distance between the cavities in
the tangential direction. Here is the expression of the obtained relation :

$$T/R = 4.5 - 1.56 \frac{\alpha_t}{\alpha_r} - 4.29 \frac{\Delta_r}{\alpha_r} + \epsilon$$

where $\Delta_r$ is in %, $\alpha_t$ and $\alpha_t$ in μ , and $\epsilon$ represents a residual.

Therefore it appears that more than half of the variance of the shrinkage
anisotropy is explained by the given characteristics of small scale structures
of wood anatomy. Now, if we recall the effect of removing the influence of the
"Commiphoras", we expect that the remaining unexplained variance is due to the
difference in species.

CONCLUSION.
In conclusion, we will emphasize two points. First, the variogram is the
only theoretical tool which allows a scanning over structures of different scales. It permits with equal ease computations either from point measurements or
from measurements between fields. Here, we covered scales from 1μ to 5 mm. Second,
to attain a very specific goal (here, the estimation of a physical property), we
have used a structural analysis which is extensive and which is also of interest
by itself. In fact, both these aspects are not at all opposite. It is precisely
by its broadness of investigation that the second point of view allowed us to lo-
cate the two structural characteristics that we needed for a good correlation.

REFERENCES.
The same as in : J. SERRA, Stochastic models
in Stereology : Strengths and Weaknesses.- Proceedings
of the 4th International Congress for Stereology, 1975
Washington D.C., U.S.A.

260

National Bureau of Standards Special Publication 431
Proceedings of the FOURTH INTERNATIONAL CONGRESS FOR STEREOLOGY
held at NBS, Gaithersburg, Md., September 4-9, 1975 (Issued January 1976)

DECISION CRITERIA IN THE QUALITY CONTROL OF STEEL WITH RESPECT TO NONMETALLIC INCLUSIONS

by Stig Johansson
*Sandvik AB, Sandviken, Sweden*

ABSTRACT

Quality Control is carried out to establish if a lot is within a certain standard regarding nonmetallic inclusions. The decision has to be based on statistics and therefore the inclusion population has to be described by the mean and the standard deviation when the metallurgical treatment is constant within controllable limits. Critical features must be recorded or predicted from something that can be recorded. Such critical features can be either agglomerations of inclusion (segregations) or individual inclusions above a critical size. It is most important that the analysis is based on a sufficiently large area to obtain even a moderate degree of accuracy in a quantitative assessment of steel cleanness, therefore modern scanning technique has to be used. Among all the various parameters that can be arrived at in image analysis three have been selected as decision criteria. These are the proportion of fields with an area fraction $A_A$ above a certain level, the average area fraction $A_A$ and the particle size distribution slope.

INTRODUCTION

In the field of metallography and cleanness assessment of steel two dimensional automatic image analysis has been used for a long time. The main advantages of this technique is the time saving factor compared to manual quantitative stereological work. All basic descriptive parameters arrived at in the latter type of work can easily be calculated from automated image analysis provided the features of interest can be selected by means of variations in optical reflectivity.

The important question is what is interesting to quantify. It is obvious that the principles in choosing descriptive parameters are different for research work and for quality control. In research and development the metallurgical parameters usually are deliberately subjected to changes for obtaining some wanted property. In many of these cases the result of the change can be seen in any position of the bulk steel as the linked change in for instance inclusion morphology and/or distribution is altered everywhere.

For quality control it is quite the opposite case. The metallurgical parameters are specified beforehand and kept constant, the inclusion morphology is known and the analysis is carried out to find unwanted features in the distribution. These unwanted features will usually not appear very frequently which means that statistical treatment is necessary to be able to assure anything. Unwanted features in quality control are usually agglomerations of inclusion (segregations) or individual inclusions above a critical size. These features can be recorded either directly or indirectly by establishing the nature of the distribution.

It has been pointed out earlier by Allmand[1] that cleanness assessment must be based on a sufficiently large area to obtain even a moderate degree of accuracy. He stated that at least 300 to 400 fields with an individual field area of 0.5 mm$^2$, making a total covered surface area of 150 to 200 mm$^2$, was needed.

In the present work the analysis has been carried out on a Quantimet 360 image analyzer. The covered surface area in each analysis was 160 mm$^2$ divided into 500 fields.

The technique is described elsewhere by Gibbons[2].

DECISION CRITERIA

Area measurement

For material release it is necessary to establish a fundamental knowledge about the inclusion population when the metallurgical treatment is constant within controllable limits. If the area fraction ($A_A$) percent of oxide is chosen as a descriptive parameter then one of the possible measures of location for the population is described, others would be the mode, the median, some centile or the maximum value. In quality control one usually would like to know the proportion or percentile of the distribution which exceeds a certain critical value (degree of segregation) and the maximum value. In other words, some characteristic of the tail of the distribution is generally of great importance in deciding whether the material is good or not.

Figure 1 shows a common frequency distribution of an oxide area measurement for a specific grade of steel. The distribution is recorded as the specific field area percent of oxide inclusions. As can be seen the distribution is very far from a standard normal distribution in a single specimen with approximately 95 percent of the fields below an oxide area of 0.1 percent. The remaining 5 percent form a tail which is of the greatest importance in quality control as this part is usually responsible for the deterioration of the material properties. As mentioned earlier the tail would be approximately characterized by a centile at any two or more chosen positions.

Knowing the normal shape of an acceptable tail it is then possible to set the acceptance limit with respect to a percentile optimising the "sellers" and "buyers risk" for any single specimen.

The spread calculated as the standard deviation would also give a good estimate of the tail. Unfortunately there are no commercial instruments available which can present standard deviation and at the same time beeing fast enough to be used for quality control of steel.

When dealing with the population mean obtained from recording a finite number of individual sample means, the standard normal distribution can be assumed and the population can be characterized by a grand mean and the standard deviation. Then each mean can be subjected to a go-no go decision based on the criterion distance from the population mean. We now have two criteria for a go-no go decision the percentile of segregated fields and the distance from the population mean.

Particle size distribution

It has been found that the particle size distribution normally can be described in a log-log diagram as a straight line if the logarithmic number of particles greater than a certain size is plotted against the logarithmic particle size as indicated. By establishing the nature of the distribution in the microscopic region it is possible to predict the probability for macroscopic features. We have in fact been able to link together two size distributions obtained from two different methods. The first distribution was obtained from image analysis of microscopic features extended from 8 μm to 126 μm in length in the rolling direction and the

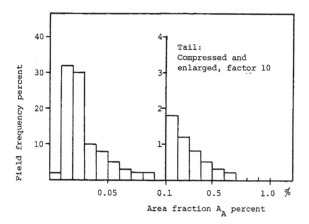

Figure 1. Distribution of specific field area
fraction for oxide inclusions in one
analysis

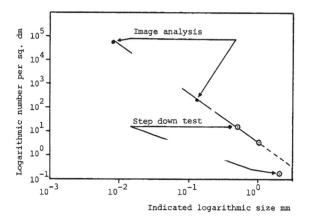

Figure 2. Particle size distribution for ductile
inclusions

second was obtained from macroscopic features extended from half a millimeter upwards in length on a step down test piece. Figure 2 shows this
relation for silicate inclusions in a rolled 18/8 stainless steel. The
slope can be used as a decision criterion in quality control. The predicted number of macroscopic features from the slope coincides with the
actual number provided the macroscopic features belong to the same population. Pieces of refractory materials or other exogeneous macroscopic
features are not constituents of this population.

This slope adds a third decision criterion to the previous two.

S, Johansson

CONCLUSION

It would be satisfactorily safe to use the percentile of segregated fields, the average area fraction $A_A$ and the particle size distribution slope as decision criteria in quality control. All other information are details which seldom have any significant influence on material properties provided the right properties have been obtained by optimising inclusion shape and acceptable level of inclusions on a research and development status of the product.

REFERENCES

1. Allmand, T.R., Coleman, D.S.
   The Microscope, Vol. 20, No 1, Jan 1972  pp 57-81

2. Gibbons, J., Soames, M.R. and Knowles, W.R.
   The Microscope, Vol. 20, No 1, Jan 1972 . pp 1-20

National Bureau of Standards Special Publication 431
Proceedings of the FOURTH INTERNATIONAL CONGRESS FOR STEREOLOGY
held at NBS, Gaithersburg, Md., September 4-9, 1975 (Issued January 1976)

AUTOMATIC RECOGNITION OF NON-METALLIC STRINGERS IN STEEL

by D. Jeulin* and J. Serra**
*IRSID Station d'Essais MAIZIERES-lès-METZ, France
**Centre de Morphologie Mathématique, Ecole des Mines de Paris, France

1 - INTRODUCTION : Two different ways may be used to give quantitative characteristics of non metallic stringers in steel : to compare stringers seen
through a microscope to standard patterns, or to make measurements with an
image analyzer. Standard patterns (S.P.), well known from metallographists,
are used to define criterions of steel quality. Image analyzers, working
automatically, allow to get objective and representative characteristics.
But measured parameters have not always an explicit meaning and their links
with the S.P. may be not obvious. So, it would be interesting to know if it
is possible to select few parameters correlated with the S.P. To answer it,
we have to survey all the kinds of patterns. In this paper, we build the
theoretical approach in a particular case, the German standard patterns for
stringers in forged or rolled steel DIEGARTEN (1). The analysis includes two
steps : a) study of each pattern to find which combination of parameters
enables to get its best characterization ; b) application of the results
to real steels jointly recognized in the same places by an observer and by
an analyzer, and test of the statistical equivalence of both estimations.
In this paper, we only carried out the first step, leaving the second one
to metallurgists.

2 - THE STANDARD PATTERNS FOR STRINGERS IN STEEL build a rectangular table.
Its columns are ten families, numbered from 0 to 9 (Threadlike sulfides for
families 0 and 1 ; oxides alumina type : 2, 3, 4 ; threadlike oxides, silicate type : 5, 6, 7 ; globular shaped oxides : 8, 9). Each family contains
9 patterns (numbered from 0 to 8 : 02 is the third of family 0, ...). The
size of stringers increases from pattern 0 to pattern 8. Stringers are black
or grey (families 0 and 1) grains on a white circular ground. Shapes of
grains change with the types of stringers (points, dashes, or globules of
different sizes). They are more or less numerous and scattered (n° 75) or
gathered (n° 37). Beside that wide range of shapes, two preferential directions may be noticed : the "vertical" direction of elongation of stringers
after forging or rolling, and the "horizontal" direction, giving the width
of details.

3 - MORPHOLOGICAL PARAMETERS : We use notions (star, erosion, closing) based
on Mathematical Morphology, exposed in (2), (3), and (5). To measure parameters defined in Morphology was conceived the texture analysis system (T.A.S.)
(6-7) manufactured by LEITZ firm (8) (Licence IRSID - Ecole des Mines). The
three more simple parameters we used are area (A), perimeter (L), and connexity number (N). Elongation and thickness of stringers are estimated with
parameters of mean intercept computed from the histograms of intercepts measured in the two directions, the stars D* (vertical) and d* (horizontal) and
with the maximal vertical and horizontal intercepts (D, d), of similar meanings
but more fluctuating. Clusters of stringers are characterized with two
parameters : hexagonal dilation and closing (4-8). Figure 1 shows the results
of these operations : the pattern (1a) is dilated (1b) or closed (1c) by an
hexagon of size 1.6 mm. The dilation regroups near grains but enlarges the
whole. The closing deletes small enclaves inside stringers and welds details
without enlarging them. After dilation is measured the connexity number, $N_D$,
giving information about the number of clusters. After closing are measured
the area $A_F$, the length of maximal horizontal intercept $d_F$, and the connexity
number $N_F$. At last, threshold variable (S) indicates whether stringers are
grey (0 for families 0 and 1) or black (1 for the others).

4 - CORRESPONDENCE ANALYSIS of J.P. BENZECRI (9) was chosen among methods
of multivariate analysis to compute the results of measurements, because it
corrects redundancies of data : two parameters giving similar information
(as A and $A_F$, D and D*, or d and d*) can be replaced by one of them in a

265

further step such as a classification, without changing the structure of the data put forward in the analysis. Three levels of correspondences can be seen : *which S.P. are similar* when reported to the measured parameters ? *Which parameters are similar* from the patterns point of view ? *Which parameters give the best characterization of each family* ? The correspondence analysis enables to visualise these 3 types of relations in order to get criterions of automatical recognition of each S.P. The method considers the 90 patterns, characterized by 12 morphological parameters, as points of a 12 dimensional space, and the parameters as points of a 90 dimensional space. The system of main axis of inertia of these points, provided with weights, is computed, and it is possible to give a simultaneous representation of the S.P. points and of the parameters points on these axis and on factor planes. Points of weight zero, called additional elements, can be projected on the axis when heterogeneous data could disturb the analysis, or when a classification is required for a sample. This principle may be used to recognize non metallic stringers in steel with the texture analyzer.

5 - RESULTS : The data, being somewhat heterogeneous (numbers, areas, perimeters), had to be balanced before processing with correspondence analysis. The threshold variable S, having no morphological meaning, is not used at that step. The family 8, where stringers are globular shaped, is particular, and thus was introduced as additional elements. The inertia ratio of the four first factors (93.89 % of the whole variation) warrants a good representation of the data in a four dimensional space. In each factor plane are noticed proximities of some parameters : A is always near from $A_F$ (the closing does not disturb the area of stringers but only deletes some enclaves). D is closed to $D^*$ and d to $d^*$ : stars and maximal intercept length have the same meaning here, yet stars are better : computed by integration of granulometry they are less sensitive to statistical fluctuations. The perimeter L, near from origin, is not important in this study. The description of the factors is given in (10) : *factor 1* opposes vertical intercepts $(D^*, D)$ to horizontal $(d^*, d)$, and to numbers $(N, N_F, N_D)$ and families 5-6 to the others. *Factor 2*, correlated with areas A and $A_F$, arranges families from pattern 0 to pattern 8, as in the table of S.P. *Factor 3* opposes horizontal intercepts $(d^*, d, d_F)$ to numbers $N_F$ and $N_D$ (Fig. 2), and separates family 8 (globular stringers) from the others. *Factor 4* correlated with $N_D$, opposite to N, is the axis of clusters : on one side families 2, 3, 4, 5 and 6 have numerous stringers gathered in one or two clusters ; on the opposite side, families 0, 1, 7 and 9 have more clusters. Figure 2 shows factor plane 3-4, where the localisation of families is caused by their number of clusters ($N_D$), of details (N) or their horizontal intercepts.

6 - AUTOMATIC CLASSIFICATION OF THE STANDARD PATTERNS : The results of the correspondence analysis showed that certain families are well detached in the different factor planes. We consider in the *3 dimensional space* (axis 1-3-4) parallelipipeds parallel to the axis containing only one family, the limits of which are determined with the help of factor planes. To class a field of measurement, we compute his coordinates in the factor space, $f_1$, $f_3$, $f_4$, from the measured values of parameters $x_j (j = 1, ... 11)$ with formula (1) given below and coefficients $a_{ij}$ given in table I ; execution of tests of table II gives the family of stringers seen in the field, while the area of stringers gives its position inside the family (axis 2).

7 - CONCLUSION : This way of automatic recognition can be performed easily and instantaneously (on one field, the parameters are measured in a fraction of second, and a small computer, connected with the T.A.S., can make the summation (1) and the discrimination of table II), and may be transposed to real samples of steel. Otherwise, few parameters are sufficient to make the recognition, since it was shown that it is not necessary to measure the perimeter L, the maximal intercepts (D,d) the area and the number of connexity after closing ($A_F$, $N_F$). In the case of other types of stringers, a preliminary correspondence analysis has to be done with the concerned table of standard patterns, to determine new coefficients $a_{ij}$ and new tests. About the precision of the results, we have to notice that it is difficult to make a discrimination between certain families ( (2 and 3),(1,0,7), (5 and 6) ),

certainly because of their great likeness. But if an objective analysis pointed out such a likeness, it certainly means that human recognition may sometimes confuse too similar families, such as 2 and 3. So, it would be good to complete our study with a comparison between visual and automatic analysis on the same samples.

### BIBLIOGRAPHY

1 - Standard patterns of non metallic stringers in steels. German standard DIEGARTEN, 1973.

2 - G. MATHERON : Eléments pour une théorie des milieux poreux. Paris, MASSON, 1967.

3 - A. HAAS, G. MATHERON, J. SERRA : Morphologie mathématique et granulométries en place. Annales des Mines, déc. 1967.

4 - J. SERRA : Introduction à la Morphologie Mathématique. Cahiers du C.M.M., fascicule 3, 1969. Ecole des Mines de Paris.

5 - J. SERRA : Stereology and structuring element Stereology 3 - 1971.

6 - Texture analyzer : French patents IRSID n° 23.273 (1965), and ARMINES-SERRA n° 70.21322 (1970).

7 - J.C. KLEIN, J. SERRA : The Texture Analyzer Stereology 3 - 1971.

8 - LEITZ - Scientific and technical information suppl. 1 - April 1974.

9 - J.P. BENZECRI : L'analyse des données PARIS DUNOD 1973.

10 - D. JEULIN : Caractérisation des chartes d'inclusions dans les aciers. Thèse D.E.A. 1974.

Table II – CLASSIFICATION OF THE FAMILIES OF STANDARD PATTERNS

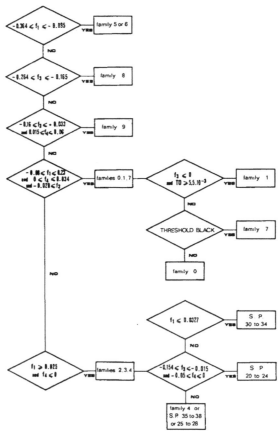

TABLE I    Coefficients aij

| xj | a 1j | a 3j | a 4j |
|---|---|---|---|
| 100 N | 0.543 | 0.119 | -0.274 |
| A | -0.103 | -0.051 | 0.108 |
| 2,5 d | 0.147 | -0.466 | 0.018 |
| 100 $N_F$ | 0.520 | 0.304 | 0.076 |
| $A_F$ | -0.044 | -0.067 | 0.029 |
| 2,5 $d_F$ | 0.093 | -0.305 | 0.016 |
| 2,5 L | 0.010 | 0.044 | -0.030 |
| 2,5 D | -0.481 | 0.137 | -0.066 |
| 100 $N_{D_*}$ | 0.371 | 0.265 | 0.468 |
| 1000 $d_*$ | 0.095 | -0.415 | 0.014 |
| 100 $D_*$ | -0.603 | 0.168 | -0.036 |

$$(1) \quad f_i = \sum_j \frac{a_{ij} \, x_j}{\sum_j x_j}$$

$$(i = 1,3,4)$$

Fig. 2   FACTOR PLANE 3-4

Fig. 1

a) Standard pattern n° 37

b) dilated picture.

c) closed picture

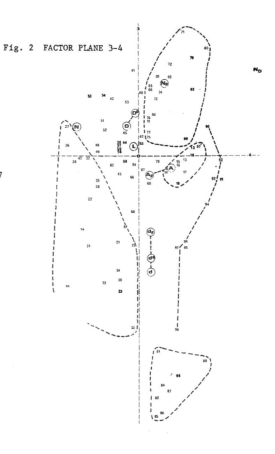

*National Bureau of Standards Special Publication 431*
*Proceedings of the* FOURTH INTERNATIONAL CONGRESS FOR STEREOLOGY
*held at* NBS, Gaithersburg, Md., September 4-9, 1975 (Issued January 1976)

METALLOGRAPHIC CHARACTERIZATION OF FRACTURE SURFACE PROFILES ON SECTIONING PLANES

by J. R. Pickens and J. Gurland
*Division of Engineering, Brown University, Providence, R.I. 02912, U.S.A.*

It is the purpose of this work to investigate certain metallurgical and geometrical features of fracture surfaces by means of measurements on metallographic sections showing a fracture surface profile.

Quantitative measurements of the linear fracture profile of SAE 4620 steels, in the quenched and tempered condition, were carried out by Shieh [1] who determined the relative concentrations of the constituents along the linear fracture profile and used them to calculate a fracture path preference index and a fracture probability index for each constituent in each fracture mode. It would be useful to extend these and other linear parameters to the characterization of the area features of the fracture surface. However, the sampling problem associated with nonrandom surfaces severely limits either the ease of measurement or the generality of the results obtained from limited measurements. Nevertheless, it will be shown that even under restricted conditions, a number of interesting results may be obtained which may contribute to the analysis and understanding of fracture problems.

I.   THEORY AND DEFINITIONS

1) Line and surface fractions of constituents

The trace of the fracture surface on a polished cross section is the lineage of interest. The basic relations between the mean length of the lineal feature per unit area of random plane sections, $L_A$; the mean area of the surface feature per unit volume, $S_V$; and the mean number of intersections between the surface or lineal features and random test lines per unit length of test line, $P_L$, for a phase $\alpha$, are:

$$\frac{S_V^\alpha}{S_V} = \frac{L_A^\alpha}{L_A} = \frac{P_L^\alpha}{P_L} \quad \text{or} \quad f_S^\alpha = f_L^\alpha \qquad (1)$$

where $f_S^\alpha$ and $f_L^\alpha$ are the surface and line fractions of the constituent $\alpha$, respectively. The preceding argument applies to any surface, random or oriented, if the test planes and test lines are randomly oriented, or if the surface elements of the surface in question are randomly oriented in space.

The conditions of random orientation are generally not present in the analysis of linear fracture traces. Equation 1 does not hold true for preferentially-oriented surfaces and nonrandomly-oriented test planes, except for the special case when the angular spatial distribution of the surface elements is the same for all the constituents.

2) Roughness

Lineal roughness is defined in terms of a reference line which is the trace of a reference plane on a metallographic section plane.    The lineal roughness, $R_L$, is defined as the ratio of the true length of the linear fracture surface profile to the projected length of this trace on the reference line. The lineal roughness can be easily measured on metallographic sections, although the apparent roughness is greatly affected by the magnification and the resolution of the optical system. The orientation of the. reference line in the specimen must be carefully defined.

### 3) Curvature

The curvature of planar curves and of curved surfaces is defined and their relationship is reviewed in Underwood [2]. For a planar curve, such as the surface trace on a section plane, the local curvature is:

$$k = \frac{d\alpha}{d\ell} \tag{2}$$

where $\alpha$ is the tangent direction angle and $\ell$ is the curve length. The numerical average of the local curvatures would be expected to differentiate between flat and curved line segments, i.e.

$$\bar{K}_n = \frac{\Sigma_1^n |\pm k|}{n} \tag{3}$$

where n is the number of measurements. The local curvature may be determined at selected points on the surface trace, by measuring the local radius of curvature r, where r is the radius of the circle of curvature tangent to the curve at a given point, and $r = 1/k$ .

## II. EXPERIMENTAL INVESTIGATION OF FRACTURE SURFACES OF 1018 STEEL

### 1) Procedures

Twenty Charpy V-notch specimens of cold-rolled 1018 steel were fractured at temperatures ranging from -193.5°C to +100°C. A plot of the impact energy versus temperature of fracture is shown in Fig. 1. The transition temperature lies between 15°C and 45°C, which is in agreement with published work on the transition temperature of similar low-carbon steels [3,4]. The average hardness of the steel is 92.5, Rockwell B, with a grain size corresponding approximately to ASTM no. 8.

The linear analysis of the fracture surface profiles was carried out on metallographic sections through the fracture surface of the broken specimens. The sections will be referred to as parallel or perpendicular cuts. The plane of the parallel cuts is perpendicular to the fracture surface and parallel to the direction of fracture propagation. The perpendicular cuts are perpendicular to the fracture surface and to the direction of fracture propagation. The fracture surfaces were nickel plated before sectioning and the metallographic sections were etched with 3% nital. Typical fracture sections of the steel broken at high and low temperatures are shown in Fig. 2. The pearlite volume fraction of specimens selected for metallographic analysis was determined by point counting on random fields well removed from the fracture. These values compare well with the calculated value of 20.5% pearlite for a nominal 0.18% carbon steel.

### 2) Pearlite fractions on linear fracture surface traces

The line segment lengths of the fracture traces were determined by moving the cross hairs of a Hewlett-Packard Digitizer, programmed to measure line length, across photographs of the fracture surface profile on the metallographic sections. For each specimen, four contiguous fields were measured at each of four locations equally spaced along the length of the fracture surface.

The total length of line segments in pearlite divided by the total fracture surface lineage measured gives the pearlite lineal fraction (Fig. 1).

It is noted that the pearlite lineal fraction of the specimens which lie on the lower shelf of the impact energy curve, is not statistically different from the pearlite volume fraction. The pearlite lineal fraction of specimens which lie on the upper shelf is significantly less than the pearlite volume fraction.

Measurements of the orientation of fracture lineage segments of pearlite and ferrite were made on parallel and perpendicular sections of 2 specimens. The method consisted of measuring the angle between the surface tangent and a horizontal reference line at random points of the surface trace. Angular distributions were plotted and the data compared by statistical methods (see Table A). As discussed in Part I, the equivalence of-lineal fraction and surface fraction of one microconstituent may be claimed only if the angular distributions are equal for the two microconstituents. Results show that this equivalence holds for the low temperature specimen but not necessarily for the high temperature specimen. Therefore, only for the brittle fracture case is the pearlite surface fraction on the fracture surface given by the fraction of pearlite on the lineal trace of the fracture surface. The relatively low pearlite lineal fractions on ductile specimens nevertheless indicate that the fracture path favors ferrite at the expense of pearlite in the ductile mode and that void growth takes place preferentially in the ferrite. No preference has been shown in the brittle mode.

According to Shieh's nomenclature [1], the fracture path preference index of the two constituents is equal for both constituents (and is therefore equal to unity) in the brittle mode, but is not equal for both constituents in the ductile mode.

3) Roughness

Lineal roughness was measured using the Hewlett-Packard digitizer. A line perpendicular to the specimen axis was used as the reference line for all roughness measurements. Lineal roughness values for 5 specimens are included in Table B. No significant differences or trends in lineal roughness could be shown between samples broken at different temperatures or between parallel and perpendicular cuts on the same sample.

4) Curvature

Local planar curvatures were computed from the output of the Hewlett-Packard digitizer. For each of two specimens, the coordinates of the linear fracture surface traces were collected on four photomicrographs at 0.01 inch intervals. At each point, the curvature was computed for line segments varied in length from 0.01 to 0.5 inch on both sides of the reference point. The radius of curvature was calculated from the equation of the circle of curvature based on the midpoint (i.e. the reference point) and the two end points of the line segments. Figure 3 shows graphically the variation of the averages of the local radii of curvature as a function of the line segment lengths. The significant results are found at the shortest line segments since they represent the true local curvature, limited only by the precision of the tracing device. With long line segments, the local curvature is obscured by the large-scale roughness of the surfaces.

At a line segment length of 0.01 inch, the average radius of curvature of the brittle fracture surface trace (Specimen No. 1) is found to be about 50% larger than the corresponding average of the ductile fracture surface trace (Specimen No. 5). The actual values for the two specimens are: 19.4, 22.7, 23.8, 28.9 inches, and 30.8, 31.0, 34.1, 35.4 inches, respectively. The lengths are reported in inches, as measured on the photomicrographs, and are not corrected for the magnification factor (1000x). These results indicate that the local planar curvature of the fracture surface traces can perhaps be used to characterize a fracture surface, although it remains to be shown whether the values obtained are indicative of the fracture mode.

Conclusion

The techniques of quantitative microscopy which enable quantitative information to be obtained from measurements made on metallographic sections of the fracture surface, were applied to Charpy V-notch specimens of 1018 steel. Statistically significant results indicate that the fracture path

271

avoids pearlite in the ductile mode but appears to have little or no preference in the brittle mode. No statistically significant difference in roughness between the two modes of fracture could be determined. However, the
average local planar curvature of brittle and ductile fracture surfaces was
found to be appreciably different.

Acknowledgements

The authors gratefully acknowledge the financial support of the U. S.
Atomic Energy Commission, Division of Physical Research, and the invaluable
help of Professor S. Burns, who initiated this work, and of Mr. Paul Williams,
undergraduate research assistant.

References

1. W. T. Shieh, Met. Trans., 5 (1974) 1069.
2. E. E. Underwood, Quantitative Stereology, Addison-Wesley Publ. Co.,
   Reading, Mass. 1970.
3. F. W. Boulger and W. R. Hansen, Trans. Met. Soc. AIME 227 (1963) 212.
4. J. A. Rineholt and W. J. Harris, Trans. ASM 43 (1951) 1175.

### Table A

| Fracture Test Temperature °C | Pearlite Average Angular Measurement | Ferrite Average Angular Measurement |
|---|---|---|
| -193.5 | 6.30° | 7.60° |
| +100 | 1.74° | 11.8° |

### Table B

| Fracture Test Temperature °C | Linear Roughness Parallel Cut |
|---|---|
| -193.5 | 2.54 |
| -73.5 | 2.18 |
| -28.5 | 2.08 |
| 0 | 2.50 |
| 100.0 | 2.30 |

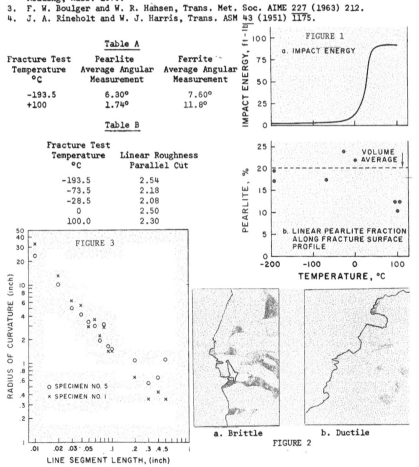

FIGURE 1

a. IMPACT ENERGY

b. LINEAR PEARLITE FRACTION ALONG FRACTURE SURFACE PROFILE

FIGURE 3

O SPECIMEN NO. 5
× SPECIMEN NO. 1

a. Brittle    b. Ductile

FIGURE 2

*National Bureau of Standards Special Publication 431*
*Proceedings of the* FOURTH INTERNATIONAL CONGRESS FOR STEREOLOGY
*held at* NBS, Gaithersburg, Md., September 4-9, 1975 (Issued January 1976)

## THE EQUIVALENT FOIL THICKNESS PRINCIPLE

by S. M. El-Soudani
*Department of Metallurgy and Materials Service, The University of Cambridge, Cambridge, England*

SUMMARY

The introduction of stereological methods to quantitative fractography is deemed possible provided certain conventions are adopted and remain consistent throughout any comparative analyses. The analogy between quantitative metallographic and fractographic relations is demonstrated using several examples whereby the analyzed features are truncated outside the test volume. These findings are formulated in terms of "the equivalent foil thickness principle."

INTRODUCTION

In a recent publication the author [1] introduced several applications of stereology to quantitative fractography. The region of a fractured metal swept by the plastic zone of a propagating crack was modelled as a thin film that was subjected to a special technique of projected-image analysis. In view of the complexity of the fracture process, it may sometimes be convenient to confine the test volume to a region of the plastic zone upon which the fractographer may wish to focus his attention. In doing so, one must keep in mind that in order to obtain useful correlations using stereological methods a selected approach must remain consistent from one specimen to another, and also, when comparing fracture behaviors in different materials. Such an approach to a rather complex problem is basically sound. However, the ultimate success of a stereological theory characterizing fractures will be assessed in terms of its consistency and ability to interpret the observed fracture behavior of materials.

In this paper we shall analyse a basic problem which simplifies the introduction of stereological principles to quantitative fractography. In singling out a test region of the plastic zone, the fractographer often finds himself confronted with the problem of truncation of features outside the region of interest. In fact, metallographers would face the analogous situation in thin film microscopy of dispersed-particle structures. A useful principle was introduced [1] without derivation for dealing with these situations. This principle, heretofore to be called "the equivalent foil thickness principle," may be formulated as follows:

(a) Metallography: "In order to account for the truncation of convex particles β, whose mean intercept length is $[\bar{L}_3]_\beta$ at the surfaces of a thin foil of thickness t, we may without changing the volume fraction of particles replace the given foil with another having a thickness $t + [\bar{L}_3]_\beta$ and solve the problem for the new foil assuming no truncation of particles."

(b) Fractography: "In order to account for the truncation of convex voids β, whose mean intercept length is $[\bar{L}_3]_\beta$ at the boundaries of the maximum-profile-height region of thickness, E, enveloping the fracture surface, we may without changing the volume fraction of voids replace this region by its equivalent of thickness $E + [\bar{L}_3]_\beta$ and solve the problem for the new fracture zone assuming no truncation of voids."

The analogy between the problems of metallography and fractography will become apparent as we treat several cases of analysis. For simplification of the notation to be used, we shall omit the subscript β throughout the derivations and include it only in the final expressions.

273

ANALOGY BETWEEN SEVERAL METALLOGRAPHIC AND FRACTOGRAPHIC PROBLEMS

A. Effect of Truncation Without Overlapping of Particles on their Projected-Image Area Fraction (See Fig. 1)

Suppose that we need to determine the total projected-image area fraction,
$[A_A^\prime]^{tot}$, of the features $\beta$ of one size class $r$. We may classify the $\beta$ particles
contributing to the projected-image area fraction into two groups: those whose
centers fall inside the foil, and those with their centers falling outside the
foil. For the former group we may write

$$[A_A^\prime]^{inn} = [N_A^\prime]^{inn} \pi r^2 = N_V t \pi r^2 \tag{1}$$

where $r$ is the sphere radius, $t$ is the foil thickness, $[N_A^\prime]^{inn}$ is the number of
inner particles per unit area of the projected image, and $N_V$ is their number per
unit volume. For the outer particles, we may also write

$$[A_A^\prime]^{out} = [N_A^\prime]^{out} [\overline{A}^\prime]^{out} = 2N_V \left[\frac{\overline{H}}{2}\right] [\overline{A}^\prime]^{out} \tag{2}$$

where $\overline{H}$ is the mean tangent diameter and $[N_A^\prime]^{out}$ is the number of outer $\beta$ particles per unit area of the projected image.

The mean projected area of the outer particles $[\overline{A}^\prime]^{out}$ may be determined
by methods of geometrical probability as follows:

$$[\overline{A}^\prime]^{out} = \int_{A_{min}^\prime}^{A_{max}^\prime} A^\prime \, p(A^\prime) dA^\prime \tag{3}$$

where the integration is confined to the outer $\beta$ particles. The product $p(A^\prime)dA^\prime$
is the probability that a given outer particle will have a projected area between
$A^\prime$ and $A^\prime + dA^\prime$. Defining an angle $\theta$ as shown in Fig. 2, and assuming that the
intersections of $\beta$ particles with the foil surface will be such that all values of
$\theta$ from 0 to $\pi/2$ are equally probable, it can be shown [2] that $p(A^\prime)dA^\prime = \sin \theta \, d\theta$.
The integration in Eq. (3) may thus be carried out in terms of $\theta$ as follows:

$$[\overline{A}^\prime]^{out} = \int_0^{\pi/2} \pi (r \sin \theta)^2 \sin \theta \, d\theta = \frac{2}{3} \pi r^2 \tag{4}$$

Substitution of Eq. (4) into Eq. (2) and the addition of Eqs. (1) and (2) yields
the total projected-image area fraction of the $\beta$ particles as follows

$$[A_A^\prime]^{tot}_\beta = [N_V]_\beta \{t + [\overline{L}_3]_\beta \} \pi r_\beta^2 \tag{5}$$

where for the spheres $\beta$ of one size, $r$, we note that the mean intercept length of
particles $[\overline{L}_3]_\beta = 2[\overline{H}]_\beta / 3 = 4r/3$ (see Ref. [2]).

In the absence of truncation, we obviously have

$$[A_A^\prime]^{tot}_\beta = [N_A^\prime]_\beta [\overline{A}^\prime]_\beta = [N_V]_\beta t \pi r_\beta^2 \tag{6}$$

Let us now assume that we have $j$ size classes. The summation may first be
carried out over each size class individually, assuming all other classes to be
absent, followed by the summation over all size classes. It can be shown that the
analogous expression to Eq. (5) takes the form

$$[A_A^\prime]^{tot}_\beta = [N_V]_\beta \pi \overline{r}_\beta^2 \{t + [\overline{L}_3]_\beta \} \tag{7}$$

and in the absence of truncation, we also have

$$[A_A^\prime]^{tot}_\beta = [N_V]_\beta \pi \overline{r}_\beta^2 t \tag{8}$$

where $\overline{r}_\beta^2 = \sum_j [N_V]_\beta^j r_j^2 / [N_V]_\beta$, and $[\overline{L}_3]_\beta = 4 \sum_j [N_V]_j r_j^3 / 3 \sum_j [N_V]_j r_j^2$

Comparison of Eqs. (5), (6), (7) and (8) shows that the effect of truncation can
be accounted for by substitution of $t + [\overline{L}_3]_\beta$ instead of $t$.

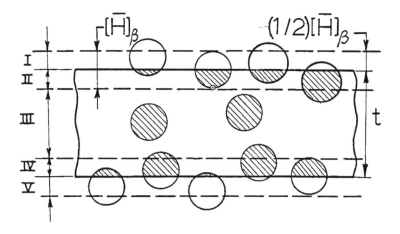

Figure 1 – A thin foil containing spherical particles of one size class truncating without overlapping at foil surfaces.

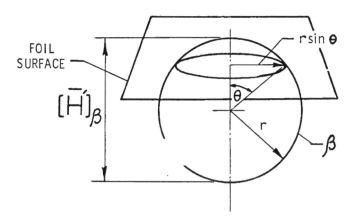

Figure 2 – Geometrical relations of spherical-particle intersections with foil surfaces.

The derivation in the case of randomly shaped convex particles may be carried out in a similar manner to the above cases and will, therefore, be omitted.

B. Effect of Truncation of Features on Volume Fraction Measurements

Several expressions have been derived [1] relating the volume and surface fractions of convex voids appearing on the fracture surface as equiaxed dimples. The following cases are of interest:

(1) Volume fraction of voids which are neither truncating nor overlapping and are intersected along their outermost perimeters by the fracture surface:

$$[V_V]_\beta = [S_S]_\beta [\bar{L}_3]_\beta / E \tag{9}$$

where, as before, E is the maximum profile height of the fracture surface and $[S_S]_\beta$ is the surface fraction of dimples.

(2) Volume fraction of voids for the case of truncation without overlapping, and the fracture surface intersects all voids along their outermost perimeters except, for physical considerations, those with centers in regions I&V(Fig 1)

$$[V_V]_\beta = [S_S]_\beta [\bar{L}_3]_\beta / \{[\bar{L}_3]_\beta + E\} \tag{10}$$

(3) Volume fraction of particles in a thin foil of thickness, t:

The volume fraction of β particles for the case of "no overlapping or truncation" is given in Ref. [2] as follows

$$[V_V]_\beta = [A_A^-]_\beta [\bar{L}_3]_\beta / t \tag{11}$$

For the case of truncation of particles without overlapping, the volume fraction is expressed by [2]

$$[V_V]_\beta = [A_A^-]_\beta - \tfrac{1}{4}[S_V]_\beta t \tag{12}$$

where $[S_V]_\beta$ is the surface area of particles β per unit volume. Substitution of the Tomkeieff Equation $[S_V]_\beta = 4[V_V]_\beta / [\bar{L}_3]_\beta$ into Eq. (12) and rearrangement yields

$$[V_V]_\beta = [A_A^-]_\beta [\bar{L}_3]_\beta / \{[\bar{L}_3]_\beta + t\} \tag{13}$$

Comparison of Eqs. (9), (10), (11) and (13) indicates that the fractographic analogues of the foil thickness, t, and projected-image area fraction, $[A_A^-]_\beta$, are respectively, the fracture-surface roughness index, E, and the true surface fraction, $[S_S]_\beta$.

DISCUSSION AND CONCLUSIONS

In the previous section we have treated two different problems: the determination of the projected-image area and volume fractions of dispersed particles which are truncating at the surfaces of a thin foil of thickness, t, and the measurement of volume fraction of voids intersected by a fracture surface having a surface roughness index, E. The analogy between the two cases, which follows directly from comparing Eqs. (5), (6), (7), (8), (9), (10), (11) and (13) is quite clear, and appears to be of a general nature. This analogy may be used primarily as a mathematical device in simplifying rather complex derivations. In treating such cases, one hypothetically replaces the test volume of thickness t or E with another of thickness (t + $\bar{L}_3$) or (E + $\bar{L}_3$), respectively, and treats the problem ignoring truncation effects. The proposed quantitative fractographic model outlined in detail in Ref. [1] is, therefore, based on the employment of stereological methods to the maximum profile-height region enveloping a fracture surface.

REFERENCES

1. S.M. El-Soudani, "Theoretical Basis for the Quantitative Analysis of Fracture Surfaces," Metallography, Vol. 7, No. 4, pp. 271-311 (1974).
2. E.E. Underwood, Quantitative Stereology, Addison Wesley, Reading, Ma. (1970).

*National Bureau of Standards Special Publication 431*
*Proceedings of the* FOURTH INTERNATIONAL CONGRESS FOR STEREOLOGY
*held at* NBS, Gaithersburg, Md., September 4-9, 1975 (Issued January 1976)

STEREOLOGICAL METHOD FOR ESTIMATING RELATIVE MEMBRANE SURFACES IN FREEZE-FRACTURE
PREPARATIONS.

by E. R. Weibel, G. A. Losa and R. P. Bolender
*Department of Anatomy, University of Berne, Berne, Switzerland*

INTRODUCTION

Freeze-fracture techniques (1,2) have become increasingly popular in the study of cellular membranes because they reveal some
aspects of internal membrane structure, particularly a characteristic distribution pattern of protein particles (3). This is due
to the tendency of membranes, in frozen samples (-100°C), to split
in half between the two phospholipid leaflets whenever they are
hit by the fracture plane. From the point of view of stereology
this means, however, that the observable "fracture surface"
follows certain elements of structure, namely the membrane "mid-
plane", and is hence not independent of the structure.

It is the purpose of this paper to show that under certain conditions a stereological analysis of the relative surface of membranes of various classes can nevertheless be performed. The method to be developed should allow us to estimate the relative surface of membranes characterized by a certain particle density.
Denoting particle density with $\alpha$ (measured in particle number per
$\mu m^2$) we will attempt to estimate by counting procedures the relative surface of "$\alpha$ membranes"

$$S_{S\alpha} = S_\alpha / S_T \qquad (1)$$

where $S_T$ is the total membrane surface.

MODEL AND METHOD

Let the material be composed of spherical membrane vesicles as
they occur, e.g., in microsomal fractions. If a horizontal fracture plane through the ice matrix of the frozen suspension hits
the vesicles above the equator the fracture surface is deflected
upwards forming a convex profile; if the vesicle is hit below the
equator a concave profile is formed. The probability that any profile is formed is proportional to the vesicle diameter 2R; with a
numerical density of vesicles $N_V$ the numerical density of profiles
is

$$N_A = N_V \cdot 2R \qquad (2)$$

It is useful to restrict the analysis to concave profiles only.
The reason is that the two membrane halves are asymmetrically
loaded with particles; there is adequate evidence that a concave
profile of a specific type of membrane will mostly if not always
display the same membrane face. It can further be shown that the
density of protein particles per unit area of concave membrane
face is characteristic for various membrane classes. The density
of particles $\alpha$ on concave faces may therefore be used as a label
for membrane types.

The particles are revealed by shadow casting the fracture surface with platinum at an angle of 45° to the normal to the frac-

ture plane. As a consequence many profiles will display a cast shadow which obscures the particles. A certain fraction of profiles will however be devoid of cast shadow, namely those generated by a fracture plane hitting the vesicle in the region marked with X in Fig. 1a, namely the cap between the lower pole and the tangent point of a 45⁰ line. Since

$$X = (1 - \sqrt{1/2}) \cdot R \qquad (3)$$

we find the numerical density of profiles without cast shadow to be

$$N_A{}^* = (1 - \sqrt{1/2}) \cdot R \cdot N_V \qquad (4)$$

It can easily be shown that, if the sphere radius R is variable, the numerical density of profiles without cast shadow depends on the mean vesicle diameter:

$$N_A{}^* = (1 - \sqrt{1/2}) \bar{R} \cdot N_V \qquad (5)$$

The number of concave profiles without cast shadow hence depends directly on the arithmetic mean radius of the vesicles. Since the total profile number is

$$N_A = 2\bar{R} \cdot N_V \qquad (6)$$

the relative frequency of profiles without cast shadow is

$$(1 - \sqrt{1/2})/2 \simeq 14.65\% \qquad (7)$$

meaning that about 15% of all profiles should be concave without cast shadow. This has been shown to be the case. This relationship is independent of vesicle size, so that the study of concave shadowless profiles affords an unbiased sample of the vesicles in the fraction.

## NUMERICAL FREQUENCY DISTRIBUTION OF VESICLES WITH RESPECT TO PARTICLE DENSITY

The first measurement that can now be performed is the estimation of the relative number of vesicles with particle density $\alpha$:

$$N_{N\alpha} = N_\alpha / N_T$$

To this end, a small test circle of known area is centered onto concave shadowfree profiles and the particle number in the test area is recorded. This leads immediately to a numerical frequency distribution of vesicles with $\alpha$ as the variable.

As shown above the restriction of the sample to shadow-free concave profiles has not introduced any bias. However, the use of a test circle of finite size g to measure particle density (Fig. 1b) has eliminated from the sample all profiles whose radius is smaller than g. This will affect small vesicles more severely than larger ones (Fig. 1b) and will thus introduce a bias. Eq. (5) becomes

$$N_A{}' = N_V \left[ \sqrt{R^2 - g^2} - R\sqrt{1/2} \right] \qquad (8)$$

It is hence not a constant function of vesicle size but depends on the difference between F and g. This may introduce a bias into the numerical frequency distribution of particle density if membranes of different values of $\alpha$ had a tendency to form vesicles of largely differing size. However, it can be shown that the choice of a small enough test circle ($R/g > 6$) reduces the sampling error to less than 5%.

A further bias may have been introduced in the estimation of particle densities on the projection of a curved surface onto a plane. Due to this $\alpha$ is an apparent particle density which is an overestimate of true density as a function of membrane curvature. It can be shown, however, that with small test circles this overestimation is inappreciable.

## MEMBRANE SURFACE DISTRIBUTION OF VESICLES WITH RESPECT TO PARTICLE DENSITY

The "surface frequency" or relative membrane surface of vesicles with particle density $\alpha$ is

$$S_{S\alpha} = N_{V\alpha} \cdot \bar{s}\alpha / N_V \cdot \bar{s} \qquad (9)$$

where $\bar{s}\alpha$ and $\bar{s}$ are the mean vesicle surface in class $\alpha$ and in the total population, respectively. The problem is to infer $S_{S\alpha}$ from the estimated numerical frequency of profiles $N_{N\alpha}$.

It is evident from the above that $N_{N\alpha}$ is an unbiased estimate of $N_{V\alpha}/N_V$ if the size of vesicles in class $\alpha$ does not deviate greatly from that in the total population. If the distribution of radii in $\alpha$ agrees well with that in the total population we also find that the mean surface areas $\bar{s}\alpha$ and $\bar{s}$ are equal. In this case

$$S_{S\alpha} = N_{N\alpha} \qquad (10)$$

If $\alpha$ vesicles show a different size distribution from that in the total population this will affect both the sampling bias and the mean surface of the vesicle, which is proportional to the second moment of the distribution. Expressing the vesicle radius R as a multiple of the test circle radius g, such that $R = q \cdot g$, we find

$$S_{S\alpha} = N_{N\alpha} \left[ \left[ 1 - E(q \sqrt{q^2 - 1}) / E(q^2) \right] \Big/ \left[ 1 - E\alpha(q \sqrt{q^2 - 1}) / E\alpha(q^2) \right] \right] \quad (11)$$

It can easily be shown that, if $q > 3$, i.e. if the test circle is smaller than one third the vesicle radius, this can be approximated by
$$S_{S\alpha} = N_{N\alpha} \cdot \left[ E\alpha(q^2) / E(q^2) \right] = N_{N\alpha} \cdot \left[ E\alpha(R^2) / E(R^2) \right] \quad (12)$$

where $E\alpha(R^2)$ and $E(R^2)$ are the second moments of the size distribution in class $\alpha$ and in the total population.

The membrane surface distribution among the various particle density classes hence follows from the numerical frequency distribution, but the ratio of second moments must be used to correct for a bias due to unequal size distribution of vesicles derived from various membrane types. Experimentally we have found that the second moments of the size distribution in the subclasses investigated deviated no more than 5 to 10 percent from that of the total population, which probably related to the fact that the microsomal preparations were obtained by differential centrifugation which selects a certain range of vesicle sizes.

## DISCUSSION

The method here developed has enabled us to quantitatively repartition the pool of membrane vesicles in a microsomal fraction to the various classes of cellular membranes from which they were

derived; this gave a very close agreement with parallel cytochemical estimates of relative membrane area (4,5). This method permits
to exploit the technique of freeze-fracturing for quantitatively
classifying membranes on the basis of their characteristic particle densities.

The only assumption made was that the vesicles were spherical,
a condition fulfilled in our specific preparation. In further development this type of analysis should be extended to become applicable to other vesicle forms.

Acknowledgement

This work was supported by grants from the Swiss National Science
Foundation.

References

(1) Moor, H.: Freeze-etching. Int. Rev. Cytol. 25, 1969

(2) Branton, D.: Exp. Cell Res. 45, 703, 1967

(3) Branton, D.: Phil. Trans. Royal Soc. London B, 261, 133, 1971

(4) Losa, G.A., Bolender, R.P., and Weibel, E.R.: Quantitative
characterization of microsomal membranes by freeze-etching.
(in preparation)

(5) Bolender, R.P., Paumgartner, D., Losa, G.A., and Weibel, E.R.:
Correlated biochemical and morphometric analysis of liver cells.
(in preparation)

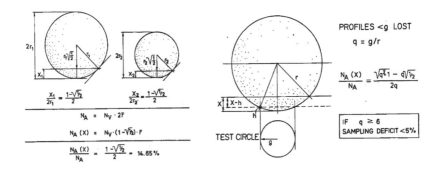

Fig. 1a                              Fig. 1b

*National Bureau of Standards Special Publication 431*
*Proceedings of the* FOURTH INTERNATIONAL CONGRESS FOR STEREOLOGY
*held at* NBS, Gaithersburg, Md., September 4-9, 1975 (Issued January 1976)

APPLICATION OF VARIOUS STEREOLOGICAL METHODS TO THE STUDY OF THE
GRAIN AND THE CRACK STRUCTURE OF CONCRETE

by Piet Stroeven
*Materials Science Group, Stevin Laboratory, Department of Civil Engineering,*
*Delft University of Technology, Stevinweg 4, Delft, Netherlands*

SUMMARY
         Various statistico-geometrical as well. as deterministic stereological meth-
ods have been applied to study the structure and the structural defects in plain
concrete, subjected to a compressive loading. Hitherto, use has been made of an
artificial spherical single-sized ceramic aggregate instead of coarse gravel for
most of the investigations. Crack patterns which have been visualized with the
help of the filtered particle method (based on fluorescence), have been analysed
by means of the method of directed secants. Apart from structural information
and results for engineering applications, this approach facilitated a mutual comparison of various stereological techniques in this field of materials science.

INTRODUCTION
         At present, though still hesitatingly, a definite interest is being exhib-
ited in the more fundamental background of concrete material behaviour. An increasing number of research projects is nowadays conducted in which the phenome-
nological approach is supplemented to a larger extent with results from structural investigations than ever before. It is found, of course, that the methods used
to tackle concrete problems are quite comparable to those used by investigators
(sometimes for decades already) in other diciplines. Analogous problems had to be
solved in fields like metallurgy, petrography, mineralogy, ceramics technology,
etc. As yet, however, the use of stereology in concrete materials science is
rather unique. As for concrete, such analyses have been handled exclusively for
petrographic investigations by Larsen and for porosity measurements by Lauer,
among others.
         It is clear that many examples can be indicated where the investigators
could have taken full advantage of the stereological techniques, e.g. in the investigation of grain anisotropy by Wu and Karl, in a study of surface cracking
by Bennet and Raju, and by the establishment of a semi-empirical relationship between $A_A$ and $V_V$ for glas marble concrete by Pigeon!
         In order to attain that stereology will be accepted, however, as a suitable
approach to the determination of the relevant features of the grain, the crack or
the pore structure in concrete, one should direct his attention to the development of labour-saving methods. For the time being, a manual step cannot be
avoided in the evaluation procedures, which prevents reaping full benefit from
the usage of (automatic) image analysers (Stroeven, 1975).
         Since various aspects of this work have been published elsewhere (Stroeven,
1973a, 1973b, 1973/74, 1975) I will concentrate at present on the newly obtained
results and the further elaborated primary data. For a good understanding, however, a rough sketch has to be presented of a more general framework for this
particular application of stereology.

STRUCTURAL EXPERIMENTS
         An artificial aggregate was used in the applied mixes instead of the coarse
gravel. The weight fraction of this single-sized spherical material (diameter D =
16 mm) has been varied between 10 and 50% (with respect to the amount of sand in
the mix).
Cubical specimens with linear dimensions of 200 mm were obtained by removing a
disturbed boundary layer of 25 mm from larger units. Then, approximately 11 mm
thick tiles were sawn from the specimens, the sawcut measuring about 5 mm. Finally, all sections were photographed, what resulted in 35 to 40 images per specimen,
all presenting an effective area element (AE) of 200 mm x 200 mm.
         The structural analysis has been performed in either of the following ways:
a. by random sampling, followed by statistico-geometrical analyses of the selec-
    ted image(s), yielding average structural information;

b. by uniform sampling and by subjecting all successive images to statistico-geometrical analyses, a procedure which reveals structural gradients in addition to overall information;

c. by serial section reconstruction performed by a computer; this method offers
insight into the complete configuration (Hammersly) apart from all information supplied by following procedure b.

Concrete is a very complex, macroscopically heterogeneous material. As a consequence, being confronted with a coarse structure, one has to select an AE of
relatively large size in order to be representative for composition. Therefore,
the AE will coincide usually with the sampled sub-area in our experiments. On
the contrary, the size of the AE will be generally inadequate to attain configuration homogeneity. This problem has to be solved by scanning a large quantity
of sub-areas.

## STEREOLOGICAL ANALYSIS

### a. *Grain structure*

An areal, a lineal, a features count and a random secants analysis have
been applied for the acquisition of volume fraction data. For the experiments
falling in category b, the last method was selected, since it is the least time
consuming operation in stereology. The other methods were used, according to procedure a, for the structural analysis of all three mixes and in accordance with
c, for the study of the grain structure of the intermediate mix. In the latter
case an IBM-1130 computer performed all operations.

Attention was also paid to characteristic distances like the nearest neighbour distance (in 2 and 3 dimensions) and the free spacing. Apart from average
stereological data, the serial section technique also yielded information with
respect to the complete configuration of the spheres.

The lineal analysis executed by the computer has proved to be convenient
to estimate the size of the RAE for composition as well as for configuration
homogeneity. On the one hand, the mean intercept length covering the mortar $L_m$,
is related to the mean free spacing of the spheres $\lambda$, and by that, to the composition of the mix, i.e. $\bar{V}_v$. On the other hand, the distribution of $L_m$ reflects
the complete range of grain distances. However, this part of our experiments was
described previously (Stroeven, 1973b, 1975).

In the near future computer generated random states of single-sized spheres
will be analogously analysed. The density will be gradually increased up to values met in the experiments. By transforming the growing amount of overlap into
displacements of the proper spheres according to prescribed rules, we will try
to approach "reality" as revealed by the stereological analyses.

### b. *Crack structure*

The lineal features in a plane, constituted by the crack pattern, have
been analysed with the help of Saltikov's method of uniformly spaced directed
secants. Sections situated parallel to the loading direction (the axis of symmetry) will reflect the representative features of the spatial crack structure as
indicated by Underwood (1968, 1970) and described by Hilliard (1962).

From an engineering standpoint this approach is promising, since the crack
development process under increasing loads can be interpreted in tangible structural terms, like crack density, or total specific crack length $L_A$ and the complete orientation distribution (or, simply, the degree of orientation $\omega$).

## RESULTS

The degree of experimental scatter has been compared with the calculated
values of the variance for methods of volume fraction analysis (Hilliard, 1968).
The data have, therefore, been subjected to a suitable regression analysis, in
order to eliminate the effect of segregation in the "rich" mix and the edge
disturbance effect in the lean one.

In accordance with systematic point count results, all applied methods have
demonstrated to be more accurate than predicted (Table 1). Moreover, in the case
of the intermediate mix, the various methods showed about the same degree of
scatter when based on the same subarea. This confirms a study of Hilliard & Cahn,
who reported that the variance of an optimized analysis is primarily determined
by the number of observations, the type of observation being of secondary importance.

Table 1. Comparison of results from various methods for volume fraction analysis

| Reference | | $V_v$ Experiment (%) | | | $CV_t$ Theory (%) | | $CV_e$ Experiment | | | $\dfrac{CV_e(D)}{CV_t(L)}$ | $\dfrac{CV_e(A)}{CV_t(A)}$ |
|---|---|---|---|---|---|---|---|---|---|---|---|
| Weight fraction (%) | Directed secants (D) | Areal (A) | Lineal (L) | Features count (F) | Areal)[2] (A) | Lineal)[3] (L) | Directed secants (D) | Areal (A) | Features count (F) | | |
| 50 | 36.89 | | | | | 8.14 | 6.50 | | | 0.80 | |
| 30 | 22.60 | 20.12 | 21.00 | 20.63 | 14.2 | 11.21 | 10.05)[1] | 11.0 | 10.4 | 0.90 | 0.77 |
| 10 | 9.25 | | | | | 16.43 | 15.28 | | | 0.93 | |

)[1] value could be biased (Moran), is confirmed, however, by an identical result (CV = 10.2) obtained from a random set of 11 out of 38 images.

)[2] $CV = (1.2/N)^{\frac{1}{2}}$ . . . . . $CV$ = coefficient of variation

)[3] $CV = (1 - V_v)(2.2/N)^{\frac{1}{2}}$ , total length of test lines :

1600 mm (50%)
2800 mm (30%)
4000 mm (10%)

To be able to interpret these results in a more general way, one can compare the formulae for the variance of an areal (A), a lineal (L) and a systematic point count (P) analysis over the complete range of volume fractions (without taking into account the possible violations of the assumptions which underlie the formulae). The following inequalities can be derived readily from Fig. 1.:

$$CV(P) < CV(A) < CV(L) \quad \text{for} \quad V_V < 0.26$$
$$CV(P) < CV(L) < CV(A) \quad \text{for} \quad V_V > 0.26$$

where CV is the coefficient of variation.

As a consequence, for practical values of $V_v$, the systematic point count is expected to show the lowest variance. It is worthy of mention that the mutual comparison is concerned with the intermediate mix, which is situated in the range where $CV(A) \simeq CV(L)$.

The line spacing δ in the lineal analysis, however, has been based on the condition that on the average every circle in the image plane should be cut by a single traverse (thus, δ = 10 mm $\simeq \bar{x}$ = $^2/3$ D). The calculated value of the variance of the lineal analysis will be only 70% of that of the areal analysis, since, in doing this, the number of intercepts will be doubled as compared to the number of features. The experimental scatter of $L_m$ has as yet not been analysed.

The variation in the material composition will manifest itself in a damped way in the scatter of the bulk modulus K. These stiffness fluctuations can be approximated by means of the Hashin & Shtrikman bounds for a two-phase model (which is in this case a reasonable assumption according to Brown). Accepting a CV of, say, 5% for mixes used in practice, an RVE with linear dimensions of 200. $(3.6/5.0)^2 \simeq 100$ mm, i.e. 6.5 times the grain size, is required, which corresponds to results of a study on the variance in strain gauge readings (Cooke & Seddon).

The configuration homogeneity, on the other hand, was studied by means of the discrete intercept length distribution $F(L_m)$, where the index m refers to the mortar. Such histograms, as shown in Fig. 2 have been obtained by subjecting 18 successive sections of a specimen of the intermediate mix to a lineal analysis with the help of uniformly spaced traverses; an operation which has been repeated for 7 different orientations $\theta_i$ of the line array. The theoretical density distribution of $L_m$ (representative for a very large AE) was approximated by calculating the average frequency distribution with respect to the location of the section and the orientation of the line array.

The differences between the "theoretical" distribution and a histogram representing an increasing number of subareas N for a given value of $\theta_i$ can be expressed by means of the $x^2$-value. Mean and standard deviation of the $x^2$-values for a given value of N facilitate the determination of the confidence limits of $x^2$. Finally, a suitable regression analysis by means of hyperbolic functions provided a series of continuously decreasing functions with rising values of N (Fig. 3. The size of the RAE for configuration homogeneity could be obtained by requiring the various curves to attain different levels of significance for the $x^2$-

283

value. The solutions for N were located in the range from 24 to 36. This means that the linear dimensions of the RAE should be 5 to 6 times larger than the sampled AE (i.e. 1000 to 1200 mm). And, as a consequence, the linear dimensions of the RVE's for structure insensitive (e.g. K) and structure sensitive proper- ties, respectively, will differ one order of magnitude.

The final conclusion, drawn from data previously published on the distri- bution of the spheres in concrete, is that the largest deviations from the random state occur locally. Particularly the nearest neighbour distribution in space was shown to be strongly influenced by this tendency to strive for (local) order. On the contrary, the free spacing was expected to be less governed by this tendency. Therefore, the average value of the mean free spacing was calculated for 33 spheres included in a hypothetical cubical element, centrally located in the re- constructed part of a specimen, containing a total number of 455 spheres.

A completely ordered system (based on a single-stagger configuration) will give occasion to 40 unobstructed free spacings per sphere, with an average length of 31.9 mm ($V_V$ = 20.6%). From the random secants analysis a value of 38.8 mm could be deduced. The "exact" value of $\bar{\lambda}$ yielded 43.8 mm on the average, being based on 97 unobstructed spheres. This latter result, particularly, demonstrates that the average deviations from the ordered state are quite large (which is also confirmed by the size distribution of the circles in the image plane). In the near future, it will become possible to analyse $\bar{\lambda}$ in the computer-generated quasi-random systems, as described before.

The analysis of the defect structure in concrete has been limited, hitherto, to the zero-state and a state of uniaxial compression beyond discontinuity (say, 75% of ultimate). The total specific crack length $L_A$ and the complete orientation distribution have been determined in sections, situated parallel to the axis of symmetry, constituted by the loading direction. Typical roses have been obtained. Indeed, perpendicular to the axis of symmetry an isometric system was revealed by the circular shape of the rose. In addition, the dependence of the properties of the roses on the orientation of the section with respect to the axis of symmetry is now being studied.

Finally, we are at the moment investigating axial sections over the comple- te stress-strain range in uniaxial compression in order to gain information with respect to the size of the RVE, the onset of structural loosening and the mecha- nisms of cracking that govern the different stages of the mechanical behaviour of concrete. A single example is presented in Fig. 5.

CONCLUSIONS

The stereological approach to the study of the grain structure and, partic- ularly, the defect structure in concrete as a function of the loading history, the mix proportions, the casting and densification techniques, etc., is very in- trigueing from a methodological standpoint, but also very promising with respect to the engineering interest. The time-consuming operations of preparing large (or a large number of) specimens and of executing the stereological analyses - accepting for the time being that a manual step cannot be avoided in the data acquisition - are a handicap for the investigator in this particular field of ma- terials science.

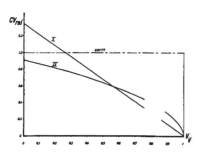

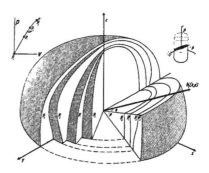

Fig. 1. Comparison of the variance of optimized lineal (L) areal (A) and systematic point count (P) analyses applied to a dispersion of single- sized spheres

I  $CV_{rel} = 1.35(1 - V_V)_1 = CV(L)/CV(A)$
II $CV_{rel} = 0.91(1 - V_V)^{\frac{1}{2}} = CV(P)/CV(A)$

Fig. 4. Combination of roses-of-the- number of intersections in a polar figure and according to the orienta- tion of the relevant section in the specimen.
Different loading stages are indi- cated.

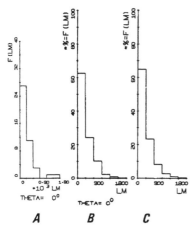

A       B       C

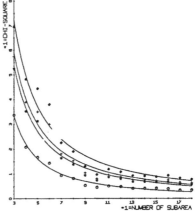

Fig. 2. Histograms for the intercept length through the mortar.
A: subarea size 200 mm x 200 mm.
B: collected values representing 15 subareas.
C: collected values for 18 subareas and 7 different orientations of the line array.

Fig. 3. Deviations between the in- tercept length distribution in the mortar for a total number of N sub- areas and for a representative area element with respect to configuration (viz. Fig. 2) quantified by means of the $\chi^2$-tests. The regression curves represent the average value with re- spect to orientation, $\overline{\chi^2}$, and the three upper confidence limits $t_{\alpha/2, n-1}$ for n = 7 and for the significant levels $\alpha$ = 10%, 5% and 1%, respec- tively. For further explanation see the text.

285

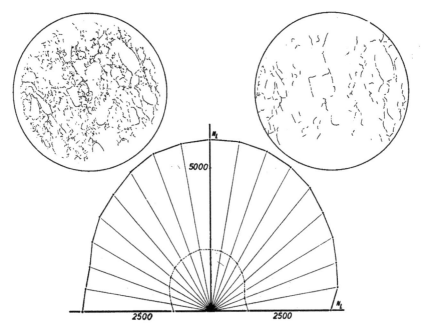

Fig. 5. Lineal features in a plane as constituted by the copied crack patterns. The patterns represent the state of disrupture at the surface on an axial section of a cubical specimen shown in Fig. 11 of Stroeven (1975). The specimen has been subjected to a relatively high uniaxial compressive load. The pattern at the right only reveals the more advanced cracks. The roses of the number of intersections at the bottom present the orientation distribution of both patterns. As can be observed, the microcracks which have joined under relatively high loadings have caused the degree of orientation $\omega$ to increase.
——·—— : $\omega= 17\%$  - - - - - - - : $\omega= 29\%$. Dimensions in picture points (Quantimet 720).

## REFERENCES

Bennet, E.W. & Raju, N.K. (1971) Cumulative fatigue damage of plain concrete in compression. In: Proc. Conf. Structure, Solid Mechanics and Engineering Design, 1096. J. Wiley & Sns.

Brown, C.B. (1965) Minimum volumes to ensure homogeneity in certain conglomerates. J. Frankl. Inst. 279, 189.

Cooke, R.W. & Seddon, A.E. (1956) The Laboratory use of bonded-wire electrical resistance strain-gauges on concrete at the Building Research Station. Mag. Concr. Res. 8, 31.

Hashin, Z. & Shtrikman, S. (1963) A variational approach to the theory of elastic behaviour of multiphase materials. J. Mech. Phys. Solids, 11, 127.

Hammersly, J.M. (1972) Stochastic models for the distribution of particles in space. In: Proc. Symp. Statistical and Probabilistic Problems in Metallurgy, 17. Suppl. Adv. Appl. Prob. Israel.

Hilliard, J.E. (1962) Specification and measurement of microstructural anisotropy. Trans AIME. 224, 1201.

Hilliard, J.E. (1968) Measurement of volume in volume. In: Quantitative Microscopy, 45. McGraw-Hill Book Co.

Hilliard, J.E. & Cahn, J.W. (1969) An evaluation of procedures in quantitative metallography for volume-fraction analysis. Trans Met. Soc. AIME. 221, 344.

Larsen, G. (1961) Microscopic point measuring. A quantitative petrographic method of determining the Ca(OH)₂ content of the cement paste of concrete. Mag. Concr. Res. 13, 71.

Lauer, K.R. (1971) The effect of the boundary conditions on the air void system of air entrained concrete. In: Proc. Res. Conf. 50-years of Building Institute, 76, Academia, Prague.

Moran, P.A.P. (1972) The probabilistic basis of stereology. In: Proc. Symp. Statistical and Probabilistic Problems in Metallurgy, 69. Suppl. Adv. Appl. Prob. Israel.

Pigeon, M. (1969) The process of crack initiation and propagation in concrete. Ph.D. Thesis, Imp. Coll. London.

Stroeven, P. (1973a) Structural investigations of concrete by means of stereological techniques. In: Proc. RILEM Symp. Fresh Concrete, 5-2-1, Leeds.

Stroeven, P. (1973b) Some aspects of the micromechanics of concrete. Ph.D. Thesis, Delft.

Stroeven, P. (1973/74) Methods for the evaluation of topological characteristics of the pore structure of concrete; application to crack porosity. In: Proc. RILEM/IUPAC Symp. Pore Structure and Properties of Materials, C529. Academia, Prague.

Stroeven, P. (1975) Stereometric analysis of structural inhomogeneity and anisotropy of concrete. In: Proc. Symp. Quantitative Analysis of Microstructures, 291. Practical Metallography Special Issue 5, Dr. Riederer Verlag, Stuttgart.

Underwood, E.E. (1968) Surface area and length in volume. In: Quantitative Microscopy, 77. McGraw-Hill Book Co.

Underwood, E.E. (1970) Quantitative Stereology, 48. Addison Wesley Publ. Co.

Wu, M. & Karl, F. (1970) Gefügeuntersuchungen an Betonen und Diskussion des Gefüge-einflusses auf die technischen Eigenschaften. Tonind. Z. 94, 449.

*National Bureau of Standards Special Publication 431*
*Proceedings of the* FOURTH INTERNATIONAL CONGRESS FOR STEREOLOGY
*held at* NBS, Gaithersburg, Md., September 4-9, 1975 (Issued January 1976)

THE USE OF QUANTITATIVE METALLOGRAPHY AND STEREOLOGICAL METHODS TO DESCRIBE COMPOSITE
MATERIALS

by Jean-Louis Chermant and Michel Coster,
*Groupe de Cristallographie et Chimie du Solide, ERA 305, Laboratoire de Chimie Minérale Industrielle,*
*Université de Caen, 14032 Caen Cedex, France.*

ABSTRACT
     This investigation indicates some  examples of using quantitative metallo-
graphy and stereology,  to investigate physical phenomena  related to the microstructure of materials.  After a critical  study of the different  methods  leading  to the choice of the most suitable,  we investigate the growth kinetics of
tungsten  carbide crystals in cobalt.  It was shown that the proposed growth law
corresponds to a kinetic which is  limited  by a second order interface reaction,
and this result was obtained  only after employing stereology.  The study terminates in an extension of stereology to quantitative fractography.  It was possible to show that  for tungsten carbide-cobalt composites, the rupture has a preferred propagation in the cobalt phase and, for the carbide phase, the transgra-
nular fracture represents only a small percentage of the total surface of  fracture.

INTRODUCTION
     Composites are an interesting example for metallographic  investigation
using  quantitative and stereological methods (1)(2).  There are several ways of
analysing stereological problems : either by generalization of known methods to
more complex geometry,  or by the use of mainly  experimental methods.  We  have
chosen in fact the latter methods to investigate a material of relatively simple
structure : tungsten carbide crystals in a cobalt matrix.
     We shall briefly give the reasons which led us to use the chosen stereolo-
gical methods (3),  afterwards we shall give  two kinds of application of  these
methods : the analysis of a growth kinetics and the  analysis of fracture paths.

CRITICAL ANALYSIS OF METHODS
     The crystals of tungsten carbide  are prismatic crystals and  their  grain
size is close to a micron.  It was  therefore necessary to choose (i) the observation procedure which is the easiest to use without introducing too large syste-
matical errors,  (ii)  the stereological method which allows the crystal size in
three-dimensional space to be determined the most accurately.
     The different methods  of observation (optical, scanning and  transmission
electron microscopy) were compared.  A comparison between scanning and transmission electron microscopy, on replicas, shows that no error is introduced during
the sampling.  The discrepancy is small between optical and electron microscopy
and is only significant for the smallest crystals.  Optical microscopy is  used
to investigate the largest particle size, while electron microscopy is used for
the smallest particle size.
     Point and linear analysis are comparable for measuring the volume fraction
according to similar kinds of observation.  The electron microscopy gives results
closer to those of chemical analysis than optical microscopy, but coincidence of
the results is obtained only if the volume ratios,found experimentally,are corrected so as to eliminate the Holmes effect (4), the importance of which is due  to
the large difference in hardness of the two phases.
     The metallographic analysis gives a measure of crystal size in two-dimensio-
nal space.  If we want know this crystal size in three-dimensional space, we must
use corrective methods which allow us to pass from two- to three-dimensional space (5)-(9). These methods have been established for spherical particles, which is

287

not the case of WC-Co. We therefore tried to find out if one of these methods could be used for non-spherical crystals.

To investigate this problem, we have used chromium powder and spheroïdal bronze, as reference specimens. By microscope observations, the true particle size distribution was established. These powders were mixed with a resin and the polished surfaces of these resin composites were then analyzed. The results of the analysis of these sections were transformed by means of the correction methods of S.A. Saltykov (5) and A.G. Spektor (9) in order to obtain the repartition in volume.

The Saltykov method is perfectly verified for spherical particles, in spite of the errors which accumulate for the smallest classes. The conformity is not so good for non spherical particles, but the existing errors are kept within reasonable limits, and the result is a good representation of the space distribution. The Spektor method does not give as good results as the Saltykov method, in particular, it is sensitive to the chosen class interval and to the non-sphericity of the particles. This is the reason why, subsequently, we used the Saltykov method whose geometric progression of classes is well adapted to the TGZ 3 Zeiss analyzer and to particle size of WC-Co.

## APPLICATION OF STEREOLOGY TO A GROWTH KINETICS

In order to apply stereological methods to the evolution of the microstructure of a material, we first investigated the growth kinetics of tungsten carbide crystals in cobalt.

Whatever the limiting mechanism, the growth of crystals is due to their difference of solubility. G.W. Greenwood, I.M. Lifshitz and V.V. Slyozov (11) and C. Wagner (12) have proposed a kinetic model limited by the diffusion, where the dispersed particles had a diameter increasing as : $D^3 = D_0^3 + kt$. Elsewhere H.E. Exner and F.H. Fischmeister (13) and C. Wagner (12) proposed another model limited by interface reactions with a diameter increasing as : $D^2 = D_0^2 + kt$. Finally, E. Hanizsch and M. Kalhweit (14)-(16) have investigated the problem of the growth of dispersed particles from a general point of view. They indicated that for an infinite time, there exist a limit distribution curve of the diameter of spherical crystals growing in a matrix. This limit curve is a function of the type of reaction. These authors have reported for a second order reaction controlled kinetic that the diameter of the crystals increases as : $D^3 = D_0^3 + kt$.

The kinetic growth has been investigated on WC-Co materials with a cobalt range from 10 to 25 wt.%, and with two WC crystal diameters : 2.2 and 0.7 µm. The growing was carried out under vacuum of $10^{-5}$ torr in graphite crucibles at temperatures between 1350 and 1700°C. The sintering time is between 2 h.30 and 100 h. After sintering the specimens are polished and etched by the standard techniques used for cemented carbides, and analyzed by point and linear manual counts and with a TGZ 3 Zeiss analyzer. Afterwards the particle size distribution was transformed into the diameter of an equivalent sphere in three-dimensional space.

The growth process by difference of solubility can be followed by the growth path method of R.T. de Hoff (17) and H.J. Woodhead (18). In a first representation for different treatment times the reciprocal of the cumulative curve of the number of crystals per unit volume is plotted as a function of their diameter. One notes the large decrease in the number of crystals and specially that of the smallest, during treatment. This indicates that it is the smallest crystal that dissolves to the profit of the large crystals. From this first series of curves, a second type of representation is derived : one plots as a time function the diameter of the smallest of the largest N'v crystals which allows to follow the evolution of families of crystals of different initial diameters (Fig. 1). This plot puts in evidence a critical diameter $D_c = f(t)$ which corresponds to crystals which neither increase nor decrease at a given moment. The ensemble of the maxima of the cur-

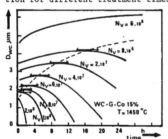

Fig. 1 : Growth-paths for the WC-Co 15 wt.%, $\bar{D}_{WC} = 2.2$ µm, for different treatment times at 1450°C.

ves at N'v = K gives the variation of this critical diameter as a function of time. This variation is given according to the law : $D_0^3 = D^3 + kt$.

With the growth path method the kinetic appears clearly, and in spite of the uncertainty of its outline this method is the only one which permits a determination of the critical diameter. The use of this method is dependant on a knowledge of the number of crystals per unit volume. This explains the great importance which must be given to the use of Saltykov's method, which allows one to obtain this parameter reasonably accurately.

The evolution of the microstructural characteristics was then investigated as a function of time. It was noticed that contiguity (19) does not constitute a determining factor of the kinetic as the following law is obtained :

$$C_{WC} = C_0 + k''t^{-0.12}$$

On the other hand the mean free path in cobalt phase and the mean diameter evolve according to the laws : $1^3_{Co} = 1_0{}^3 + k't$ ; $D^3_{WC} = D_0{}^3 + kt$.

The plots relative to the two particle sizes are fairly parallel, the speed constants are therefore pratically constant. On the other hand, the fact that the law in $D^3$ is satisfied is not significant enough for the search for a model.

The evolution of the size distribution curves of equivalent sphere diameters is given by the cumulative representation. For each given treatment time, on the same graph were plotted (Fig. 2) :

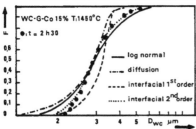

Fig. 2 : Theoretical limit curves according to the different growth processes (K: Kalhweit, R : Reaction) and experimental points for WC-Co 15 wt.%, $\bar{D}_{WC}$ = 2.2 μm.

- the experimental data obtained after Saltykov corrections.
- the log-normal distribution curve having the same median and the same standard deviation as the experimental points.
- the limit curve in diffusion-controlled growth.
- the limit curve in reaction controlled growth of the first and second orders.

The comparison of the different curves with the experimental results (Fig. 2) indicates that the limit curve of second order interface reaction is the closest to our experimental points. The evolution of the size distribution curves and of the main characteristics as a function of treatment time indicates that these curves had the same limit form for every treatment time. It should be noted that such a result can only be obtained by using a stereological corrective method. Indeed the surface analysis without corrections gives a curve close to the log-normal curve and this is not interpretable from the kinetic point of view for dispersed systems.

The whole lot of obtained results (20) has allowed us, thanks to stereology, to find that the growth kinetic of WC in Co is governed by a second order interface reation which is : $W_{Co} + C_{Co} \rightarrow WC_{crystal}$, and the rate is in accordance with $v = k(W_{Co}).(C_{Co})$.

APPLICATION OF STEREOLOGY TO FRACTURE INVESTIGATION

Most methods of quantitative metallography can be transposed to the investigation of fracture paths in scanning and transmission electron microscopies. Nevertheless some of these methods must be adapted in order to define new parameters describing the fracture feature.

The measure of fracture surface occupied by each type of rupture (trans- or inter-granular) is achieved in the same way as the ratios of polished surface occupied by each phase. In this case linear analysis seems to be the most effective (20)(22). The crystal distribution curve along the fracture path can also be obtained by linear or surface analysis. Nevertheless it would be interesting to calculate the size distribution at the surface of fracture path in terms of the diameters of the equivalent spheres, rather than the size distribution in volume. It was shown that the Saltykov method can be used for this type of calculation after some small corrections (20)(22).

By way of example, we chose to investigate the fracture of WC-Co plates by three-point bending. Two types of analysis were performed on each fractured specimen : first a polished surface was analyzed in order to determine the grain size and the principal microstructural parameters, and secondly the same kind of study was undertaken on the fractured faces.

Point analysis has allowed us to define, on thé one hand thè fracture ratios in the cobalt and, on the other hand, the ratios of inter- and trans-granular fracture in the tungsten carbide phase (Table I). Surface analysis allowed to es-

| | Microstructural parameters | | | | Fractured surface ratios | | |
|---|---|---|---|---|---|---|---|
| | $V_{V(Co)}$ % | $\bar{D}_{WC}$ µm | $C_{WC}$ | $\sigma_d$ µm | $S_{S(Co)}$ % | trans-gra-nular % | inter-gra-nular % |
| un-notched plate | 20 | 2.26 | 0.37 | 0.73 | 36 | 6 | 58 |
| | 20 | 1.05 | 0.34 | 0.33 | 45 | 1 | 54 |
| | 20 | 0.67 | 0.30 | 0.20 | 46 | 0 | 54 |
| notched plate ($K_{IC}$ specimen) | 20 | 2.25 | 0.37 | 0.73 | 30 | 10 | 60 |
| | 9 | 2.25 | | 0.73 | 12 | 12 | 76 |

Table I : Values of microstructural parameters and of ratios of fractured surface for different composites WC-Co fractured by three point bending : $V_V(Co)$ : cobalt volumic ratio ; $\bar{D}_{WC}$ : mean diameter of tungsten carbide crystals ; $C_{WC}$ : contiguity of carbide crystals ; $\sigma_d$ : standard deviation.

tablish the apparent outlines of the crystals concerned by an intergranular fracture surface (23). The results in this table indicate that the fracture travels preferentially through the cobalt phase and that the fracture is essentially intergranular along the carbide phase. A significant difference is observed between the size distribution curves determined from the polished surfaces and those determined from the fracture path. The variation of this difference indicates that the fractured surface contains larger crystals than the polished surface for a rather small grain size, and the larger the difference is the smaller the grain size. We can thus conclude: for this kind of materials the fracture is predominantly in the cobalt and the essentially intergranular fracture travels along the largest crystals of the grain size.

CONCLUSION

We were not attempting to present, in an exhaustive way, all the possibilities offered by stereology. We wished to show that its use entails a judicious choice of methods and, once this achieved, it could help to solve many problems, as much in the field of microstructural changes, as in the field of mechanical properties. Thus quantitative metallography and stereology can contribute largely to the characterization of materials, and to find correlations between mechanical, physical and chemical characteristics as a function of microstructural parameters.

REFERENCES

1. E.E. UNDERWOOD - Quantitative Stereology Addison Wesley, 1970.
2. R.T. DE HOFF, F.N. RHINES - Quantitative Microscopy, Mc Graw Hill Book, 1968.
3. J.L. CHERMANT, M. COSTER, A. DESCHANVRES - Metallography, 8, 1975, 121.
4. J.W. CAHN, J. NUTTING - Trans. AIME, 215, 1959, 526.
5. S.A. SALTYKOV - Stereometric Metallography, Metallurgizdat, Moscou, 1958, 2nd Ed.
6. E. SCHEIL - Z. Metallkde, 27, 1935, 199.
7. H.A. SCHWARTZ - Metals Alloys, 5, 1934, 139.
8. W.A. JOHNSON - Metal. Progr., 49, 1946, 87.
9. A.G. SPEKTOR - Zavod. Lab., 16, 1950, 173.
10. G.W. GREENWOOD - Acta Met., 4, 1956, 243.
11. I.M. LIFSHITZ, V.V. SLYOZOV - J. Phys. Chem. Sol., 19, 1961, 35.
12. C. WAGNER - Z. Electrochem., 65, 1961, 581.
13. H.E. EXNER, F.H. FISCHMEISTER - Z. Metallkde, 57, 1966, 187.

14. E. HANITZSCH, M. KAHLWEIT - Z. Phys. Chem. N.F., 57, 1968, 145.
15. E. HANITZSCH, M. KAHLWEIT - Z. Phys. Chem. N.F., 65, 1969, 290.
16. E. HANITZSCH, M. KAHLWEIT - Symposium on Crystallization, 1969, p. 130.
17. R.T. DE HOFF - Met. Trans., 2, 1971, 521.
18. J.W. WOODHEAD - Metallography, 1, 1968, 35.
19. J. GURLAND - Trans. AIME, 1958, p. 452.
20. M. COSTER - Thèse de Docteur ès-Sciences, Caen, juillet 1974.
21. J.L. CHERMANT, M. COSTER, G. HAUTIER, P. SCHAUFELBERGER - Powd. Met., 17, 1974, 85.
22. J.L. CHERMANT, M. COSTER - Metallography, to be published.
23. J.L. CHERMANT, M. COSTER, A. DESCHANVRES, A. IOST - 4ème Symposium Européen de Métallurgie des Poudres, Grenoble, France, 13-15 mai 1975.

*National Bureau of Standards Special Publication 431*
*Proceedings of the* FOURTH INTERNATIONAL CONGRESS FOR STEREOLOGY
*held at* NBS, Gaithersburg, Md., September 4-9, 1975 (Issued January 1976)

## A DEFINITION OF THE TWO-DIMENSIONAL SIZE OF IRREGULAR BODIES

by G. M. Timčák
*Mineral Research Laboratory, Košice Technical University, Košice, Czechoslovakia*

ABSTRACT

A method of deriving 2D size ("shape") of irregular, curvi-angular bodies
is given. In this method a statistically adequate number of random planar
sections subjected to threefold processing are used. The section shapes are
converted to "polygons" using the principle of Phines Tangent Count for the
generation of the "polygon" vertices; then the frequency distribution of the
sections with different number of vertices is determined; for the "polygons"
with the greatest frequency of occurrence the area distribution of their
parent sections is established. The most frequently occurring area value of
the "polygon" with the greatest frequency of occurrence is considered to be
the "true 2D size" of the given body.

The 2D size ("shape") of regular polyhedrons has been found to be definable as the most frequently observed area of the most frequently occurring
polygonal random sections (Timčák 1973/74) using a two-step sizing procedure
(cf. Fig.1). This method has been suitable also for the determination of the
true form of sectioned regular polyhedrons.

The above definition was, however, not directly applicable to irregular,
curvi-angular bodies (such as real mineral grains) due to the difficulty to get
a sufficiently distinctive description of the section shapes of such bodies.
It appears, however, that this problem could be overcome if the section shapes
could be approximated to some angular shapes or defined by descriptors of two-
dimensional character. A method suitable for this purpose was found to be the
Rhines Tangent Count (DeHoff 1967). In this counting, points of tangent of a
unidirectionally sweeping line with the perimeter of the particle sections are
generated at the loci where the character of the particle section curvature
allows such event. When joined within every particle section, these points
form a network which can be treated as polygons using the method suggested
earlier (cf. Fig.2). In the first stage of this sizing procedure, the number
of tangent points (TP) can generate "polygons" with different number of edges.

The generated "polygons" do not describe the original section shape fully,
but if the sectioning planes are randomly distributed, they provide an unbiased
approximation of the original shape. In case that bodies distributed in space

are considered, the random distribution of both, the bodies and the planes
is important (cf. Timčák 1975).

The procedure of 2D sizing would thus consist of three steps: 1) The
conversion of the original shape to a "polygonal" one (cf. Fig.2a, 2b). This
procedure may be performed either using a suitable TV display system fitted
with a light pen, or an acoustic tracing system, linked to a digital computer
(cf. Dvorak et al. 1974, Cowan&Wane 1973) or on a suitably modified TV based
image analyser (cf. Fisher 1971, Müller 1973). 2) The determination of the
frequency distribution of the generated sections with various number of TPs
(i.e. edges) (Fig.2c). At present, this procedure can be performed semiman-
ually or on a computer based image analysing system capable of complex pattern
recognition (cf. Grasseli 1969). 3) The area analysis of the "polygons" with
the most frequently occurring number of TPs (Fig.2d). For this measurement,
the original areas of the sections are taken and not those covered by the
generated "polygon". This type of measurement can be performed by any of the
methods mentioned in point 2 or using a suitable TV–based image analyser
(cf. Gibbard et al. 1972, Müller 1973).

The example in Fig.2a shows pagioclasses from an East Slovakian andesite
(C.No. 5001/1) obtained after the image analysis of a thin section. The
mineral is chemically eroded and has developed a hypidiomorphic form. The
transformed image is shown in Fig.2b. The frequency distribution of features
with different number of TPs is given in Fig.2c. A dominance of features
with 4 TPs is evident, so the areas of the parent shapes of these features were
analysed next (Fig.2a). The result (Fig.2d) shows that assuming that the
grains belong to a monodispersed population, the grain section having the
"true 2D size" of the analysed mineral is the one with 4 TPs and an area of
approximately 40 units$^2$. Grain sections satisfying these conditions were
marked by black triangles (Fig.2a). As, however, in truth the sections arose
by sectioning a polydispersed population, the marked sections represent a
weighted average 2D size of the sectioned minerals.

The stereological reconstruction of polydispersed systems would be
analogous to that suggested for polyhedral bodies (Timčák 1973/74).

Contrary to the polyhedral bodies, the determination of the true form of
irregular bodies from the above determined data would be difficult. It may be
described, however, in terms of largest and smallest 2D parameters and of
partial surface curvatures. This limitation, however, is not critical in cases
when only a statistical size (or the distribution of this size) of a body,
or a population of bodies is required.

LITERATURE:

COWAN W.M., WANN D.F. (1973): A computer system for the measurement of cell and nuclear sizes; J. Microscopy, Vol.99, Pt.3, Dec., pp. 331-348

DE HOFF R.T. (1967): The relationship between mean surface curvature and the stereologic counting measurements; In: H. Elias (Ed.): Stereology; Sprinver V., Berlin; pp.95-105

DVORAK J.A. et al. (1974): A simple video method for the quantification of microscopic objects; J. Microscopy, Vol.102, Pt.1, Sept., pp.71-78

FISHER C. (1971): The new Quantimet 720; Microscope, Vol.19, 1st q., pp.1-20

GIBBARD D.W. et al. (1972): Area sizing and pattern recognition on the Quantimet 720; Microscope, Vol.20, 1st q. pp.37-50

GRASSELI A. (ed.) (1969): Automatic interpretation and classification of images; Acad. Press, N. York, 430 pp.

MÜLLER W. (1973): Das Leitz Textur-Analyse-System; Leitz Mitteilungen; Suppl. I., 4; pp. 101-116

TIMČÁK G.M. (1973/74): A probabilistic definition of the 1 and 2 dimensional "size"; Proc. Pore Symp. RILEM/IUPAC; Academia, Prague, Vol.3, pp.A 33-45

TIMČÁK G.M. (1975): Stereological analysis of the size and spatial distribution of idealised random mineral systems; Manuscript; 120pp.

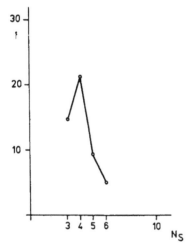

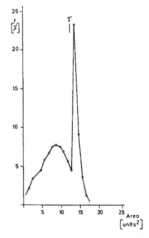

Fig.1a. Frequency distribution of cube sections with different number of sides. Quadrangular sections showing greatest incidence.

Fig. 1b. Area frequency distribution of quadrangular cube sections. Quadrangular sections with approx. 13 sp. units represent the "true 2D size" of the cube (1' is the face area of the cube).

293

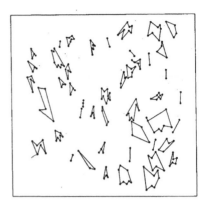

Fig.2a. Thin section of andesite with discriminated plagioclasses. Dots show the generated TPs. Grains with superimposed shading have 4 TPs; those marked with triangles represent the weighed "true 2D size" of the sectioned grain population.

Fig.2b. The same image with the natural boundaries supressed, displaying the "polygons" formed by the joining of the TPs of the individual sections.

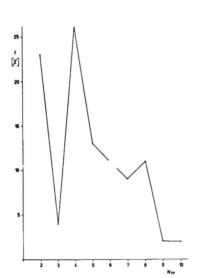

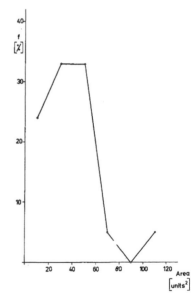

Fig.2c. Frequency distribution of grain sections with different number of TPs.

Fig.2d. Area frequency distribution sections with 4 TPs.

*National Bureau of Standards Special Publication 431*
*Proceedings of the* FOURTH INTERNATIONAL CONGRESS FOR STEREOLOGY
*held at* NBS, Gaithersburg, Md., September 4-9, 1975 (Issued January 1976)

STEREOMETRIC MICROSTRUCTURE AND PROPERTIES OF TWO PHASE MATERIALS

by Gerhard Ondracek

*Institut für Material- und Festkörperforschung, Kernforschungszentrum Karlsruhe, Germany*

The properties of materials are governed by the variables of state, e.g. temperature and chemical composition, their x-ray structure and their microstructure. It is the matter of the present paper to consider the effect of microstructural parameters of two phase matrix materials on their properties at constant state conditions and unchanged x-ray structure.

The dependence of the properties on the microstructure of a two phase matrix
material concerns the geometry and geometrical arrangement of its continuous
(matrix) and discontinuous (dispergent) phases. It leads to the variation of
each property between a lower and upper bound, which correspond - stereologically - to the well-known cases of parallel and series array of the materials
phases. There are three stereometric factors completely describing the microstructure quantitatively which determine the special property value of given
real two phase materials in between the bounds: the shape factor, orientation
factor and concentration factor of the dispersed phase (fig. 1).- The width of

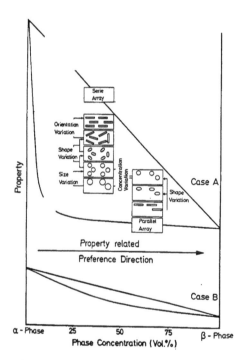

fig.1: Property bounds and stereometric factors of two phase materials

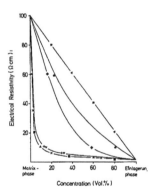

fig. 2: Theoretical shape of the
electrical resistivity of two
phase matrix materials ($p_M:p_D=10^2$)
parallel array of the phases ■
disc inclusions perpendicular to
the electric field ▲
fiber inclusions parallel to the
electric field X
disc inclusions statistically
oriented +
fiber inclusions stat.oriented ▼
spherical inclusions ◆
fiber inclusions perpendicular to
the electrical field ●
disc inclusions parallel to the
electrical field ✶
series array of the phases ◀

295

the variational region between the bounds increases with increasing difference between the property values of the phases (compare phases A and B, fig.1). Extended variational regions therefore occur in porous materials considering pores as the dispersed phase. But also if both phases are solid, this difference may cover several orders of magnitude as for example in the case of cermet resistivities ($\rho(Al_2O_3/\rho$ (Mo)$>10^{10}$). In such cases, the calculation of the property values of a given two-phase material as a function of its microstructure is technically relevant. Due to that statement, the electrical resistivity of two phase materials and its dependence on stereometric factors will be the subject treated as an example here. - Equations have been derived treating the two phase material as quasi-homogeneous continuum which provide the theoretical tool to calculate the changes in the electrical resistivity by variation of the orientation, shape or concentration of the dispersed phase /1-5/. The most general form of the <u>stereometric function of the electrical resistivity of two phase matrix materials</u> is

$$
1-c_D = \frac{\rho_C-\rho_D}{\rho_M-\rho_D}\frac{\rho_M}{\rho_C}\left[\frac{\rho_C}{\rho_M}\right]^{f(F,\cos^2\alpha)}\left[\frac{\rho_D+\left(\dfrac{1}{(1-F)\cos^2\alpha + 2F(1-\cos^2\alpha)} - 1\right)\rho_C\ \rho_M}{\rho_D+\left(\dfrac{1}{(1-F)\cos^2\alpha + 2F(1-\cos^2\alpha)} - 1\right)\rho_M\ \rho_C}\right]^{\psi(F,\cos^2\alpha)} \tag{1}
$$

$$
f(F,\cos^2\alpha) = \frac{F(1-2F)}{1 - (1-F)\cos^2\alpha - 2F(1-\cos^2\alpha)} \tag{2}
$$

$$
\psi(F,\cos^2\alpha) = \frac{F(1-2F)}{1 - (1-F)\cos^2\alpha - 2F(1-\cos^2\alpha)} + \frac{2F(1-F)}{(1-F)\cos^2\alpha + 2F(1-\cos^2\alpha)} - 1 \tag{3}
$$

Simplifications of equation 1 follow for extreme difference between the resistivities of the phases

$$
\frac{\rho_M}{\rho_D}\gg:\rho_C=\rho_M(1-c_D)\frac{1-\cos^2\alpha}{F} + \frac{\cos^2\alpha}{1-2F}; \tag{4} \qquad \frac{\rho_M}{\rho_D}\ll1:\rho_C=\rho_M(1-c_D)\frac{1-\cos^2\alpha}{F-1} - \frac{\cos^2\alpha}{2F} \tag{5}
$$

and also in particular cases of shape and orientation /5/. Using these equations, resistivity curves as shown in fig.2 are obtained for an assumed two phase material pointing out graphically the influence of stereometric microstructure. - It is the aim of the indirect stereometric factors introduced in equations 1-5 to take into account theoretically the effect of the microstructure on the properties concerned. Their definition therefore follows obligatory from the derivation of the equations. As usual, if real structures and relationships have to be described quantitatively, assumptions are required leading to approximate results only. Defining the stereometric factors due to equation 1, these assumptions include that
- the actual particles of the dispersed phase differentiated by size, shape and orientation can be substituted theoretically by the same number of average sized, shaped and oriented particles, where the irregular particle shape is characterised by an adequate spheroid.

Following this line,
- the phase concentration factor is simply the volume content of the dispersed phase
- the shape factor is directly related to the deelectrization factor which follows from fig.3 as a function of the axial ratio /6-9/,
- the orientation factor is the average of the cosinus squares of those angles which are formed by the rotation axes of the phase particle substituting spheroids and the direction of the electric field. It is also a function of the mean axial ratio of the spheroids /8,10/:

$$
\overline{\left[\frac{(\frac{z}{x})^2_{\shortparallel}\overline{(\frac{b'}{a'})}^2-1}{(\frac{z}{x})^2_{\shortparallel}-1}\right]_A} \quad \begin{array}{c}\text{prolate}\\ \hline \text{spheroids}\end{array}\ \cos^2\alpha\ \begin{array}{c}\text{oblate}\\ \hline \text{spheroids}\end{array}\quad \overline{\left[\frac{(\frac{z}{x})^2_{=}\overline{(\frac{a'}{b'})}^2-1}{(\frac{z}{x})^2_{=}-1}\right]_A} \tag{6}
$$

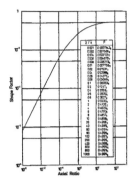

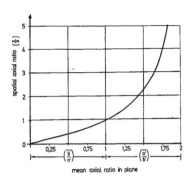

fig.3: Shape factor and axial
ratio of spheroids /5,6/

fig.4: Mean axial ratio of intercept ellipses in plane and axial
ratio by the respective spheroid$(\frac{z}{x})$

The phase concentration factor immediately follows by measuring the mean areal
concentration of the dispersed phase in section planes according to the principle of Delesse. In order to determine the shape and the orientation factor,
it is necessary to measure the following quantities stereometrically in plane
sections of the material investigated:
- the largest extension of each area of sectioned dispersed particles which is
  taken to be the large axis of the substituting sectional ellipse (a'),
- the area of each section of dispersed particles itself (A'),
- the mean area concentration of the dispersed phase ($c_D$),
- the average number of sectioned areas of dispersed particles per unit area of
  plane ($\overline{N_A'}$),
- the total number of section areas of dispersed particles (N').
Using fig.3 and equation 6 as well as the following relationships /4,7,8,10,11,
12/, these measurable stereometric quantities lead to the indirect shape and
orientation factor:

$$\widetilde{(\frac{z}{x})} = (1-n_=) \, (\frac{z}{x})_{,,} + n_= (\frac{z}{x})_= \tag{7}$$

$$(\frac{z}{x})_{,,} = f(\overline{\frac{a'}{b'}}) \; ; \quad (\frac{z}{x})_= = f(\overline{\frac{b'}{a'}}) \quad (s.fig.4) \tag{8}$$

$$n_= = \frac{c_D - N_i V_{V_{,,}}}{V_= - V_{,,}} = \frac{6c_D[(\overline{\frac{b'}{a'}}) \, a_{,,} + (\overline{\frac{a'}{b'}})b_=] - 2\overline{N_A'} \cdot \P \cdot a_{,,} b_{,,}^2}{\P[(\overline{\frac{b'}{a'}})a_{,,} + (\overline{\frac{a'}{b'}})b_=][a_=^2 b_= - a_{,,} b_{,,}^2]} \tag{9}$$

$$\frac{b'}{a'} = \frac{4A'}{\P(a')^2} \; ; \quad \frac{a'}{b'} = \frac{\P(a')^2}{4A'} \tag{10}$$

$$(\overline{\frac{b'}{a'}}) = \frac{1}{N'} \Sigma \frac{b'}{a'} \neq \frac{\overline{b'}}{\overline{a'}} \; ; \quad (\overline{\frac{a'}{b'}}) = \frac{1}{N'} \Sigma \frac{a'}{b'} \neq \frac{\overline{a'}}{\overline{b'}} \tag{11}$$

$$\overline{a'} = \frac{1}{N'} \Sigma a' \; ; \quad a_= = \frac{\overline{4a'}}{\P} \; ; \quad a_{,,} = \frac{4\overline{a'} \cdot (\frac{z}{x})_{,,}}{\pi \cdot (\overline{\frac{a'}{b'}})} \tag{12}$$

Comparisons between measured electrical resistivities of cermets and calculated
curves, taking into account the stereometric effects by using the above equa-

tions, are given in fig.5 and 6. Additional information about such comparisons
including other properties as thermal conductivity, Youngs modulus and thermal
expansion coefficients,is already available in the literature /4,5,12,13,14/.
Summarizing the results, the comparisons point out that the derived equations are
not only a sufficient engineering approach to reality, but also
- provide a better insight into the relationships between properties and microstructure,
- offer one way to taylor made materials by precalculating optimalized microstructures causing desired property values,
- may be used instead of data collections by calculating the properties of two-
phase materials, the phase properties of which are known,
- substitute the direct property measurement by a stereometric microstructural
analysis or extrapolation in such cases where the first is either less accurate,
very difficult or even impossible.
An impressive example in which direct measurements have been substituted successfully by stereometric microstructural analysis is the determination of the thermal
conductivity of porous nuclear fuels which cannot be measured directly by the
disturbing irradiation influences /14-16/.-The extrapolation of conductivities
of intermetallic compounds may serve as another demonstration in that context.
Due to their brittleness, pure intermetallic compounds often cannot be classified
satisfactorily by powder metallurgical methods and also cannot be prepared single-
phased by melting techniques. This is the reason why properties of pure intermetallic phases are frequently unknown. But such phases are available in dense
two-phase materials. Measuring the conductivities of these materials and their
nonintermetallic phase the equations considered here allow the calculation of
the conducticity of the intermetallic phase /17/.

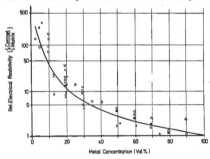

fig.5:Relative el.resistivity of metal-
matrix UO$_2$-cermets [5]

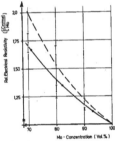

fig.6:Relative resistivity metal-
matrix UO$_2$-Mo-Cermets. x statistically oriented UO$_2$ discs
o UO$_2$ Spheres [5]

List of symbols and literature: a' = larger axis of
sectional ellipse in plane; A' = area in plane of
sectioned dispersed particle; α = angle between the
rotation axis of a spheroid and the field direction;
(a'/b'), (b'/a') = mean values of axial ratios of
the sectional ellipses measured in planes taken
statistically through the specimen; (a'/b')$_A$ ,
(b'/a')$_A$ = mean values of axial ratios of the sectional ellipses measured in planes perpendicular
to the direction of the electric field; b' = smaller axis of the sectional ellipse in plane; c =
phase content; C = index for composite, cermet, two
phase material; cos$^2$α = orientation factor; D =
index for dispersed phase; F = (indirect) shape
factor (concerning field properties); M = index for
matrix material; n$_o$= relative amount of oblate
spheroids; N' = total number of section areas of
dispersed phase particles; N'$_A$ = average number of
section areas of dispersed phase particles per
unit area of plane; ρ = electrical resistivity;
x = a$_n$ = b$_n$ = minor axis of spheroid; z = a$_n$ =
b$_n$ = rotation axis of spheroid; $\frac{a}{b}$ = axial ratio of
prolate (index n) or oblate (index ") spheroids.

Literature
/1/ Niesel W., Ann.Phys. 6-10(1952)336
/2/ Ondracek G., Schulz B., Ber.Dtsch.Keram.Ges.
48-10(1971)427
/3/ Ondracek G., Schulz B., Ber.Dtsch.Keram.Ges.
48-12(1971)525
/4/ Schulz B., Dissertation Universität Karlsruhe (1974)
/5/ Ondracek G., Z.f.Werkstofftechnik 5-8(1974)
416
/6/ Stille U., Arch.Elektrotechnik 38-3/4(1944)91
/7/ Ondracek G., Schulz B., Practical Metallography (1973)16
/8/ Ondracek G., Schulz B., Practical Metallography (1973)67
/9/ Ondracek G., KFK-Ext. 6/73-2(1973)40
/10/ Hoff R.T.de, Rhines F.N., Trans.Met.Soc.AIME
221(1961)975
/11/ Bakke T., Nazaré S., The Microscope Fourth
Quarter (1975)
/12/ Nazaré S., Ondracek G., Powder Metallurgy Int.
6-1(1974)8
/13/ Nazaré S., Ondracek G., Ceramurgia (1975)
to be published
/14/ Ondracek G., Schulz B., J.Nucl.Mat.46 3(1973)
253
/15/ Ondracek G., Schulz B., KFK 1274/1(1974)112
/16/ Ondracek G., Schulz B., KFK 1999(1974)43
/17/ Schmitt R., Diplomarbeit Universität Karlsruhe, Institut für Werkstoffkunde II (1975)

*National Bureau of Standards Special Publication 431*
*Proceedings of the* FOURTH INTERNATIONAL CONGRESS FOR STEREOLOGY
*held at* NBS, Gaithersburg, Md., September 4-9, 1975 (Issued January 1976)

USE OF MATHEMATICAL MORPHOLOGY TO ESTIMATE COMMINUTION EFFICIENCY

by Ph. Cauwe
*Société Wild Leitz-Paris, 86, Avenue du 18 juin 1940, 92504 Rueil-Malmaison, France*

By using the rigorous theoretical background of Mathematical Morphology of G. MATHERON and J. SERRA and with the associated electronic device, the Texture Analysing System (LEITZ T.A.S.), we have tried to study comminution problems with a new point of view. A first attempt of a model uses the net of Poisson flats in Euclidean spaces as described by R.E. MILES. Due to bad agreement with experimental data we have built a new model of Poisson twin flats in Euclidean spaces. Experimental values performed on comminuted products by use of the Texture Analysing System are in good agreement with the model and lead to easy interpretations of milling conditions.

1. DEFINITION: A comminution is a size reduction of large particles to finer ones. The comparison of size distributions before and after milling is the more usual way to estimate the action of the operation.

2. THE SIZE OF A PARTICULE: In the field of comminuted products, the size of a particule or grain is much more difficult to define than these of a circle or a cube. In practice, the mesh sizes of sieves are used: this method does not refer to any true concept of size. Another manner is to build up a concept of size independently of the technological ways used to obtain size distributions of particles. By doing so, each physical method can refer to a same principle and can be compared. This way is the one used by the Mathematical Morphology developed by G. Matheron and J. Serra /6/. Using the well known structuring element, the Mathematical Morphology gives the geometrical interpretation of the studied structure. Using the theoretical background and the associated Texture Analysing System (LEITZ T.A.S.) we try to build up a model to estimate the efficiency of comminution processes.

3. THE WAY TO A MATHEMATICAL MODEL OF COMMINUTION: Like the electronics engineer looking for the equation describing the signal transform when going through a "black box", we try to find the transfer function of a milling product going through a comminution machine. We try to combine tools given by the Mathematical Morphology to describe the transformation of a material of large particles to finer ones. By doing so, we are independant of experimental conditions and the parameters involved in the model have their own signification. We can expect to have a sharp description of a size distribution.

4. POISSON FLATS: A FIRST MODEL OF COMMINUTION. With reference to the literature, one can admit that the fracture propagation within ores and minerals fragments follows a net of existing cracks. We can imagine that comminuted particles have statistically equivalent geometric properties as these of a net of polyhedra in the three dimensional space. Our first attempt is to describe the fracture process by a net of Poisson flats in the 3-dimensional space yet described by R.E. MILES /5/. Using Poisson flats in the 3-dimensional space we obtain a random partition of the space into polyhedra. The parameter of the model is $\lambda_3$ (independent of direction $\alpha$): the density of the Poisson law. Morphologically speaking, in the 1-dimensional space, the erosion by a linear structuring element of a segment defined by two consecutive points of the Poisson process and containing the origin gives an eroded segment, if not empty, having the same properties as the original one. This fact is equivalent to the lack of ageing. G. Matheron generalises this property to Poisson flats of space $R^n$ : the probabilist properties (volume,...) of Poisson polyhedra containing the origin are not modified after erosion by a convex compact on condition the eroded polyhedra are not empty /4/. Knowing that Poisson flats in the 3-dimensional space induces Poisson lines and Poisson points on planes and lines going through them, very interesting relations between model's parameters in the different spaces can be calculated /1/,/4/,/5/. For example, the relation between the Poisson densities $\lambda_i$ in space i : $\lambda_3 = 2\lambda_2/\pi = {}^-\lambda_1/\pi$   (1).

With this kind of relation, experimental measures on lines and areas provide two different ways to compute the parameters of the model in the 3-dimensional space. One of the basic relation used is the linear granulometry as defined by G. Matheron and J. Serra /6/. The practical equation is : $1-G(\tau) = (1+\pi\lambda_3\tau)$. Exp $(-\pi \cdot \lambda_3 \cdot \tau)$ where $\tau$ is the length of a segment included in a polyhedra and $1-G(\tau)$ the cumulative size distribution according $\tau$. Figure 1 shows equation (2) computed for different values of $\lambda_3$ where $\lambda_3$ is a measure of the fineness. Comparison of theoretical equation with experimental values performed by the Texture Analysing System on polish sections of comminuted products have shown general good agreement (figure 2). But theoretical equation has shown also a systematic underestimation of the amount of fine particles. This fact led us to modify the model to a new original distribution of flats in Euclidean space.

5. THE MODEL OF POISSON TWIN FLATS. Roughly, one can say that each plane of the Poisson flats in the 3-dimensional space is replaced by two planes. The distance between the planes of a twin is given by a probability law $F_3(x)$. Here also the model of the 3-dimensional space induces Poisson twin lines and twin points on planes and lines going through it. Same kind of relations between parameters of the model in different spaces can be calculated /1/. One can easily understand that internal region of each pair of planes is cut by a higher number of planes giving a greater proportion of small polyhedra. The equivalent relation to equation (2) can be calculated /1/. By comparing this equation with experimental results we have found much better agreement, specially for the part of fine particles. To verify we have also used the probability $P(C_R)$ for having a circle $C_R$ of radius R inside a polygon of the Poisson twin flats in space 2.

6. CONFIRMATION OF THE MODEL OF POISSON TWIN PLANES. Using the linear and the two dimensional logics facilities of the Texture Analysing System (LEITZ T.A.S.) computation of equations leads to experimental values for parameters $\lambda_i$ and $d_i$ (i = 1,2) ($d_i$ is the parameter of $F_i(x)$ ).

SAMPLE PREPARATION: Experimental measures using optical microscope and TV camera can only be performed on flat surfaces as polish sections. By mixing the comminuted products with a Teflon powder (10 $\mu$m) homogeneous distribution of ores particles is obtained. After compression, sintering and polishing measures are performed on the grains sections. According that the properties of the Poisson polyhedra depend only on themselves and not on their mutual disposition in space; the absence of contact between two grains in a sample is the sole condition to reach valuable experimental data. 180 polished sections from 30 different tests of milling conditions have been done and measured by the Texture Analysing System. Two kinds of products have been studied: a copper-lead-zinc ore with a complex mineralogic texture and a highly homogeneous magnetic concentrate of iron.

MODEL'S PARAMETERS ESTIMATION. Experimental values of $P(\ell)$ and $P(C_R)$, obtained by the Texture Analysing System (LEITZ T.A.S.), have been smoothed (quadratic equation) before calculations of logarithms and derivatives. These numerical treatments lead to values of $\lambda_1$ and $d_i$ (i=1,2) reported on Table I with the values of $\lambda_2^*$ calculated from experimental values of $\lambda_1$ according equation (1). One can easily test that the ratio $\lambda_2^*/\lambda_2$ is equal to one. More precisely it can be seen on table I that the mean value of the ratio is 1,023 with a standard deviation of 0.085. This result is also presented on figure 3. Table II presents similar results: estimation of $d_3$ from experimental values of $d_1$ and $d_2$. Tables I and II are only the more demonstrative results proving the validity of the Poisson Twin Planes as a model for comminuted products. Other relations between the 1 and 2-dimensional spaces have been verified as shown in /1/.

USE OF PARAMETERS TO ESTIMATE COMMINUTION CONDITIONS. Model's parameters and comminution conditions have been obtained from a Magnetite concentrate milled in a pilot ball mill under different experimental conditions: weight of concentrate, size of balls. For tests I, II, III various samples have been taken at different times of milling and then analysed. In the following, results are presented versus time (minute) and power (KWH) consumption.

THE PARAMETER d: On figure 4 it can be seen that the values of d stabilise at a constant value after a given time. As discussed in /1/, the parameter d is related to the smaller size of particules and the steady stade value is a measure of the smallest grain size obtainable for given grinding conditions.

THE PARAMETER $\lambda$ : For example figure 5 presents $\lambda_1$ and $\lambda_2$ versus time for Test I.

From this kind of diagrams one can appreciate how grinding action is going for given conditions. Comparison of λ-curves of various milling condition leads to the better grinding effect.
POWER CONSUMPTION: Knowing that the maximum power efficiency of a comminution is 1%, correlations between power and model's parameters must be studied. Diagrams of power versus time of milling is of general use in comminution studies (fig.6): very poor information is given by these diagrams. On the other hand, curves of λ versus power consumption lead to easy appreciation of the better comminution conditions. From figure 7, it is obvious that the better conditions are these of test III: – up to 20 KWH/T tests I, II, III are equivalent; – for power higher than 20 KWH/T, higher power consumption does not increase fineness of the products in test I conditions; – this situation is better for test II but even better for test III. Another kind of result can be read on this figure: for the same fineness of product (same λ) conditions of test III comparing to those of test II, save 20% of energy.
7. CONCLUSIONS. – For the first time, sharp relations have been found between geometrical characteristics of particules and comminution conditions. Using the powerfull "tools" given by the Mathematical Morphology this study gives easy and clear view of the comminution process.
– Knowing that fractures propagation follows pre-existing cracks it is remarkable that to adjust a model to a milling process it is not necessary to know the net of cracks in situ. This fact is specially interesting when thinking of the not visible cracks! – It is well known that equations with many parameters can always fit experimental data but these parameters are hardly related to physical realities. One can remark that our model has only two parameters. We have tested /1/ that it enables to cover the wider range of comminution conditions: crushers, grinders, ball mills minerals and ores of various types. – Furthermore, the model can help for interpretating physical phenomena involved in comminution processes. As discussed in /1/ the parameter λ can be interpretated as a direct measure of milling conditions efficiency and parameter d can be correlated to a practical limit for the minimum size of grains obtainable for given experimental conditions.
8. BIBLIOGRAPHY.
/1/ CAUWE PH.: Morphologie Mathematique et Fragmentation-Thèse-Univ.Cath.Louvain-Lab. Traitement des Minerais-Belgique.
/2/ HADWIGER H.: Vorlesungen über Inhalt, Oberfläche und Isoperimetrie-Springer Verlag-1957.
/3/ KLEIN J.C.-SERRA J.: The texture analyser-Stereology 3, 1971.
/4/ MATHERON G. : Random sets and integral geometry-Intersciences-Wiley & S. 1975.
/5/ MILES R.E.: A synopsis of Poisson flats in Euclidean space-Izv. Akad. Nauk. armianskoi SSR, 3, 1970, 263-285.
/6/ SERRA J. : Introduction à la Morphologie Mathématique-Cahiers du Centre de Morphologie Mathématique-Fascicule 3-1969.

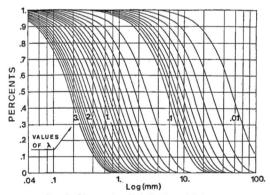

Fig. 1: Linear size distribution of Poisson planes in space 3.

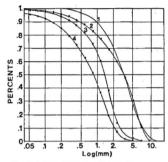

Fig. 2: 1-Theoretical size distribution; 2-Quartz, roll crusher; 3-Dolomite, jaw crusher; 4-Galena, roll crusher. Points are experimental, curves follow eq.(2)

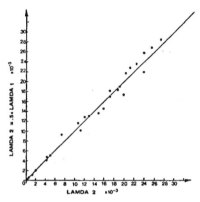

Fig. 3: Relation between $\lambda_2$ and $\lambda_2^o$.

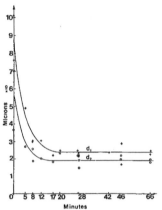

Fig. 4: $d_1$ and $d_2$ vs. time of milling.

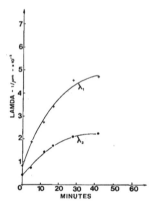

Fig. 5: $\lambda_1$ and $\lambda_2$ vs. time of milling, test I.

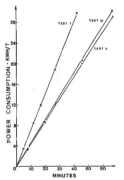

Fig. 6: Power consumption vs. time of milling.

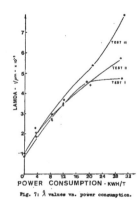

Fig. 7: $\lambda$ values vs. power consumption.

| | TABLE I | | | | TABLE II | | | | |
|---|---|---|---|---|---|---|---|---|---|
| Test number | $\lambda_1$ $1/\mu m$ | $\lambda_2^*$ $1/\mu m$ | $\lambda_2$ $1/\mu m$ | $\lambda_2^*/\lambda_2$ | $d_1$ $/\mu m$ | $d_2$ $/\mu m$ | $d_{31}$ $/\mu m$ | $d_{32}$ $/\mu m$ | $d_{31}/d_{32}$ |
| 6737 | 0.182 | 0.091 | 0.091 | 1.001 | 760 | 595 | 507 | 505 | 1.004 |
| 6738 | 0.266 | 0.133 | 0.144 | 0.923 | 760 | 595 | 507 | 505 | 1.004 |
| 6739 | 0.372 | 0.186 | 0.202 | 0.920 | 760 | 595 | 507 | 505 | 1.004 |
| 6740 | 0.658 | 0.329 | 0.298 | 1.104 | 760 | 595 | 507 | 505 | 1.004 |
| 6741 | 1.401 | 0.700 | 0.658 | 1.064 | 760 | 595 | 507 | 505 | 1.004 |
| 6742 | 2.399 | 1.200 | 1.315 | 0.912 | 500 | 370 | 333 | 314 | 1.061 |
| 6743 | 4.008 | 2.004 | 2.054 | 0.975 | 220 | 130 | 147 | 110 | 1.336 |
| 6744 | 9.230 | 4.515 | 4.236 | 1.090 | 99 | 57 | 66 | 48 | 1.375 |
| 6745 | 25.702 | 12.851 | 12.291 | 1.046 | 53 | 36 | 35 | 31 | 1.129 |
| 6746 | 36.436 | 18.218 | 17.077 | 1.067 | 35 | 24 | 23 | 20 | 1.150 |
| 6747 | 43.701 | 21.851 | 20.127 | 1.086 | 28 | 21 | 19 | 18 | 1.056 |
| 6748 | 51.847 | 25.924 | 23.789 | 1.090 | 26 | 20 | 17 | 17 | 1.000 |
| 10701 | 8.229 | 4.115 | 4.115 | 0.999 | 83 | 60 | 55 | 51 | 1.784 |
| 10702 | 18.759 | 9.380 | 7.495 | 1.251 | 42 | 31 | 28 | 26 | 1.071 |
| 10703 | 27.303 | 13.652 | 14.657 | 0.931 | 31 | 20 | 21 | 17 | 1.235 |
| 10704 | 34.187 | 17.094 | 17.092 | 1.001 | 24 | 19 | 16 | 16 | 1.000 |
| 10705 | 45.841 | 22.921 | 21.072 | 1.088 | 24 | 19 | 16 | 16 | 1.000 |
| 10706 | 47.300 | 23.650 | 22.353 | 1.058 | 24 | 19 | 16 | 16 | 1.000 |
| 10707 | 20.353 | 10.177 | 11.127 | 0.915 | 32 | 24 | 21 | 20 | 1.050 |
| 10708 | 26.174 | 13.087 | 12.805 | 1.022 | 24 | 19 | 16 | 16 | 1.000 |
| 10709 | 34.840 | 18.920 | 18.920 | 1.000 | 24 | 19 | 16 | 16 | 1.000 |
| 10710 | 44.094 | 22.047 | 23.884 | 0.923 | 24 | 19 | 16 | 16 | 1.000 |
| 10711 | 57.351 | 28.675 | 27.151 | 1.056 | 24 | 19 | 16 | 16 | 1.000 |
| 10713 | 10.013 | 5.007 | 5.030 | 0.995 | 83 | 60 | 55 | 51 | 1.784 |
| 10714 | 23.294 | 11.647 | 10.715 | 1.087 | 32 | 24 | 21 | 20 | 1.050 |
| 10715 | 29.300 | 14.650 | 15.787 | 0.928 | 24 | 19 | 16 | 16 | 1.000 |
| 10716 | 36.830 | 18.415 | 18.650 | 0.987 | 24 | 19 | 16 | 16 | 1.000 |
| 10717 | 34.944 | 17.472 | 19.738 | 0.885 | 24 | 19 | 16 | 16 | 1.000 |
| 10718 | 53.716 | 26.858 | 25.618 | 1.049 | 24 | 19 | 16 | 16 | 1.000 |
| 10719 | 78.290 | 39.145 | 32.022 | 1.222 | 24 | 19 | 16 | 16 | 1.000 |
| Mean value of the ratio | | | | 1.023 | | | | | 1.067 |
| standard deviation | | | | 0.085 | | | | | 0.208 |

$\lambda$ values have to be multiplied by $10^{-3}$

$\lambda_2 = 0.5\lambda_1$ ; $d_{31} = 2d_1/3$ ; $d_{32} = 8d_2/3\pi$

*National Bureau of Standards Special Publication 431*
*Proceedings of the* FOURTH INTERNATIONAL CONGRESS FOR STEREOLOGY
*held at* NBS, Gaithersburg, Md., September 4-9, 1975 (Issued January 1976)

MICROSCOPY, RADIAL DISTRIBUTION ANALYSIS, AND SORTABILITY

by Robert M. Doerr
*U.S. Bureau of Mines, Rolla Metallurgy Research Center, Rolla, MO 65401, U.S.A.*

A method, based on a radial distribution analysis, was developed for determining the spatial distribution of a phase in an aggregate. Two components of
the determination are (1) the concentration of the phase and (2) the enclosed
proportion of the total quantity of the phase present, each as a function of the
radius of an appropriately centered enveloping sphere. A measuring microscope
and computer analysis are used to obtain the radial distribution of the phase in
plane sections of the aggregate. The method was developed to quantify statistically the dependency of the grade and recovery potentials on the fragment size
to which a copper ore is broken for sorting.

INTRODUCTION
    This is a report on a method (1) to determine the sizes of regions of se-
lected target grade in an ore and (2) to estimate the potential recovery of the
desired mineral when the ore, broken to such sizes, is processed to yield a concentrate of that grade. When coupled with cost data, mean grade of ore, and
product value, such information is useful in deciding the feasibility of sorting
and the fragment size at which sorting would be most effective.
    Modern digital electronic circuitry has made high speed serial automatic
sorting practical for various separations, including minerals beneficiation.[1-17]
In some methods of separation, such as flotation and by magnetism, the separating action is a direct response to the properties on which the separation is
based. However, the two essential steps of sorting, detection and separation,
can occur at different times and/or places. Thus, for example, if fragments
that possess a characteristic (such as above-threshold radioactivity, induction,
fluorescence, or some combination) are readily and instrumentally identifiable
as different from those not exhibiting the characteristic, a distinct, powered
unit can be triggered to separate the individual fragments of one class from the
stream. For minerals beneficiation, the significant difference is usually that
of grade, or proportion of a valuable species, between fragments. Effective application of such sorting depends on the spatial distribution of the valuable
species within the host rock and breakage to fragments to 'liberate' regions of
sufficient grade difference. The economics of sorting are adversely affected by
over-breakage of the ore.[1] The importance of fragment size on sortability has
been recognized and the effects researched.[15]
    To visualize the radial distribution analysis,[18] consider as a specimen a
plane section of an ore fragment, with a vanishingly small circle centered on an
appropriate copper particle in the plane surface. The grade of copper within
the circle is unity. As the radius of the circle is increased, areas of rock
devoid of copper, and other copper particles, are enveloped; the overall fraction of copper within the circle generally declines. With allowance for areas
of the circle that extend beyond the edges of the specimen, the grade within the
circle ultimately reaches the specimen grade. The largest radius at which the
grade within the circle is equal to a chosen target grade, taken to be greater
than the mean, is determined from an empirical equation relating the grade to
the radius. The fraction of total specimen copper within that radius is taken
as the potential recovery for the target grade chosen. The most appropriate
central particle to choose is the one that leads to the highest potential recovery at the target grade, given an acceptable fit of the equation to the
data. It has been shown that grade (area fraction) analyses made in two
dimensions are statistically applicable to three dimensions (volume fraction).[19-22]

PROCEDURE

The sample used was drawn from fragments of conglomerate native copper ore that had been crudely preconcentrated by testing each fragment for copper by an induction bridge coupling method and retaining those for which the presence of metal was indicated. Raw ore would have necessitated too much data taking and would not have helped elucidate the application of the method. The ore fragments were sectioned by diamond sawing and the resulting plane faces were polished. These faces constituted the sample. By use of a microscope with a stage vernier and a grid reticle, data on the size and location, relative to an arbitrary origin, of each copper particle in the plane were recorded, as were data on the location of the perimeter of the specimen and of any pores therein. In the nine-specimen sample, totaling 7 260 mm², the plane area and location of 34 660 copper particles were measured and recorded.

The grade vs. radius data for the radial distribution analyses (about appropriate 'central' particles) were developed and a forward stepwise linear multiple regression analysis routine[23] was modified and used to fit the radial distribution curve. For each specimen, the radius and potential recovery for each of the concentrate grades 2 to 12 area % copper were tabulated and the radial distribution curve plotted.

The central particle for each specimen was chosen for reasons of economy from among the approximately 50 largest copper particles in the specimen. Temporarily ignoring all the smaller particles, the grade within a certain distance of each particle was determined, and the particle chosen to be the central one was the one that led to the highest grade within that distance. Two distances were tried for each specimen: half the distance to the most remote of these largest particles and 6 mm. In most cases, the two distances led to different central particles; the one used was the one that led to the higher indicated potential recovery. The 6-mm distance was found better only for lean, probably reject, specimens and the half-maximum distance for the others.

The model used for the radial distribution relationships was:

$$y = \sum_{i=0}^{7} \beta_i x_i + \varepsilon$$

where y is ln (grade within circle of radius r), $\beta_i$ is the ith regression coefficient, $x_0 = 1$, $x_1 = $ radius r, $x_2 = x_1^2$, $x_3 = x_1^3$, $x_4 = 1/x_1$, $x_5 = x_4^2$, $x_6 = x_4^3$, $x_7 = $ ln $(x_4)$, and $\varepsilon$ is random error. For another ore, a different model may be required. A grade-vs.-radius point was calculated for each copper particle, not just for the approximately 50 largest.

RESULTS AND DISCUSSION

Characterization of the sample is presented in Table 1, in which are shown the rock and copper contributions of each specimen and selected totals and means. Most of the copper is contained in two rich specimens which aggregate 65% of the copper in 24% of the rock. These and all percentages in this report are area-percent. Owing largely to the two rich specimens, concentrates with 5% copper and fairly high recovery are seen to be possible from sorting the sample without further crushing. This occurrence presumably stems from the small sizes of the specimens.

The results from fitting the data to the model are also shown in Table 1. For no specimen did all the terms of the model prove significant. The resulting radii for a target concentrate grading 5% copper, recoveries, and a measure of the goodness of fit of the model to the data are presented. Three specimens are seen to have regions of 5% copper with radius not less than 12 mm.

Crushed to a 12- to 15-mm radius range, the sample should be sortable to a concentrate with a grade over 5% and with recovery of about 62% of the copper in about 18% of the feed. All of specimen 1 would be included in a 5% concentrate, even without further crushing. Crushed pieces from specimens 6, 7, and 9, aggregating 34% of the feed would be rejected if the value of the copper requires a concentrate grading 5%, on the basis of the low potential recovery at 5%, and the pieces from specimens 3 and 8 would probably be rejected if the cost of sorting at a 6-mm radius is uneconomic.

TABLE 1. – Sample Data and Results from Radial Distribution Analyses

| Specimen* | Size, $mm^2$ | Cu area % | (Correlation coefficient)$^2$ | Radius for 5 area % target concentrate, mm | Recovery At This Radius, % |
|---|---|---|---|---|---|
| 1 | 946.2 | 7.79 | .999 | 32.5*** | 100.0*** |
| 2 | 757.9 | 4.48 | .886 | 15.4 | 56.4 |
| 3 | 671.3 | 2.02 | .983 | 3.8 | 17.7 |
| 4 | 250.4 | 1.37 | .954 | 6.0 | 57.6 |
| 5 | 1365.1 | 1.15 | .946 | 12.0 | 71.4 |
| 6 | 1230.7 | 1.08 | .991 | 2.7 | 8.6 |
| 7 | 655.5 | 0.927 | .991 | 1.7 | 6.2 |
| 8 | 772.4 | 0.868 | .987 | 3.3 | 20.9 |
| 9 | 610.8 | 0.049 | .998 | 0.1 | 2.1 |
| Totals | 7260.3 | 19.734 | | | |
| Means | 806.7 | 2.193** | .971 | | |

\* Arranged in order of descending grade.
\*\* Weighted mean is 2.29 area %.
\*\*\* Specimen grade is about 7.8%.

TABLE 2. – Radial Distribution Analysis for Specimen 2

| Target Concentrate Grade, Area % | Radius for Target Grade, mm | Recovery At This Radius, % |
|---|---|---|
| 5 | 15.4 | 56.4 |
| 6 | 2.7 | 1.9 |
| 7 | 2.3 | 1.9 |
| 8 | 2.1 | 1.5 |
| 9 | 1.9 | 1.2 |
| 10 | 1.8 | 1.2 |
| 11 | 1.7 | 1.2 |
| 12 | 1.6 | 1.2 |

TABLE 3. – Radial Distribution Analysis for Specimen 5

| Target Concentrate Grade, Area % | Radius for Target Grade, mm | Recovery At This Radius, % |
|---|---|---|
| 2 | 25.2 | 94.7 |
| 3 | 18.9 | 91.2 |
| 4 | 15.0 | 83.4 |
| 5 | 12.0 | 71.4 |
| 6 | 9.4 | 67.7 |
| 7 | 7.0 | 51.5 |
| 8 | 3.8 | 10.5 |
| 9 | 1.6 | 3.8 |
| 10 | 1.3 | 3.6 |
| 11 | 1.1 | 3.6 |
| 12 | 1.0 | 1.8 |

As examples of the radial distribution analysis, tabular results are presented for two specimens. For specimen 2 these results are presented in Table 2. The recovery falls off abruptly from 56 to 2% upon changing the target concentrate from 5 to 6%, and strongly suggests that the specimen contains two or more rich regions which, together with the intervening rock, all in one relatively large fragment, could be acceptable in the 5% concentrate but not in a 6% concentrate. Thus, the potential recoveries indicated for the 6% and richer cases are low if one or more additional comparably large regions are of target grade. It is thus necessary to use specimens of size appropriate to the target grade and to the distribution of the valuable mineral. The most remote copper particle is about 39 mm from the central one; this suggests that the central one is not near the center of the specimen, i.e., that fracture occurred through or near the richest copper region. Results for specimen 5 are presented in Table 3. Here the recovery falls off slowly when the target concentrate grade is increased from 2 to 7%, while the radius drops from about 25 mm to about 7 mm. Thus, much of the specimen copper is in the region that includes the central particle.

The method of radial distribution analysis enables the determination of key data needed for judging the feasibility of sorting an ore and the size at which to sort. However, manual recording of the voluminous data proved too tedious to permit working a sample comprised of enough specimens to perform needed statistical analyses and provide confidence limits on the results. Specimens large enough to permit full determination of the copper distribution were similarly precluded. Thus it is recommended that for implementation, the data be recorded by a computerized microscope[24-26] with the computer controlling the stage drives. Such apparatus is within the state of the art. Further, with an automated microscope, it would be feasible simply to assay each hexagonal field of view, record its position, and develop an assay-vs.-location data point for each such field rather than for each particle. This approach, which is not feasible with manual data acquisition, would reduce computer memory requirements and enable simplified programming. It should allow application of cluster analysis, and with it, treatment of the shape factor and independence from limitations on the shapes of rich regions and from the problem of multiple rich regions, so that large enough specimens can be evaluated.

CONCLUSIONS

The method of radial distribution analysis provides data that are useful, when combined with economic factors, for judging whether or not a mineral deposit lends itself to sorting as a means of concentration or preconcentration, providing the valuable species is detectable. These data relate the effect of the rock fragment size at which sorting is performed, and selected target grades to the potential recoveries at such grades.

ACKNOWLEDGMENT

This report is modified from the author's thesis submitted to the University of Missouri-Rolla in partial fulfillment of the requirements for the degree Master of Science in Ceramic Engineering.

REFERENCES

1. Wyman, R. A., Selective Electronic Mineral Sorting to 1972, Mines Branch Monogr. No. 878, Canada, Dept. of Energy, Mines, and Resources, Ottawa, 1972, 67 pp.
2. Newman, F. C. and P. F. Whelan, Photometric Separation of Ores in Lump Form, Chapt. in Recent Developments in Mineral Dressing, Inst. of Min. and Met., London, 1953, pp. 359-383.
3. Carson, Robert W., Sorter Goes Solid State Now. That Price Is Right, Prod. Eng., Vol. 39, Feb. 12, 1968, pp. 72-74.
4. Harris, Virgil, Electro-Optics Sort Products by Color, Automation, Vol. 17, Jan. 1970, pp. 53-57.
5. Sorts Potatoes Electro-Optically, Food Eng., Vol. 41, Feb. 1969, pp. 122-128.
6. Sclater, Neil, Electronic System Sorts Silver From Clad Coins, Prod. Eng., Vol. 39, Nov. 18, 1968, pp. 62-63.
7. Mariacher, Burt C., The Challenges of Mineral Separation and Beneficiation, Fine Grind Col., Min. Eng., Vol. 21, June 1969, pp. 49-50.
8. Sorting Limestone Electronically, Rock Prod., Vol. 73, March 1970, pp. 76-79.
9. Hintikka, Ossi V. I. and Andrew Balint, Optical Separation of Limestone in Southern Finland by Lohjan Kalkkitehdas Oy, Trans. SME/AIME, Vol. 250, Sept. 1971, pp. 203-207.
10. Automatic Optical Sorters Separate Premium White Marble, Rock Prod., Vol. 74, June 1971, pp. 52-56.
11. Zeising, Gordon F., William S. Hannan, and William A. Griffith, Concentration, Min. Eng., Vol. 24, Feb. 1972, pp. 91-94.
12. Discrimination in Diamonds, Engineering, Vol. 207, May 9, 1968, p. 708.
13. X-rays Sift Diamonds From Gravel Mix, Mach. Design, Vol. 41, Aug. 7, 1969, p. 112.
14. Colborne, G. F., Electronic Ore Sorting at Beaver Lodge, Canada, Min. & Met. Bull., Vol. 56, No. 616, 1963, pp. 664-668.
15. Carlson, David H., "Some Characteristics

Affecting the Mechanical Sortability of Native Copper Ores", SME-AIME Preprint No. 70-B-47, Feb. 1970, 9 pp.
16. Carlson, D. H. and M. E. Volin, "Electromechanical Sorting of Poor Rock From Michigan Copper Mining Operations", Paper Presented at AIME Upper Peninsula Section Spring Technical Meeting, May 22-23, 1968.
17. Miller, Vernon R., Robert W. Nash, and Alfred E. Schwaneke, "Detecting and Sorting Disseminated Native Copper Ores." U.S. BuMines Report of Invest. 7904, Washington, 1974, 14 pp.
18. Doerr, Robert M., Radial Distribution Analysis For Sorting, Trans. SME/AIME, Vol. 255, 1974, pp. 4-9.
19. Underwood, Erwin E., Quantitative Metallography, Metals Eng. Quart., Vol. 1, Aug. 1961, pp. 70-81.
20. Delesse, M. Procede Mechanique Pour Determiner La Composition Des Roches, Ann. Des Mines, Vol. 13, 4th Series, 1848, pp. 379-388.
21. DeHoff, Robert D., and Frederick N. Rhines, Eds., Quantitative Microscopy, McGraw-Hill, N.Y., 1968, 422 pp.
22. Pellisier, G.E., and S. M. Purdy, Chairmen, Stereology and Quantitative Metallography, STP 504, Amer. Soc. for Testing and Materials, Philadelphia, 1972, 182 pp.
23. Efroymson, M. A., "Multiple Regression Analysis", Chapt. 17 in Mathematical Methods for Digital Computers, Ed. by Anthony Ralston and Herbert S. Wilf, Vol. 1, John Wiley & Sons, Inc., New York, 1960, pp. 191-203.
24. Cole, Michael, Image Analysis, Amer. Laboratory, Vol. 3, June 1971, pp. 19-28.
25. Gray, R. J., "The Present Status of Metallography", Chapt. in Fifty Years of Progress in Metallographic Techniques, STP 430, Am. Soc. for Testing Materials, Philadelphia, 1968, pp. 17-62.
26. Schuck, John, TV Analyzing Computer Expands Scope of Microscopic Studies, Res. & Develop., Vol. 17, April 1966, pp. 28-31.

*National Bureau of Standards Special Publication 431*
*Proceedings of the* FOURTH INTERNATIONAL CONGRESS FOR STEREOLOGY
*held at* NBS, Gaithersburg, Md., September 4-9, 1975 (Issued January 1976)

SIZE DISTRIBUTION ANALYSIS IN SITU ON INDIVIDUAL OR INTERCONNECTED PHASES BY IMAGE ANALYSES

by C. Gateau and J. M. Prévosteau
*Service Géologique National (B.R.G.M.), B.P 6009, 45018 Orléans Cédex, France*

ABSTRACT

We present the analytical conditions which must be respected for the different size distribution possibilities by image analyser ; size distributions in numbers or in measurements, all or nothing size distributions, sizing by openings according to a structural element.

Using examples of applications taken from the study of natural materials, and in particular of ores, we illustrate the choice of size distribution analysis modes selected which depends on the texture examined, the type of problem posed, the stereological interpretation and, finally, on the analytical efficiency.

In addition, we show how several modes of size distribution analysis can be used together in order to define supplementary morphological criteria which are extremely interesting for the three-dimensional restitution of the distributions measured in one or two dimensions.

A-CHOICE OF A SIZE DISTRIBUTION TYPE

The idea of size distribution supposes a classification of different parts of a whole according to size criteria. The nature of this geometric parameter must obviously be adapted to the texture of the material to be analysed.

The least favourable case corresponds to an interconnected medium (e.g. fissural or porous network). Size distribution in such a material can only be attempted by the intermediary of its opening (defined by Matheron) according to a structural element, of variable size and possibly form ; the dimension of this structural element then serves as a size distribution parameter.

In addition to the preceding analysis, which can always be applied, the individual particles may be characterised by new parameters. In particular, they may be classified according to their specific area, to their perimeter or their vertical and horizontal projections. In addition to these size criteria, we have also developed new modes of distribution analysis according to parameters of form and structure.

When the particles are in the form of aggregates, it is necessary to use the intermediary of structural elements, preferably bi-dimensional (octogon, hexagon, rectangle).

When the size parameter has been chosen, then the method of representing the size distribution must be selected. This may be expressed either as a number or as a measurement. In the later case the number of particles belonging to each class is balanced by their average dimension.

The determination of all the above parameters can only be made rapidly and automatically by an image analysis procedure.

The choice of the type of measurement must therefore be adapted to the problem posed. Several factors must be considered in this choice :
.the type of texture : individual or interconnected medium-grains with or without inclusions - morphology of grains ;
.the nature of the problem : size distributions in numbers or in measurements - need to measure certain parameters in view of a stereological interpretation ;
.the analytical efficiency conditioned by the size of the analysis frame and the magnification used.

B-SIZE DISTRIBUTION ANALYSIS BY LINEAR MEASUREMENTS OR BY OPENINGS ACCORDING TO
A STRUCTURAL ELEMENT

*1-RECALL OF THE MATHEMATICAL DEFINITION OF THE MOST USED SIZE DISTRIBUTIONS*
$P$ (h)=probability that a segment of length h is entirely included in the grains
$F_1$(h)=probability that a chord of grains belongs to a chord of less than h (size
distribution in number)
$G_1$(h)=probability that a point of grains belongs to a chord of less than h (size
distribution in measurement)
$A_1$(h)=probability that an analysis point belongs to a chord of grains greater or
equal to h.
The quantities P(h), $F_1$(h) and $A_1$(h) are not independent parameters, but
are accessible by different measurement functions at the level of the image analyser. $A_1$(h) = p {1-$G_1$(h)} = P(h) – h P'(h)  where p is the percentage of grains

$$1 - F_1 (h) = \frac{P'(h)}{P'(0)}$$

$\Omega_2$(s)=probability that an analysis point belongs to the area scanned by a
structural element S, entirely included in the grains.
This structural element S may, in our case, have very varied forms : octogon, hexagon, square, rectangle, segment, etc.
Grains size distribution analysis of the linear type or by openings according to a bi-dimensional structural element may be applied to all textures
whether they be individual or interconnected.

*2-RECALL OF THE ANALYTICAL CONDITIONS TO BE RESPECTED*
The size distributions defined from one single image are the most frequently erroneous owing to the intersection of the phase considered with the edges of
the field of analysis. If the number of the analysis field is high, this error
is statistically eliminated, on condition, nevertheless, that a guard region is
used whose dimensions are greater than the largest erosion step employed : $h_{max}$.
It is the case for an interconnected medium P(h) ; $A_1$(h) *(fig. 1a)*
Automatic linear opening analysis $\Omega_1$(h) requires a double guard region :
$2h_{max}$ *(fig. 1b)*. Size distribution analysis by bi-dimensional opening $\Omega_2$(s) involves new restrictions on the dimensions of the guard region depending on the
form and the maximum size of the structural element *(fig. 2-3)*.
All these specific conditions for the QTM 720 are summarized in *table 1*.

*3-STUDY OF THE SIZE DISTRIBUTION OF ALLUVIAL PRODUCTS*
All the various possibilities of size distribution analysis may be applied
to this very simple case (isolated convex grains).
In the absence of any other restrictions, the method giving the best results is therefore chosen. In this case, linear measurements from P(h) would be
used, since it is this method which allows the largest analysis frame.

*4-STUDY OF A MASSIVE MINERALIZATION*
The size distribution study of an interconnected medium, whether this be a
porous or fissural network, a gangue or a massive mineralization, obviously pre-
cludes the all or nothing methods. In this case, techniques of openings according to a structural element (linear $\Omega_1$(h) or bi-dimensional $\Omega_2$(s)) will be
used. More particularly, for some ore-dressing applications which are connected
with problems of comminution, of mineral liberation and of fissural networks
which may be clogged, it is necessary to take into account all the previsional
factors which are responsible for the quality of the products obtained and to
acquire them by a bi-dimensional analysis.

C-ALL OR NOTHING SIZE DISTRIBUTIONS

*1-DEFINITION*
The all or nothing size distribution analysis, which can only be applied to
individual media, can be constructed according to size parameters, and also form
(F) or structure (m) criteria.
The general definition of such a size distribution may be expressed as

follows :
$A_2(P)$=probability that an analysis point belongs to a grain whose parameter is
more than a value P.

### 2-RECALL OF THE ANALYTICAL CONDITIONS TO BE RESPECTED

The all or nothing size distribution analysis requires that the guard region be chosen so that its dimension is greater than the largest size of the
particles : $F_{max}$ *(fig. 4)*.

### 3-STUDY OF A ZIRCON-RUTILE ORE

As in the first example, rutile has the form of individual particles. However, the presence of inclusions greatly disturbs the measurements of size distributions whether these be by    linear methods or by bi-dimensional openings.
As these inclusions are much smaller compared with the grains, the all or nothing area sizings $A_2(a)$ are perfectly adapted to the problem.

### 4-PARTICULAR CASE OF THE QUANTITATIVE EVALUATION OF MIXED GRAINS

This new kind of all or nothing size distribution analysis makes it possible to establish a distribution of mixed grains (two phases) according to a parameter of structure on which the relation may be calculated at the level of
each particle of the area of a phase over the total area of the two phases :
$A_2(m)$. The practical adaptation of such a study has made it necessary to work on
two phases X and Y which are simultaneously detected and differentiated.
m = {area X} / {area (X + Y)}  for each particle.
Using the example of an ore containing two sulphides : pyrite and chalcopyrite, it is possible to count the grains either as a whole in the usual way or
to isolate those which contain a certain percentage of pyrite. Also we can measure the area concerning only the mixed grains. *Photograph 1* gives the number of
mixed grains and *photograph 2* shows us the area of mixed grains which are bigger in area than 800 $\mu^2$ and which contain more than 30 % of pyrite. It should be
remembered that the grains measured may be identified by the presence of a luminous dash at their lower extremity.
The study of mixed grains of this ore is in fact translated by a distribution curve $A_2(m)$ in number or in measurement according to the degree of mixture
m which varies from 0 to 1.
An important problem still remains for the stereological interpretation.
Certain causes of error in this domain are mentioned by JONES (5). The estimation of mixed grains is, in fact, biased, because the informations from the
image analyser at the level of an analysis plan underestimates their importance
owing to the textural arrangement specific to each grain.

### D-COMBINED USE OF SIZE DISTRIBUTION : CHARACTERIZATION OF THE ELLIPTICITY OF GRAINS

All size distributions depending on size parameters are obtained in one or
two dimensions. We have determined three-dimensional restitution formulae, particularly for $A_1(h)$ (ref. 1) and for $A_2(a)$ (ref. 2), and have applied Matheron's
results which however are only applicable to spherical grains. Thus it is very
important to determine with what degree of accuracy these formulae may be
applied.
Using the QTM 720 pattern recognition system we have carried out automatic
morphological analyses of particles by the combination of several size factors
with which a form factor can be defined. The choice of this form factor is however limited to combinations of simple functions. In particular, it is possible
to establish a distribution of particles according to the relation :
$K = \dfrac{area}{(perimeter)^2}$ of each particle.
This parameter, which in fact is an index of circularity, has the advantage
of being non-dimensional. Its use is nevertheless limited to convex particles.
Another disadvantage is seen with the magnification of the object studied in the
revelation of certain details, in particular jagged border edges which increase
the value of the perimeter without much modifying the area.
The system may be improved by the simultaneous use of an all or nothing

311

size distribution of the area $A_2$(a) and an octogonal opening size distribution for example $\Omega_2$(s). From amongst the particles of a given area (chosen by the all or nothing technique) it is possible to retain only those which are likely to contain a structural element also of a given size.

In practice, if the classes are fine enough, the selection of particles can be made according to values obtained by the relation :

$$\alpha = \frac{\text{area of largest structural element contained in the grain}}{\text{actual area of the grain}}$$

We have just defined a form coefficient which characterizes the difference from a given geometric figure.

The major disadvantage of this method is that it does not use non-dimensional parameters. If the distribution of particles according to their form is sought then the dimension of the structural element must be varied for each grain size. Such a technique would quickly lead to a prohibitive number of measurements.

However, let us suppose that the grains are to be separated into two classes only : for example, those for which $\alpha > 0.75$ (i.e. those which are the most nearly octogonal) and all the others. In practice, at the QTM 720 scale and with only one programmer it will be possible to obtain in one single operation, an area size distribution in 16 classes for each of the 2 sub-divisions defined by the value of $\alpha$.

E-CONCLUSION

A large number of possibilities of size distribution analysis methods are now available so that all types of individual or interconnected textures may be dealt with. The success of these automatic analyses will be variable if the most suitable mode of distribution is not chosen and the corresponding analytical precautions are not taken.

A particular and original type amongst these distribution modes is the size distribution of mixed grains which we propose as a criterion of quality appreciation. This type of characterization of mixed grains has the advantage of offering new and very promising application perspectives for ore dressing but at present suffers from the stereological bias introduced into its estimation. It is therefore important to develop a mathematical support which will make it possible to correct this bias by taking into account the effects of size, localisation and form of the components of a mixed grain spatially.

REFERENCES

1-J.L. PINAULT, J.M. PREVOSTEAU, D.G. WILLIAMS
Perspectives of application of quantitative
television microscopy to earth sciences.
*Journal of microscopy* (1972), 95, 357-365.

2-C. GATEAU, J.M. PREVOSTEAU
Exploitation d'un analyseur d'images à l'étude
de la structure des pores d'un matériau.
*Congrès RILEM-IUPAC (PRAGUE 1973)*. Comptes-
rendus IV, C 517-527.

3-C. GATEAU, J.M. PREVOSTEAU, F.X. VAILLANT
Correlations between texture of rock samples
and induced polarization reaction.
*Congress of Petrophysics (NEWCASTLE 1974)*.

4-J.M. PREVOSTEAU, Z. JOHAN

Étude texturologique et minéralogique appliquée
à la valorisation des grès à zircon et rutile.
*Journées des sciences de la Terre (MONTPELLIER 1975)*.

5-M.P. JONES, J.L. SHAW
Automatic measurement and stereological assess-
ment of mineral data for use in mineral techno-
logy.
*Xe International mineral processing congress,
LONDON 1973*.

6-G. BARBERY
Determination of particle size distribution
from measurements on sections.
*Powder technology*, 9 (1974) 231-240.

Position of the analysis zone according to the type of size distribution

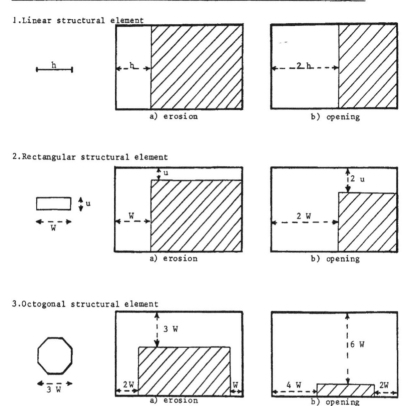

1.Linear structural element

h

a) erosion         b) opening

2.Rectangular structural element

a) erosion         b) opening

3.Octogonal structural element

a) erosion         b) opening

4. All or nothing area size distribution

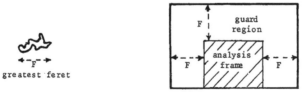

greatest feret

guard region

analysis frame

TABLE 1

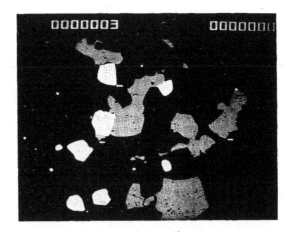

1 – Detection of two phases (white : pyrite –
grey : chalcopyrite)

– Counting mixed grains.

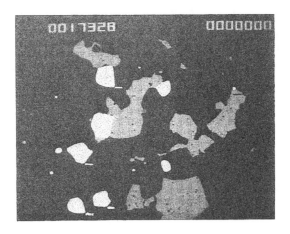

2 Area measurement with both criteria for
each grain :
– size : more than 800 $\mu^2$
– mixity : more than 30 % pyrite.

*National Bureau of Standards Special Publication 431*
*Proceedings of the* FOURTH INTERNATIONAL CONGRESS FOR STEREOLOGY
*held at* NBS, Gaithersburg, Md., September 4-9, 1975 (Issued January 1976)

STEREOLOGICAL DETERMINATION OF THE PROPORTIONAL VOLUME OF DENDRITES IN
CAST IRON BY MEASURING PLAN MICROGRAPHS

by Robert Wlodawer
*Sonnenblickstrasse 8,. CH-8404 Winterthur, Switzerland*

SUMMARY
    As is generally known, the determination of the true proportional
volume of oriented structural constituents, in this case of rectangular
dendrites, is a rather difficult task (i.e. see bibliography in
UNDERWOOD 1970 and in GAHM 1971).- The present paper describes a method
that allows the volume of rectangular dendrites to be determined, and
that even in the case of oblique and randomly cut sections.
    By way of verbal simplification, dendrites of this kind may be des-
cribed as "millimeter paper translated in space" (see Tab.1). Oblique
sections through such cubes result in cutting triangles with a large
variety of angles. These angles may be finally described by spatial
superposition of trigonometric functions. By means of nomograms it is
possible to recalculate the true relations in the rectangular system
for the triangle angles measured in the micrograph and thus to
determine the proportional volume of the dendrites.

INTRODUCTION

Table 1 : Derivation of equations and
representation of simple
geometrical relations.

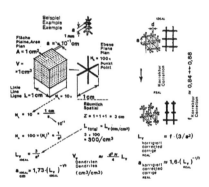

First we determine the relations in rectangular
systems, considering the
proportional volume of the
rods (=cylinders) and the
junctions($\approx$ spheres).Intro-
ducing the quotient $Q = d/a$
we determine the dendrite
volume $V_V$ immediately
(Tab.1 and Figs. 1 and 2).
    The relations, the terms
and calculations in randomly cut or all-over oblique
sections are shown in
Fig.3, resulting in Fig.4.
    The correctness of the
method has been verified:
(1) On dendrite model lattices with known dendrite
volume $V_V$(WLODAWER,1970,71)
(2) by reciprocal checking
of the results with rectangular and all-over oblique-
angled sections (BÜCHI,
SOLTERMANN,WLODAWER,1973)
in actual structures.

USING OF THE NOMOGRAM FOR OBLIQUE SECTIONS AND EXAMPLES
    First we measure the angle of the cutting triangle and determine PJ.
Then we count the number of sectional dendrite axes ($N_A$) within the triangle and rectify this by means of the quotient PJ, i.e. we imaginarily
set the oblique lying dendrite lattice parallel to the basal planes.Con-
sequently we determine the axis density .DA in the rectangular system.
    Fig.5 illustrates an all-over oblique section through dendrites in
cast iron. The cutting ellipses are oriented in markedly different directions. The very simple computations are arranged in tabular form. There
resulted a proportional volume $V_V$ = 0,23 or 23 % dendrites etc.
    Illustrative is the comparison of the real state with the equilibri-
um state (in this alloy approx. only 3 % Dendrites). The actual proper-

315

ties of the material depend on the actual structure. Hitherto unexplained or puzzling problems associated with defects may be consequently solved by stereological methods (WLODAWER 1976).

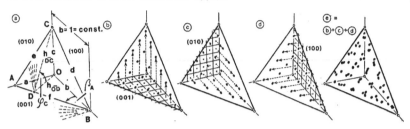

Fig. 3 : Graphic elucidation of the term "projection density" PJ :
Projection of dendrite axes on the cutting triangle, sum of the partial pictures.- The summary picture would be rather confusing without construction aids.
Example of the trigonometric computation of PJ.

$$PJ = \cos \varphi_C \cdot \left[ 1 + \cos \beta_A \cdot tg\, \varphi + \sin \beta_A \cdot tg\, \varphi_C \right]$$

$$\beta_A = 18°; \quad \sin \beta_A = 0,3090$$
$$\cos \beta_A = \frac{0,9511}{1,2601}$$

| $\varphi_C°$ | $tg\, \varphi_C$ | $1 + tg\, \varphi_C \cdot \left[ \sin \beta_A + \cos \beta_A \right]$ | $\cos \varphi_C$ | PJ |
|---|---|---|---|---|
| 27 | 0,5095 | 1,5430 | 0,8910 | 1,463 |
| 30 | 0,5774 | 1,7280 | 0,8660 | 1,496 |

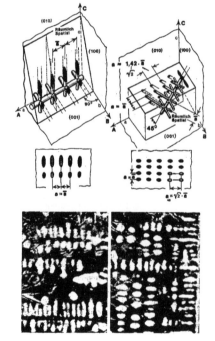

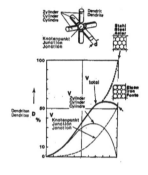

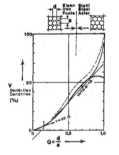

**Fig. 1**
Principle and secondary axes parallel to edges a,b and c Comparison of the perspective sketch with real sectional figures.

**Fig. 2**
A secondary axis parallel to edge b.- All other axes are oblique.- Comparison of the perspective sketch with real section figures.

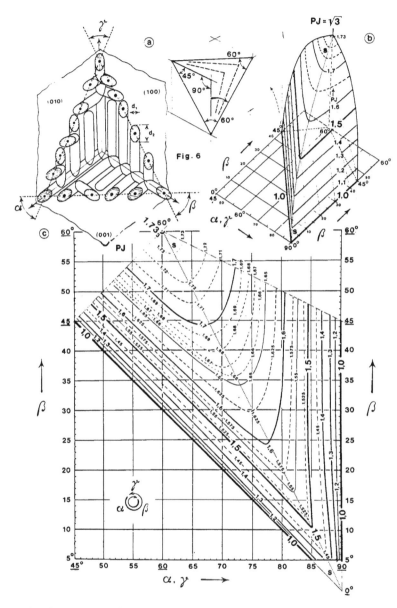

Fig. 4 : Nomograms for the determination of the projection PJ from
the angles of the ellipse chains. Triangle chain formed of
cutting ellipses. The ellipses are from dendrite axes originating from different directions. Representation of PJ
as a spatial diagram and as a plane nomogram.

Fig. 5 : Example for the determination of an
obliquely sectioned dendritic structure (cast iron, etching Marble,
10 X).The calculations are tabulated.

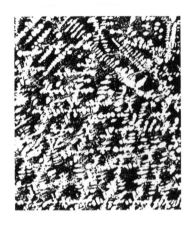

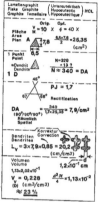

REFERENCES

UNDERWOOD, E.E.

Quantitative Stereology
(Addison-Wesley Publishing Company,
Reading, Mass., 1970 ).

GAHM, J.

Handbuch für die visuelle Messung mit dem
quantitativen Fernseh-Mikroskop VIDEOMAT

(Firmenschrift, Carl ZEISS, W-Germany,
Oberkochen 1971)

WLODAWER, R. (1970)

Giesserei-Rundschau, 17 , 23

BÜCHI, R. ; SOLTERMANN, B. (1973)

Basisversuche zur quantitativen Erstarrungs-
lenkung in hochtemperierten keramischen
Feingussformen für dick-und dünnwandige
Stücke aus Stählen Kobaltbasis-Legierungen.

(Ingenieurarbeit, Fachhochschule Duisburg )

WLODAWER, R.

Directional Solidification of Iron Castings

(book in preparation, Giesserei-Verlag GmbH.
Düsseldorf, 1976 ).

WLODAWER, R. (1971)

J. Microscopy 95 , 285

ACKNOWLEDGEMENTS

The author wishes to express his thanks to Prof E.R.Weibel, Department of Anatomy, Berne University, for the valuable discussion bearing
on this paper.

318

# 6. APPLICATIONS IN BIOLOGY

*Chairpersons:*

H. Elias
E. R. Weibel
W. S. Tyler
G. A. Meek
A. M. Carpenter
H. Haug
J. L. Binet

*National Bureau of Standards Special Publication 431*
*Proceedings of the* FOURTH INTERNATIONAL CONGRESS FOR STEREOLOGY
*held at* NBS, Gaithersburg, Md., September 4-9, 1975 (Issued January 1976)

MORPHOMETRY, STEREOMETRY AND STEREOLOGY APPLIED TO CANCER DEVELOPMENT IN THE LARGE INTESTINE OF MAN

by Hans Elias, D. Bokelmann and R. Vögtle
*Anatomic Institute, University of Heidelberg, Germany*

The three terms which appear in the title of this presentation are often used as synonyms; but they are not. Morphometry
is the measurement and counting of structures, regardless of the
methods employed. Measuring tapes, tachometers, telemeters, etc.
can be employed to obtain morphometric information. Stereometry
is the same as solid geometry. It is one of the methods employ-
able for morphometry.

Stereology is three-dimensional interpretation of flat
images by criteria of geometric probability (extrapolation from
two- to three-dimensional space). As Professor Haug aptly put it,
stereology as contrasted with photogrammetry is based on uniaxial viewing in cases where parallax can not be employed. Stereology is the most frequently used method in the search for the
solution of problems in morphometry. It can also answer questions
of shape-determination, of orientation in space and of density
gradients.

Anisotropic objects can usually be adapted to stereological
investigation by random distribution of fragments of the object
or by random sectioning.

In the following pages I will present a pathological object
which exhibits stages from normal intestinal mucosa, through
various transitional forms, each of specific properties, to the
fully developed cancer. In cases of this kind stereology can be
applied only to the terminal stage of transformation, namely to
the fully developed malignant tumor.

Figure 1 is a schematical presentation of the marginal area
of a typical ring-shaped adeno-carcinoma of the colon. Beginning
with normal gut wall (A), we encounter in the direction toward
the tumor: an area of hypotrophy and hypoplasia (B), a trench
with strongly hypertrophic and hyperplastic branched glands (C),
an area of renewed hypotrophy with nuclear multiplication and
hyperchromasia (D) and almost abruptly following the fullfledged
adenocarcinoma (E). Toward the center of the ring-shaped lesion
is an area of ulceration (F).

Confronted with specimens of this kind, randomization of the entire piece can not lead to a significant result, since morphometric information is needed for each specific area. Fig. 7 shows how heterogeneous the zone of transformation is.

## Stereological approach

Only in the cancer area itself, in which the development is completed, are stereological procedures applicable. For example, the nucleus/cytoplasmic volume ratio has been determined by superimposing optically (see demonstration by Elias and Botz) a grid of lines with their intersection points. This can not be done in conventional paraffin sections, but only in semithin sections, where the thickness of the slice is negligible (Fig. 2). Result in case 1:

$$V_{nuclei}/\ V_{cytoplasm} = 0.32 \ ,$$

using the formula $V/V_T = P/P_T$.

The surface area of the nuclear envelope is determined by the intersection method, $S/V_T = 2\ P/L_T$ (Fig. 3). It was found to be approximately 69 $cm^2/mm^3$ of cancer epithelium and 193 $cm^2/mm^3$ of nuclear volume.

## Direct morphometry

To determine the lengths of the glands for each developmental area, the specimen must be oriented carefully to permit cutting exact longitudinal sections through the tubular glands. The length of individual glands is measured directly by an optically superimposed scale (see demonstration by Elias and Botz, Fig. 4). This and the next step involve direct morphometry. And this next step is the measurement of the diameter (2r) of the gland (Fig. 5, right). These two steps permit to determine the surface area of each gland.

## Stereometry

From the length $\ell$ of the gland and its diameter, its surface area can be determined by the well-known formula

$$S = \ell \ 2r\,\pi$$

In tangential sections of the gland (and in serial sections tangential sections can be found for every longitudinally cut gland) one counts the number of nuclei per unit area $A_T$ (Fig. 5, left).

If the average number of nuclei per test area $\dot{A}_T$ is n, then the number of cells in the entire gland is

$$N = \ell \; 2r \, \pi \, \frac{n}{A_T} \qquad \text{(Fig. 6)}$$

This numerical information enables the observer to determine whether a gland is enlarged by hypertrophy or hyperplasia or by both, or if its size is reduced by hypoplasia or hypotrophy or both.

Table 1 shows preliminary results in case 2.

Table 1

Comparison of morphometric parameters in the marginal area of a typical, ring-shaped colon adenocarcinoma (case 2). Lines 1, 4, 5 and 7 of the last column contain estimated values, since the tumor extends beyond the edge of the specimen.

| Table 1 | Normal | near Tumor | Trench near tumor | Adenocarcinoma |
|---|---|---|---|---|
| 1. Length of glands | 470 $\mu$ | 333 $\mu$ 70% | 2 021 mm 430% | 4 000 $\mu$ 850% |
| 2. Diameter of glands | 90 $\mu$ | 57 $\mu$ 63% | 105 $\mu$ 116% | 200 $\mu$ 222% |
| 3. Radius of glands | 45 $\mu$ | 28.5 $\mu$ 63% | 52 $\mu$ 116% | 100 $\mu$ 222% |
| 4. Surface area of glands | 132 540 $\mu^2$ | 59 600 $\mu^2$ 44% | 666 930 $\mu^2$ 503% | 2 512 000 $\mu^2$ 1903% |
| 5. Volume of glands | 2 988 260 $\mu^3$ | 849 303 $\mu^3$ 28% | 17 178 500 $\mu^3$ 574% | 120 000 000 $\mu^3$ 4016% |
| 6. Number of cells per mm$^2$ | 16 000 | 19 200 120% | 16 000 100% | 10 000 62% |
| 7. Number of cells per gland | 2 120 | 1 142 53% | 10 672 503% | 25 120 1184% |
| 8. Volume of cells | 2 120 $\mu^2$ | $1040^3$ 49% | 3 563 $\mu^3$ 168% | 800–8000 $\mu^3$ 37%–377% |

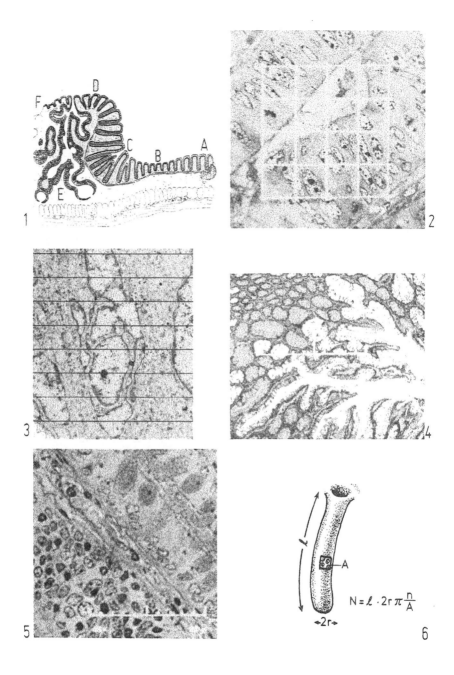

$$N = \ell \cdot 2r\,\pi\,\frac{n}{A}$$

H. Elias, D. Bokelmann and R. Vögtle

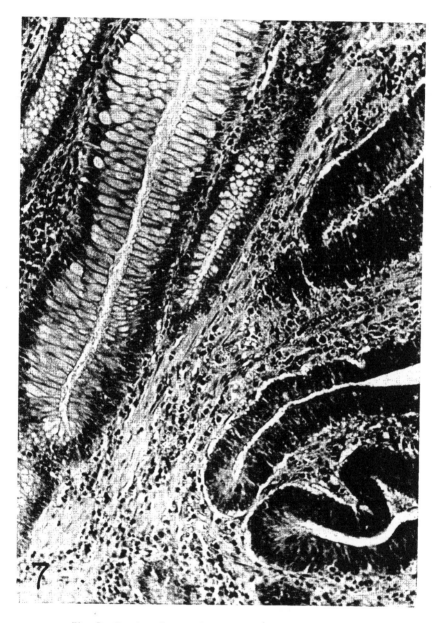

Fig. 7 : Boundary Between Precancerous Glands and Carcinoma

National Bureau of Standards Special Publication 431
Proceedings of the FOURTH INTERNATIONAL CONGRESS FOR STEREOLOGY
held at NBS, Gaithersburg, Md., September 4-9, 1975 (Issued January 1976)

ANALYSIS OF HETEROGENEOUS COMPOSITION OF CENTRAL NERVOUS TISSUE

by S. Eins and J. R. Wolff,

Max-Planck-Institute for biophysical Chemistry, Dept. Neurobiology/Neuroanatomy, Göttingen, Germany

We should like to discuss two points of methodological importance in stereological analysis of brain structures.

The first problem is connected with the calculation of numeric density parameters, which require the absolute values of that quantity to which the density relates (volume, area, length). Especially in biological objects dimensional changes occur which are caused by preparative influences involving chemical and physical alterations of the object. The resulting changes are strongly dependent on the particular preparative procedure and small variations in the method may result in serious differences of preparative behaviour. There - fore it seems indispensable to assess the effect of a given method on the same object which has to be analysed by morphometry. For this purpose, automatic image analysis is a very convenient instrumentation method. Area, perimeter and coordinates can be measured separately for each brain slice out of a given number of samples by an image analysing computer (Quantimet 720, IMANCO, Ltd.). The frequency of measurements may be varied by an appropriate program and prints similar to that of figure 1 are obtained. Here the relative volume changes of 8 frontal rat brain slices are compared, which underwent exactly the same preparative procedure in a perfusion chamber. Differences between the final volumes of individual slices in the order of 10 % and more are usual. This may be partly due to the variable composition of slices according to the inhomogeneous structure of central nervous tissue (gray and white matter, layers, barrels, columns and not yet identified patterns). According to our experience (1) with central nervous tissue it is impossible to predict the final volume change of a given slice even after a systematic investigation of the particular preparative method used. This is a strong argument for following the behaviour of single slices instead of averaging larger samples and calculating mean correcting factors. Nevertheless, there are typical effects of the various preparative procedures on tissue volume.

Our second point is to demonstrate a further special aspect in the analysis of heterogeneous material, i.e. the necessity of appropriate sampling methods. Random sampling, although a very useful method for general statistics in homogeneous populations is less appropriate, if the detailed structure of an object is blurred by averaging over the inhomogeneous composition of the specimen. In these cases only measurements from comparable object regions should contribute to a data set. As this pattern of homologous start points may not be known before the measurements it must be detected by systematic sampling. This first step leads to a defined space standard for the distribution of homologous points in the object to which further measurements should be related. Of course, within homologous regions random sampling must be applied. We have found that a special set of blood vessels vertically penetrating the cerebral cortex can be used as a marker for sampling. Applying this sampling pattern a systematic variation of packing density of synapses and myelinated nerve fibres in Lamina I has been detected.

Vascular pattern: According to the ontogenetic development of the cerebral cortex, the intracerebral vessels form a typical distribution and branching pattern (see fig. 2). This is due to the fact

327

that the cortical plate (anlage) develops by consecutive apposition of the more superficial layers on top of the deeper ones (2). In addition to thickening the cortex also increases its area. Thickening is accompanied by elongation and terminal branching of pre-existing vessels, while the increase in area evokes the formation of new penetrating vessels which supply the superficial layers. The complete (mature) vascular pattern is characterized by typical mean distances between old, large penetrating vessels, which may be used as starting points for the structure specific sampling in the cerebral cortex.

Degenerating fibres: After small intracortical lesions degenerating fibres appear in Lamina I (fig. 3). Since this type of degeneration can also be induced after undercutting, these nerve fibres must originate from local neurons. Their distribution is cluster-like and cannot be related to the position of the inducing lesion. However, maxima and minima of degeneration are distributed in a manner similar to that of the large penetrating cortical vessels.

If a regular (linear or 2-dimensional) sampling is applied instead of random sampling, the mean synaptic density with large standard deviations is transformed into a periodic variation which is strongly correlated to the position of penetrating vessels (3).

In conclusion: Ontogenetically old, long penetrating vessels can be used as starting points for systematic sampling in the cerebral cortex. It can be demonstrated by the distribution of other structures (synapses and special nerve fibres) that the vascular system in the cortex is not only a practical, but also a biologically relevant spatial standard. Further studies are required to search for similar ontogenetical spatial standards in other organs and tissues.

References

(1) S. EINS and E. WILHELMS, Fortschritte der automatischen Bild-analyse, IMANCO-Symposium, Frankfurt 1975.

(2) M. BERRY and A.W. ROGERS, J. Anat. 99, 691 (1965).

(3) J.R. WOLFF, this conference.

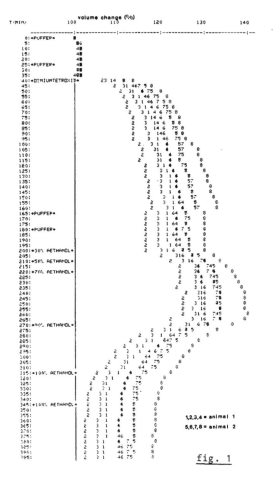

fig. 1

INCREASE IN CORTICAL SURFACE AREA~NUMBER OF PENETRATING VESSELS

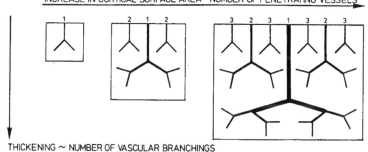

THICKENING ~ NUMBER OF VASCULAR BRANCHINGS

fig. 2

S. Eins and J. R. Wolff

HETEROGENEOUS DISTRIBUTION OF DEGENERATING FIBERS IN LAMINA I OF THE VISUAL CORTEX OF CAT

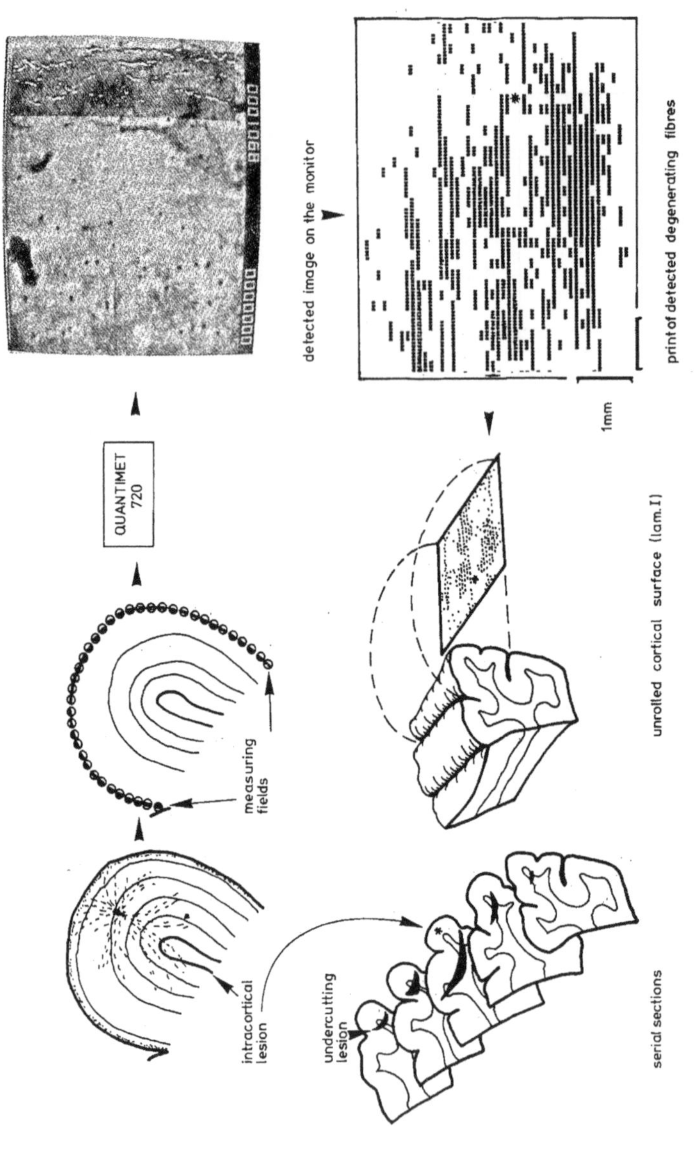

fig. 3

330

National Bureau of Standards Special Publication 431
Proceedings of the FOURTH INTERNATIONAL CONGRESS FOR STEREOLOGY
held at NBS, Gaithersburg, Md., September 4-9, 1975 (Issued January 1976)

STEREOLOGICAL ANALYSIS OF THE HETEROGENEOUS COMPOSITION OF CENTRAL NERVOUS TISSUE:
SYNAPSES OF THE CEREBRAL CORTEX

by J. R. Wolff
Max-Planck-Institute for biophysical Chemistry, Dept. Neurobiology/Neuroanatomy, Göttingen, Germany

Central nervous tissue represents a complex 3-dimensional network which is heterogeneous, being composed of many interdigitating neuronal and glial cell processes. This inhomogeneity has several consequences: 1.) The typical "dendritic" shape of neurons is caused by polar, arborised cell processes. Processes of different polarity (axons and dendrites) make specific contacts (synapses) with each other. Consequently the size, orientation and arrangement of the processes of a neuron may define the probability of synapse formation with other neurons. 2.) Each neuron represents only a small fraction of the space, in which its processes are distributed. Therefore normal neurons may be forced to change their shape, if the tissue shrinks after the degeneration of surrounding neurons. 3.) Random sampling is unsuitable for detecting variations in small tissue fractions within complex tissues of heterogeneous composition. Consequently stereological analysis relies on sampling and statistical methods which do not blur the observed heterogeneity (1).

In the cerebral cortex various types of heterogeneities are known, such as vertically oriented barrels, stripes and columns, as well as bundles of dendrites and axons which penetrate a horizontal set of layers (see fig. 3). Varying numbers of laminae and sublaminae have been defined in relation to the distribution of different components. Obviously none of these structural characteristics can be used as a space standard or as starting points for sampling in the cerebral cortex.

All available data on the synaptic density in the cerebral cortex show relatively high standard deviations of 20 - 25% about the mean (2). In the present study we tried to determine whether this high variability is caused by non-random, systematic variations of the synaptic density in the cortical tissue.

Therefore a systematic sampling has been applied in the present study consisting of vertically and horizontally oriented equidistant sequences of electronmicrographs (magnification 15000 : 1). In addition, all structures which do not contain or make synapses (cell nuclei, blood vessels) were subtracted from the micrograph area.

Thus the packing density of synapses has been estimated in comparable regions of the "neuropil".

In all vertical sample sequences the synaptic density decreases from the superficial to the deeper cortical layers. However, the mean level of the absolute values may differ from one sequence to another (fig. 1). This difference increases with the horizontal distance between the vertical lines of measurements.

Horizontal sequences of samples show a periodic variation (increase and decrease) of the packing density of synapses. The maxima are correlated with the position of rather thin, vertically orientated blood vessels (arteries?) branching in the superficial two-thirds of the cortex. In contrast, the minima surround wide blood vessels (venes) which penetrate most of the cortex.

In agreement with the variable distance between the vertical blood vessels in sections through the cortex, the periodicity of the synaptic density is not a regular one. However, in the visual cortex of rats a minimum frequently occurs about 300-500 μm from the next maximum (fig. 2 and 3).

The distribution of synapses has been found to be similar in various parts of the cortex and in different species: rats, cats, rhesus monkeys. In addition, we compared the synaptic density in the normal visual cortex with that after undercutting of the posterior lateral gyrus of cats (fig. 2) and of area 17 of rats. Although, after undercutting all afferent nerve fibres degenerate in the cortex, the periodic variations were essentially retained(fig. 2). These results suggest that the periodic variations of the packing density of synapses, which depend on the position of vertically orientated blood vessels, represent a pattern which is characteristic for the neocortex.

The cortical vascular pattern becomes complete during the early postnatal life (3) and is retained even after heavy degeneration in the undercut cortex (unpublished). Therefore the intracortical vascular system may be used as a space standard (number of penetrating or vertical vessels per surface area X total thickness of the cortex), which indicates all variations of the tissue volume caused by growth or retardation, swelling or atrophy. In addition, the vertical vessels can be used as starting points for systematic sampling in the horizontal level, i.e. parallel to the surface.

A study on the distribution in lamina I of myelinated axons originating from neurons in the deep layers of the cortex demonstrates that variations in their packing density seem also to be related to the position of long penetrating vessels (1). Further studies are required to determine whether these topological relations between different structural components depend on common developmental conditions.

By simultaneous estimation of the volume fractions of blood vessels, cell nuclei, glial processes and myelinated axons, the number of synapses could be related to various types of tissue volumes(cortex volume; neural tissue volume; "neuropil"; neuronal neuropil; collective volume of dendrites, unmyelinated axons and boutons). These calculations show that neither the gradient from superficial to deeper layers nor the periodic variations of the synaptic density parallel to the surface of the cortex vanish, if any of the volumes is applied. Consequently, all the intracortical variations of the synaptic density depend on real variations of the number of synapses per dendrite or axon. Since variations of about 100 % occur, there must be neurons, e.g. pyramid cells in the same Lamina, which are only about 300 to 500 μm apart, but receive very different amounts of synaptic input. These results may demonstrate the dangers of random sampling and averaging in a stereological analysis of complex and heterogeneously composed tissues.

## References

(1) S. EINS and J.R. WOLFF (this meeting)
(2) B.G. CRAGG: Brain 98, 81 - 90 (1975)
    G. VRENSEN and D. De GROOT: Brain Res. 58, 25 - 35 (1973)
(3) Th. BÄR and J.R. WOLFF: Z. Anat. Entw.-Gesch. 141,207-1221(1973)

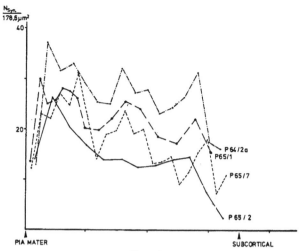

fig. 1

The values of synaptic density are connected by lines along vertical sampling sequences. post. lat. gyr. of cat. Note: different
sample sequences may be separated over their total length, although
all show the same gradient (slope) towards the subcortical white
matter.

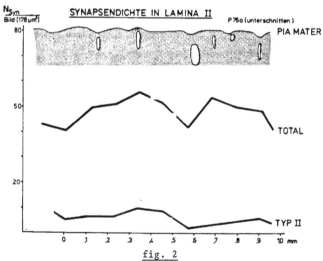

fig. 2

The density of synapses varies periodically in a sequence of samples which have been taken from lamina II of the rat's cerebral cortex (area 17) 3 months after undercutting. Gray-Type II synapses
show similar, but smaller variations.

J. R. Wolff

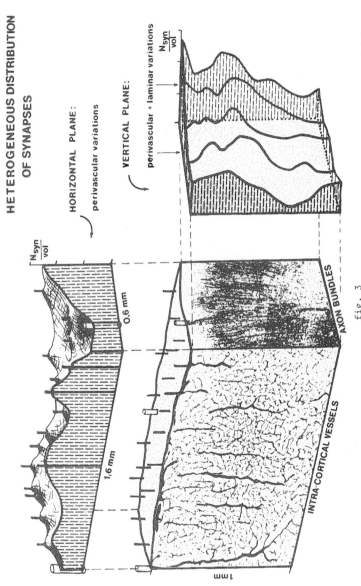

fig. 3

Synopsis of the horizontal and vertical variations of the packing density of synapses, which are related to the position of blood vessels penetrating the cerebral cortex. Note: Vertical axon bundles show different periodicity.

*National Bureau of Standards Special Publication 431*
*Proceedings of the* FOURTH INTERNATIONAL CONGRESS FOR STEREOLOGY
*held at NBS, Gaithersburg, Md., September 4-9, 1975 (Issued January 1976)*

## THE DEVELOPMENT OF MYELINATED FIBERS IN THE VISUAL CORTEX OF CATS

by H. Haug
*Medical School Lübeck - Department of Anatomy, D-2400 Lübeck, Ratzeburger Allee 160, Germany*

The development of myelinated fibres has been studied with stereological procedures. Earlier quantitative investigations of these fibres have been carried out more in the peripheral nervous system (1, 3, 6, 9, 11, 12 etc.) than in the central nervous system (2, 4, 5, 10). The results have been gained by very different procedures and samples of various sizes. In the following it will be reported on an investigation, which is based on 15 animals of different ages counting 44000 myelinated fibres. The length per unit volume has been measured by the number of intercepts per area (8) and in the same step the diameters have been estimated with a TGZ 3 Particle Size Analyser in low-power electron micrographs of the visual cortex (4). From each electronmicrograph the distance from the surface of the cortex has been registered in order to estimate the different developmental speed in the different cortical layers. The procedure of evaluation in rows and layers was similar to the above mentioned method in light microscopy (7).

Our investigations have produced the following results:

1) The myelination in the visual cortex of the cat begins in the $30^{th}$ day of life. This fact indicates that the function of the visual cortex begins in about 30 days after birth. After the opening of the eyelids it must be assumed that the visual function is directed by lower visual centers of the brain.

2) The first fibres which appear are input fibres of greater diameters (1,5 µm). They grow only up to the lower third of the cortex and have developed nearly completely by the $60^{th}$ day.

3) The average diameter will be smaller with increasing age (0,9 µm) in an adult cat , but this is due to the predominant development of smaller fibres and not with a loss of thicker fibres which are growing, as well, but at a slower rate.

4) After the $60^{th}$ day the first internal system of myelinated fibres can be observed. It lies below the surface and has almost no connection to the deeper layer of fibres for in- and output.

5) The size of brain of the cat has been fully developed with about the $100^{th}$ day. However, the last fibre systems of very small fibres are growing after the $164^{th}$ day. The oldest brain of an advanced adolescent cat has only two third of the adult cat fibres.

335

6) The small fibres which appear after the 1oo[th] day belong to the internal systems of the visual cortex and are lying mostly in the middle of the cortex. Their diameters are small, below 1 µm.

7) Our evaluation shows that the storing of the original values of the fibre diameters has a great advantage in respect to different combinations of results for different kinds of distribution eg. in rows perpendicular to the surface or in layers parallel to the surface.

## Summary

The development of the myelinated fibres was studied by means of optomanual stereological procedures (TGZ 3). The following results are important. The first input fibres appear at the 3o[th] day of life. The last internal fibres grow after the 16o[th] day at a time in which the cat is in the transition phase between adolescence and adult stage. Most of the fibres have a diameter smaller than 1 µm.

## References

1) Berthold, C.-H. and Carldstedt, T., Neurobiol. 3 (1973), 1 - 18.

2) Bishop, G. H. and Smith, J. M., Exp. Neurol. 9 (1964) 483 - 5ol.

3) Donovan, A., J. Anat. lol (1967) 1 - 11.

4) Haug, H., Stereology, Proceedings of the Second International Congress for Stereology, Chicago.1967, Springer New York Inc. 66 - 67.

5) Haug, H., Proceedings 25. Anniversary Meeting Electron Microscopy Society of America 1967 b, Ed. C. Arceneaux, Claitoc's Book Store, Baton-Rouge, 56 - 57.

6) Haug, H. und Rast, A., Microsc. Acta 72 (1972) 136 - 146.

7) Haug, H., in Quantitative Analyse von Gefügen in Medizin, Biologie und Materialentwicklung, Ed. H. E. Exner, Dr. Riederer GmbH, Stuttgart (1975) 91.- 1o2.

8) Hennig, A., Z. wiss. Mikrosk. 65 (1963) 193 - 194.

9) Landolt, J. P., Topliff, E. D. L. and Silverberg, J. D. Brain Res. 54 (1973) 31 - 42.

lo) Samorajski, T. and Friede, R. L., J. Comp. Neur. 134 (1968) 323 - 338.

11) Sunderland, S. and Roche, A. F., Acta anat. 33 (1958) 1 - 37.

12) Young, R. F. and King, R. B., J. Neurosurgery 38 (1973) 65 - 72.

National Bureau of Standards Special Publication 431
Proceedings of the FOURTH INTERNATIONAL CONGRESS FOR STEREOLOGY
held at NBS, Gaithersburg, Md., September 4-9, 1975 (Issued January 1976)

THE VOLUMES OF ELASTIC TISSUE AND SMOOTH MUSCLE CELLS IN THE ARTERIAL WALL

by Vladimír Levický
Institute of Normal and Pathological Physiology, Slovak Academy of Sciences, 884 23 Bratislava,
Czechoslovakia

Most studies dealing with the amount of elastin, collagen and other components of the blood vessels wall have been performed by means of biochemical methods /Lowry et al.,1941; Neuman and Logan,1950; Harkness et al.,1957; Banga,1969/. Studies concerning the volume of individual tissues and thus directly morphological parts of the wall are substantially limited in number and concentrate mainly on aorta while to other arteries they pay only a little attention /Hürthle,1920; Lang,1965/.

This work is intended to broaden and specify existing knowledge concerning the volume of two important structural components of the arterial wall-elastic tissue and smooth muscle cells in the walls of arteries of different diameter. At the same time it is an attempt at correlating the results with the function of the arterial wall.

Method

For taking the blood vessels excisions in the experiment were used 39 dogs.

The point-counting procedure /Underwood,1970/ was used to find out the data on the volume of both elastic tissue and smooth muscle cells. The histological section of the arterial wall was observed by means of a test grid inserted in the eyepiece of the microscope. The total magnification was 1250x. Through moving the microscope stage the vessel wall section is covered, beginning with the outside boundary of the adventitia towards the intima, until the number of about 500 or more test points is reached. The distance between test points was 100 μm.

The relative values of errors for the structures observed were calculated from formulas and tables presented by Haug/1962/ and oscillated within 5-15 %.

In the first stage of the observation the relative volume relationship was determined between the smooth muscle cell and its nucleus. The coefficient that was obtained served to calculate the volume of total smooth muscle on the basis of the volume of nuclei of smooth muscle cells. This procedure had been chosen in view of the fact that the used method requires a precise delimitation of the evaluated structures that may not be achieved through the simultaneous staining of elastic tissue and smooth muscle cells. In the next stage the relative volumes of elastic tissue and nuclei of smooth muscle cells in the arterial walls of different areas in the direction from aorta towards the vascular bed periphery were determined which enables making comparisons of measured values in the arteries of different structure, diameter and function. The following staining methods were used: elastic tissue was stained by the resorcin – fuchsin, the smooth muscle cells nuclei by the Mayer's haematoxylin and the smooth muscle cells for the calculation of the cell to its nucleus volume relation by the Heidenhain's haematoxylin.

Results

A. The ratio of nucleus volume to that of a whole smooth muscle cell equals 1:14.5, i.e. the nucleus volume accounts for 6.9%

of the total smooth muscle cell volume at the average. No
significant differences in this ratio were found on the sections of arteries taken from different areas of arterial bed.

B. Figs. 1-6 represents the volume of elastic tissue and smooth
muscle cells nuclei in the aorta and the arteries of different
consecutive regions. These measurements could be summarized
in the following way:

1/ The elastic tissue volume decreases from the proximal part
of aorta where it achieves values from 22.0-25.0% in the
distal direction where its values decrease down to 4.0-6.0%
in the smallest arteries /palmar branch of radial artery,
the arteries of the arterial circle of the brain/ while in
the other arteries /dorsal pedal artery, facial artery, arcades of the jejunal arteries/ they are somewhat higher:
8.0-11.0%.

2/ The volume of smooth muscle cells nuclei is the same in all
the arteries. The average values move in the range of 3.0
to 4.0%. As the volume of smooth muscle cell is a 14.5
multiple of that of its nucleus, the total volume of smooth
muscle calculated on the basis of the nuclei volume fluctu-
ates in most arteries between 45.0-55.0%.

## Discussion

In our study we put under examination arteries of considerably differing size - from aorta with the diameter of 11.0 mm until arteries with the diameter around 0.5 mm. In its course we
examined the volume of smooth muscle cells in the arteries of
both elastic and muscular type. In all the cases we arrived at
equal values, in other words we found out that the relative
volume of smooth muscle cells is constant and does not depend on
the arterial wall structure.

From the functional point of view the main function of elastic tissue consists in effecting the elastic tension in response
to the vessel distension with lower and normal transmural pressures /Burton,1965/. Its presence makes also possible a graded control of vessel diameter exerced by smooth muscle. Because in the
smaller arteries the vessel wall tension decreases and the diameter is being more and more controlled by the smooth muscle
cells, the decrease in the elastic tissue volume in smaller arteries appears to be an accompanying phenomenon of functions
changed in this way.

On the basis of presented findings it is possible to draw
the conclusion that the property of arteries of muscular type to
actively change the blood vessel diameter by means of muscular
contraction - in comparison with arteries of elastic type - is
made possible by decreasing in volume of elastic tissue with
keeping the same relative volume of smooth muscle - i.e. active
contractile component of the vascular wall.

Bibliography

Banga,J., in Int.Symp.Biochem.Vascular. Wall, Fribourg 1968 Part II, pp.18-92, Ed.by M.Comél and L.Laszt, S.Karger, Basel-New York 1969

Burton,A.C., Physiology and Biophysics of the Circulation, Year Book Medical Publishers, Chicago,1965

Harkness,M.L.R., Harkness,R.D., McDonald,D. A., Proc.Roy.Soc.Lond.Ser.B. 146:541-551, 1957

Haug,H., in Med.Grundlagenforschung 4:299-344, 1962

Hürthle,K., Pflügers Arch. 183:253-270,1920

Lang,J., Angiologica 2:225-284,1965

Lowry,O.H., Gilligan,D.R., Katersky,E.M., J.Biol.Chem. 139:795-804,1941

Neuman,R.E., Logan,M.A., J.Biol.Chem. 186:549-556,1950

Underwood,E.E., Quantitative Stereology, Addison-Wesley Publ.Co., Reading,Mass,1970

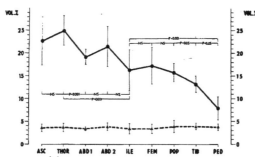

Fig.1. Aorta and arteries of the hind-limb: ASC — ascend.aorta, THOR
— thor.aorta, ABD 1 — abdom.aorta before the arising of cran.mesent.
art., ABD 2 — abdom.aorta before the arising of renal art., ILE —
ext.iliac art., FEM — femoral art., POP — popliteal art., TIB — cran.
tibial.art., PED — dors.pedal art.
Symbols: ●—● elastic tissue, ▲—–▲ smooth muscle cells nuclei,ver-
tical lines = standard deviations, P = the degree of statistical
significancy, NS = a non-significant difference

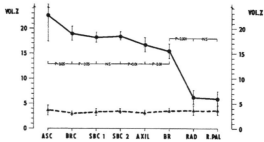

Fig.2. Brachiocephalic artery, subclavian artery and arteries of the
fore-limb: BRC — brachiocephal.art., SBC 1 — subclav.art./ the beginning /, SBC 2 — subclav.art./ the end /, AXIL — axillary art.,
BR — brachial art., RAD — radial art., R.PAL — palmar branch of radial art.

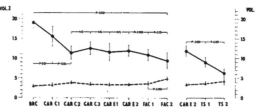

Fig.3. Common carotid artery, external carotid artery and its ex-
tracranial branches: CAR C 1 — comm.carot.art./ the beginning /,
CAR C 2 — comm.carot.art.before the arising of cran.thyreoid.art.,
CAR C 3 — comm.carot.art./ the end /, CAR E 1 — extern.carot.art.
/ the beginning /, CAR E 2 — extern.carot.art.before the arising of
facial art., FAC 1 — facial art./ the beginning /, FAC 2 — facial
art.before the arising of submental art., TS1— superfic.temporal
art./ the beginning /, TS 2 — superfic.temporal art./ the end /

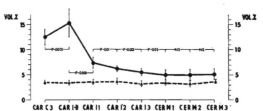

Fig.4. Internal carotid artery and middle cerebral artery: CAR I-B –
bulb of intern.carot.art., CAR I 1 – intern.carot.art.just after the
bulb, CAR I 2 – intern.carot.art.before getting into carot.canal,
CAR I 3 – intern.carot.art.after getting out of carot.canal, CER M 1
– midd.cerebr.art.before the first branching, CER M 2 – midd.cerebr.
art.before the second branching, CER M 3 – midd.cerebr.art.before
the third branching.

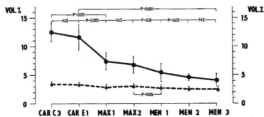

Fig.5. External carotid artery, maxillary artery and middle meninge-
al artery: MAX 1 – maxill.art./ the beginning /, MAX 2 – maxill.art.
before the arising of midd.mening.art., MEN 1 – midd.mening.art.be-
fore getting into oval foramen, MEN 2 – midd.mening.art.after getting
out of oval foramen, MEN 3 – midd.mening.art.in the temporal region.

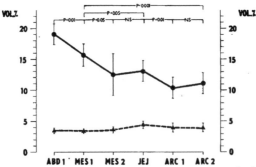

Fig.6. Mesenterial arteries: MES 1 – cran.mesenter.art./ the beginning /, MES 2 – cran.mesenter.art.before the arising of 6th jejunal
art., JEJ – 6th jejunal art., ARC 1 and ARC 2 – primary and secondary arcades of jejun.art.

National Bureau of Standards Special Publication 431
Proceedings of the FOURTH INTERNATIONAL CONGRESS FOR STEREOLOGY
held at NBS, Gaithersburg, Md., September 4-9, 1975 (Issued January 1976)

PROGRESS, SUCCESS AND PROBLEMS IN APPLYING STEREOLOGY IN BIOLOGICAL RESEARCH

by Ewald R. Weibel
Department of Anatomy, University of Berne, Berne, Switzerland

Exactly 50 years have passed since, for the first time, a stereological principle was derived with a biological application in mind:
in 1925 the Swedish mathematician S.D. Wicksell published his now
classical memoir entitled "The corpuscle problem: a mathematical
study of a biometric problem" (1). It is only slightly more than a
decade, however, that the stereological approach is making progress
in becoming an established method of biological investigation. The
number of biological papers making successful use of stereological
methods has been rapidly increasing over the past years. However,
one of the remaining problems is the lack of awareness for the potential of a stereological analysis among still a large fraction of
biomorphologists; indeed, in perusing some of the leading biological
journals one is astounded to find a large number of papers in which
a quasi-quantitative analysis of structure was used to produce ambiguous data, whereas solid morphometric data could have been easily
produced - often with even less effort.

Inspite of considerable progress - and even some success - it is
clear, however, that the application of stereology in biological research still poses a number of very serious problems. These become
particularly prominent as one attempts to proceed from descriptive
morphometry to the use of stereology in an analytical framework in
combination with a physiological or biochemical analysis. These
problems derive (a) from the nature and complexity of biological
specimens, (b) from the difficulty of controlling artefacts, (c)
from the necessity to work with a variety of morphological methods
to reveal the properties of structural elements, and (d), last but
not least, from the persisting lack of adequate technology to efficiently produce the large volume of data often required. I shall not
deal with this latter aspect as this has been done elsewhere (2,3),
but shall concentrate on the other problems, particularly focussing
on those arising in attempts to quantitate the structural make-up of
cells by electron microscopy.


1. Identification of cell constituents

Cells are subdivided into various functional compartments by membranes. These membranes are sheet-like aggregates of proteins and
lipids; in their basic constitution they are all alike and consequently look very similar in an electron micrograph (Fig. 1), but
they differ by the function of the proteins most of which are highly
specific enzymes. Thus various types of membranes can be distinguished; but they are recognised in electron micrographs essentially
by the configuration of the membrane-bounded space and by their context. The major problem resulting from this situation is that image
analysis depends on interpretation of the constituents by trained
investigators, and that automatic image analysis is still not feasible, except under very special circumstances. Hence, stereological
analysis of electron micrographs of cells remains a laborious enter-
prise, although the proper planning of point counting procedures,

. R. Weibel

the use of a variety of test systems, and computer assistance can greatly alleviate the chore (4,5).

In recent years considerable progress has been made in specifically "tagging" membranes and cell compartments by cytochemistry, i.e. by allowing an enzyme to react in situ with a suitable substrate so as to produce an insoluble electron-opaque precipitate. Fig. 2 shows such a preparation in which the black reaction product marks sites of glucose-6-phosphatase, a marker enzyme for endoplasmic reticulum. The problem with these methods lies in the difficulty to adequately control the reaction quantitatively and to assure even penetration of the substrate. Some progress is currently being made by using isolated cells in suspension (6).

A further new technique for tagging specific protein components is that of immunocytochemistry (7), but here the pitfalls of the method are even greater. If these methods of specific tagging of cell constituents are further improved automatic image analysis may eventually become accessible on a broader scale.

The method of freeze-fracturing offers a further way of identifying different membrane types by the pattern, size, and density of intramembranous particles; these are displayed on the fracture surface because the fracture has a tendency to follow the mid-plane of membranes and hence splits them in half (8). The main problem for using stereology on such preparations is that we are dealing with a non-flat surface, and that the "section" (= fracture surface) leads to a biased display of structure elements. We shall show elsewhere, however, how non-conventional stereology can be used to quantify membrane systems on freeze-fracture preparations (9).

## 2. Biological organisation and sampling for stereology

Stereological methods are derived on the basis of probabilistic models: the structural elements are supposed to be randomly distributed, randomly oriented, and to be randomly intercepted by a stereological test procedure (section, test lines, test points, etc.). Biological structure is, however, characterised by an extraordinarily high level of organisation: cells of different function are clearly separated and often arranged in a strictly hierarchical order. Many cells, tissues or organs also show a pronounced polarity or anisotropy. This appears to exclude a stereological approach to quantitation of such structural systems.

Indeed, this is one of the more fundamental problems in biomorphometry. We shall consider three situations.

### 2.1. Gland-like structures

A gland-like structure is characterised by a system of branching ducts which are organised according to a strict hierarchy which establishes a strong order and polarity of the elements (Fig. 3). A morphometric analysis of such a hierarchical system cannot be based on classical stereological methods, but must use systematic "hierarchical" sampling procedures (10,11).

In most gland-like structures the "parenchyma" or functional complex of cells is found at the termination of the last branches of the duct system (Fig. 4). The units attached to each terminal branch show a remarkable uniformity: they occur in very large number, and appear furthermore rather "randomly" oriented in space because the duct system branches in all directions like a tree. A plane section will therefore hit these parenchymal units randomly; a stereological analysis hence becomes possible, but it must exclude "duct elements" from the sample. In general, gland-like structures must therefore be analysed by multiple stage sampling procedures. Such an approach has been successfully used for example in studying lung (10,12), liver (13,14,15), Salivary glands (16), and pancreas (17).

As the demands on the reliability of the stereological analysis increase it will however be mandatory to examine whether a probabilistic structural model is indeed satisfactory or whether more stringent model conditions must be introduced.

## 2.2. Polarised stratified structures

Such structures are exemplified by skin or mucous membranes: several layers of cells extend from the base to the surface, and there is a definite gradient in the structural and functional property of the cells. Such systems clearly call for systematic stratified sampling procedures; whilst some progress has been made (18) a large number of problems remain to be solved.

## 2.3. Anisotropic and periodic structures

The extreme case of anisotropic and periodic structure is certainly skeletal muscle (Fig. 5): the strict parallelism of the straight myofibrils renders these cells anisotropic; the subdivision of the myofibril into sarcomeres of identical length makes a perfect periodic structure. It is found that all other cell constituents comply to this strict order and are likewise anisotropic and periodic. Eisenberg et al. (19) have proceeded to a thorough analysis of the conditions for performing a stereological analysis on such cells.

An anisotropic system could actually be analysed on a strict cross-section, perpendicular to the anisotropy axis; periodicity of the elements however would introduce considerable sampling errors. It has been shown that an oblique section yields a representative sample and allows measurement of both $V_V$ and $S_V$ of various components (20). Alternatively, one can use sections parallel to the anisotropy axis and orient a test line system at a fixed angle (19,21). By this approach one has however abandoned to a large measure the fundamental probabilistic basis on which stereological methods have been derived.

## 3. Preparation artefacts

## 3.1. Processing artefacts

Before a section can be viewed in the (electron) microscope the tissue must be fixed, stained and processed through a considerable number of steps. Cells are osmometers and may swell or shrink, depending on the osmotic properties of the processing solutions. Compression of the section by the microtome knife is another important artefact. Although some studies of such effects have been made (22), in most instances control of these artefacts is still rather intuitive.

## 3.2. Section thickness and Holmes effect

Basic stereological methods are developed for true two-dimensional sections of no thickness. The systematic errors introduced by using them on sections of finite thickness are well-known (23-26) but nevertheless difficult to control in practice: they depend on the relative dimension of section thickness and object size, but are heavily dependent on the shape of the objects. As the demands on the reliability and reproducibility of stereological data increase with the sophistication of biomorphometric studies, such effects - or rather errors - must find a reliable compensation. One way is to develop correction coefficients which adequately consider the properties of the specimen and the measuring condition. In view of correcting volume and surface density estimates on cellular membrane systems in the framework of an analytical study (27) we have recently re-examined this problem systematically.

The cytoplasmic membrane system of cells occurs in essentially three forms (Fig. 1): (a) as spherical vesicles, (b) as narrow cisternal spaces (rER), and (c) as tubular networks. Approximating these shapes with (a) spheres, (b) thin flat discs, and (c) thin cylindrical tubules we derived a set of correction coefficients for $V_V$ and $S_V$ estimates which consider the total projection of the section content, leading to overestimation, as well as the loss of profiles produced by grazing sections, leading to underestimation. In the case of spherical vesicles the effect of size distribution on the error due to finite section thickness was also considered. Table I presents a synopsis of the formulas developed for the three basic shapes; these correction factors, by which the raw estimates must be multiplied, are easily determined by means of graphs, as shown in Figs. 6-9.

The analysis of these three basic shapes has shown that section thickness leads to very different errors in the estimation of $V_V$ and $S_V$ of various cell components. For endoplasmic reticulum of the rat liver, for example, it was found that raw estimates of $S_V$ had to be corrected by a factor of 0.95 for cisternal rER, whereas a factor of 0.70 was required for the tubular sER. The surface area estimate of spherical vesicles as they are found in microsomal fractions from the same tissue required correction by a factor of 0.72. Whilst the surface area estimates of rER cisternae was affected by a small error only, the estimate of $V_V$ of rER required correction by a factor of 0.54; this is due to the fact that the length of the membrane trace, on which the estimation of $S_V$ is based, is little affected by varying inclination of the cisterna to the section plane, whereas the area of the projected band, the basis for the $V_V$ estimate, increases both with section thickness and with inclination of the cisterna.

As this example shows, consideration of section thickness effects can significantly improve the validity of stereological data. The use of correction factors together with basic stereological methods is, however, only partially satisfactory. It is to be hoped that the development of new stereological principles will allow us, in the future, to obtain our measurements with methods that account *a priori* for finite section thickness.

## 4. Stereology as a tool in analytic cell biology: the beginnings of a future

The foundations of modern cell biology have essentially been laid by Albert Claude, Christian de Duve, and George Palade, last year's Nobel Laureates. Their concept was that the many functions of the cell must be associated with a particular structural element, and that it should be possible to identify that association (28). They proceeded to a combined use of cell fractionation (based on size and density of the components), biochemistry, and electron microscopy, partly combined with cytochemistry and autoradiography.

In this work biochemistry was quantitative, allowing the establishment of balance sheets. The structural analysis however remained qualitative, so that a true quantitative structure-function correlation could not be established. Instead of relating the activity of membrane-bound enzymes to the area of membrane present, they used the protein content as a biochemical reference for structure.

Only rather recently have Baudhuin (29) and Wibo (30) from de Duve's group proceeded to a stereological analysis of the membrane vesicles present in the various fractions by using a Wicksell analysis; this resulted in a good correlation between structural and biochemical data.

Over the past years, Bolender et al. (27) have conducted a systematic study in which they attempted a quantitative correlation between biochemical determinations of enzyme activities measured in fractions and a stereological analysis of intact cells; this required an intermediate step, namely the stereological analysis of the fractions. The difficulty of identifying the origin of membrane vesicle in a fraction required the use of methods of ·"tagging" by cytochemistry and a stereological evaluation of freeze-fracture preparations. The final attempt to establish a correlation further necessitated sound correction of artefacts, such as compression and section thickness. It had further to consider that the cell population of the liver is heterogeneous, and that the various cell types may contribute differently to the homogenate and to the stereological analysis which focussed predominantly on hepatocytes - hence a sampling problem was prominent (31).

The approach developed in this context is very involved and laborious; if it can be rendered more efficient this may however prove to be a profitable field for applying stereology.

## 5. Conclusions

The application of stereological methods in biological research has undoubtedly made considerable progress. The biologist now possesses a set of methods that lend themselves to a large variety of applications, from basic research in cell biology to more practical aspects of pathology (32-34). In many fields the use of stereology has been undoubtedly successful. Nevertheless, a large number of problems persist or even increase as the stereological approach becomes more sophisticated. The main residual problems are (a) an improvement in the efficiency of the methods, (b) a better control of artefacts, and (c) development of methods that ensure the validity of stereological measurements inspite of non-ideal conditions of the specimen. These problems need to be solved in the near future.

345

References

(1) Wicksell, S.D.: Biometrica 17, 84, 1925

(2) Haug, H.: Experiences with opto-manual automated evaluation systems. (This volume)

(3) Weibel, E.R., Fisher, C., Gahm, J., and Schaefer, A.: In: Stereology 3. J. Microscopy 95, 367, 1972

(4) Weibel, E.R.: Stereological techniques. In: Principles and techniques of electron microscopy. M.A. Hayat, ed., Van Nostrand-Reinhold, New York, 1973

(5) Rätz, H.U., Gnägi, H.R., and Weibel, E.R.: J. Microscopy 101, 267, 1974

(6) Wanson, J.C., Drochmans, P., May, C., Penasse, W., and Popowski, A.: J. Cell Biol. 66, 23, 1975

(7) Kraehenbühl, J.P., and Jamieson, J.D.: Review of Exp. Pathology 13, 1, 1974

(8) Branton, D.: Phil. Trans. Royal Soc. London B, 261, 133, 1971

(9) Weibel, E.R., Losa, G.A., and Bolender, R.P.: Stereological method for estimating relative membrane surfaces in freeze-fracture preparations. (This volume)

(10) Weibel, E.R.: Morphometry of the human lung. Berlin, Springer, 1963

(11) Horsfield, K., and Cumming, G.: J. appl. Physiol. 24, 373, 1968

(12) Weibel, E.R.: Physiol. Reviews 53, 419, 1973

(13) Loud, A.V.: J. Cell Biol. 37, 27, 1968

(14) Weibel, E.R., Stäubli, W., Gnägi, H.R., and Hess, F.A.: J. Cell Biol.: 42, 68, 1969

(15) Bolender, R.P., and Weibel, E.R.: J. Cell Biol. 56, 746, 1973

(16) Albegger, K.W., and Müller, O.: Arch. klin. exp. ONK-Heilk. 204, 27., 1973

(17) Bolender, R.P.: J. Cell Biol. 61, 269, 1974

(18) Schroeder, H.E., and Münzel-Pedrazzoli, S.: J. Microscopy 92, 179, 1970

(19) Eisenberg, B.R., Kuda, A.M., and Peter, J.B.: J. Cell Biol. 60, 732, 1974; J. Ultrastr. Res. 51, 176, 1975

(20) Weibel, E.R.: J. Microscopy 95, 229, 1973

(21) Sitte, H.: In: Quantitative Methods in Morphology, p. 167, E.R. Weibel and H. Elias, eds., Berlin, Springer, 1967

(22) Weibel, E.R., and Knight, B.W.: J. Cell Biol. 21, 367, 1964

(23) Holmes, A.H.: Petrographic Methods and Calculations, London, Murby, 1927

(24) Chayes, F.: Mineral Magazines 31, 276, 1956

(25) Hennig, A.: Z. wiss. Mikr. 63, 67, 1956; Mikroskopie 12,7,1957 25,154,1969

(26) Cahn, J.W.: Am. Mineralogist 44, 435, 1959

(27) Bolender, R.P., Paumgartner, D., Losa, G.A., and Weibel, E.R.:
     Correlated biochemical and stereological analysis of
     liver cells (in preparation)

(28) De Duve, Ch.: Science <u>189</u>, 186, 1975

(29) Baudhuin, P., and Berthet, J.: J. Cell Biol. <u>35</u>, 361, 1967

(30) Wibo, M., Amar-Costesec, A., Berthet, J., and Beaufay, H.:
     J. Cell Biol. <u>51</u>, 52, 1971

(31) Blouin, A., Bolender, R.P., and Weibel, E.R.: The distribution
     of organelles and membranes between hepatocytes and non-
     hepatocytes in the rat liver parenchyma (in preparation)

(32) Thurlbeck, W.M.: Thorax <u>22</u>, 483, 1967

(33) Kapanci, Y., Weibel, E.R., Kaplan, H.P., and Robinson, F.R.:
     Lab. Invest. <u>20</u>, 101, 1969

(34) Bachofen, M., and Weibel, E.R.: Chest <u>65</u>, 14, 1974

## Acknowledgements

This work was supported by grants from the Swiss National Science Foundation.

## Note added after the Congress:

The reader may be intrigued by the fact that in Fig. 7 the curves intersect the dashed line for discs of infinite diameter, implying that as the section thickness increases truncation leads to greater overprojection, which is evidently not well possible. Using the formulas for overprojection developed by Miles (1975) and presented at the Congress, one finds, however, that the curves for both $K_V$ and $K_S$ become steeper as the relative diameter $\delta$ decreases. They start at 1.0 and converge with the curves presented in Fig. 7 as section thickness increases.

## Reference:

Miles, R. E. : On estimating aggregate and over-all characteristics from thick sections by transmission microscopy. This volume p. 3, 1975.

347

## TABLE I

| CORRECTION COEFFICIENTS FOR SECTION THICKNESS AND TRUNCATION | | NO TRUNCATION |
|---|---|---|
| $V_V = V_V' \cdot K_V$ | $S_V = S_V' \cdot K_S$ | |
| **SPHERES** $\begin{array}{c}\longmapsto D \longrightarrow \\ d_o \end{array}$ T $T/D = g$ $d_o/D = \rho$ | $K_V = \dfrac{2}{(2 + \rho^2)\sqrt{1 - \rho^2} + 3g}$ $K_S = \dfrac{\pi}{2\left(\rho\sqrt{1 - \rho^2} + \sin^{-1}\sqrt{1 - \rho^2} + 2g\right)}$ | $\dfrac{2}{2 + 3g}$ $\dfrac{\pi}{\pi + 4g}$ |
| **SPHERES** $F(D)$ $m_n = \dfrac{E(D^n)}{E^n(D)}$ $\gamma = h_o/D$ | $K_V = \dfrac{2m_3}{2m_3 + 3gm_2 - 3\gamma^2 + \gamma^3}$ $K_S = \dfrac{\pi m_2}{2\left\{(\rho + \sin^{-1}\sqrt{1 - \rho^2})m_2 + 2g - \rho - \rho\sqrt{1 - \rho^2}\right\}}$ | $\dfrac{2m_3}{2m_3 + 3gm_2}$ $\dfrac{\pi m_2}{\pi m_2 + 4gm_1}$ |
| **DISCS** $\begin{array}{c} d \\ D \end{array}$ T $T/d = g$ $D/d = \delta$ | $K_V = \dfrac{\delta}{\delta - 1 + g\left(g + \frac{\pi}{4} + \frac{1}{2}\delta\right)}$ $K_S = \dfrac{\frac{\pi}{2}\left(\frac{1}{2}\delta^2 + \delta\right)}{\frac{\pi}{4}\delta^2 + \left(\frac{4}{\pi} - \frac{1}{2}\right)\delta - \frac{4}{\pi} + 2g\left[\frac{2}{\pi}g + \left(\frac{1}{\pi} + \frac{1}{2}\right)\delta + 1\right]}$ | $\dfrac{1}{1 + \frac{1}{2}g}$ $1$ |
| **CYLINDERS** $\begin{array}{c} d \\ L \end{array}$ T $T/d = g$ $L/d = \lambda$ | $K_V = \dfrac{\lambda}{\lambda - 2 + \frac{4}{\pi}g\left(\frac{\pi}{4}\lambda - 2g + \frac{1}{2}\right)}$ $K_S = \dfrac{\pi\lambda}{\pi\lambda + 4g\left(\frac{1}{2}\lambda - \frac{4}{\pi}g + \frac{1}{\pi}\right)}$ | $\dfrac{1}{1 + g}$ $\dfrac{\pi}{\pi + 2g}$ |

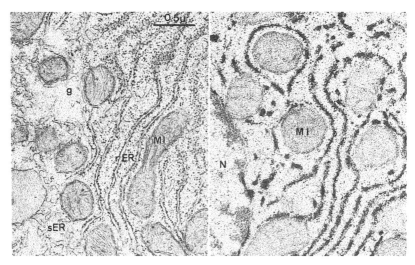

Fig. 1: Electron micrograph of rat liver hepatocyte showing various forms of cytoplasmic membrane system. rER: ribosome-studded cisternae of endoplasmic reticulum, sER: tubules of smooth ER among glycogen granules (g), MI: mitochondria, N: nucleus.

Fig. 2: Comparable region of hepatocyte cytoplasm after glucose-6-phosphatase reaction in situ. ER profiles are marked by black precipitate. Preparation and micrograph courtesy of Dr. D. Paumgartner.

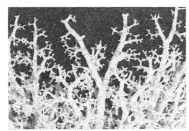

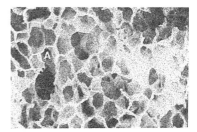

Fig. 3: Resin cast of peripheral airway tree from human lung. Note branching of ducts in gland-like structures. Preparation Dr. h.c. W. Weber.

Fig. 4: Scanning electron micrograph of human lung parenchyma. Alveoli (A) as the terminal units of airways are densely packed. Courtesy Dr. P. Gehr.

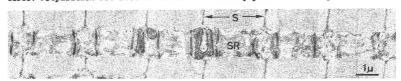

Fig. 5: Longitudinal section of skeletal muscle showing anisotropy and periodicity of myofibrils and organelles. S: Sarcomere, SR: sarcoplasmic reticulum.

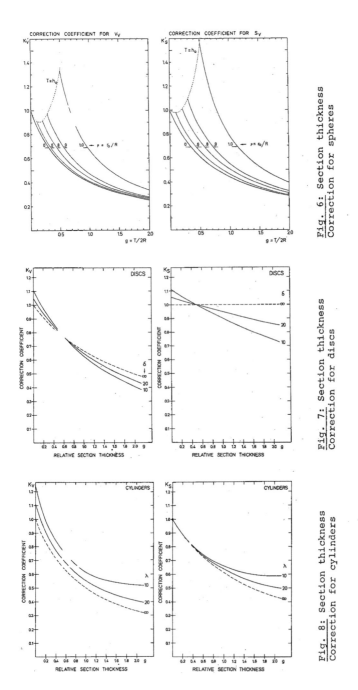

Fig. 6: Section thickness
Correction for spheres

Fig. 7: Section thickness
Correction for discs

Fig. 8: Section thickness
Correction for cylinders

National Bureau of Standards Special Publication 431
Proceedings of the FOURTH INTERNATIONAL CONGRESS FOR STEREOLOGY
held at NBS, Gaithersburg, Md., September 4-9, 1975 (Issued January 1976)

HEXACHLOROBENZENE-INDUCED SMOOTH ENDOPLASMIC RETICULUM IN RAT LIVER: A CORRELATED
STEREOLOGIC AND BIOCHEMICAL STUDY.

by T. Kuiper-Goodman, D. Krewski, H. Combley, M. Doran and D. L. Grant
Health Protection Branch, Bureau of Chemical Safety, Toxicology Research Division, Tunney's
Pasture, Ottawa, K1A 0L2, Canada

## Introduction

Hexachlorobenzene (HCB) is a fungicide and an industrial pollutant.
It causes porphyria cutanea tarda in man (1) and rats (2,3). In the rat HCB
also induces the activity of drug metabolizing enzymes (3) and a proliferation
of smooth endoplasmic reticulum (4). Because of its low biodegradability the
accumulation of HCB in the biochain is of concern to health and environmental
scientists. The present study was conducted to relate the morphological and
biochemical changes induced by HCB in order to better assess their toxicological
significance.

## Methods

Groups of six-week old male and female Charles River (COBS) rats
(70 per group) were fed diets containing HCB at daily doses of 0, 0.5, 2.0,
8.0 or 32.0 mg per kg body weight in ground Master Fox cubes with 5% corn oil.
Four rats from each dose group were sacrificed at 21, 42, 62 and 89 days for
measurement of morphological and biochemical parameters and for electron
microscopy. To assess the reversibility of the changes noted the remaining rats
were returned to an HCB-free diet after 103 days, and 4 rats per group were
sacrificed at 7, 14, 28, 49, 112 and 231 days after cessation of treatment.

The rats were anesthetized with ether and a small slice of the large
lobe of the liver was excised for electron microscopy. Small blocks of less than
1 mm$^3$ were fixed for 3 hrs in 2% phosphate buffered glutaraldehyde (ph 7.3, 425
mOsm) at 4°C, post fixed in 1% osmium tetroxide, dehydrated through alcohols and
propylene oxide, and embedded in Epon. Sections stained with uranyl acetate
and lead citrate were examined at 60 kV (for better contrast) in a Siemens
Elmiskop 1A. Photographs were printed on Kodak Polycontrast rapid RC paper,
which does not stretch during processing. For comparison with biochemical data,
stereologic analyses were performed on 4 treated animals and 2 out of 4 control
animals per group. For each animal 2 blocks were selected at random, and 9
photographs were taken both at 1500 x (one grid square of 300 mesh uncoated grid)
and at 10,000 x (specified corner of each grid square) magnification, from which
7 at each magnification were selected at random. For each block, estimates of
the morphometric parameters for the liver as a whole were made by applying the
point counting procedures described by Weibel (5) to the seven photographs
treated as one unit (a set of IBM Fortran IV computer programs was developed for
this purpose), and estimates for each animal were obtained by averaging the two
block values. The size of the block was such that usually 2 out of 3 zones of
the liver acinus were included (6).

Drug metabolizing enzyme activity and porphyrin levels were measured
using methods described previously (3).

Mean values of morphometric and biochemical parameters for each treated
group were compared with those of the control group at each time point. A pooled
estimate of error variance was used where appropriate with an overall level of
significance of less than 0.05 (7).

## Results and Discussion

Relative liver weight was significantly increased ($p < 0.05$) at the two
highest dose levels for both sexes, and slowly returned to normal after the
animals were returned to an HCB-free diet (Fig. 1). Examination of semi-thin
sections of liver by light microscopy showed enlargement of hepatocytes in
zones 2 and 3 at the two highest dose levels in both males and females (Fig. 2
a,b). In the enlarged cells there were large pale areas, which by electron
microscopy, were found to consist of proliferated smooth endoplasmic reticulum
(SER) (Fig. 3). In males large whorls of compacted membranes surrounding lipid
droplets often accompanied the proliferated SER. The nuclei of enlarged hepato-

351

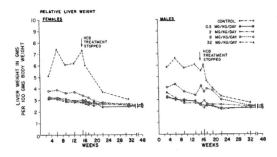

*Figure* 1. Relative liver weight in rats fed HCB in the diet for 15 weeks followed by control diet. The standard error of the difference between two means is indicated on the abscissa for each time interval.

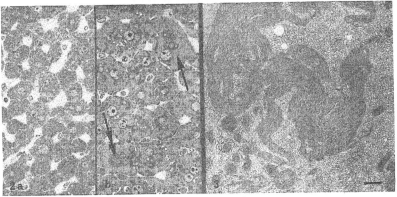

*Figure* 2. Semi-thin sections of rat liver stained with toluidine blue: *a*. zone 2 of a control rat *b*. zone 2 and 3 of a male rat fed HCB at 32 mg/kg for 15 weeks. Pale SER area (single arrow), membranous whorls (double arrow). x 400.

*Figure* 3. Electron micrograph of a zone 3 cell from a male rat fed 32 mg/kg HCB for 12 weeks showing proliferated SER and membranous whorls.(length of bar = 1μ).

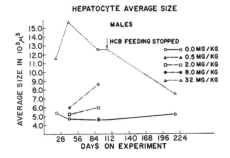

*Figure* 4. Average size of hepatocytes in male rats fed HCB for 15 weeks followed by a recovery period. Data not corrected for zone distribution.

cytes also appeared enlarged. Mitochondria in enlarged hepatocytes were very small and sparse. Cells in zone 1 appeared unaffected by HCB.

Stereologic analysis of the liver showed some changes with time in control animals. These changes were probably related to changes in the relative liver weight shown in Figure 1. In general the results agreed well with those reported by others (5,8). In control males the total volume of liver per 100 gm body weight declined from $3.62 \pm 0.01$ ml at 9 wks of age to $2.09 \pm 0.13$ ml at 37 wks of age; females declined from $3.07 \pm 0.06$ ml to $2.62 \pm 0.07$ ml at 28 weeks. Total SER volume varied between 0.06 and 0.25 ml/100 gm body weight in males and between 0.04 and 0.37 ml in females throughout the experiment.

The total volume of liver per 100 gm body weight ($V_l$) was significantly increased ($p < 0.05$) at the two highest dose levels in males and at the highest dose level in females mainly due to an increase in total hepatocyte volume ($V_h$). There was no apparent change in the number of hepatocytes per 100 gm body weight ($N_h$), so that the increase in $V_h$ was only due to an increase in hepatocyte size (Fig. 4). Total nuclear volume ($V_n$) increased slightly at the highest dose level in males and females; this appeared to be due to an increase in nuclear size, as evidenced by an increase in nuclear diameter. Fig. 5 shows that the percentage of SER in hepatocyte cytoplasm increased in a dose related manner until the cytoplasm in males at the highest dose level contained about 60% SER. Most of the increase in $V_h$ was, therefore, due to an increase in $V_{ser}$. Similar changes of a lesser magnitude were observed in females.

Approximately 80% of the hepatocytes were found in zones 2 + 2,3 + 3, where the effects of HCB were most pronounced (53-77% SER in cytoplasm at the highest dose level). Cells in zone 1 + 1,2 contained about the same amount of SER (10-12% of cytoplasm) as control hepatocytes. On this basis post-stratified estimates by zone (9) could be prepared and an estimate for the liver as a whole obtained by weighting the zone estimates in the above proportions. Because of the high degree of homogeneity within zones the stratified estimates will be considerably more precise than estimates which ignore zone differences.

Assuming that the drug metabolizing enzymes are located on the surface of the SER membranes, the surface area of SER should be related to drug metabolizing enzyme activity. There was a dose related increase in SER surface area/gm liver ($S_{ser}$) (up to 10-fold at the highest dose-level in males). In males the increase in $S_{ser}$ was associated with the increase in drug metabolizing enzyme activity (Fig. 6), but at the highest dose level there was an excess of SER, suggesting the formation of membrane without enzyme. This was also accompanied by the formation of whorls of excess membrane. In females at the highest dose level, $S_{ser}$ was significantly increased ($p < 0.05$) at 3 weeks and subsequently declined somewhat. This was accompanied by an initial increase in drug metabolizing enzyme activity followed by a 50% inhibition. At this time the liver porphyrin level was greater than 700 nM/g liver (10).

Most parameters had recovered to normal values after males had been placed on a recovery diet for 16 weeks. Females also tended to recover, but to a lesser extent due to the shorter recovery period of only 7 weeks. It is likely that the recovery could be attributed to a gradual removal of excess membranes either by autophagy as suggested by Bolender *et al* (11) or through myelin whorl formation or both, rather than to a replacement of hepatocytes. There was no evidence of cell division, and in the normal liver there is no turnover of liver cells.

The changes in SER, with the accompanying increase in drug metabolizing enzyme activity appear, therefore, to be adaptive rather than toxic, in that liver cells seem to recover to normal hepatocytes, without morphological signs of permanent injury due to proliferated SER. The long term implications of chronically elevated levels of these membranes and enzymes are, however, not known.

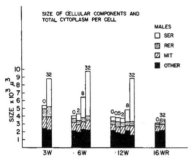

*Figure* 5. Size of cellular components and total cytoplasm per cell in male rats, based on an estimate of cell number pooled over all groups and time period. (W = weeks).

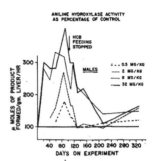

*Figure* 6. Aniline hydroxylase activity in male rats. Values are expressed relative to control values.

## References:

1. Cam, C. and G. Nigogosyan. Acquired toxic porphyria cutanea tarda due to hexachlorobenzene. J. Amer. Med. Assoc. 183, 88, 1963.

2. Ockner, R.K. and R. Schmid. Acquired porphyria in man and rat due to hexachlorobenzene intoxication. Nature 189, 499, 1961.

3. Grant, D.L., F. Iverson, G.V. Hatina and D.C. Villeneuve. Effects of hexachlorobenzene on liver porphyrin levels and microsomal enzymes in the rat. Environ. Physiol. Biochem. 4, 159, 1974.

4. Medline, A., E. Bain, A.I. Menon and H. Haberman. Hexachlorobenzene and rat liver. Arch. Pathol. 96, 61, 1973.

5. Weibel, E.R., W. Stäubli, H.R. Gnägi and F.A. Hess. Correlated morphometric and biochemical studies on the liver cell. I. Morphometric model, stereologic methods, and normal morphometric data for rat liver. J. Cell. Biol. 42, 68, 1969.

6. In this study we used the 3 circulatory zones of liver acini as suggested by Rappaport. Rappaport, A.M., Z.J. Borowy, W.M. Lougheed and W.N. Lotto. Subdivision of hexagonal liver lobules into a structural and functional unit; role in hepatic physiology and pathology. Anat. Record 119, 11, 1954.

7. Miller, R.G. Simultaneous Statistical Inference. McGraw Hill, Toronto, 1967, p67, p76.

8. Loud, A.V. A quantitative stereological description of the ultrastructure of normal rat liver parenchymal cells. J. Cell. Biol. 37, 27, 1968.

9. Raj, D. Sampling Theory. McGraw Hill, Toronto, 1958.

10. Grant, D.L., G. Hatina, I.C. Munro. Environmental Quality and Safety. Thieme, Stuttgart, 1975. in press

11. Bolender, R.P. and E.R. Weibel. A morphometric study of the removal of phenobarbital - induced membranes from hepatocytes after cessation of treatment. J. Cell Biol. 56, 746, 1973.

*National Bureau of Standards Special Publication 431*
*Proceedings of the* FOURTH INTERNATIONAL CONGRESS FOR STEREOLOGY
*held at* NBS, Gaithersburg, Md., September 4-9, 1975 (Issued January 1976)

AUTOMATED MEASUREMENT OF AIRSPACES IN LUNGS FROM DOGS CHRONICALLY EXPOSED TO AIR POLLUTANTS

by D. Hyde,* A. Wiggins,* D. Dungworth,* W. Tyler,** and J. Orthoefer***

*School of Veterinary Medicine, University of California, Davis, California 95616, U.S.A.*
**California Primate Research Center, University of California, Davis, California 95616, U.S.A.*
***U. S. Environmental Protection Agency, 1055 Laidlaw Avenue, Cincinnati, Ohio 45237, U.S.A.*

SUMMARY

Morphometric data were obtained by automated analysis from the lungs of 66 dogs which had been subjected to long-term exposure to a variety of air pollutants. FORTRAN IV (level H) programs allowed data integrity checking, editing, and reduction before computing various parameters. A distribution of percent cumulative frequency of airspace chords in 14 size groups was computed and fitted by a two parameter cumulative probability law of the form $F(x) = P(X \leq x) = (1 - e^{-\alpha x})^{\beta}$. The estimated mean chord length ($\hat{\mu}_x$) was obtained by calculating the area above the curve $(1 - e^{-\alpha x})^{\beta}$. The $\mu_x$ was found to be a sensitive indicator of group differences in sizes of the distal airspaces. Lungs of dogs exposed to oxides of nitrogen, irradiated automobile exhaust, or oxides of sulfur alone and with irradiated automobile exhaust had significantly enlarged distal airspaces by all methods of evaluation.

INTRODUCTION

Stereological methods were used to evaluate lungs from groups of dogs exposed to a variety of air pollutants. Quantitation of histological sections by manual methods is time-consuming and the probability of operator error increases with the number of slides processed. To avoid these limitations, we employed automated analysis in which optical images were analyzed by a television scanner coupled to a computer (Cole, 1966). Thurlbeck (1974) commented that automated measuring microscopes offer a reasonable compromise between the tedium and error of the human mode and the complexity and great expense of computer pattern recognition techniques which utilize binary image digitization and analysis (Levine et al., 1970).

Our primary objective was to determine whether there were significant increases in sizes of airspaces distal to the end bronchi in the lungs of any of the groups of dogs and thus provide quantitative assessment of changes indicative of early emphysema. This was achieved by using a mathematical model based on the distribution of chord lengths of distal airspaces.

METHODS

Sixty-six dogs, distributed among control and treatment groups (Table 1), were exposed to a variety of air pollutants for 61 months (Hinners et al., 1966). A wide variety of physiological parameters were determined before, during, and after exposure (Lewis et al., 1974; Bloch et al., 1972, 1973; Vaughan et al., 1969). Two years after cessation of exposures, the dogs were weighed, measured, and killed by an intravenous injection of pentobarbital. The lungs, trachea, and attached structures were removed from the thorax and trimmed of extraneous tissue. The lungs were weighed, and subsequently the left lung was removed at the left primary bronchus for biochemical analysis. The trachea was cannulated and the right lung was fixed in the normal dorsoventral orientation by intratracheal perfusion with dilute Karnovsky's fixative (Karnovsky, 1965) at 30 cm $H_2O$ pressure in a fixative bath for 16 to 18 hours. The volume of the fixed right lung was then measured by fixative displacement. After samples were taken for scanning electron microscopy (SEM) and transmission electron microscopy (TEM), the lobes of the right lung were cut into 1 cm thick slices (Fig. 1). The fraction of non-parenchyma (bronchi and blood vessels down to 2 mm diameter) and parenchyma were determined using the point-count method of Dunnill (1962).

Nine 2 to 3 $cm^2$ blocks of tissue were cut from each right lung (Fig. 1) and measured. Paraffin sections $7\mu$ thick were cut, stained with hematoxylin and eosin, and measured microscopically using a calibrated eyepiece reticle. The linear correction factor (p) for shrinkage of the fixed tissue due to processing was calculated from these measurements.

The stained slides were examined at a magnification of 125X on a Quantimet 720 image analysing computer (Cole, 1966). The instrument was set to detect the airspaces, and the detection level was standardized by use of a preset level of detection on a wire mesh grid. Measurements were made at 20 stratified random points on each slide (Thurlbeck, 1967). At each point the volume fractions of airspace and tissue, the number of alveolar intercepts, and the number of intra-alveolar chord lengths in 14 size classes were determined, printed on a teletype, and punched on papertape. The punched papertape was read into a Burroughs 6700 for computation of the 201,960 measurements. FORTRAN IV (level H) programs allowed data integrity checking, editing, and reduction before computing and graphically displaying various stereologic parameters.

The fractions of tissue and airspace were corrected for the Holmes effect (Weibel, 1963). The total length of scanning lines multiplied by the number of fields and by the fraction of the corrected mean airspace per field, divided by the number of intra-alveolar chords $>12\mu$ gave the mean chord length of the processed tissue ($L_{c1}$). Those chords $<12\mu$ were rejected because they represented mostly structures not normally air-filled, such as capillaries and post-capillary venules.

The estimated mean chord length ($\hat{L}_c$) of unfixed tissue was calculated from the processed to unfixed tissue by formula 1 (Thurlbeck, 1967):

$$\hat{L}_c = L_{c_1} \; (p) \left(\frac{TLC_R}{V_{LR}}\right)^{1/3} \tag{1}$$

The total lung capacity of the right lung ($TLC_R$) was determined by multiplying TLC by 0.59 which is the fractional volume of the right lung (Cree et al., 1968). TLC was measured on these dog lungs by plethysmographic methods (Lewis et al., 1974). The volume of the fixed right lung ($V_{LR}$) was determined by fixative displacement. The internal surface area of the unfixed right lung ($S_{LR}$) was calculated as described by Weibel (1963) and Thurlbeck (1967).

A distribution of percent cumulative frequency of airspace chords in 14 size groups was computed for each dog lung. A two parameter cumulative probability law of the form $F(x) = P(X \leqslant x) = (1 - e^{-\alpha x})^\beta$ was found to provide the best fit to that distribution. The parameters $\alpha$ and $\beta$ were estimated using a combination of minimum modified chi-square (Cramer, 1946) and the Newton-Raphson procedures (Froberg, 1965). The estimated mean chord length $\hat{\mu}_X$ was obtained by calculating the area above the curve $(1 - e^{-\alpha x})^\beta$ using the 6th Order Newton-Cote formula for numerical integration, formula 2 (Froberg, 1965):

$$\xi(X) = \int_0^\infty [1 - F(x)]dx \tag{2}$$

The variance of X was determined by formula 3 (Froberg, 1965):

$$\hat{\sigma}_X^2 = 2\int_0^\infty x[1 - F(x)]dx - [\int_0^\infty 1 - F(x)dx]^2 \tag{3}$$

$\hat{\mu}_X$ and $\hat{\sigma}_X^2$ were corrected for the Holmes effect and to unfixed dimensions by formula 1. Since the average n per group was large (1347 fields), the differences between $\mu_X$ and $S_{LR}$ of the control group and exposure groups were tested using the standard normal test statistic (Snedecor and Cochran, 1967).

RESULTS

The $\hat{\mu}_X$ calculated by formula 2 had a standard error of 2.76, while $L_c$ calculated by formula 1 had a standard error of 4.14. Because of the lower standard error for all groups, $\hat{\mu}_X$ was used to calculate $S_{LR}$. The corrected $\hat{\mu}_X$ of groups 4, 6, 7, and 8 demonstrated significant increases at the 0.1% level while group 3 demonstrated a significant increase at the 1% level relative to the control group (Table 1). The $S_{LR}/TLC$ ratio of groups 2-4 and 6-8 were also significantly decreased at the 0.1% level when compared to the control group (Table 1).

The most to least significant increase in $\mu_X$ occurred in groups 7, 4, 6, 8, and 3, respectively. The group order of significant differences determined by $S_{LR}/TLC$ was 4, 7, 6, 3, 8, and 2. The TLC to body weight (BW) ratio was significantly increased at the 1% level only in group 4.

DISCUSSION

The automated analysis method of estimating $L_c$ was compared to the manual method of linear integration for estimating $L_c$ (Weibel, 1963) in a study of the distal airspaces in a series of young and aged dogs (unpublished). The slopes of the regression lines of age plotted against $L_c$ by automated analysis and $L_c$ by the manual method were tested to be equal at the 5% level of significance (Snedecor and Cochran, 1967). The SE of $L_c$ by automated analysis was less than that of $L_c$ by the manual method. The y-intercept of the regression line of $L_c$ by automated analysis was slightly smaller than the regression line of $L_c$ by the manual method, because automated analysis counted some blood vessel walls as interalveolar septa.

The $\hat{\mu}_x$ values of our distribution formula are sensitive to subtle shifts in airway chord lengths. There were highly significant increases in the sizes of the distal airspaces in lungs of dogs in groups 3, 4, 6, 7, and 8 with complete agreement between the two parameters used to indicate enlargement ($\hat{\mu}_x$, $S_{LR}$/TLC). The degree of enlargement was greatest in groups 7 and 4 and slightly less in groups 6, 8, and 3.

Scanning electron microscopic observations of these same dogs provided confirmation that destruction of tissue accompanied dilatation; therefore, the lesion represented a true emphysematous process. However, in group 4, hyperinflation accounts for the significant increase in the TLC/BW ratio in this group.

The conclusions with regard to the effects of the various mixtures of air pollutants tested are that 32–36 months after the end of an approximate 5-year exposure period, mild emphysema was present in the lungs of dogs exposed to oxides of nitrogen, irradiated automobile exhaust, or oxides of sulfur alone or with irradiated automobile exhaust. Evaluation of possible regional differences among various lobes and regions of the lung is underway, as is correlation of the total spectrum of morphologic changes observed in lungs of exposed dogs with pulmonary function data.

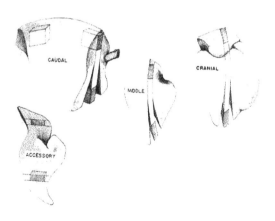

FIGURE 1. Blocks of tissue were selected from sites in the cranial and middle (transverse plane), caudal (sagittal plane), and accessory lobes (frontal plane). The cranial, middle, and caudal lobes are depicted from the lateral view, while the accessory lobe is depicted from the caudal view.

D. Hyde, A. Wiggins, D. Dungworth, W. Tyler and J. Orthoefer

EXPOSURE GROUP

|  | 1 | 2 | 3 | 4 | 5 | 6 | 7 | 8 |
|---|---|---|---|---|---|---|---|---|
|  | CA | R | I | $SO_x$ | $R + SO_x$ | $I + SO_x$ | $NO_2$ high | NO high |
| No. Dogs | 11 | 9 | 4 | 7 | 9 | 11 | 6 | 9 |
| $\hat{\mu}_x \times 10^2$ (microns) | 1.44 | 1.50 | 1.58 | 1.69 | 1.45 | 1.64 | 1.77 | 1.60 |
| SE | 2.06 | 2.50 | 3.90 | 2.90 | 2.20 | 2.31 | 3.73 | 2.51 |
| P | ---- | ---- | 1.0% | 0.1% | ---- | 0.1% | 0.1% | 0.1% |
| $S_{LR}/TLC \times 10^1$ ($m^2/1$) | 1.85 | 1.69 | 1.59 | 1.49 | 1.79 | 1.60 | 1.50 | 1.68 |
| SE $\times 10^{-1}$ | 2.60 | 2.84 | 3.83 | 2.51 | 2.66 | 2.22 | 3.09 | 2.60 |
| ?. | ---- | 0.1% | 0.1% | 0.1% | ---- | 0.1% | 0.1% | 0.1% |

TABLE 1. The dogs were distributed among a control group (CA) and treatment groups of raw automobile exhaust (R), irradiated automobile exhaust (I), oxides of sulfur ($SO_x$), raw automobile exhaust and oxides of sulfur ($R + SO_x$), irradiated automobile exhaust and oxides of sulfur ($I + SO_x$), nitrogen dioxide high/ nitric oxide low ($NO_2$ high), and nitric oxide high/nitrogen dioxide low (NO high). The mean chord length ($\hat{\mu}_x$) and internal surface area of the unfixed right lung to total lung capacity ratio ($S_{LR}/TLC$) are recorded along with their standard errors (SE) and tested level of significance (P) in eight groups of dogs.

ACKNOWLEDGEMENTS
    This study was supported by EPA Contract No. 68-02-1732. We would like to thank Dott Morrissey for typing and editing the manuscript. We acknowledge the assistance of Don Hallberg and Mike Vaida for computer programming, and Janet Mumford for sketching Figure 1.

REFERENCES

Bloch, W.N., Lewis, T.R., Busch, K.A., Or- thoefer, J.G., and Stara, J.F. (1972) Cardiovas- cular status of female beagles exposed to air pollutants. Arch. Environ. Health 24, 342.

Bloch, W.N., Lassiter, S.G., Stara, J.F., and Lewis, T.R. (1973) Blood rheology of dogs exposed to air pollutants. Toxicol. Appl. Pharmacol. 25, 576.

Cole, M. (1966) The metals research Quanti- met. Microscope 15, 148.

Cramer, H. (1946) Mathematical Methods of Statistics. Princeton Univ. Press, Princeton, New Jersey.

Cree, E.M., Benfield, J.R., and Rasmussen, H.K. (1968) Differential lung diffusion, capillary volume, and compliance in dogs. J. Appl. Physiol. 25, 186.

Dunnill, M.S. (1962) Quantitative methods in the study of pulmonary pathology. Thorax 17, 320.

Froberg, C.E. (1965) Introduction to Numer- ical Analysis. Addison-Wesley, Reading, Massachu- setts.

Hinners, R.G., Burkart, J.K., and Conner, G.L. (1966) Animal exposure chambers in air pollution studies. Arch. Environ. Health 13, 609.

Karnovsky, M.J. (1965) A formaldehyde-glutar- aldehyde fixative of high osmolality for use in electron microscopy. J. Cell Biol. 27, 136A.

Levine, M.D., Reisch, M.L., and Thurlbeck, W.M. (1970) Automated measurement of the internal surface area of the human lung. IEEE Trans. Bio- med. Eng. BME-17, 254.

Lewis, T.R., Moorman, W.J., Yang, Y., and Stara, J.F. (1974) Long-term exposure to auto ex- haust and other pollutant mixtures. Arch. Environ. Health 29, 102.

Snedecor, G.W., and Cochran, W.G. (1967) Sta- tistical Methods. The Iowa State Univ. Press, Ames, Iowa.

Thurlbeck, W.M. (1967) The internal surface area of nonemphysematous lungs. Am. Rev. Respir. Dis. 95, 765.

Thurlbeck, W.M. (1974) Measuring tissues. Chest 65, 1.

Vaughan, T.J. Jr., Jennelle, L.F., and Lewis, T.R. (1969) Long-term exposure to low levels of air pollutants: Effects on pulmonary function in the beagle. Arch. Environ. Health 19, 45.

Weibel, E.R. (1963) Morphometry of the Human Lung. Academic Press, New York.

*National Bureau of Standards Special Publication 431*
*Proceedings of the* FOURTH INTERNATIONAL CONGRESS FOR STEREOLOGY
*held at* NBS, Gaithersburg, Md., September 4-9, 1975 (Issued January 1976)

VOLUMETRIC DETERMINATIONS OF CELLS AND CELL ORGANELLES FROM TWO-DIMENSIONAL
TRANSSECTIONS

by L. G. Lindberg
*Cytodiagnostic Department, University Hospital of Lund, Lund, Sweden*

One of the most important problems in stereology is to estimate how well a
purely mathematical model used in calculations conform to the actual problem
when it is applied e.g. in biological work.
To calculate volumes of cells or cell organelles from area measurements in
transsections has since long been a frequent problem. - Sections of many cells
or cell organelles fairly well conform to the shape of wellknown geometrical
figures and presumedly their three-dimensional counterpart to the shape of
bodies that can be described in relatively simple mathematical terms.
In general terms the formula

$$V = \beta \times \bar{a}^{-3/2}$$

describes how the mean transsectioned area $\bar{a}$ can be used to calculated the
volume V of a body from which the sections were obtained. There is one difficulty, however, and that is to estimate $\beta$ in the expression above - a coefficient which is dependent on the shape of the transsected body only.

The most common shapes of cells in mammalian tissues are ellipsoid or cylinder-
like shapes and generally there is a rotational symmetry, i. e. two of the
three axis are the same. In this case a relation between $\beta$ and the eccentricity
$\epsilon$ (the relation between the perpendicular axis to the axis of rotation) can be
found.[1]
The general formula is

$$\bar{a} = V \bigg/ \!\!\int_{o}^{\pi/2} \cdot H_\theta \cdot \sin\theta \cdot d\theta$$

where $\bar{a}$ is the average transsectional surface area of the body and $H_\theta$ is the
projected length of the body and $\theta$ is its angle of inclination and $V^\theta$ is the
volume. When the volume of the ellipsoid body or cylinder is used the following
two relations are found between the eccentricity $\epsilon$ and the coefficient $\beta$
(fig 1).

$$\beta_{ellipsoid} = \sqrt{6/\pi} \cdot h(\epsilon) \cdot \sqrt{h(\epsilon)} / \epsilon$$

where

$$h(\epsilon) = 0.5 + (0.5\,\epsilon^2 / \sqrt{1 - \epsilon^2})\; {}^e\!\log\,((1 + \sqrt{1 - \epsilon^2}) / \epsilon) \qquad \text{for}\,\epsilon < 1 \quad (1)$$

$$h(\epsilon) = 1 \qquad \text{for}\,\epsilon = 1 \quad (2)$$

$$h(\epsilon) = 0.5 + (0.5\,\epsilon^2 / \sqrt{1 - \epsilon^2})\,(\pi/2 - \text{arctg}\,(1/\sqrt{\epsilon^2 - 1})) \qquad \text{for}\,\epsilon \sim 1 \quad (3)$$

and

$$\beta_{cylinder} = (2 + \pi \cdot \epsilon) \cdot \sqrt{2 + \pi \cdot \epsilon} / (4\,\epsilon \cdot \sqrt{\pi})$$

Thus the practical problem of volume estimation from transsections has been
reduced to find a good method of obtaining the numerical value of $\epsilon$.
Theoretically this is very simple - make enough transsections and calculate
the ratio between the shortest and longest axis - the smallest of these
values can be taken as a good approximation of $\epsilon$.

However, in practical work it is impossible to make transsections of biological
material without a section compression i.e. the cutting knife compresses the

section as it cuts through the embedding media and tissues - this causes a
shortening of all lengths along the cutting direction (in the order of 10-30%).
Thus length measurements cannot be used.
Another way can be used - the largest possible area of a transsectioned ellipsoid body is $\pi b^2/\epsilon$ (if the axes are a, b and b). This area can be approximated by the largest observed area, and thus $\epsilon$ can be calculated.

However, the use of extreme values can be dangerous as these are most likely
to be influenced by methodological errors in processing or preparation of
tissues, e.g. swelling or shrinking due to osmotic effects, toxic agents,
traumatic damages etc.
It was found that the distribution of the transsected areas is dependent only
on $\epsilon$ (fig 2).[2] (For $\epsilon < 1$ the density is increasing in (0,1), decreasing
in (1,1/$\epsilon$) with a peak in y = 1).

From this figure it can be seen that using two different moments of the distribution of the areas, $\epsilon$ can be calculated. A requirement that must be fulfilled
is that the size of the transsected bodies is fairly constant. One approach is
too choose the average (expected) area E(A) and the variance V(A) and compute
the square of the coefficient of variation

$$R^2 = V(A)/E(A)^2$$

The exact mathematical relation between $\beta(\epsilon)$ and $R^2(\epsilon)$ can thus be given[2].

$$E(A) = 2\pi b^2/(3h(\epsilon))$$

and

$$V(A) = 16\pi^2 \cdot b^4 (h(\epsilon) - 0.5)/(15\epsilon^2 \cdot h(\epsilon)) - E(A)^2$$

where $h(\epsilon)$ is given above (1). Figure 3 gives a diagram of this relation.

$R^2(\epsilon)$ is estimated by $s^2/\bar{a}^2$, where S and $\bar{a}$ are observed standard deviation
and mean area respectively. For values of $\epsilon$ over 0.3 up to 1 a simple approximation of $\beta(\epsilon)$ can be used.
This approximation is exact for $\epsilon = 1$ and having an error that is slowly
increasing as $\epsilon$ decreases. (For $\epsilon > 0.3$ this error is below 3 per cent).

Thus $V_{est} = (1.382... + 3.4 (s^2/\bar{a}^2 - 0.2)) \cdot \bar{a}^{3/2}$

A practical application of the present principles has been undertaken when
different varieties of a malignant experimental tumor - Rous sarcoma - induced
in rats - were studied. This tumor is composed of elongated cells irregularily
arranged. - The cells fairly well conform to ellipsoid bodies as seen under the
light microscope and even the electron microscopic picture is in agreement with
this assumption.
Transsections of cells and their nuclei are arranged according to size in a
histogram and superimposed upon this the theoretical distribution of areas
is given (fig 4). As is seen the practical application conform rather well to
the theoretical basis although there must be some variation both in size and
eccentricity of the transsected cells. Thus the sampling of tissues, the
measurements and all calculation methods seem to be reliable as the obtained
results when controlled against the theoretical model fairly well conform
to what is expected.[3,4]

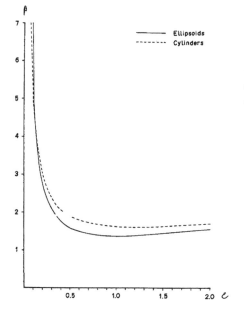

Fig 1  Relation between
the coefficient $\beta$ and the
eccentricity $\epsilon$ of rotatory
ellipsoids or cylinders.

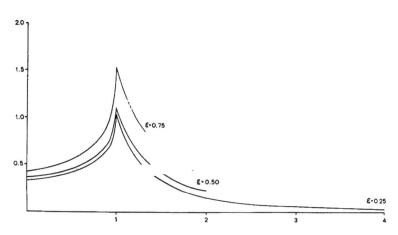

Fig 2  Theoretical distribution (density function) of areas for ellipsoids
with different eccentricity.

2.000

0.200      0.400      0.600      0.800      1.000      1.200      1.400   R²(ε)

Fig 3   Relation between the coefficient $\beta$
and the coefficient of variation $R^2(\epsilon)$.

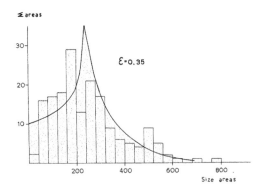

$\Sigma$ areas

$\epsilon = 0.35$

Fig 4   Observed distri-
bution of areas (histo-
gram) plotted against
theoretical distribution
of areas for ellipsoid
tumor cells from a Rous-
virus induced rat sarcoma.

Size areas

REFERENCES

1. Lindberg, L.G. & Vorwerk, P.: On calculating
volumes of transsected bodies from two-
dimensional micrographs.
Lab Invest 23:315, 1970

2. Lindberg, L.G. & Vorwerk, P.: Estimation of
volumes of ellipsoid bodies in conglomerates
through random plane sections.
Lab Invest 27:384, 1972

3. Lindberg, L.G., Mitelman, F. & Vorwerk, P.:
A comparative morphologic study of Rous rat
sarcoma with normal and abnormal chromosomal
pattern.
Lab Invest 27:387, 1972

4. Lindberg, L.G. & Mitelman, F.: Morphometric
analysis of sarcomas with different karyo-
types. Lab Invest 31:90, 1974

362

National Bureau of Standards Special Publication 431
Proceedings of the FOURTH INTERNATIONAL CONGRESS FOR STEREOLOGY
held at NBS, Gaithersburg, Md., September 4-9, 1975 (Issued January 1976)

MORPHOMETRIC ANALYSIS OF THYMIC CELLS DURING MURINE RADIOLEUKEMOGENESIS

by J. Boniver, R. Courtoy and L. J. Simar
Institute of Pathology, University of Liege, Belgium

Summary : By using stereological methods, the following quantitative changes of the blastic population in the thymic subcapsular
zone during radiation induced lymphomagenesis in C57BL mice were
demonstrated :
1) Apparent accumulation of blast cells, mainly due to a simultaneous decrease of the small lymphocyte number;
2) Modifications of the number of the three subcapsular blast cell
aspects (lymphoblasts, X cells, RSN cells) : accumulation of RSN
cells and of X cells during post-RX regeneration in the late atrophic preleukemic thymuses;
3) A few variations of the cellular structure, except for a reduction of the surface density of the Golgi membranes in X cells and
RSN cells of the atrophic and lymphomatous thymuses.
The relation between these changes and kinetic and functional modifications of thymic blast cells during leukemogenesis is discussed.
RSN : ring-shaped nucleolus cells.

Introduction : Murine radiation-induced lymphomas result likely
from the cancerous transformation of thymic subcapsular blast cells
by an oncornavirus (11). At the ultrastructural level, this morphologically heterogeneous blastic population contains two peculiar
cell aspects respectively named "X cells" (13) and "ring-shaped nucleolus cells" (5), besides typical lymphoblasts (fig. 1). A morphometric analysis (4) was performed to determine quantitatively
the eventual morphological changes of this population during radioleukemogenesis (5).

Results : 1. Numerical density : The numerical density of blast
cells and small lymphocytes in the subcapsular zone, determined by
using the Abercrombie's method (1), and the absolute blast cell num
ber, based on a evaluation of the subcapsular volume (4) show parallel variations (fig. 2). After each X ray induced thymic involu-
tion, regeneration is characterized by a transient accumulation of
blast cells, the number of which does not depend on the number of
X ray doses. However, after the fourth irradiation, the number of
small lymphocytes does not return rapidly to the normal level;
then, the blast cell proportion becomes higher than normally. Later
on, the single changes occur on day 90, in the late atrophic thymuses in 30% treated mice; the small lymphocyte number is statistically smaller than in the other thymuses, inducing an apparent accumulation of blast cells. The numerical density of each blast cell
type in the subcapsular zone is determined by multiplying their respective percentage determined with electron microscope by the total
numerical density described above (fig.3). Each 150 R dose induces
a transient increase of the density of lymphoblasts and X cells.
After the last irradiation, X cells and ring-shaped nucleolus cells
become more numerous than normally. In the late atrophic thymuses,
the numerical density of lymphoblasts and ring-shaped nucleolus
cells is higher than in non atrophic and in untreated thymuses.

2. Nuclear and cytoplasmic constituents : The volumetric and surface densities are obtained by point and intersection counting methods (6, 14, 15, 16) and the numerical density by Weibel and
Gomez method (22).

Dense chromatin (fig.4) : Its volumetric density in the nucleus is
higher in ring-shaped nucleolus cells and mainly in X cells than
in lymphoblasts. It does not change during leukemogenesis.

Nucleolus (fig.5,6) : There is no significant difference of its volumetric density in the three blast cell types; the respective proportions of fibrils, granules and fibrillar centre inner the nucleolus are not identical since the fibrillar centre is peculiarly
well developed in ring-shaped nucleoli.

Cytoplasmic organelles : The cell content in mitochondria, endoplasmic reticulum and dense granules is similar in the three cell
types and shows few variations during leukemogenesis. The surface
density of Golgi membranes (fig.7) is decreased in X cells and
mainly in ring-shaped nucleolus cells in late atrophic and lymphomatous thymuses.

Discussion : The apparent accumulation of blast cells in the subcapsular zone of the thymus during radioleukemogenesis is mainly
due to a decrease of the number of small lymphocytes. This phenomenon is simultaneous with changes of the thymic blast cell composition. As according to recent kinetic studies (5), the three blast
cell aspects correspond to different stages in the cell cycle, it
is likely that these morphological modifications are related to
changes of the kinetics of thymic lymphopoïesis at some important
phases of radioleukemogenesis as regeneration (1) and late atrophy
(10, 12). The single cellular change observed during the preleukemic period is a reduction of the Golgi zone in some blast cells.
This modification could be correlated with a variation of the poly-
saccharide production since it is well admitted that the Golgi zone
plays a role in the synthesis of the cell coat material (2, 8, 9)
which is frequently modified in cancerous cells (7).

References :
1) ABERCROMBIE, M., Anat. Rec., 1946, 94, 239-251.
2) BENNETT, G., LEBLOND, C.P., HADDAD, A., J. Cell Biol.,1974, 60,
   258-284.
3) BONIVER, J., COURTOY, R., SIMAR, L.J., J. Microscopie, 1975, 22,
   57a.
4) BONIVER, J., COURTOY, R., SIMAR, L.J., BETZ, E.H., in preparation
5) BONIVER, J., DELREZ, M., SIMAR, L.J. and HAOT, J., Beitr. Path.,
   1973, 150, 229-245.

6) DE HOFF, R.T. and RHINES, F.N., Quantitative microscopy, Mc Graw
   Hill Book Co, New York, 1968.
7) EMMELOT, P., Europ. J. Cancer, 1973, 9, 319-334.
8) FAVARD, P., The Golgi apparatus. In : Handbook of Molecular Cytology, A. Lima de Faria, Ed., North Holland, Amsterdam, 1969,
   1130-1155.
9) ITO, S., Fed. Proc., 1969, 28, 12-25.
10) JARPLID, B., Acta Radiol., supplt. 279, 1968.
11) KAPLAN, H.S., Nat. Cancer Inst. Monographs, 1964, 14, 207-217.
12) SIEGLER, R., HARRELL, W. and RICH, M.A., J. Nat. Cancer Inst.,
    1966, 37, 105-121.
13) SIMAR, L.J., HAOT, J. and BETZ, E.H., Europ. J. Cancer, 1968, 4,
    529-535.
14) UNDERWOOD, E.E., Quantitative stereology, Addison Wesley Ed.,
    Reading, Mass, 1970.

15) WEIBEL, E.R., Int. Rev. Cytol., G.H. Bourne and J.F. Danielli
    Ed., 1969, <u>26</u>, 235-302.
16) WEIBEL, E.R. and ELIAS, H., Quantitative Methods in Morphology,
    Springer Verlag, Berlin, 1967.
17) WEIBEL, E.R. and GOMEZ, D.M., J. Appl. Physiol., 1962, <u>17</u>, 343-
    365.

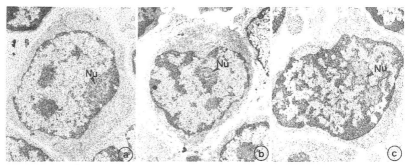

<u>Figure 1</u> : Ultrastructural aspects of thymic subcapsular blast cells
in C57BL mouse (x6650) : a) <u>typical lymphoblast</u> : few abundant dense chromatin and compact nucleolus (Nu); b) <u>ring-shaped nucleolus
cell</u> : granular and fibrillar components encircle the fibrillar cen
tre; c) <u>X cell</u> : abundant dense chromatin irregularly distributed,
compact nucleolus.

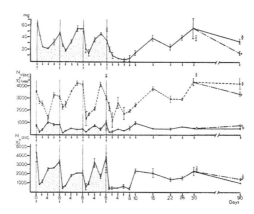

<u>Figure 2</u> : Variations of the thymic weight (in mg), numerical densi-
ty ($N_V$(zsc)) of blast cells (●——●) and small lymphocytes (●- -●)
and absolute number of blast cells ($N_{zsc}$) in the subcapsular zone
of the thymus of C57BL mice which have received 1, 2, 3 or 4 irra-
diations of 150 R (arrows).
Φ control; o——o——o : evolution to late atrophy (A).

|              | Control 35 d. old | 1 Rx D.4 | 4 Rx D.8 | 4 Rx J.90 | 4 Rx J.90 atrophy | Control 6 months |
|--------------|-------------------|----------|----------|-----------|-------------------|------------------|
| Lblasts      | 438÷98            | 713÷111  | 251÷90   | 101÷25    | 230÷15            | 104÷27           |
| RSN cells    | 99÷26             | 68÷ 26   | 194÷42   | 70÷42     | 235÷34            | 119÷41           |
| X cells      | 94÷23             | 261÷121  | 380÷48   | 155÷35    | 90÷23             | 75÷18            |

Fig.3 : Numerical density of blast cells in the subcapsular zone
during radioleukemogenesis.

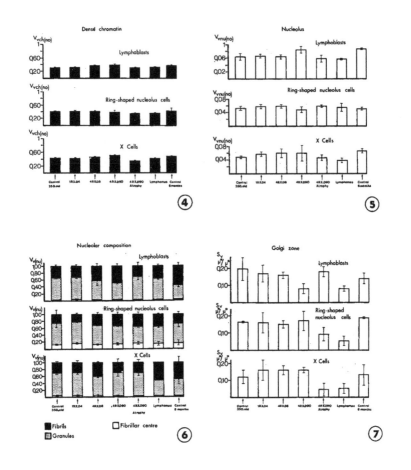

Figures 4-7 : Quantitative changes of thymic blast cells during radioleukemogenesis. Volumetric density of dense chromatin (fig.4) and
nucleolus(fig.5) in the nucleus; volumetric density of various components in the nucleolus (fig.6) and surface density of Golgi membranes in the cytoplasm (fig.7).

*National Bureau of Standards Special Publication 431*
*Proceedings of the* FOURTH INTERNATIONAL CONGRESS FOR STEREOLOGY
*held at* NBS, Gaithersburg, Md., September 4-9, 1975 (Issued January 1976)

## STUDIES ON THE PARTIALLY-ORIENTED SURFACES OF SKELETAL MUSCLE MITOCHONDRIA

by N. T. James and G. A. Meek
*Department of Human Biology & Anatomy, The University, Sheffield S10 2TN, United Kingdom*

The mitochondria of skeletal muscle fibres are usually elongated and lie with their long axes parallel with that of the fibre. Consequently, their surfaces form a partially orientated system. In a partially orientated system there is a *linear* component which is formed by the surfaces of mitochondria parallel with the long axis of the fibre and an *isometric* component which is formed by the rounded ends of the mitochondria.

Several quantitative techniques are available for analysing partially orientated systems (Underwood, *Quantitative Stereology*, Addison-Wesley, 1970). They are widely used in non-biological subjects, for example, in analysing the partially orientated crystals in metal alloys. They have not previously been used to analyse the properties of orientated organelles in markedly anisotropic muscle fibres. The use of such techniques could provide much data on the organization of muscle fibres which might otherwise be unobtainable.

The general relationship of any system of organelles, including mitochondria, is

$$S_v = 2 P_L \ \mu m^2 \ \mu m^{-3}$$

where $P_L$ is the number of intersections of organelle surface per unit length of test line in a stereological grid. This formula *could* be used to analyse orientated structures provided that a sufficient number of section planes were used in order to satisfy the necessary criterion of complete randomness. The data so obtained, however, cannot be used to derive further information on the surface areas of organelles.

The theory for analysing partially orientated surfaces conveniently requires that only *one* plane of section be used for analysis. On each longitudinal section of a skeletal muscle fibre a linear test array is applied sequentially, first in a plane parallel with the long axis of the fibre and then perpendicular to this axis. When the lines of the test array are parallel with the long axis of the fibre (Fig. 1) only the terminal isometric surfaces of the mitochondria intersect the test lines. The surface area of the isometric regions $(S_v)_{isom}$ is given by

$$(S_v)_{isom} = 2(P_L)_\parallel \ \mu m^2 \ \mu m^{-3} \quad \dots\dots\dots\dots\dots\dots\dots\dots\dots \ (1)$$

where $(P_L)_\parallel$ is the number of surface intersections per unit length of the parallel test array.

When the test array is perpendicular (Fig. 2) to the long axis of the fibre both the linear orientated and the isometric surfaces of the mitochondria

367

intersect the test lines. The surface area of the orientated regions of mitochondria $(S_v)_{lin}$ is given by

$$(S_v)_{lin} = \frac{\pi}{2}\left((P_L')_{\perp} - (P_L')_{\shortparallel}\right) \mu m^2 \; \mu m^{-3} \; \ldots\ldots\ldots\ldots\ldots \; (2)$$

Total surface area of the mitochondria $(S_v)$ is given by the formula:

$$S_v = (S_v)_{isom} + (S_v)_{lin} \; \ldots\ldots\ldots\ldots\ldots \; (3)$$

Addition of (2) and (3) gives

$$S_v = \frac{\pi}{2}(P_L')_{\perp} + 2(P_L')_{\shortparallel} - \frac{\pi}{2}(P_L')_{\shortparallel} \; \mu m^2 \; \mu m^{-3}$$

The degree of linear orientation of any partially orientated system of surfaces $(\omega_{lin})$ can be defined as $(S_v)_{lin} \Big/ S_v$ whereby

$$\omega_{lin} = \frac{(S_v)_{lin}}{(S_v)_{lin} + (S_v)_{isom}}$$

which reduces to:

$$\omega_{lin} = \frac{(P_L')_{\perp} - (P_L')_{\shortparallel}}{(P_L')_{\perp} + \frac{4}{\pi}(P_L')_{\shortparallel} - (P_L')_{\shortparallel}}$$

The pectoralis major muscle of the pigeon was selected for analysis since it contains two highly different types of muscle fibre. These are arbitrarily designated type I and type II (George & Berger, *Avian Myology*, Academic Press, 1966). Type I fibres are relatively narrow in diameter ($\sim 30 \; \mu m$) and contain many mitochondria. They are slow twitch fibres and they derive their energy aerobically by the oxidation of lipids. Type II fibres are relatively broad ($\sim 70 \; \mu m$) and contain few mitochondria. They are fast twitch fibres and they derive their energy by anaerobic glycolysis. Some typical values for $S_v$ and $\omega_{lin}$ are recorded in Table 1.

It is possible that estimates of $S_v$ and $\omega_{lin}$ could be useful in quantifying alterations in structure which occur in skeletal muscles following experimental procedures, for example, during muscular hypertrophy.

Table 1

Stereological parameters calculated for
avian type I and II fibres

| Parameters Calculated | Muscle Fibres | |
|---|---|---|
| | Type I | Type II |
| $(P_L)_{\perp}$ $\mu m^{-1}$ | 1.76 | 0.06 |
| $(P_L)_{\parallel}$ $\mu m^{-1}$ | 0.64 | 0.04 |
| $S_V$ $\mu m^2$ $\mu m^{-3}$ | 3.04 | 0.11 |
| $\omega_{lin}$ | 0.58 | 0.29 |

Values obtained by applying $\sim$ 500 $\mu$m test line length to each of 20 type I fibres and 20 type II fibres.

A value of $\omega_{lin}$ close to zero indicates there is no linear orientation of the mitochondrial surfaces. Maximum orientation occurs at $(P_L)$ = 0 when $\omega_{lin}$ = 1.

Fig. 1. A randomly selected longitudinal section of a type I muscle fibre in the pectoralis major muscle of the pigeon. A linear test array has been applied parallel with the long axis of the fibre. The number of intersections of mitochondrial surfaces with the test lines provide values for $(P_L)$ . x 35 000

Fig. 2. The same electron micrograph as in Fig. 1. The same linear test array has been applied perpendicular to the long axis of the fibre. The number of intersections of mitochondrial surfaces with test lines provide values for $(P_L)$ .

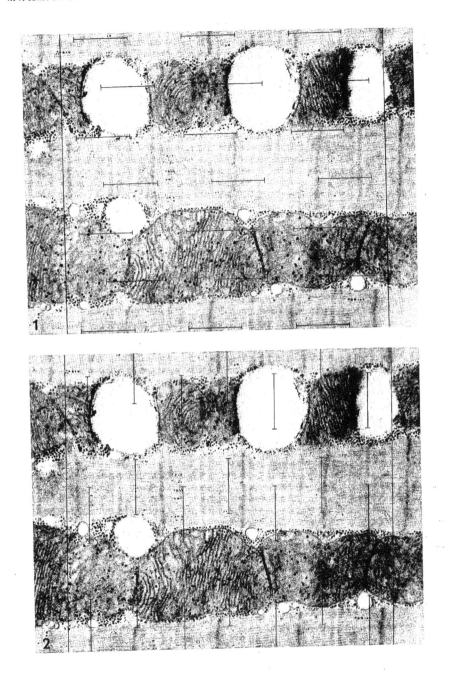

National Bureau of Standards Special Publication 431
Proceedings of the FOURTH INTERNATIONAL CONGRESS FOR STEREOLOGY
held at NBS, Gaithersburg, Md., September 4-9, 1975 (Issued January 1976)

DETERMINATION OF THE NUMBER OF CELLS WITH MULTIPLE NUCLEOLI IN HISTOLOGICAL SECTIONS

by A. Schleicher, H. -J. Kreschmann, F. Wingert, and K. Zilles

Department of Anatomy, Medical University Hannover, and Institute of Medical Information Science and Biomathematics, University of Münster, Germany

For certain neuroanatomical investigations the number of neurons in different parts of the nervous system has to be estimated. The neuronal cell body, its nucleus, or its nucleolus may be used as the unit to indicate the presence of a cell in the histological section, but the larger the unit used, and the thinner the section, the greater the counting error due to multiple counting of units split by cutting. Therefore, the most widely used method is that of counting the nucleoli of the neurons as the smallest, but still well distinguished cellular constituent (1).

The method for counting neuronal nucleoli is based on the assumption that each neuron has one nucleus with only one nucleolus. Counting samples in the brain stem of Tupaia belangeri, we found mononucleated neurons with one, two and three nucleoli.

In order to estimate the total number of neurons as well as the number of neurons with one, two and three nucleoli, two aspects must be considered:

- When cutting the structure, nuclei may be split in such a way that each fragment has one or more nucleoli. Thus nucleoli belonging to the same neuron may appear in adjacent sections (Fig. 1a).

- Smaller fragments of split nucleoli are not seen or recognized, but fragments exceeding a certain diameter are also counted as nucleoli (Fig. 1b).

Counting nuclei or nuclear fragments with at least one nucleolus as neurons in the sections we find $x'_1$ neurons with one nucleolus, $x'_2$ neurons with two and $x'_3$ neurons with three nucleoli:

(1) $$n = x'_1 + x'_2 + x'_3$$

The true numbers are denoted by $x_1$, $x_2$ and $x_3$:

(2) $$N = x_1 + x_2 + x_3$$

The relations between the numbers of neurons counted and the true numbers are formulated as a system of linear equations:

371

(3)
$$a_{11} \cdot x_1 + a_{21} \cdot x_2 + a_{31} \cdot x_3 = x_1'$$
$$a_{22} \cdot x_2 + a_{32} \cdot x_3 = x_2'$$
$$a_{33} \cdot x_3 = x_3'$$

For instance, the number of neurons counted with one nucleolus is the sum of nuclei with one nucleolus, of nuclear fragments with one nucleolus from nuclei with two nuleoli and of nuclear fragments with one nucleolus from nuclei with three nucleoli.

b

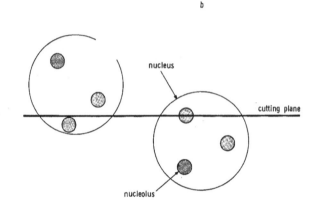

Fig.                    Schematic illustration of counting errors caused
                        by split nuclei and nucleoli.

The calculation of the coefficients is based on the following assumptions:

- Both the nuclei and the nucleoli are spheres.
- The nuclear diameter is smaller than or equal to the thickness of the section.
- The nuclei are uniformly distributed perpendicular to the cutting plane.

By using geometrical probability assumptions one finally obtains the following expressions:

(4)  $\quad (1-(1-2p_{11})\cdot\frac{D}{T})\cdot x_1 \;+\; 2p_{21}\cdot\frac{D}{T}\cdot x_2 \;+\; 2p_{31}\cdot\frac{D}{T}\cdot x_3 \;=\; x_1'$

$\qquad\qquad (1-(1-2p_{22})\cdot\frac{D}{T})\cdot x_2 \;+\; 2p_{32}\cdot\frac{D}{T}\cdot x_3 \;=\; x_2'$

$\qquad\qquad\qquad\qquad (1-(1-2p_{33})\cdot\frac{D}{T})\cdot x_3 \;=\; x_3'$

D is the nuclear diameter and T is the thickness of the sections.
Nuclear fragments with j nucleoli or visible nucleolar fragments
from nuclei with i nucleoli have the probability $p_{ij}$ to be counted in the section, assuming that all nuclear splits are equally
probable and that the nucleoli are equally distributed throughout
the nucleus. The values of these probabilities depend on

(5)  $\qquad\qquad b = \dfrac{d}{D} \quad$ and $\quad c = \dfrac{\sqrt{d^2 - d^2_{min}}}{d} \; ,$

where d is the nucleolar diameter and $d_{min}$ is the diameter of the
smallest nucleolar fragments counted.

Table of the probabilities $p_{ij}$ for c = o

| b | $p_{11}$ | $p_{21}$ | $p_{22}$ | $p_{31}=p_{32}$ | $p_{33}$ |
|------|-------|-------|-------|-------|-------|
| 0.00 | 0.500 | 0.257 | 0.371 | 0.193 | 0.307 |
| 0.05 | 0.500 | 0.245 | 0.377 | 0.183 | 0.317 |
| 0.10 | 0.500 | 0.234 | 0.382 | 0.176 | 0.325 |
| 0.15 | 0.500 | 0.224 | 0.387 | 0.169 | 0.331 |
| 0.20 | 0.500 | 0.218 | 0.390 | 0.163 | 0.337 |
| 0.25 | 0.500 | 0.216 | 0.391 | 0.162 | 0.338 |
| 0.30 | 0.500 | 0.216 | 0.392 | 0.161 | 0.329 |
| 0.35 | 0.500 | 0.218 | 0.390 | 0.160 | 0.340 |
| 0.40 | 0.500 | 0.225 | 0.386 | 0.164 | 0.336 |

The probabilities depend on c by

(6)  $\qquad p_{ij}(b,c) = p_{ij}(b,c=o) + \dfrac{1}{2}\cdot b\cdot c \quad$ for $\quad i = j .$

The calculation of some of these probabilities may be done explicit-
ly ($p_{11}$; $p_{ij}$ and d = 0), others are calculated using a Monte-Carlo-
procedure on a computer where the probabilities are estimated by
their frequencies occurring in a simulation procedure (2).

The following example demonstrates the differences between counted
and corrected numbers. In the nucl.n.oculomotorii of a Tupaia belangeri we measured: thickness of section 20 μm, nuclear diameter
10.6 μm, nucleolar diameter 2.6 μm and the diameter of the smallest
nucleolar fragments counted 2.0 μm. The results are as follows.

| counted numbers | corrected numbers |
|---|---|
| $x_1' = 7426$ | $x_1 = 6176$ |
| $x_2' = 3078$ | $x_2 = 3168$ |
| $x_3' = 74$ | $x_3 = 81$ |
| $n = 10578$ | $N = 9425$ |

References

1. B.W. KONIGSMARK, "Methods for the counting of neurons",
   Contemporary Research Methods in Neuroanatomy,
   edited by W.J.H. NAUTA and S.O.E. EBBESSON.
   New York: Springer-Verlag (1970) 315

2. A. SCHLEICHER, K. ZILLES, H.-J. KRETSCHMANN, F. WINGERT,
   "Bestimmung der Anzahl der Zellen mit mehr als einem
   Nucleolus im histologischen Schnittpräparat", Acta
   microsc. (in press).

*National Bureau of Standards Special Publication 431*
*Proceedings of the* FOURTH INTERNATIONAL CONGRESS FOR STEREOLOGY
*held at* NBS, Gaithersburg, Md., September 4-9, 1975 (Issued January 1976)

COMBINED LIGHT AND ELECTRON MICROSCOPY OF SERIALLY SECTIONED CELLS AND TISSUES

by Yrjö Collan,
*Second Department of Pathology, University of Helsinki, Finland*

ABSTRACT

By studying herring eggs sectioned in series it was shown that corresponding images of cells in sections could no more be identified reliably after the distance between sections increased above 0.13 times the diameter of the egg (psychological safety limit). For round cells the probability for a cell to be present in one section if it is found in the other is $2r-t$ / $2r+s_1+s_2+t$, where $2r$ is the diameter of the cell, $t$ the distance between the sections, and $s_1$ and $s_2$ the thicknesses of the sections. To pick up those cells common in both sections two types of criteria were recommended, one based on the occurrence of the nucleus within the images of the cell in both sections, the other on the diameter of the cell or nuclear image. Formulas for the application of these criteria were developed. Applicability of the method was studied in the laboratory mouse. Brain glial cells (ample material between cells), liver parenchymal cells and kidney proximal tubule cells (large cell size) could easily be studied with this method. Small intestinal gland cells (tall but thin cells), and spleen lymphocytes or spermatogonia (small cell size) needed special precautions that when taken secured reliable matching. Based on experiments on the sensitivity of autoradiography it was concluded that combined light microscope autoradiography and electron microscopy under suitable conditions will be about 50 times more sensitive than conventional electron microscope autoradiography.

INTRODUCTION

In electron microscope laboratories sections stained with various basic dyes are found useful in determining the area under cutting. However, the expansion of methodology offered by sectioning electron and light microscope sections in series is far from exploited. Aspects related to this question have been studied in the author's laboratory (1, 2).

REVIEW OF PRINCIPLES

When combining light and electron microscopy the object to be studied must be present in both light microscope (LM) and electron microscope (EM) sections. In the present method 1 - 5 EM sections are cut first and immediately after these a light microscope section, 0.5-1.0 μm thick. While studying these sections it is important to minimize the risk of erroneously matching two different cells, one present in the EM- and the other in the LM-section, but in relation to other cells in corresponding location in each section. It was shown that both psychological and geometrical factors are involved.

The psychological factors were studied by cutting densely packed paraffin embedded herring eggs in series. Pairs of sections were cut 4 - 1200 μm apart from each other. Each section was photographed and the pairs of photographs were shown to 15 persons. These were asked whether they were able to identify corresponding cells in the picture pairs. The results showed that above the distance of 100 μm matching was found difficult or impossible. The limiting distance was called "the psychological safety limit (PSL)". For general use on round objects it was presented as the ratio between the distance of the sections and the diameter of the eggs, giving the value 0.13.

Application of the PSL does not guarantee that the images of cells in corresponding location in sections are images of the same cells. Even immediately adjacent sections contain a certain number of cells that are present in only one of the sections. The probability for a cell or any tissue structure to be present in two sections provided it is found in one of them can be determined by the formula

$$p = \frac{2r - t}{2r + s_1 + s_2 + t} \qquad (1)$$

where $2r$ is the diameter of the tissue structure perpendicular to the plane of sectioning, $t$ is the distance between the sections, and $s_1$ and $s_2$ are thicknesses of the sections.

To find the cells common to both sections three parameters have been investigated: 1) the occurrence of the nucleus in the cell in the section, 2) the diameter of the cell in the section, 3) the diameter of the nucleus in the section.

Of these, the first one is that most easily applicable to practical situations; only cells are studied that show nucleus in both sections. When applying this criterion for round cells, to avoid mismatching one of the following formulas should hold

$$2r \, (1 - \sin a) > t \quad ; \quad a < 60° \qquad (2)$$

$$2v - 2r \, (1 - \sin a) > s_1 + s_2 + t \qquad (3)$$

(where $r$ is the radius of the cells under study, $a$ is the angle between the cutting plane and the line connecting the centres of the two cells, $t$ is the distance of the nucleus from the cell surface, $s_1$ and $s_2$ are thicknesses of the sections).

After characterization of the cell population under study, the conditions necessary for applying the method can be determined. This is most easily done by drawing the two sine curves presented by the left sides of the formulas and adjusting the parameters $s_1$, $s_2$ and $t$ so that under any given test conditions at least one of the formulas holds. The two other parameters are applied in a corresponding fashion (2).

APPLICATION TO VARIOUS TISSUES

To determine to what extent the method is applicable, tissues of an adult white laboratory mouse weighing 20 g were investigated. Samples were taken from the small intestine, testis, kidney cortex, liver, spleen and brain, fixed in 3% glutaraldehyde and embedded in Epon. The small intestinal gland cells, spermatogonia, proximal tubule cells, liver parenchymal cells, lymphocytes and glial cells were studied. Cell diameter, nuclear diameter, cell surface - nuclear surface distance and the distance between the nuclei of two adjacent cells were measured (Table 1).

The table also contains other data, such as values of the formula (1) and the PSL for the smallest cell diameters in each group for immediately adjacent light and electron microscope sections. Under these conditions the most unfavourable situation in tissues is combatted by the most favourable way of matching sections. In the following various tissues are discussed in detail; the presence of the nucleus of the cells in both sections being the criterion of sameness.

Small intestinal glands

If studied in sections cut perpendicular to their long axis the cells cause no problems. However, gland cells often appear longitudinal in sections. The cell nuclei are long and the distance between the nuclei is a limiting parameter ($s_1 + s_2 + t < 0.7$ µm) because mismatching is improbable when the cells are cut with a plane not roughly parallel to their long axis. Further, less than 90 % of cells are present in both sections and the PSL reaches high values easily. This is why the use of immediately adjacent sections only is recommended. The thickness of the light microscope section should not exceed 0.5 µm.

Testis

These cells vary in size the smallest ones being the size of the lymphocytes and the largest about double the diameter. More than 90 % of cells are common to both sections. Also the PSL is small allowing comparison of 1 - 4 EM sections with an 0.5 µm LM section. The formulas (2) and (3) recommend the use

TABLE 1.

Dimensions (μm) of cells in various organs in a laboratory mouse after fixation in cacodylate buffered 3 % glutaraldehyde and embedding in Epon.

| | Small intestine gland cells | Sperm- ato- gonia | Kidney proximal tubules | Liver parench. cells | Spleen lympho- cytes | Brain glial cells |
|---|---|---|---|---|---|---|
| **Cell diameter** | | | | | | |
| Minimum | 4.3 | 6.8 | 10.2 | 20.3 | 5.1 | 10.3 |
| Maximum | 31.0 | 15.3 | 23.7 | 39.0 | 11.0 | 15.5 |
| **Nuclear diameter** | | | | | | |
| Minimum | 3.4 | 5.1 | 5.1 | 10.2 | 4.2 | 6.9 |
| Maximum | 15.5 | 11.8 | 9.3 | 15.3 | 8.5 | 11.2 |
| **Cell surface – nuclear surface distance** | | | | | | |
| Minimum | 0.3 | 0.5 | 0.8 | 3.4 | 0.2 | 1.0 |
| Maximum | 15.5 | 5.9 | 6.8 | 16.9 | 4.2 | 3.4 |
| **Distance between nuclei in adjacent cells** | | | | | | |
| Average | 0.9 | 3.4 | 4.2 | 16.9 | 0.7 | –[xx] |
| Minimum | 0.7 | 1.7 | 1.7 | 8.5 | 0.5 | – |
| Maximum | 3.0[x] | 6.8 | 8.5[x] | 30.5 | 5.9 | – |
| **Material between cells** | | | | | | |
| | ± | ± | ± | | | |
| **Probability from the formula (1)[xxx]** | | | | | | |
| | 0.88 | 0.92 | 0.94 | 0.97 | 0.89 | 0.94 |
| **PSL** $(s_1 + s_2 + t / 2r)$[xxx] | | | | | | |
| | 0.14 | 0.08 | 0.06 | 0.03 | 0.12 | 0.06 |

[x]Nuclei of another tubulus or gland often are still farther, up to 10 μm
[xx]not determined
[xxx]$2r$ = smallest measured, $t = 0$, $s_1 = 0.5$ μm, $s_2 = 0.1$ μm.

of 1 – 3 EM sections with an 0.5 μm LM section. This recommendation should be followed to secure reliable matching of the smallest cells also.

Proximal tubule of the kidney
Here the cells are cubical and have a large amount of cytoplasm around the nucleus. The formulas (2) and (3) show that with an 1 μm LM section three most adjacent EM sections can be reliably matched, with an 0.5 μm LM section more than ten such sections.

Liver parenchymal cells
These cells were large and had so rich cytoplasm that no problems could develop in applying the method.

Spleen lymphocytes
The distance between the nucleus and the cell surface is so small that the use of a thin (0.3 – 0.5 μm) LM section and only the adjacent EM section is recommended. Under these circumstances the comparison is reliable also for the smallest cells. Both the PSL and the values from formulas (2) and (3) are critical if these limits are exceeded. This is due to the dense packing of lymphocytes in the spleen. For these cells one could also use nuclear or cell diameter as a criterion. Then, only cells with nuclei above a certain diameter in both sections are matched. Under those conditions care must be taken not to rule out a true cell population with a small nuclear diameter.

*Yrjö Collan*

## Brain glial cells

These cells gave data comparable to those of the kidney proximal tubule cells. However because of the thick layer of cytoplasm around the nucleus they allowed liberal application of the method.

Table 1 also reports interstitial material between cells. In liver and spleen cells are practically without any intercellular material at all. If the same cells are found more loosely distributed - e.g. lymphocytes in the connective tissue - the applicability of the method increases. Small intestine, kidney and testis form concentrates of cells surrounded by a basement membrane. Between glands and tubules there is some connective tissue and this is why there is little danger in mixing data between two glands or kidney tubules. In the brain intercellular material is ample and the method is even more applicable than appears from the tabulated size estimates.

## SPECIAL APPLICATIONS

In addition to allowing comparison of various light microscopic staining characteristics (3,5) with ultrastructure and to bridging the gap between paraffin sections and electron microscopy (4) the method is applicable to autoradiography (2). Because autoradiography is done on the LM section the sensitivity of the method is better than that of EM autoradiography. To test this samples of mouse testis were studied. The animal was injected with 1 µCi/g body weight of tritiated thymidine intraperitoneally and killed one hour later. From one Epon block numerous silver EM sections and 0.5, 1.0, 1.5 µm thick LM sections were cut. LM autoradiography was carried out by using Ilford K5 emulsion as described earlier (2). For electron microscope samples the same emulsion, diluted 1:4 was used and spread on the grids with a steel wire loop. Exposure time was 3 weeks, and samples were developed with Kodak D 19b for 2 minutes. In screening the EM sections only occasional cells were found labelled with one grain above the nucleus. The LM sections gave following data

| Section thickness | Mean grain count of 100 cells $\pm$ SD |
|---|---|
| 0.5 µm | 7.7 $\pm$ 2.7 |
| 1.0 " | 10.2 $\pm$ 3.8 |
| 1.5 " | 12.1 $\pm$ 5.5 |

Although definite counts could not be made from the EM sections there is great difference in sensitivity. Difference in the thickness of sections may be 12 fold (or even more) in these methods (1.0 µm and 0.08 µm). In addition, the emulsion being thin and diluted in EM autoradiography may cause further 1.5 - 5 fold decrease in sensitivity. So it turns out that under suitable conditions the combined autoradiography method will be about 50 times more sensitive than EM autoradiography.

## REFERENCES

1. Collan Y, Collan H: Interpretation of serial sections. Z Wiss Mikrosk 70: 156-167, 1970
2. Collan Y: Combining light and electron microscopic findings on individual cells: A theoretical and methodological study exemplified by combined electron microscopy and light microscope autoradiography. Microscopica Acta 75:48-60, 1973
3. Kaufmann P, Stark J: Enzymdarstellungen im Semidünnschnitt. Acta Histochemica (Jena) 42:178, 1972
4. Lynn JA: "Adjacent" sections - a bridge in the gap between light and electron microscopy. Hum Pathol 6:400-402, 1975
5. Snodgree AB, Dorsey CH, Bailey GWH, Dickson LG: Conventional histopathologic staining methods compatible with Epon-embedded,osmicated tissue. Lab Invest 26:329, 1972

*National Bureau of Standards Special Publication 431*
*Proceedings of the* FOURTH INTERNATIONAL CONGRESS FOR STEREOLOGY
*held at* NBS, Gaithersburg, Md., September 4-9, 1975 (Issued January 1976)

MORPHOMETRIC "ORGANELLE PROFILES" AS METABOLIC INDICATORS

by Sasha Malamed and Lawrence C. Zoller
*Department of Anatomy, College of Medicine and Dentistry of New Jersey, Rutgers Medical School,
Piscataway, New Jersey, 08854, U.S.A.*

SUMMARY

Two systems are under morphometric study in order to detect possible differences in cell organelles pointing to specific metabolic alterations. Electron micrographs of normal and experimental or diseased cells are compared by using morphometric "organelle profiles" (MOP's), i.e., histograms of volume density, numerical density, etc. of mitochondria, smooth endoplasmic reticulum (SER), etc. MOP's of dissociated adrenocortical cells: The lack of an ACTH effect on mitochondrial volume and surface density is consistent with the view that ACTH increases steroid output by increasing the activity, but not the amount of the mitochondrial enzyme which participates in pregnenolone synthesis. In contrast, the increases in the SER parameters induced by ACTH suggest a concurrent increase in the amount of enzymes associated with the SER: $3\beta$ -hydroxysteroid dehydrogenase and 21-hydroxylase. The roughly equal decrease in lipid droplet volume density and numerical density indicates a depletion almost entirely due to disappearance of lipid droplets, rather than to their diminution in size. MOP's of cultured skin fibroblasts from normal and homozygotic familial hypercholesterolemic (FH) patients: The almost twofold increase in the SER volume and surface density of FH cells agrees with reports of their increased cholesterol production.

INTRODUCTION

Our recent work has been based on the rationale that structural changes in cell organelles that can be detected by morphometry may reflect quantitative changes in metabolic activities characteristic of those organelles. We hope that such implied metabolic alterations will aid in the understanding of certain cellular mechanisms and thus lead to the design of therapeutic measures for diseased cells.

We have used morphometry as an indirect metabolic probe in our studies of the effect of ACTH on rat adrenocortical cells and our studies on the possible changes associated with familial hypercholesterolemia (FH) in human skin fibroblasts.

RAT ADRENOCORTICAL CELLS

Trypsin-dissociated (Sayers, Swallow and Giordano, 1971) rat adrenocortical cells were incubated for two hours with and without ACTH and then prepared for transmission electron microscopy and morphometry (Zoller and Malamed, 1975; Zoller, Gibney and Malamed, 1975). A section through part of a typical cell incubated without ACTH is shown in Fig. 1. The cells appeared qualitatively the same whether or not ACTH was present.

We studied three cell organelles associated with steroidogenesis: The mitochondria and the smooth endoplasmic reticulum (SER) both of which bear enzymes and the lipid droplets which supply cholesterol from which the steroid hormones are made (cf. Malamed, 1975). The results of morphometric analyses (Weibel, 1966, 1969) are shown in Fig. 2. Of the eleven properties measured, five showed significant (Student's T test; $P < 0.05$) ACTH effects. Diameter measurements in the light microscope and similar methods using electron micrographs showed that the volume of cells with ACTH was no different than the volume of the cells without ACTH. Thus the ACTH-induced per-unit-volume changes of the cell organelles were due to changes in the organelles themselves.

ACTH increased mitochondrial numerical density by 20 percent over the control (no ACTH) value, but had no effect on the volume density or surface density of their inner or outer membranes. The SER of ACTH-treated cells was 41 percent higher than controls in volume density and 68 percent higher than controls in surface density. In contrast, lipid droplets were depleted by ACTH: 41 percent in volume density and 34 percent in numerical density.

What do these results mean in terms of the secretory function of the adrenocortical cell? Because ACTH markedly enhances steroid hormone production in preparations such as ours (Malamed, Sayers and Swallow, 1970), the following interpretations are possible:

Mitochondria: The increase in number of mitochondria may be necessary for later increases in mitochondrial volume as reported for long term ACTH-treatment of adreno-cortical cells in situ (Nussdorfer, Mazzocchi and Rebonato, 1971). The lack of an ACTH effect on mitochondrial volume and surface density is consistent with the view that ACTH increases steroid output by increasing the activity, but not the amount of the mitochondrial enzyme which participates in pregnenolone synthesis (Koritz and Kumar, 1970).

SER: The increases in volume density and surface density induced by ACTH suggest a concurrent increase in the amount of enzymes associated with the SER: $3\beta$ -hydroxy-steroid dehydrogenase and 21-hydroxylase (Inano, Inano and Tamaoki, 1969).

Lipid droplets: The 41 percent decrease in volume density and numerical density induced by ACTH agrees well with earlier qualitative observations with the light microscope (cf. Malamed, 1975). Because 75-80 percent of the adrenocortical choles-terol in the rat is in the lipid droplets (Moses, et al., 1969) and cholesterol is the raw material from which steroid hormones are made (cf. Malamed, 1975), this acute deple-tion in lipid droplet volume was expected.

HUMAN NORMAL AND FAMILIAL HYPERCHOLESTEROLEMIC (FH) FIBROBLASTS

The cells used in this study were provided by Dr. Avedis K. Khachadurian from his cultures of skin fibroblasts obtained from four patients with FH and from four normocholesterolemic siblings of these patients or unrelated individuals. FH is a hereditary disorder characterized by elevation of plasma cholesterol and beta lipo-proteins. The FH cells we studied were from FH homozygotes who characteristically succumb to atherosclerotic heart disease at 21 years of age.

The cells were cultured in monolayers in a lipid-rich conventional medium containing 10 percent calf serum and harvested by trypsinization (Khachadurian and Kawahara, 1974). Preparations for transmission electron microscopy and morphometric and statistical methods were similar to those used for adrenocortical cells as described above.

Fig. 3 shows a section through part of a typical skin fibroblast from a normocholes-terolemic individual. Qualitatively it is indistinguishable from a typical skin fibroblast from an FH patient. Morphometric analyses, however, detected differences between normal and FH cells in three of the thirteen organelle properties measured: volume density of the mitochondria, volume density of the SER, and surface density of the SER. Because the total volumes of normal and FH cells were the same, the mitochondrial and SER differences in volume and surface were absolute as well as relative to cell volume.

The mitochondrial volume density of FH cells was 59 percent higher than that of normal cells, thus suggesting an increase in any of the metabolic activities performed by these organelles as for example, electron transport and the synthesis of ATP. Hence it would be tempting to conclude that the FH cells are "burning more fuel" than are normal cells. This may turn out to be true, but other morphometric data argue against this view because all the other mitochondrial properties measured were the same in normal and FH cells. Thus, the significance of the increase in mitochondrial volume of the FH cells is unclear.

The changes in the SER are easier to interpret. FH cells overproduce cholesterol (Khachadurian and Kawahara, 1974) and cholesterol is synthesized by the SER (cf. Malamed, 1975). Thus the 71 percent higher volume density and 90 percent higher surface density of FH cells as compared to normal cells was expected. Of course, these data alone tell us nothing about any possible causal relationships between the structural and functional changes in the SER of FH cells.

## VALUE OF MORPHOMETRIC "ORGANELLE PROFILES" AS METABOLIC INDICATORS

Because the rate-limiting enzymes for steroidogenesis (Karaboyas and Koritz, 1965) are thought to be mitochondrial (cf. Malamed, 1975), ACTH might be expected to produce some increase in mitochondrial properties. Yet our morphometric data show no increase in mitochondrial parameters associated with ACTH-enhanced steroid output. Thus it appears that the "resting" or unstimulated adrenocortical cell has adequate mitochondrial "apparatus" to respond to acute ACTH stimulation and the mechanism of the acute ACTH effect does not appear to involve the production of new mitochondrial material, i.e., enzymes. Rather, our findings in agreement with those of Koritz and Kumar (1970), indicate that ACTH stimulates steroid secretion by activating already available mitochondrial enzymes, perhaps by removing some inhibitory factor(s).

The ACTH-induced changes in the SER and in the lipid droplets as shown by morphometry are interpreted as indices of chemical events. The increases in SER parameters indicate synthesis of steroidogenic enzymes such as $3\beta$ -hydroxysteroid dehydrogenase and 21-hydroxylase (Inano, Inano and Tamaoki, 1969) and the decreases in lipid droplet parameters reflect the depletion of cholesterol used for the formation of the steroid hormones.

The similar decreases in lipid droplet volume density (41 percent) and numerical density (34 percent) induced by ACTH permit an interesting inference about how the cell mobilizes its lipid droplet cholesterol when called upon to produce steroid hormones. It appears that some lipid droplets are depleted entirely and disappear before the choles - terol in other droplets is drawn upon. Were this not the case, there would be no decrease in number of lipid droplets along with the decrease in volume density. Why the cell should use up some droplets completely before starting to use others is puzzling and raises other questions as, for example, how the cell chooses which lipid droplets to use first.

The results of studies on the FH cells so far tell us far less than do those on the adrenocortical cells. At first we thought that FH cells might differ from normal cells in that they were less able to destroy excess cholesterol via action by lysosomes. Accordingly we studied spheromembranous bodies, one form of these cell organelles. However, we found no difference between FH cells and controls in volume or number of spheromembranous bodies.

The large increases in SER parameters of FH cells is to be expected in cells producing excess cholesterol, but the significance of the increase in mitochondrial volume density is unclear. This mitochondrial change revealed by morphometry does, however, suggest that mitochondrial oxidative phosphorylation and other functions be investigated in the FH cells.

Although our morphometric studies with these two cell systems are still at their earliest stages, the results so far indicate to us that morphometry is useful as a kind of indirect cytochemistry. Ultimately it is hoped that this use of morphometry may have value as a diagnostic tool in early detection of specific diseases.

This work was supported by NIH grant AM 16833, AHA grant GR-74-SOM-2, and National Foundation March of Dimes grant 1-352. Ms. Jean Gibney provided expert technical assistance.

REFERENCES

Inano, H. , Inano, A. & Tamaoki, B. (1969) Submicrosomal distribution of adrenal enzymes and cytochrome P-450 related to corticoidogenesis. Biochim. Biophys. Acta 191, 257.

Karaboyas, G. C. & Koritz, S. B. (1965) Identity of the site of action of 3', 5'-adenosine monophosphate and adrenocorticotrophic hormone in corticosteroidogenesis in rat adrenal and beef adrenal cortex slices. Biochemistry 4, 462.

Khachadurian, A. K. & Kawahara, F. S. (1974) Cholesterol synthesis by cultured fibroblasts: decreased feedback inhibition in familial hypercholesterolemia. J. Lab. Clin. Med. 83, 7.

Koritz, S. B. & Kumar, A. M. (1970) On the mechanism of action of the adrenocorticotrophic hormone. J. Biol. Chem. 245, 152.

Malamed, S. (1975) Ultrastructure of the mammalian adrenal cortex in relation to secretory function. In: Handbook of Physiology; Endocrinology VI (Ed, by G. Sayers, H. Blaschko and A. D. Smith), pp. 25-39. American Physiological Society, Washington.

Malamed, S., Sayers, G. & Swallow, R. L. (1970) Fine structure of trypsin-dissociated rat adrenal cells. Z. Zellforsch. 107, 447.

Moses, H. L., Davis, W. W., Rosenthal, A. S. & Garren, L. D. (1969) Adrenal cholesterol: localization by electron microscope autoradiography. Science 163, 1203.

Nussdorfer, G., Mazzocchi, G. & Rebonato, L. (1971) Long term trophic effect of ACTH on rat adrenocortical cells. Z. Zellforsch. 115, 30.

Sayers, G., Swallow, R. L. & Giordano, N. D. (1971) An improved technique for the preparation of isolated rat adrenal cells. Endocrinology 88, 1063.

Weibel, E. R. (1966) Practical stereological methods for morphometric cytology. J. Cell Biol. 30, 23.

Weibel, E. R. (1969) Stereological principles for morphometry in electron microscope cytology. Int. Rev. Cytol. 26, 235.

Zoller, L. C. & Malamed, S. (1975) Acute effects of ACTH on dissociated adrenocortical cells: quantitative changes in mitochondria and lipid droplets. Anat. Rec. 182, 473.

Zoller, L. C., Gibney, J. & Malamed, S. (1975) Acute effects of ACTH on dissociated adrenocortical cells: quantitative changes in mitochondria and smooth endoplasmic reticulum. Anat. Rec. 181, 517.

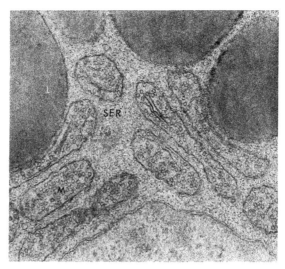

Fig. 1. Electron micrograph of a section through a trypsin-associated rat adreno-cortical cell. M, mitochondrion; SER, smooth endoplasmic reticulum; L, lipid droplet; N, nucleus. X 23,500.

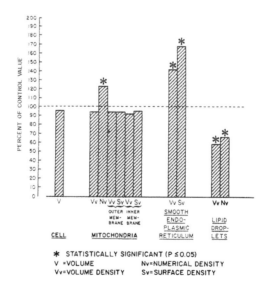

Fig. 2. Morphometric organelle profiles of trypsin-dissociated rat adrenocortical cells incubated with ACTH (1000 microunits; 2 hours).

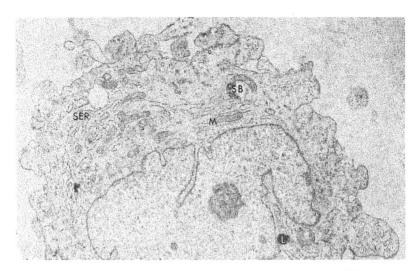

Fig. 3. Electron micrograph of a section through a human skin fibroblast. M, mitochondrion; SER smooth endoplasmic reticulum; L, lipid droplet; SB, sphero-membranous body. X 7200.

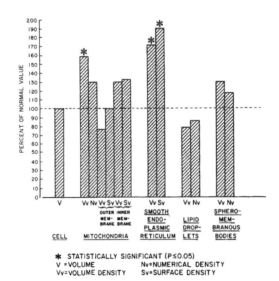

Fig. 4. Morphometric organelle profiles of cells from homozygotes with familial hypercholesterolemia.

*National Bureau of Standards Special Publication 431*
*Proceedings of the* FOURTH INTERNATIONAL CONGRESS FOR STEREOLOGY
*held at* NBS, Gaithersburg, Md., September 4-9, 1975 (Issued January 1976)

A QUANTITATIVE ANALYSIS OF SOME ULTRASTRUCTURAL ASPECTS OF SEED DEVELOPMENT

by Colin E. Hughes and Lewis G. Briarty
*Botany Department, School of Biological Sciences, University of Nottingham, Nottingham NG7 2RD, England*

SUMMARY

The application of straightforward stereology techniques in the analysis of developing seed structure provides accurate information on morphogenesis. Studies on the endosperm of developing wheat seeds show that the amyloplasts, the organelles in which starch granules are deposited, divide only during the very early development of the tissue, and that subsequently the amyloplasts are partitioned among the still dividing cells. Some information on starch granule synthesis is provided by the correlation of starch build-up with membrane changes within the amyloplasts.

Changes are also noted in the S/V of the RER in relation to reserve protein synthesis in both wheat and French bean; a high value for S/V is achieved before protein synthesis begins, this then drops fairly rapidly as the protein is produced. Nuclear volume remains a constant proportion of the cell volume, even though the latter changes markedly.

The significance of these points is discussed in relation to the use of stereology in the early analysis of the effects of genetic manipulation.

INTRODUCTION

The fine structure of developing seeds of economic importance, in particular cereals and legumes, is a research area receiving increasing investigation as the importance of such seeds as primary sources of nutritional protein, carbohydrate and lipid is realised. In this context we have been working on two seed types, the French bean (*Phaseolus vulgaris*: Briarty, 1973) and wheat (*Triticum aestivum*) in an attempt to quantify the changes in ultra-structure which occur in the cells of the seed storage tissues throughout the entire period of seed development. Both of these seeds produce significant amounts of starch and protein, and it was with the hope of obtaining a better understanding of the mechanism by which these materials are synthesised that the studies were undertaken.

The results to be described were all obtained using material fixed in glutaraldehyde/OsO$_4$ and embedded in epoxy resin using standard procedures: quantitative data was obtained by point-counting with light and electron microscopy using techniques described by Weibel (1969), and all data was processed using the Pocoster program (Gnägi *et al.*, 1970).

WHEAT GRAIN DEVELOPMENT

In both bean and wheat the storage tissue development follows a similar pathway, there is an initial phase of cell division during which the cells are fairly vacuolate, and this is followed by a phase in which the bulk of the reserves is laid down. Starch is deposited in amyloplasts, membrane-bound organelles similar to chloroplasts, while protein is deposited in small vacuoles in the cytoplasm. During this phase of reserve deposition there is massive cell enlargement and the phase is terminated at maturity by dehydration of the seed.

Starch is the major reserve in the cells of the wheat grain, and the greater part of the starch is present at maturity in the form of large, lenticular granules (A-type), each granule lying within its own amyloplast. Thus the number of granules present in the mature grain, and the consequent yield of the seed, is a function of the number of amyloplasts produced in each cell, and of course the number of cells in the grain. One of the factors affecting starch yield in the mature grain is therefore the number of amyloplasts present in the individual cells during the phase of cell division, when the cell contents are divided up between the daughter cells. It is generally accepted that amyloplasts themselves undergo division, thus restoring the organelle

385

numbers to some extent after cell division.

A plot of the number of A-type starch granules per cell (Fig.1) shows a steady drop to a plateau after the seed reaches an age of 12 days. This corresponds to the age at which cell division stops in the grain: the indication is therefore that after this time there is no increase in the number of A-type amyloplasts, and thus of starch granules, in each cell of the grain - a suggestion which has been made on other evidence by.Evers (1974). At what stage of development, then, do the amyloplasts stop dividing? Do they them-. selves divide as the cells divide, maintaining the number per cell to some extent, or are they distributed passively to the dividing cells?

A plot of the total number of amyloplasts present per grain (Fig.2) shows a fairly constant value as early in development as our data so far extends, some 6 days after fertilisation, even though cell division is continuing fairly rapidly. It therefore appears that, over this period at least, there is no multiplication of the amyloplasts, but that these organelles are simply distributed to the daughter cells as the latter divide. There is little A-type starch granule initiation after about 6 days after fertilisation: the number of A-type starch grains which a cell produces is presumably fixed in the very early stages of grain development.

*Starch Synthesis*

The mechanism of starch granule deposition is the subject of a great deal of debate (Evers,1974: Geddes,1969). While some of the biochemistry is well known, the sites of action of the responsible enzymes within the amyloplast are unknown. A study of the changes in volume of the various components of the amyloplast during starch synthesis provides indications on this latter point.

The major component of the organelle, apart from the starch, is a system of tubules which run around the periphery of the granule (Fig.3). A plot of starch granule volume against the volume of the tubules shows a straight line relationship throughout the developmental stages examined (Fig.4), and it is tentatively suggested that some of the enzyme systems responsible for starch synthesis may be associated with the tubular membrane system of the amyloplast. (The S/V of the tubules remains virtually constant over the period examined.)

*Protein Synthesis*

In both wheat and bean seeds the synthesis of the protein reserves begins somewhat later than starch synthesis. It is likely that in both species the reserve protein is synthesised on the rough endoplasmic reticulum (RER) membranes, and this idea is strengthened by the apparent increase in surface area of the RER which parallels the development of storage protein (Fig.5).

A parameter which has not previously received attention, however, is the S/V of the RER in relation to its role in protein synthesis. The data presented above (Fig.5) indicate that, prior to the onset of protein synthesis, the S/V increases rapidly then drops as protein synthesis commences. The increase in S/V indicates a flattening of the RER cisternae while the decrease denotes a swelling, perhaps as a result of the accumulation of. synthesised protein within the membranes. The path by which synthesised protein moves to its site of deposition within the cell is unknown in these species, and this work provides at least circumstantial evidence that part of that path lies in the RER cisternae.

The similarity of the conformational changes in the RER during protein synthesis in these widely different species suggests that the phenomenon is perhaps more widespread, and deserving of further study.

Another parameter of interest in these developing cells is the $V_v$ of the cell occupied by the nucleus. In wheat the cell volume increases 6-fold, and in bean 43-fold over the periods studies. In spite of these large variations, however, the $V_v$ occupied by the nucleus remains fairly constant at $3 \pm 1\%$ in wheat and $4.5 \pm 0.7\%$ in bean: the nuclear volume thus maintains a fairly close relationship with the cell volume. The immediate interpretation of this is that an increase in nuclear volume, and therefore presumably in activity, is required to produce and/or to maintain control over a cell of increasing volume. A similar relationship between nuclear DNA content and cell volume has been noted by Smith (1973).

CONCLUSIONS

The results described above provide data on a number of parameters of developing seeds, relating in particular to the mechanisms of cell growth. They also suggest possible points in the growth cycle at which outside influences might effect some degree of crop improvement. It appears that the overall starch yield in wheat, for example, is a function of many variables including the number of cell divisions occurring in the seed, the number of times that individual amyloplasts divide, and the number of amyloplasts inherited by the gametes from the parents, as well of course as the nutritional state of the parent plant during seed development. Morphometric stereology, providing quantitative data on cellular morphogenesis, could be used to monitor changes produced by genetic manipulation at a level relatively close to the controlling DNA.

ACKNOWLEDGEMENTS

Financial support from the Science Research Council and the Flour Milling & Baking Research Association is gratefully acknowledged. The authors wish to thank Dr.H-R.Gnägi for material help in application of the Pocoster program, and Tony Evers (F.M.B.R.A.) for valuable advice and discussion.

REFERENCES

Briarty, L.G., 1973. *Stereology in Seed Development studies: Some preliminary work.* Caryologia 25 supp., 289-301.

Evers, A.D., 1974. *The development of the grain of wheat.* Proc. 4th Int. Cong. Food Science and Technology. Madrid, 1974. (In press.)

Geddes, R. 1969. *Starch Biosynthesis.* Quarterly Reviews. Chem. Soc. London 23, 57-72.

Gnägi, H-R., Burri, P.H. & Weibel, E.R., 1970. *A multipurpose computer program for automatic analysis of stereological data obtained on electron micrographs.* 7 ème Cong. Internat. de Micr. Electronique, Grenoble, Vol.I, 443-444.

Smith, D.L., 1973. *Nucleic acid, protein and starch synthesis in developing cotyledons of Pisum arvense L.* Ann. Bot. 37, 795-804.

Stamberg, O.E., 1939. *Starch as a factor in dough formation.* Cereal Chem. 16, 769.

Weibel, E.R., 1969. *Stereological principles for morphometry in electron microscopic cytology.* Int. Rev. Cytol. 26, 235-302.

FIGURE LEGENDS

Fig. 1 Starch granule numbers per cell in developing wheat seed.

Fig. 2 Total number of cells (•) and amyloplasts (o) in developing wheat seed.

Fig. 3 Amyloplast in developing wheat seed showing starch and membrane tubules (Scale = 0.5μm).

Fig. 4 Starch granule volume per amyloplast plotted against tubule volume per amyloplast in developing wheat seed. (Numbers = age, days after fertilisation)

Fig. 5 Changes in S/V (•) and area (▲) of RER, and protein content (■) in developing seeds of (a) bean and (b) wheat.

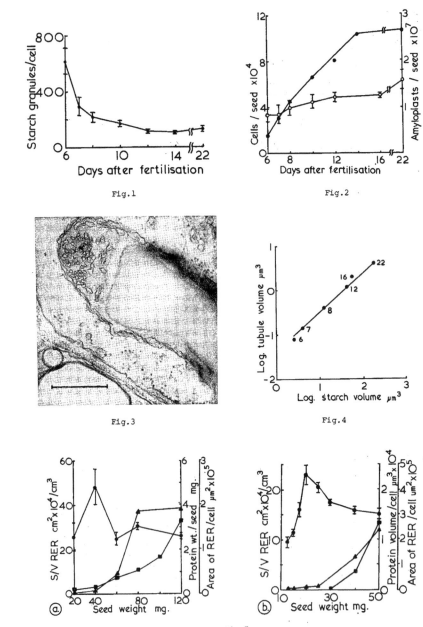

Fig.1

Fig.2

Fig.3

Fig.4

Fig.5

*National Bureau of Standards Special Publication 431*
*Proceedings of the* FOURTH INTERNATIONAL CONGRESS FOR STEREOLOGY
*held at* NBS, Gaithersburg, Md., September 4-9, 1975 (Issued January 1976)

THE BRAIN AS A "STEREOLOGICAL DEVICE"

by T. Radil-Weiss and J. Radilová
*Institute of Physiology, Czechoslovak Academy of Sciences, Prague, Czechoslovakia*

It is probable that stereological ideas will be soon adopted
in the study of visual perception as the brain is functioning
in a sense as a "stereological device" by which three-dimensional judgments can be performed on the basis of two-dimensional
representations of the external world upon the retina. The aim
of the present paper is to report on the results of experiments
in which three-dimensional perceptual interpretation of plane
geometrical figures by humans (i.e. a subjective phenomenon)
is studied by means of objective psychophysiological and quantitative techniques.


METHODS
Reversible figures (Fig.1) well known in experimental
psychology are used as stimulus material. Their basic feature
is that they may be interpreted perceptually in two different
ways. In case these patterns are constantly illuminated both
interpretations alternate spontaneously (the instant of change
is signaled by the subject by operating a switch). When they
are shown for a very short time, for instance by means of a
flash light, the pattern is seen in any trial in one of the two
possible ways only (signaled by pushing one of a pair of knobs).
Time series analysis is performed with this data. In the case
stimulus pattern is illuminated by flash electroencephalographic
evoked potentials from the surface of the skull can be also
recorded, classified into two groups according to the subjective
interpretation of the stimulus (which remains the same physical-
ly) and separately processed for increasing the ratio between
"signal" - the evoked brain potential and the "noise" - the
spontaneous electroencephalographic activity.


RESULTS
The time course of alternation of pairs of perceptual
interpretations of the constantly illuminated reversible figure
shown in Fig.1 C have been analysed under two different conditions (Radilová et al., 1974): either when perceived as a reversible three-dimensional concave or convex object, or when it
was seen (in another experimental session after a different
verbal instruction) as a two-dimensional reversible figure -
background pattern. Although the perceptual process develops
unconsciously, the reversal rate was about two time slower in
case the drawing was interpreted as three-dimensional (Fig.2).
Three-dimensional interpretation of a plane figure requires
probably more elaborate nervous processing needing more time.
Rational explanation of the principle of three-dimensional
interpretation of a more complex two-dimensional reversible
object (Schröder staircase - Fig.1 B) makes the spontaneous
reversal rate faster showing that learned cognitive factors play
an important role in the development of three-dimensional interpretation of plane objects (Radilová, Radil-Weiss,1975).
When a simple reversible figure the Necker cube (Fig.1 A)
was constantly illuminated (Radilová et al., 1972) both types

of intervals were of the same length and the types of interval distributions were almost equal demonstrating that both perceptual interpretations having in this case little symbolic meaning are equivalent. Neighbouring intervals either of the same type or of different type tend to be of similar duration what means that the actual mode of perception depends to some degree from the previous one. In the case of repeated exposure of the same pattern by flash subjective interpretations of the stimulus in consecutive trials were not independent and underlying process could be described on the basis of Markov chain model (Radilová et al., 1973 a). Thus in the case when both three-dimensional interpretations of the two-dimensional pattern are equivalent, the actual way of perceiving the object depends very much from the mode of seeing it in the previous trial.

Evoked brain potentials have been recorded after the repeated presentation of this pattern and separately averaged in two groups according to its actual subjective interpretation (Fig.3). The curves of averaged evoked brain potentials have been found different in a part of subjects showing that different brain processes might be responsible for the two types of three - dimensional interpretations of the same two-dimensional geometrical pattern (Radilová et al., 1973 b, 1975).

REFERENCES

Radilová,J., & Radil-Weiss,T. (1975) Two and three dimensional figure reversals. Activ. nerv. super. in press.
Radilová,J., Radil-Weiss, T. & Krekule, I. (1972) Psychophysiological analysis of the perception of reversible figures. Physiol. bohemosl. 21, 429.
Radilová, J., Radil-Weiss,T. & Havránek, T. (1973 a) Quantitative description of the Necker cube. Physiol. bohemosl. 22, 427-428.
Radilová,J., Radil-Weiss,T. & Špunda, J. (1973 b) Evoked responses induced by presentation of the Necker cube. Physiol. bohemosl. 22, 427.
Radilová, J., Radil-Weiss, T. & Špunda, J. (1974) Objective analysis of subjective perception of reversible figures. Activ. nerv. super. 16 , 297.
Radilová, J., Radil-Weiss, T., Špunda, J. & Indra, M. (1975) Electroencephalographic correlates of the perception of visual pattern. In: Biokybernetik (Eds by H.Drischel, & N. Tiedt), pp 268-272. Gustav Fischer Verlag, Jena.

TEXT TO FIGURES:
Fig.1: Two-dimensional reversible visual patterns used in psychophysiological experiments.
Fig.2: Histograms of intervals corresponding to pairs of three (A) and two-dimensional (B) interpretation of the same reversible figure. Histograms 1 and 2 represent pairs of intervals of both types. On the horizontal axis intervals of different length, on the vertical axis their incidence.
Fig.3: Typical example (at regular stimulus presentation with 5 sec intervals) of the average EEG activity. A - 500 msec before and B - 500 msec after the presentation of the pattern of the Necker cube by means of flash light. Curves (1) and (2) represent averaged bioelectrical activity in cases of the two different subjective three-dimensional perceptual interpretations of the same two-dimensional geometrical pattern. Curve (3) is the arithmetical difference between both.

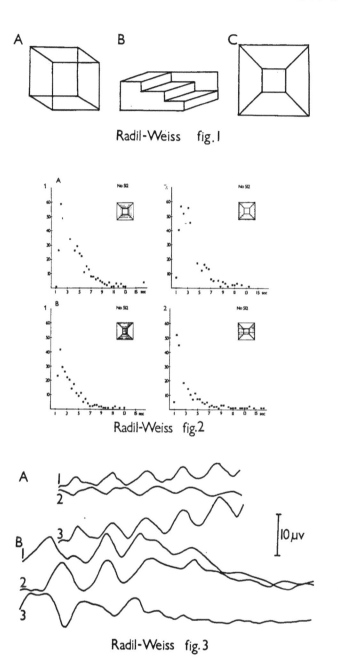

Radil-Weiss   fig. 1

Radil-Weiss   fig. 2

Radil-Weiss   fig. 3

*National Bureau of Standards Special Publication 431*
*Proceedings of the* FOURTH INTERNATIONAL CONGRESS FOR STEREOLOGY
*held at* NBS, Gaithersburg, Md., September 4-9, 1975 (Issued January 1976)

STEREOLOGICAL ANALYSIS OF NEURAL ORGANOGENESIS IN THE CHICK EMBRYO

by O. Mathieu and Paul-Emil Messier
*Departement d'Anatomie, Université de Montréal, C.P. 6128, Montréal, Canada*

ABSTRACT

Stereological methods were used to study the organogenesis of the neural tube in chick embryos at stages 5 to $10^+$. Transverse sections from various levels along the cephalo-caudal axis were analysed. The point-counting method was used to quantify, in light microscopy, the nucleo-cytoplasmic ratio and the volumetric density of the intercellular spaces, nuclei and cytoplasm of cells in the neural tube. Using the electron microscope we have studied the volumetric density and the surface ratio of mitochondria as well as the surface density of the endoplasmic reticulum in the cytoplasm.

INTRODUCTION

During embryonic development undifferentiated cells associate to form definite cellular layers, which are precisely shaped to form specific organs. Stereology is useful in the study of cytodifferentiation as it permits to quantify cellular structures at various stages of development and to follow their fate in time. This paper presents the morphometric analysis of cells of the neurulating chick embryo.

MATERIAL AND METHODS

Chick embryos (Gallus domesticus) were prepared for electron microscopy using usual procedures. Several stages of development from the neural plate to the closed neural tube (stage 5 to $10^+$ according to Hamburger and Hamilton, 1951) were analysed. Transverse sections were cut at defined levels along the neural tube of each embryo, considering the sequential development from cephalic to caudal region. For embryos at stage $8^+$ or older these levels were:
  level 1: anterior to the head      level 2: posterior to the head
  level 3: the first somite          level 4: the last somite
  level 5: the node of Hensen
Table 4 shows which levels were analysed in our younger less differentiated specimens.
The section area of the neuroepithelium was determined by cutting and weighting drawings made by projections of the sections on a constant density paper. The point-counting method of Glagolev (1933), applied on micrographs taken at a magnification of 1000 X with the light microscope, was used to determine 1) the nucleo-cytoplasmic ratio 2) the volume fraction of intercellular spaces, nuclei and cytoplasm in the neural tube.
In electron microscopy, we determined 1) the volume density of mitochondria in the cytoplasm 2) the surface density of the granular endoplasmic reticulum in the cytoplasm 3) the surface to volume ratio of the mitochondria

These parameters were determined according to Glagolev (1933) for volumetric density and according to Saltykov (1958) for surface density.

RESULTS

Table 1 shows the value of the mean $\pm$ standard error obtained at each level of the six embryos of each stage for the area of the section of the neural tube. The statistical analysis (Friedman's m ranking and variance analysis) of these data has shown that there are differences ($p \leq .05$) 1) along the cephalo-caudal axis of embryos of the same stage, stage 5 excepted 2) from early to later stages of development, at any level along the cephalo-caudal axis except for level 3. Table 1 illustrates that at stage $8^+ - 10^+$ the section area of the neural tube decreases from level 1 to level 4 and then increases as the node of Hensen is approached. Between these stages the section area 1) increases anterior to the head (level 1) 2) decreases both posterior to the head (level 2) and in the last somite region (level 4) 3) does not change in the first somite region (level 3).

Furthermore our results indicate that
1) the nucleo-cytoplasmic ratio (Table 2) does not change at any stage along the cephalo-caudal axis, except at stage 7 where it increases ($p \leq .01$) from the back of the head to the node of Hensen. Also,the value of the nucleo-cytoplasmic ratio changes ($p \leq .05$) from early to later stages, at each of the five levels studied.

2) the volume fraction of the intercellular spaces in the neural tube (Table 3) decreases from the cephalic region to the node of Hensen at stages 6 to 8-. It also decreases from level 1 to level 4 whereupon it increases approaching level 5 in embryos aged $8^+$ to $10^+$. Moreover the volume fraction of the intercellular spaces in the neural tube does not change from younger to older specimens at level 1 and 5, while from stage 8- to $10^+$ it decreases ($p \leq .01$) at levels 2, 3 and 4.

3) limited space does not allow us to present our results in a table form, but we found that the volumetric densities of cytoplasm and nuclei both increase ($p \leq .05$) from the head to the node of Hensen at each stage. Moreover from stage 7- to 10- the volumetric density of nuclei in the neural tube increases ($p \leq .01$) at all levels; while the volumetric density of the cytoplasm in the neural tube does not change, except at level 5.

4) none of the parameters considered change along the cephalo-caudal axis of the stage 5 embryos.

Table 5 shows the results obtained at level 2 for the volume fraction of mitochondria in the cytoplasm ($V_V$ mito/$_{cyto}$), the surface density of the endoplasmic reticulum in the cytoplasm ($S_V$ R.E./$_{cyto}$) and the surface to volume ratio of mitochondria ($R_{mito}$). As mentioned above the volume fraction of the cytoplasm in the neural tube did not change at level 2 from stage 5 to $10^+$. The volume fraction of mitochondria and the surface density of the endoplasmic reticulum in the cytoplasm did not change at level 2 from one stage to another. The surface to volume ratio of mitochondria decreases at level 2 along the stage series. This decrease indicates that the size of the mitochondria increases from stage to stage, as there is a negative correlation between the size of a particle and its surface to volume ratio. However, the volumetric density of mitochondria does not change from one stage to another suggesting that mitochondria are less numerous from stage 5 to $10^+$.

DISCUSSION

Our results point to changes in relative volume of some cellular structures as they evolve during neurulation. We find that all parameters remain constant along the cephalo-caudal axis of young specimens (stage 5). When comparing data of a given level from one stage to another, it must be kept in mind that the position of level 2, 4 and 5 moves posteriorly within the embryos from stage 5 to $10^+$. At level 3, we show that from stage 8- to $10^+$ the variations in the total volume of the intercellular spaces does not interfere with the unchanging area of the section of the neural tube. Yet as the neural tube

Table 1: Area $(mm^2)$ of the neural tube section at different stages

| Level | 5 | 6 | 7 | $8^-$ | $8^+$ | 9 | $10^-$ | $10^+$ |
|---|---|---|---|---|---|---|---|---|
| 1 | .0286 | .0217 | .0178 | .0305 | .0375 | .0467 | .0465 | .0596 |
|   | .0034 * | .0025 | .0019 | .0019 | .0011 | .0050 | .0068 | .0056 |
| 2 | NA | .0342 | .0286 | .0402 | .0342 | .0305 | .0226 | .0196 |
|   |   | .0036 | .0014 | .0021 | .0024 | .0014 | .0017 | .0017 |
| 3 | .0293 | .0362 | .0220 | .0218 | .0174 | .0204 | .0173 | .0184 |
|   | .0020 | .0021 | .0026 | .0007 | .0005 | .0019 | .0006 | .0016 |
| 4 | NA | .0329 | NA | NA | .0135 | .0108 | .0082 | .0075 |
|   |   | .0021 |   |   | .0011 | .0004 | .0005 | .0007 |
| 5 | .0248 | .0291 | .0218 | .0219 | .0228 | .0183 | .0156 | .0174 |
|   | .0044 | .0033 | .0029 | .0018 | .0022 | .0014 | .0018 | .0020 |

Table 2: Nucleo-cytoplasmic ratio at different stages

| Level | 5 | 6 | 7 | $8^-$ | $8^+$ | 9 | $10^-$ | $10^+$ |
|---|---|---|---|---|---|---|---|---|
| 1 | .26 | .30 | .27 | .28 | .33 | .32 | .38 | .32 |
|   | .01 | .03 | .02 | .01 | .02 | .01 | .02 | .01 |
| 2 | NA | .28 | .26 | .28 | .31 | .36 | .34 | .35 |
|   |   | .02 | .02 | .01 | .02 | .01 | .02 | .02 |
| 3 | .28 | .28 | .29 | .31 | .31 | .35 | .37 | .36 |
|   | .01 | .01 | .02 | .00 | .01 | .01 | .01 | .01 |
| 4 | NA | .27 | NA | NA | .33 | .35 | .39 | .34 |
|   |   | .02 |   |   | .02 | .02 | .03 | .02 |
| 5 | .28 | .28 | .32 | .29 | .35 | .36 | .36 | .39 |
|   | .02 | .02 | .02 | .02 | .02 | .03 | .02 | .02 |

Table 3: Volume fraction of the intercellular spaces in the neural tube at different stages

| Level | 5 | 6 | 7 | $8^-$ | $8^+$ | 9 | $10^-$ | $10^+$ |
|---|---|---|---|---|---|---|---|---|
| 1 | .18 | .19 | .23 | .21 | .21 | .20 | .21 | .21 |
|   | .02 | .03 | .03 | .01 | .01 | .01 | .01 | .01 |
| 2 | NA | .17 | .20 | .22 | .18 | .17 | .14 | .13 |
|   |   | .01 | .03 | .01 | .01 | .02 | .01 | .01 |
| 3 | .13 | .16 | .17 | .15 | .13 | .14 | .10 | .11 |
|   | .00 | .02 | .01 | .01 | .01 | .01 | .01 | .01 |
| 4 | NA | .14 | NA | NA | .14 | .11 | .08 | .09 |
|   |   | .01 |   |   | .01 | .01 | .01 | .01 |
| 5 | .12 | .13 | .13 | .11 | .12 | .14 | .11 | .14 |
|   | .01 | .01 | .01 | .01 | .01 | .01 | .01 | .02 |

*In each of the above table, the first line of figures= mean of 6 embryos; second line= ± standard error;  NA= value not available (level not considered)

closes, neuroepithelial cells get closer and closer so as to potentiate surface interactions that may be modified from stage to stage. Concomitantly
the nucleo-cytoplasmic ratio increases as a consequence of the increase in the
volume fraction of the nuclei. This may reflect an enlargement in euchromatin
content and therefore point to an increase in DNA replication process during
neurulation.

At level 2 the decrease in the neural tube section area is approximately
50% between stages 8⁻ and 10⁺. This decrease cannot be caused only by the
reduction of the volume fraction of the intercellular spaces as these occupy
respectively 22% and 13% of the volume of the neural tube at stage 8⁻ and 10⁺.
Then it would seem that fewer cells may be implicated in the maturation of the
neural tube at level 2 along the stage series studied. Moreover the nucleo-
cytoplasmic ratio and volumetric density of the nuclei increase in the neural
tube. The increase in DNA replication referred to earlier is not coupled with
an increase in endoplasmic reticulum-related synthesis of proteins since we
have shown that the surface density of the endoplasmic reticulum in the cytoplasm remains unchanged.

In conclusion, our data allow us to draw a curve of the characteristics
of the epithelial cells in the chick embryo undergoing normal neural organogenesis. Use of this curve will enable us to estimate the effect of various
treatments on the chick embryo neuroepithelium provided we know 1) the stage
of the embryo 2) the level which is analysed along the embryo's axis.

This work was supported by the Medical Research Council of Canada and the
Department of Health and Welfare, RODA division.

Table 4: Transverse levels analysed in embryos of stage 5 to 8⁻

| Level | 5 | 6 | 7 | 8⁻ |
|---|---|---|---|---|
| 1 | middle of the head | anterior to the head | | |
| 2 | NA | posterior to the head | | |
| 3 | equidistant be-tween 1 and 5 | 1/3 the distance between 2 and 5 | somite | second somite |
| 4 | NA | 2/3 the distance between 2 and 5 | NA | NA |
| 5 | node of Hensen | | | |

NA: not available (level not considered)

Table 5: Volumetric density of the mitochondria in the cytoplasm,
surface density of the endoplasmic reticulum and surface to volume
ratio of the mitochondria at level 2 in embryos of different stages

| Parameter | 5 | 6 | 7 | 8⁻ | 8⁺ | 9 | 10⁻ | 10⁺ |
|---|---|---|---|---|---|---|---|---|
| $V_V$ mito/cyto | 5.8 .6 | 6.2 .4 | 6.2 .3 | 6.1 .4 | 6.1 .2 | 7.2 .3 | 6.6 .3 | 6.7 .1 |
| $S_V$ R.E./cyto | .37 .06 | .39 .05 | .37 .04 | .35 .05 | .35 .03 | .51 .04 | .37 .02 | .48 .11 |
| $R$ mito | 10.3 .4 | 9.4 .4 | 8.7 .2 | 8.9 .5 | 8.7 .5 | 8.4 .1 | 8.6 .2 | 8.0 .3 |

For each parameter, first line of figure = mean of 6 embryos, second
line = ± standard error

BIBLIOGRAPHY
Glagolev, A.A. (1933): Trans. Inst. Econ. Min. Moscow, 59
Hamburger, V. and H.L. Hamilton (1951): J. Morphol., 88: 49
Saltykov, S.A. (1958): "Stereometric Metallography", second edition, 446 p.,
Moscow

National Bureau of Standards Special Publication 431
Proceedings of the FOURTH INTERNATIONAL CONGRESS FOR STEREOLOGY
held at NBS, Gaithersburg, Md., September 4-9, 1975 (Issued January 1976)

APPLICATION OF STEREOLOGICAL METHODS TO THE STUDY OF PRE-IMPLANTATION EMBRYOGENESIS IN MICE

by Russell L. Deter,
*Department of Cell Biology, Baylor College of Medicine, Houston, Texas 77025, U.S.A.*

## ABSTRACT

Application of stereological methods to the study of mammalian preimplantation embryogenesis requires control of the morphogenic process, visual access to the embryos during development, identification of individual embryos and adequate evaluation of sampling procedures. In preparation for a study of early mouse embryogenesis in vitro, methods have been developed which achieve these objectives. Control of embryogenesis through the blastocyst stage has been obtained by use of a perfusion culture system in which the parameters affecting development are defined. The culture chamber of this system permits continuous observation of embryo development with any type of optical microscope. In this system, up to 97% of 2-4 cell embryos have developed into morulae or blastocysts in 48 hours. Culturing of embryos in compartments formed by fibers of a nylon mesh has permitted identification of individual embryos by position. A method for embedding embryos in plastic while they are still in the mesh has been developed. Because individual embryo identification and orientation are conserved, evaluation of the response of two cell embryos to the preparative procedure has been possible. This in turn has led to selection of conditions which minimize dimensional distortion. Specification of section samples has been made possible by the definition of a coordinate system based on the structure of the nylon mesh. This coordinate system has been used to design a sectioning procedure which permits location of individual sections within the embryo. These sections contain a complete cross-section of the embryo and can be studied in their entirety with either the light or electron microscope. Analysis of any level of structural organization and evaluation of sample adequacy can be carried out using the information contained in these sections.

## INTRODUCTION

Mammalian embryogenesis prior to the time of implantation in the uterus is characterized by multiple cell divisions, cavitation of the resulting cell mass (blastocyst formation) and differentiation of the first specific cell population the trophoblasts. These events, which usually occur within the lumen of the oviduct, have been the subject of many morphological studies with both the light and electron microscopes. Such studies have been greatly facilitated by the development of methods which can support embryogenesis to the blastocyst stage in vitro.

As a consequence of previous investigations, an outline of the morphogenetic events responsible for the transformation of a fertilized zygote into an implanting blastocyst has been obtained. However, this outline is essentially descriptive as there has been no previous attempts to quantitate morphological changes or assess their variability. To make possible a quantitative morphological analysis of embryo development, the following series of investigations were undertaken.

## OBJECTIVES

Quantitative morphological studies of pre-implantation embryogenesis require that the following four objectives be realized:
1) embryogenesis must occur in a controlled environment to reduce the number of variables affecting development
2) visual access to embryos during their maturation must be possible so that the morphogenic process can be monitored and appropriate stages selected for further study

3) individual embryo identification must be maintained during all stages
   of the investigation to allow assessment of variability in the developmental process and correlation of information obtained from both living
   and sectioned embryos
4) evaluation of the adequacy of samples used in stereological analyses
   must be possible

Procedures developed in our laboratory have achieved these objectives.

## METHODS

### Embryos

The mouse embryos used in the development of the system to be described were obtained by inducing superovulation in the immature BALB/C mice with
pregnant mare's serum (PMS) and human chorionic gonadotropin (HCG), followed by
mating with adult 129 males (Gateş, A.H. Methods in Mammalian Embryology, 1972
p. 62). Two to four cell embryos were collected by flushing the oviduct with
Whitten's defined medium (Whitten, W.K., Adv. Biosci. 6:131, 1970), equilibrated
with 5% $O_2$ - 5% $CO_2$ - 90% $N_2$, 44 hours after HCG injection.

### Culture System

To obtain embryo development under controlled conditions and constant
visual access, a perfusion culture system built around the chamber designed by
Dvork and Stotler (Dvork, J.A. and Stotler, W.F., Exptl. Cell Res. 68:144, 1971)
has been developed. Embryo culture is carried out in a small space (volume:0.2
ml) formed by two coverglasses separated by a spacer ring which is located in
the center of the chamber. This space is perfused (1 ml/hr) with Whitten's medium by means of a peristaltic pump. The chamber is placed on the stage of bright
field microscope and its temperature maintained at $37^{o}C$ using a Nicholson air-
stream stage incubator (Model C300).

As embryos can be swept out of the chamber by the moving medium, a
means for restraining them must be used. For this purpose a piece of nylon mesh
(1.0 x 0.5 cm), bonded to a polystyrene or polyester coverslip by heating, has
been found to be most satisfactory. The mesh fibers form small compartments
(100 x 100 u) into which individual embryos fall when placed over the mesh in
suspension. Friction between the zona pellucida and the fibers of the mesh prevent both rotation and mechanical displacement if the mesh is stretched sufficiently.

### Preparation of Embryos for Light and Electron Microscopy

High resolution light and electron microscopy can be carried out only
on specimens considerably thinner than single embryos or even one blastomere.
Therefore, microscopic study of the embryogenic process requires sectioning.
This is possible if embryos are fixed and embedded in plastic. Fixation of
mouse embryos has been carried out using 4% gluteraldehyde followed by 1% osmium
tetroxide, both buffered in 0.1 M cacadylate buffer, pH 7.3. Embryos were then
stained with uranyl acetate, dehydrated in ethanol and embedded in the low vis-
cosity epoxy resin described by Spurr (Spurr, A.R. J. Ultrastruc. Res. 26:31,
1969). These procedures were carried out on embryos confined to compartments of
a nylon mesh bonded to a polyester coverslip. An evaluation of the effects of
this procedure on two cell embryos was made by determining the perpendicular distance from the center of the cleavage plane to the periphery of one blastomere
after each processing step. These measurements, made on photomicrographs, were
compared and the percent change calculated.

### Sectioning

The key to sample evaluation lies in the process of sectioning. An
important requirement for such evaluations is the ability to specify section
position within the intact embryo. This can be done with the following sectioning procedure by definition of an appropriate coordinate system. If one

mesh fiber of the square containing the embryo is considered to lie on the X axis, a line perpendicular to that fiber (usually parallel to another mesh fiber) will represent the Y axis. This X-Y plane can be made congruent with the coverslip underlying the mesh and the origin placed under one corner of the square. A line perpendicular to this plane and passing through the origin is then the Z axis. Using an LKB Pyramitome (Model 11800), it is possible to construct a cube whose face lies in the X-Z plane and whose right side lies in the Y-Z plane. The other three sides are perpendicular to the face and cut to the point where sections parallel to the face (X-Y plane) are just large enough to contain the maximal embryo profile and the polyester-epoxy resin junction. For electron microscopy, ribbons of such sections are first picked up on an uncoated slot grid (sections float on a water droplet) which is then inverted and touched to the Formvar coated surface of a second grid. This procedure avoids section folding which is often found following section transfer with usual techniques. Sections on the coated slot grid can be seen in their entirety if the slot has dimensions of 0.2 x 1.5 mm.

### Microscopy
Sections prepared as just described can be studied with both the light and electron microscopes following staining. A RCA-3G electron microscope, modified to operate between 700X and 10,000X and having a 350 negative roll film camera, has been used to obtain negatives of sections at both low and high modifications. Stereological procedures are carried out on the projected image of these negatives using a specially designed projection system (Deter, R.L., J. Microsc. 100:341, 1974).

### RESULTS AND DISCUSSION

#### Embryo Development
With this perfusion culture system, approximately 80% of the 2-4 cells embryos develop into morulae or blastocysts when all conditions investigated are considered together. This value can be increased to 97% with selection of appropriate parameters such as duration in culture, embryo stage at beginning of culture, etc. No attempt has been made to maximize blastocyst development but 70% of the embryos reached this stage in some instances. The use of a defined medium, which is replaced completely every 12 minutes, and an accurately controlled environmental temperature permit pre-implantation embryogenesis to proceed in a very reproducible manner.

#### Visual Access
Because of the special design of the Dvork-Stotler chamber, any type of optical microscopy can be used to study embryos as they develop in culture. Polystyrene coverslips do not interfer with phase, interference contrast or bright field microscopy. The polyester coverslips, on the other hand, cannot be used with phase objectives. Because the spacer ring and the two coverglasses are less than 1.2 mm thick, high magnification objective lens can be used.

#### Individual Embryo Identification
The sequestration of single embryos within compartments formed by the fibers of nylon mesh provides a means for permanently separating embryos from one another. As the embryo's position in the mesh usually remains unchanged throughout the culture period and even during fixation and embedding, data on individual embryos can be collected by both direct observation and from studies on sections. This attribute allows an evaluation of variability in the morphogenic process at all levels of organization.

#### Sampling
The ability to keep track of individual embryos throughout their prep-

aration for sectioning allows correlation of information obtained from studies of sections with that obtained from living embryos. However, sectioning is a sampling procedure and is carried out after the embryo has been subjected to extensive chemical treatment. The usefulness of such correlations therefore depends upon knowledge of the effects of the preparative procedure (which could introduce distortion) and the representativeness of the section sample taken for study. Examination of our procedure for fixing and embedding embryos has revealed a consistent embryo shrinkage during uranyl acetate staining and alcohol dehydration. This effect could only be counteracted by allowing swelling to occur during fixation. With proper balance of the swelling and shrinking artifacts, embryos having dimensions similar to those in the living state (% change for unfixed vs. embedded: $-3.2[\pm 2.4$ S.D.$]$%) can be obtained. The variability among embryos indicates that evaluation of the response of each embryo to the preparative procedure is required if major distortions are to be avoided.

Evaluating the representativeness of samples requires information about the size, shape, orientation and spacial distributions of the objects being studied and therefore must be decided in each individual case. However, specification of the location of sections taken for study is an essential part of this process. With the fixed coordinate system and sectioning procedure previously described, such specification is possible because each individual section can be related to the Y axis. By keeping a record of section numbers and thicknesses (determined from interference colors) starting with the first section to contain a part of the embryo, a Y coordinate value can be obtained for each section in the sample. These can be located on a Y axis superimposed on a micrograph made of the embedded embryo prior to sectioning. Perpendiculars drawn through each of these points intersect the image of the embryo and indicate where the embryo was sampled. If some specific distribution of sections is desired, appropriate Y coordinates could be chosen prior to sectioning.

With the section sample selected, subsequent sampling and measurement problems become those common to all stereological studies carried out on sections. Since the entire cross section of the embryo can be seen in the electron microscope, any level of embryo organization can be sampled, with the added advantage of high magnification for identification of boundaries or specific structures. X and Z coordinate values can be determined for any part of the section by measuring the perpendicular distance to the right hand side of the section (X) or to the polyester-epoxy resin junction (Z). With the flexibility in sampling provided by these procedures, many types of quantitative morphological studies should be possible.

*National Bureau of Standards Special Publication 431*
*Proceedings of the* FOURTH INTERNATIONAL CONGRESS FOR STEREOLOGY
*held at* NBS, Gaithersburg, Md., September 4-9, 1975 (Issued January 1976)

THE DETERMINATION OF SIZE DISTRIBUTION ON LYMPHOBLASTS IN ACUTE LEUKEMIA

by E. Feinermann and G. A. Langlet
*Centre d'Etudes Nucléaires de Saclay, Division de Chimie, Service de Chimie-Physique, B. P. 2*
*91190-Gif-sur-Yvette, France*

Stereology is the study of the structure of matter in three dimensions at the microscopic level, based on the examination of two-dimensional sections through the material. According to this definition (E. Underwood), we have entered upon the size distribution of leucocytes within the study of acute lymphoblastic leukemia.

The stereological study of the leucocytes assumed spherical, for a group of patients suffering from this disease has enabled the population distribution of the diameter to be determined. Thin sections obtained by classical procedures of embedding, cutting and staining , have been examined with a Quantimet.

Several procedures have been suggested for determining true-size distributions of spherical particles. Many biologists and metallurgists concerned with the third dimension of cells and materials, approached this question with stereological parameters (Saltykov, 1958; Fisher and Cole, 1968; Underwood, 1969; Weibel, 1969).

First of all, cells having diameters inferior to $D_i = 5.2$ microns are not counted. Let $D_r$ be the real diameter and $D_m$ the apparent mean diameter of spherical cells.

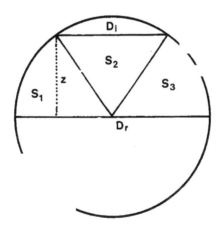

FIG. 1

Dm is given by ratio $S/z$ if S is the area of the circular segment limited by Dr and Di on Fig. 1 , and z the height of this segment.

Area S is the sum of S1 , S2 and S3 ;

$$S1 = S3 = \pi \frac{Dr^2}{4} \frac{\alpha}{2\pi} \qquad\qquad S2 = \frac{Di\ z}{2}$$

then : $S = \dfrac{Di\ z}{2} + \dfrac{Dr^2 \alpha}{4}$

and : $Dm = \dfrac{S}{z} = 0.5\ (Di + \dfrac{Dr^2 \alpha}{2\ z})$

If one sets : $x = Di/Dr$ , one gets : $\alpha = \cos^{-1} x$

and : $z = 0.5\ Dr\ \sin\alpha = 0.5\ Dr\ \sqrt{1-x^2}$

so that the formula becomes :

$$Dm = 0.5\ (x\ Dr + \frac{Dr^2\ \cos^{-1} x}{Dr\ \sqrt{1-x^2}}\ )$$

i. e. : $Dm = 0.5\ Dr\ (x + \cos^{-1} x\ /\ \sqrt{1-x^2}\ )$

From Dm and Di, one can deduce Dr. The analysis of this function by a program written in APL (A Programming Language) shows that a good approximation is obtained when using the following empirical formula :

$$Dr = (Dm - 0.24\ Di)\ /\ 0.73$$

with $6.5 < Dm < 10.5$ and $4.5 < Di < 6.5$ (microns), so that :

$$Dm = 0.73\ Dr + 0.24\ Di$$

Di would be equal to zero if all the cells could be counted. Then the formula would become : $Dr = 4\ Dm\ /\ \pi$

From the apparent diameters which have been measured, one may get the histogram of frequencies. A triple smoothing on three values at each time provides a histogram which obviously exhibits a bi-modal distribution. The observed maxima of the peaks are supposed to be close enough to mean apparent diameters to allow the derivation of the real diameters and hence of the volumes of the cells. The bi-modal distribution of the volumes seems to consist in the sum of two gaussians. It has been shown that, for all sufferers of Acute Lymphoblastic Leukemia whose cells have been examined, the apparent diameters of the majority of cells correspond to one of the two peaks at 9.2 and 10.9 microns, which leads to real volumes of 740 and 1280 cubic microns and real diameters of 11.2 and 13.4 microns respectively.

R E F E R E N C E S

1.    S.D. WICKSELL
      "The Corpuscle Problem. I. Case of Spherical Corpuscles".
      Biometrika, 18, (1926).

2.    S.A. SALTYKOV
      Stereometrie Metallography, second edition, Moscow
      Metallurgizdat (1958).

3.    S.A. SALTYKOV
      "The Determination of the size Distribution of Particles
      in an opaque Material from a Measurement of the size
      Distribution of their Sections".
      Stereology, edited by H. Elias, Proc. Second. Int.
      Cong. for Stereology, New York. Springer - Verlag
      (1967), 163.

4.    E.E. UNDERWOOD, Particle - Size Distribution, Chapter 6
      p. 149, Quantitative Microscopy (DeHoff and Rhines,
      eds) 1970.

5.    C. FISHER and M. COLE
      The metals research Image analyzing computer, the
      Microscope 16 p. 81 (1968).

6.    E.R. WEIBEL
      Stereological principles for morphometry in electron
      microscopic stereology
      Int. Rev. Cytol. 26, 235.

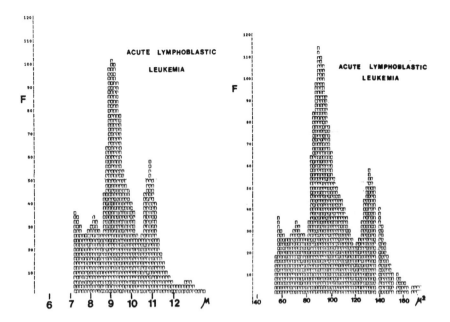

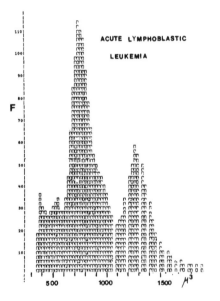

Fig. 2.

Diameter, Surface and Volume
Distribution Curves.

National Bureau of Standards Special Publication 431
Proceedings of the FOURTH INTERNATIONAL CONGRESS FOR STEREOLOGY
held at NBS, Gaithersburg, Md., September 4-9, 1975 (Issued January 1976)

A MORPHOMETRIC STUDY OF HUMAN PLACENTAE OF DIFFERENT GESTATIONAL AGES FROM NORMAL AND
TRIPLET PREGNANCIES

by Indra Bhargava,* K. Kamashki** and Yadollah Dodge,*
*Jundi Shapur University, Ahwaz, Iran
**Jawaharlal Institute of Postgraduate Medical Education and Research, Pondicherry, India

Bhargava (1969) has postulated a selective elongation of either component
of the fetal blood vessels of the placenta, as a common denominator in many
abnormal states of gestation and development. The present study is directed
towards the search of a morphometric confirmation of the above postulate.

MATERIAL AND METHODS:- The material for study comprises 10 placentae
consisting of early premature, - 10 weeks - 2, premature, non viable, - 24
weeks - 2, premature, viable, - 26 weeks - 2, fullterm 38 weeks - 2 and fullterm triplets 38 weeks - 2.

These placentae were collected at cesarean section and fixed in 10 %
formaline for 24 hours. Each placenta was divided into 1 c.c. blocks - 250
to 300. Out of these, 20 blocks were selected at random and refixed for
further 48 hours. These were embedded and sectioned at 5 u. 25 sections were
randomly selected from each block and were stained with H & E.

Observations regarding volume, area and length per unit volume - c.mm.,
of different structural elements of fetal placenta, namely villi and their
components - blood vessels, trophoblast and stroma, intervillous space and
fibrin, were recorded on 5 fields from each section by point count volumetry
and analysis of linear intercepts, using coherent multipurpose test system of
Weibel (1966).

Significance of differences between placentae of different groups, in
relation to various parameters of each structural element, was assessed by
means of the Least Significant Difference Test (Miller 1966), consisting of
two stages
   1. Testing Null hypothesis by appropriate F test
   2. Testing each single comparison with significant
    F value, by appropriate T test.

Presence of a significant correlation between the measurements of
different structural elements of the same placenta was investigated by
computing the coefficient of correlation 'r' from covariance and variance of
the two variables. The critical value of r for rejection of the null
hypothesis was taken as 0.445 at 5 %, and 0.564 at 1 % level of significance.

OBSERVATIONS AND COMMENTS:- The mean values for volume, area and length
of different structural components of placentae at different stages of a
normal pregnancy and in fullterm triplet pregnancy, per unit volume - c.mm.
of placental tissue, are presented in Table I. These values will provide the
base line for further studies on placentae from abnormal gestation.

A statistical evaluation of the differences between the parameters of
different structural elements of placentae of different groups shows that
significant differences are shown by:-
   1. 10 week placentae in relation to $S_V$ villi, $S_V$ intervillous
    space and $L_V$ blood vessels;
   2. 10 and 24 week placentae in relation to $L_V$ villi, $S_V$ stroma,
    $S_V$ fibrin and $V_V$ trophoblast.

405

TABLE I SHOWING MEAN VALUES OF PARAMETERS OF PLACENTAE IN DIFFERENT STAGES
OF GESTATION

| PARAMETER | | AGE | | | |
|---|---|---|---|---|---|
| | 10 Wk | 24 Wk | 26 Wk | 38 Wk | Triplet 38Wk |
| Villi $V_v$ | 0.57+ 0.04 | 0.52+ 0.03 | 0.65+ 0.04 | 0.52+ 0.03 | 0.50+ 0.04 |
| $S_v$ | 12.86+ 2.09 | 26.99+ 4.71 | 24.69+ 2.39 | 24.31+ 4.18 | 29.72+ 4.38 |
| $L_v$ | 3.48+ 1.04 | 11.27+ 1.92 | 14.74+ 1.60 | 15.08+ 2.29 | 22.22+ 4.44 |
| Blood $V_v$ | 0.41+ 0.05 | 0.36+ 0.03 | 0.42+ 0.03 | 0.43+ 0.07 | 0.33+ 0.04 |
| Vessels $S_v$ | 47.38+ 6.90 | 53.66+ 5.44 | 49.82+ 3.19 | 30.51+ 4.39 | 48.78+ 9.09 |
| $L_v$ | 87.30+ 11.76 | 71.10+ 12.81 | 70.84+ 10.48 | 55.99+ 15.87 | 78.53+ 18.14 |
| Troph. $V_v$ | 0.14+ 0.02 | 0.14+ 0.02 | 0.14+ 0.04 | 0.09+ 0.01 | 0.18+ 0.03 |
| Stroma $V_v$ | 0.19+ 0.05 | 0.21+ 0.03 | 0.21+ 0.03 | 0.39+ 0.02 | 0.17+ 0.07 |
| $S_v$ | 56.49+ 15.06 | 70.84+ 6.87 | 60.68+ 3.88 | 33.83+ 3.44 | 85.22+ 15.09 |
| I.V. $V_v$ | 0.20+ 0.03 | 0.27+ 0.04 | 0.19+ 0.02 | 0.30+ 0.03 | 0.25+ 0.05 |
| Space $S_v$ | 12.91+ 2.09 | 26.74+ 4.38 | 24.09+ 2.42 | 24.31+ 4.18 | 29.72+ 4.38 |
| Fibrin $V_v$ | 0.04+ 0.02 | 0.06+ 0.02 | 0.05+ 0.01 | 0.093+ 0.03 | 0.063+ 0.03 |
| $S_v$ | 2.17+ 1.09 | 3.72+ 1.15 | 1.86+ 0.55 | 5.18+ 1.72 | 4.74+ 2.24 |

$V_v$ – c.mm./c.mm., $S_v$ – sq.mm./c.mm., $L_v$ – mm./c.mm.

From the preceding analysis, the trend of morphometric changes in placentae
with advance of gestation, can be stated as:
    A.  Increase up to 24 weeks, thereafter
         1. Maintained      – $S_v$ intervillous space
         2. Decreased       – $S_v$ villi, $S_v$ blood vessels, $S_v$ stroma
         3. Increased further – $L_v$ villi, $V_v$ stroma, $S_v$ fibrin, $V_v$
                          blood vessels

    B.  Decrease up to 24 weeks, and decrease further – $L_v$ blood vessels
        and $V_v$ trophoblast.

The trend of changes becomes rather abrupt at 26 weeks, the age of
viability of the fetus and correspondingly the period of considerable
adjustments for the placenta. This is shown by significant differences of 26
week placentae in relation to:-
         1. An increase in $V_v$ intervillous space, $V_v$ stroma, $S_v$ fibrin
            and $L_v$ villi.
         2. A decrease in $V_v$ villi, $V_v$ trophoblast, $S_v$ stroma, $S_v$ blood
            vessels and $L_v$ blood vessels.

Out of these, the differences in relation to $L_v$ villi, $S_v$ stroma, $S_v$
fibrin and $V_v$ trophoblast are not shared by 24 week placentae.

The abruptness of these changes is associated with a marked decline in
correlations between different structural elements, as well as their magnitude,
thereby indicating that at this stage, the coordinated process of maturation
of placenta has started making way for the process of senescence.

Correlations among different structural components of placentae of various
groups show a complex pattern and are shown in Table II. It appears that these
correlations and consequently coordinated growth reaches a peak by 24 weeks,
followed by a gradual decline. At 10 weeks, outstanding correlations are seen
in relation to $V_v$ villi, $V_v$ blood vessels, $V_v$ stroma, $S_v$ blood vessels and
$S_v$ intervillous space. At 24 weeks, these are seen in relation to $V_v$ blood
vessels, $V_v$ intervillous space, $S_v$ villi, $L_v$ villi, $S_v$ blood vessels, $L_v$ blood
vessels and $S_v$ stroma. Regarding $V_v$ fibrin, $S_v$ fibrin and $V_v$ trophoblast,
significant correlations are seen at 10 and 24 weeks.

TABLE II SHOWING 'r'-COEFFICIENT OF CORELATION, BETWEEN PARAMETERS OF STRUCTURAL ELEMENTS OF PLACENTAE OF DIFFERENT GROUPS

| | VILLI | | | BLOOD VESSELS | | | TROPH. | STROMA | | I.V.SPACE | | FIBRIN | |
|---|---|---|---|---|---|---|---|---|---|---|---|---|---|
| | $V_V$ | $S_V$ | $L_V$ | $V_V$ | $S_V$ | $L_V$ | $V_V$ | $V_V$ | $S_V$ | $V_V$ | $S_V$ | $V_V$ | $S_V$ |
| | 1 | 2 | 3 | 4 | 5 | 6 | 7 | 8 | 9 | 10 | 11 | 12 | 13 |
| | | | | B0.45 | A0.65 | | A.045 | A0.72 | | B0.57 C0.55 | | A0.52 | |
| 2 | | | B0.56 C0.50 | C0.46 | D0.48 E0.58 | B0.55 | A0.50 B0.48 | A0.45 | E0.74 | | A0.99 B0.84 C0.49 D0.89 | B0.59 | |
| 3 | | B0.56 C0.50 | A0.51 B0.55 | | | | | | | | B0.63 E0.60 | | |
| 4 | | | A0.51 B0.55 | A0.45 B0.55 C0.50 D0.50 E0.49 | | | | | | B0.70 D0.51 | | | A0.57 |
| 5 | A0.65 | C0.46 D0.48 E0.58 | | A0.45 B0.55 C0.50 D0.50 E0.49 | | E0.52 | A0.46 | A0.74 | B0.74 C0.49 E0.90 | | D0.49 E0.58 | | |
| 6 | | B0.55 | | | E0.52 | | | | | | B0.56 | B0.56 D0.45 | |
| 7 | A0.45 | A0.50 B0.48 | | | | | | | A0.57 E0.64 | | A0.51 B0.75 | B0.55 | B0.57 |
| 8 | A0.72 | A0.45 | | | A0.74 | | A0.57 E0.64 | | | | A0.46 | B0.73 E0.45 | B0.65 C0.46 |
| 9 | | E0.74 | | | B0.74 C0.50 E0.90 | B0.52 | | | | | E0.74 | | |
| 10 | B0.57 C0.55 | | E0.60 | B0.70 D0.51 | | D0.60 | | | | | B0.47 | | |
| 11 | | A0.99 B0.84 C0.49 D0.89 | B0.63 E0.60 | | D0.48 E0.58 | | A0.51 B0.56 | A0.46 B0.75 | | B0.47 | | B0.69 | B0.59 |
| 12 | A0.52 | B0.59 | | | | | B0.56 D0.45 | B0.55 | B0.73 E0.45 | | B0.69 | | A0.58 B0.87 D0.65 E0.88 |
| 13 | | | A0.57 | | | | B0.57 | | | | B0.59 | A0.58 B0.87 D0.65 E0.88 | |

A - 10 weeks, B - 24 weeks, C - 26 weeks, D - 38 weeks single and E - 38 weeks triplet placenta

Throughout the gestation, a significant correlation is observed between
1. $V_v$ and $S_v$ of blood vessels,
2. $S_v$ villi, and $S_v$ intervillous space, and
3. $V_v$ fibrin and $S_v$ fibrin.
$S_v$ villi and $S_v$ blood vessels show significant correlations only in the later part of gestation.

Thus, in the placenta, villi and their blood vessels appear to have the most coordinated growth. In general, 24 week placenta shows the most coordinated growth, with 10 week, triplet, 26 and 38 week placentae following in a descending order.

Triplet placentae of 38 weeks show significant differences in relation to all the structural elements, when compared to a single placenta of similar age. These placentae have similarities with
1. 10 week placenta in relation to $S_v$ villi, $S_v$ intervillous space and $L_v$ blood vessels.
2. 10 and 24 week placentae in relation to $V_v$ trophoblast, $S_v$ fibrin and $L_v$ villi.

Different parameters of various structural elements of a triplet placenta display more significant correlations of a larger quantum among themselves, when compared with their counterparts in a placenta from single pregnancy of a similar duration. These facts indicate that in comparison to a placenta from a single pregnancy a triplet placenta shows a more coordinated growth of its elements, and continues to be in the process of maturation for a longer time.

A study of the interrelationship of different parameters of a structural element indicates that after the initial bulk or volume is laid down in the first trimester, area becomes the most significant parameter, on account of its key role in governing the physiological exchanges.

The above study, and the inferences therein, can be regarded as a preliminary study of the quantitative aspects of growth, maturation and senescence of the structural elements of the placenta, which may define the norms for further studies in this direction.

REFERENCES:

1. Bhargava Indra and P.T.K. Raja,
   Fetal blood vessels on the chorial surface of the human placenta in abnormal development, Experientia, 25: 520-522; 1969.

2. Miller, R.G., Jr.,
   Simultaneous statistical inference, McGraw Hill, New York, 1966.

3. Weibel, E.R., G.S. Kistler and W.F. Scherle,
   Practical stereological methods for morphometric cytology, J. Cell. Biol. 30: 23-38; 1966.

*National Bureau of Standards Special Publication 431*
*Proceedings of the* FOURTH INTERNATIONAL CONGRESS FOR STEREOLOGY
*held at* NBS, Gaithersburg, Md., September 4-9, 1975 (Issued January 1976)

## THE EFFECTS OF OPTICAL RESOLUTION ON THE ESTIMATION OF STEREOLOGICAL PARAMETERS

by H. J. Keller, H. P. Friedli, P. Gehr, M. Bachofen and E. R. Weibel
*Department of Anatomy, University of Berne, Berne, Switzerland*

It is generally assumed that the estimation of parameters such as volume and surface densities should be independent of the optical magnifications at which the measurements are performed. In comparing estimates of alveolar surface area of human lungs obtained by light microscopy (1) and by electron microscopy (2) systematic differences were however observed whose magnitude could not be explained by the differences in specimen preparation alone. It was concluded that the higher resolving power of the electron microscope, together with the smaller section thickness, allowed more surface detail to be sensed by the test system. Such systematic differences became particularly critical since the use of automatic image analysis was envisaged, and Bignon and André-Bougaran (3) had already noticed systematic differences to point counting methods to occur with these instruments.

The resolution of fine detail of an object is limited by two effects:
1) An object point is "blurred" by the analytic system to a circle of non-zero diameter on the image.
2) The dimensions of the analyzing structure (a picture point on a TV tube or the thickness of a test line on a grid) are finite.
Overall resolution is expected to be roughly the sum of these effects. Its influence on the measurement of area and boundary length should be different: the resolution of greater contour detail results in an increased boundary length whereas the enclosed area should remain the same, at least as long as the resolved detail is small compared to the dimensions of the object.

To test this hypothesis we have performed comparative measurements of 40 samples from a human lung using (a) the Quantimet 720 combined with a light microscope operated at magnifications of 45x, 110x, 440x, and 1100x, and (b) point counting on electron micrographs magnified 1000x, 4300x, and 11'000x using established methods (4).

The lung had been fixed in situ by instillation of glutaraldehyde (4); samples from various regions were obtained following a systematic scheme and processed for embedding in Epon 812. Sections about 2μm thick were cut with glass knives and stained with hematoxylin-eosin for light microscopy. From the same block 70nm sections were cut for electron microscopy.

Fig. 3 shows that the estimation of volume density of these septa in the unit lung volume is a stable parameter, showing no systematic differences with magnification, except at very low powers, where the diameter of a picture point approaches the septum thickness indicating poor overall resolution. From Fig. 3 it is seen that the surface density systematically increases with magnification, both in the Quantimet-light microscope combination, and in the electron microscope preparations evaluated by intersection counting. There are also systematic differences at the same magnification between Quantimet measurements and intersection

*H. J. Keller, H. P. Friedli, P. Gehr, M. Bachofen and E. R. Weibel*

counting estimations; one possible reason is that the two methods should be compared at the same resolution rather than magnification, but it is difficult to establish the former for the point counting method. An upper limit for our system is indicated in Fig. 3.

It is concluded that systematic differences in the estimation of certain stereological parameters, such as surface densities, may result from the use of varying magnifications. This is of particular concern if absolute values of a certain parameter are to be obtained, or if results from various studies are to be compared. Studies of this kind need to be extended to establish resolution-dependent factors for correlating stereological estimations of various parameters.

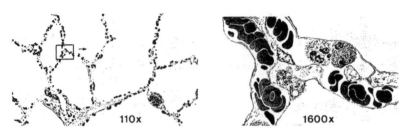

Fig. 1           Fig. 2
Representative lung fields as light and electron micrographs

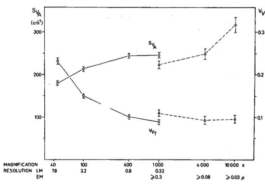

Fig. 3
Tissue volume density $V_{VT}$ and alveolar surface density $S_{VA}$ of human lung at different magnifications.
(o=light microscopy, Δ=electron microscopy)

Acknowledgement
This work has been supported by a grant from the Swiss National Science Foundation.

References
(1) Weibel, E.R.: Morphometry of the human lung. Springer, Berlin, 1963
(2) Weibel, E.R., Bachofen, M. and Gehr, P.: Prog. Resp. Res. 1975 (in press)
(3) Bignon, J., André-Bougaran, J.: C.R. Acad. Sci. 269, 409, 1969
(4) Weibel, E.R.: Respir. Physiol. 11, 54, 1970

National Bureau of Standards Special Publication 431
Proceedings of the FOURTH INTERNATIONAL CONGRESS FOR STEREOLOGY
held at NBS, Gaithersburg, Md., September 4-9, 1975 (Issued January 1976)

QUANTIFICATION OF RARELY OCCURRING STRUCTURES IN ELECTRON MICROSCOPY

by Ulrich Pfeifer
Pathologisches Institut der Universität Würzburg, Germany

In electron microscopy micrographs have always played a more important role as a vehicle of information than in light microscopy. This is particularly clearly seen in the case of electron microscopic morphometry. Here the random samples in the form of micrographs are evaluated by the superimposition of various types of lattices - a procedure which allows the determination of a great number of diverse parameters (1). There is no need to elaborate on the merits of this method. Certain limitations, however, must not be disregarded. They stem from the fact that some structures, which are known to occur regularly, are found only rarely, for instance once in a hundred or more micrographs taken at random. Under these circumstances the number of micrographs required for a statistically significant evaluation can exceed the limits of practicability.

In such a situation one would do well to recall that in light microscopy quantitative evaluations are performed mainly by counting and measuring the structures directly with the microscope. The mitotic index in the normal liver, for instance, is known to be in the range of one or two per ten thousand. Nobody would have considered determining it by taking light micrographs at random, and then evaluating them. The question now arises whether the method of direct evaluation with the microscope can and should be adapted for electron microscopy. In the present paper some methods are described which we have found of use in our work on cellular autophagy (2,3) for the quantitative evaluation of rarely occurring structures with the electron microscope. These methods have been developed for the Siemens Elmiskop I A, but may have to be modified in some details, when other types of microscopes are used.

As a first step for the evaluation desired one has to define area units which can be examined for the occurrence of the scantily represented structures of interest. As area units in our work the square openings of the copper grids supporting the specimen (100 mesh grid, Veco, Solingen, Germany) were used. It was, however, found that the size of the squares was not constant; measurements of the area units had, therefore, to be included in the evaluation. Two different methods can be used for this purpose. In the first one the lengths of a certain number of squares were measured in the light microscope by means of a micrometer screw in the ocular. Since the actual squares, which had been evaluated in the electron microscope, can subsequently not be found again in the light microscope, only the mean values of the area of squares for one grid, or for a whole experimental series, can be determined. This disadvantage is avoided by the second method, in which each square area unit can be measured directly in the electron microscope by means of a special driving mechanism for the specimen stage (Hilde Haag, Heppenheim, Germany). The digital counters of this driving mechanism indicate the actual position of the specimen stage. A calibration of the movement of the stage was performed in the following way (fig. 1):

The distance $d$, when the stage is moved from one digit of the counter to the next one, was measured by using a carbon grating replica (28 800 lines per inch, Fullam, Connecticut, U.S.A.) with the distance $i$ between

This work was supported by Deutsche Forschungsgemeinschaft, Sonderforschungsbereich 105, Dr. K. Aterman gave valuable help in preparing the English text.

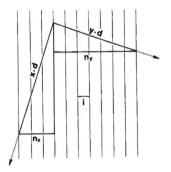

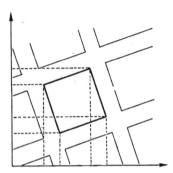

Fig. 1 Showing the geometrical basis for determining the distance d by which the specimen stage is moved, when going from one digit of the counter to the next one (see text). The thin vertical lines represent the lines of the carbon grating replica with the distance i.

Fig. 2 Showing the squares of the copper grid situated within a system of rectangular ordinates. These ordinates are represented by the two directions of movement of the specimen stage. For each of the four angle points of the square outlined by thick lines a pair of values on these ordinates can be determined.

two lines. The stage was moved in one direction by $x$ digits and the number $(n_x)$ of replica interspaces passing in center of the screen during this movement was recorded. The desk was then moved in the other direction by $y$ digits and again the number $(n_y)$ of replica interspaces was recorded. By a geometrical calculation derived from fig. 1 it can be shown that

$$d = \sqrt{\left(\frac{n_x}{x}\right)^2 + \left(\frac{n_y}{y}\right)^2} \cdot i$$

In the replica used, $i$ was 0.883 $\mu$m. The distance d was found to be 2.975 $\mu$m; it remained constant, whether the central or the more marginal areas of the specimen were examined.

The two directions in which the specimen stage is moved can be viewed as rectangular ordinates. For each of the four angles of the square area unit the values on the ordinates for the point of the angle were determined (fig. 2) by bringing the four points successively into the center of the screen; the two values on the ordinates were then read off the counters. The area of the square was then calculated from these four paired values by means of a computer program for rectilinear surface areas (Monroe 1766, punched card program 1001 ENQS).

Since the total area evaluated is not always the appropriate parameter of reference, in most instances fractional areas have to be determined. Hence

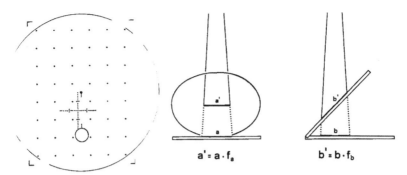

Fig. 3 Showing the fluores-
cent screen of the electron
microscope equipped with an
imprinted point lattice for
the determination of fractional areas. In addition
there is present a crossed
mm scale for the determination of two diameters of a
sectioned profile.

Fig. 4 Showing the distortion of
the two diameters of a profile
because of the oblique position
of the screen. The diameters a'
and b' read off the crossed
scale of the screen (see Fig. 3)
are divided by the factors $f_a$ and
$f_b$ in order to obtain the true
diameters a and b.

conventional morphometry with micrographs taken at random was used in our
studies as an indispensable additional method. A rough estimate of certain
parameters, however, can also be performed directly with the electron microscope. The fractional areas of parenchymal and non-parenchymal components of
a tissue, for example, were determined by the point counting method, by projecting the square area unit as a whole at low magnification onto the screen
which, as shown in fig. 3, was equipped with an imprinted point lattice. In
a study of the exocrine pancreas of the rat specimens of 24 animals were evaluated in this way, six square area units from each specimen. Without taking
any micrograph, the fractional area of parenchymal cells could be determined
with sufficient exactness; the mean was found to be 0.705, the standard deviation 0.043, and the standard error of the mean 0.009.

After these determinations the scantily represented structures of interest could be searched for. Each area unit of the specimen was now viewed
at an appropriate primary magnification, which was calibrated by means of
the carbon grating replica mentioned earlier on. A stereoscopic lens, magni-
fying 9 times, was used in addition. The electron beam was maximally focussed
by the double condenser in such a way that in the main only that area of the
screen was illuminated which was covered by the field of vision of the stereoscopic lens. Under these conditions a well illuminated image is obtained,
even if the current of the beam is kept very low. This is important both for
careful treatment of the specimen and of the cathode during the long periods
needed for the examination. When structures of interest were encountered
during the systematic scanning of the specimen by a meandering movement, the
number, the type, and the size of these structures had to be registered. This
can be done by taking a micrograph from each structure at a magnification

413

suitable for determining later on the area of the sectioned profile by means of. the point counting method. Since, however, the number of even these selected micrographs may be considerable, it was thought .advantageous to note down the required parameters directly from the image on the screen. For measurements of the area of sectioned profiles having an approximately ovoid or round shape, two diameters were taken by means of a crossed mm scale printed onto the screen as shown in fig. 3. Due to the oblique position of the screen during the measurement under the stereoscopic lens (fig. 4) the two values $\underline{a}'$ and $\underline{b}'$ obtained had to be corrected by factors $f_a$ and $f_b$ in order to determine the true diameter $\underline{a}$ and $\underline{b}$. By measuring a set of profiles first in the horizontal, and then in the oblique, position of the screen $f_a$ was found empirically to have a value of 0.906, whereas the value of $f_b$ was 1.357. The area A of a profile of interest can now be calculated from the following formula:

$$A = \frac{a'}{f_a} \cdot \frac{b'}{f_b} \cdot \frac{\pi}{4} = a' \cdot b' \cdot \frac{\pi}{4 \cdot f_a \cdot f_b} = a' \cdot b' \cdot 0.639$$

In practice it was, therefore, not necessary to register every time the two individual diameters, but only the product of these two diameters. In order to get the area of the profile in $mm^2$ at the chosen magnification, this product had to be multiplied by the factor of 0.639.

In a study of the diurnal rhythm of cellular autophagy in the proximal tubules of the rat's kidney (3) it has been shown that it is possible, with the help of the method outlined here, to examine in one experiment specimen areas amounting to 100 . $10^4$ $\mu m^2$ and more, in order to search for scantily occurring autophagic vacuoles. Their fractional volume, for example, expressed as the mean value of the diurnal cycle was found to be 2.1 . $10^{-4}$.

As always, when quantitative data in morphology are required the evaluation remains laborious. By avoiding taking micrographs, it is possible, however, with the method described here to venture into fields morphometry which so far have been hardly accessible.

References

1) Weibel, E.R.: Stereological principles for morphometry in electron microscopic cytology. Int. Rev. Cytol. 26, 235-302 (1969)

2) Pfeifer, U.: Cellular autophagy and cell atrophy in the rat liver during long-term starvation. Virchows Arch. Abt. B Zellpath. 12, 195-211 (1973)

3) Pfeifer, U., Scheller, H.: A morphometric study of cellular autophagy, including diurnal variations, in kidney tubules of normal rats. J. Cell Biol. 64, 608-621 (1975)

*National Bureau of Standards Special Publication 431*
*Proceedings of the* FOURTH INTERNATIONAL CONGRESS FOR STEREOLOGY
*held at* NBS, Gaithersburg, Md., September 4-9, 1975 (Issued January 1976)

THE MORPHOMETRY AND STEREOLOGY OF CEREBRAL ARTERIAL BIFURCATIONS

by Peter B. Canham, James G. Walmsley and J. F. Harold Smith
*Department of Biophysics, University of Western Ontario, London, Ontario, Canada*

SUMMARY

Segments of human middle cerebral arteries were fixed under pressure in 10% formalin. Serial sections were prepared and stained either with hematoxylin and eosin in order to study the arrangement of smooth muscle in the tunica media or with silver inpregnation in order to study the arrangement of collagen in the tunica adventitia. Composite photomicrographs of whole arterial segments were prepared. The darkly stained muscle nuclei were used as vector indicators of cellular orientation. The nuclei were digitized. Trigonometric transformation, modelling and projection techniques were used. In contrast to the simple cell pattern in straight portions,the cells in the bifurcation region were multi-directional and arranged in bundles. Collagen was found to be arranged circumferentially (for the single specimen studied) using the Leitz polarizing microscope and universal stage.

INTRODUCTION

Saccular aneurysms, which are balloon-like enlargements of the vascular wall, are found in the major cerebral arteries associated with the circle of Willis at the base of the brain. The fact that aneurysms occur nearly always at the apex of the bifurcation (Hassler, 1961) has stimulated us to study the architecture of the blood vessel wall. In the region of the apex (Fig. 1) where the curvature in the plane of the bifurcation does not help support transmural pressures, there seems to be no increased wall thickness (Stehbens, 1974; Macfarlane, 1975). However, the composition of the wall is different at the apex with a reduced amount of medial smooth muscle and an increased amount of adventitial collagen compared to the non-branching regions. We undertook to see if the organization of the structural elements might also show differences in the region of the apex. Vasospasm, characterized by an active or passive decrease in arterial caliber, causes reduction in regional cerebral blood flow. Our working hypothesis is that the organization of fibers of both smooth muscle and collagen are important in relation to saccular aneurysm formation, vasospasm and atherosclerosis.

Two complementary studies are in progress, the one focussing on the organization of the smooth muscle, and the second, the organization of collagen. Smooth muscle cells are long tapered cylinders, each with one cylindrical darkly staining nucleus (for sections stained with hematoxylin and eosin). Collagen is the main acellular component of the adventitia of cerebral arteries.

SMOOTH MUSCLE

We used the technique of formalin fixation of isolated vessel segments under a maintained transmural pressure of 100 ± 10 mm Hg, paraffin embedding, serial sectioning and staining with hematoxylin and eosin. We have concentrated on middle cerebral arteries, a common site for aneurysm formation in the anterior circulation. Vessels were obtained at autopsy. Certain specimens were sectioned longitudinally and others transversely. Because bifurcations are frequently not planar, a system of three cylindrical nerve fibers were embedded into the paraffin block (method of Burston & Thurley, 1957) and sectioned along with the vessel material, the nerve fibers serving as a reference coordinate system. This method involves infusion of the nerves with a higher melting point wax to allow implantation at right angles to the plane of sectioning.

Our findings from longitudinal sections cut in the bifurcation region, but away from the midplane, were suggestive of smooth muscle cell groups. A test of the continuity of the cell groups, or bundles, was made using the technique of counting the number of nuclear intersections with a line grid superimposed

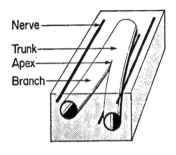

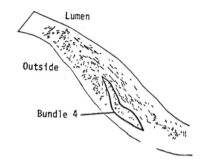

Fig. 1. A bifurcation in sectioning
block with reference nerves.

Fig. 2. Section from branching region
with bundle outlined (#4 of Table 1)

on a composite photomicrograph (described as the method of roses, Underwood,
1970). A digitized and replotted sample composite is shown in Figure 2.
Table 1 summarizes the results for five cell bundles which were identifiable
through 80μm of sectioning (each slide had three sections sliced at 6.9μm).
The parameter θ, is the mean angle (in the section plane) of the nuclei, and
Ω is a measure of the extent of orientation, being zero for complete randomness. Similar analysis repeated on other serial sections showed similar
results and we concluded that the medial smooth muscle above and below the
apex of middle cerebral arteries was arranged in groups, each group being
identifiable for several adjacent sections. The total effect was that of a
coarsely woven cellular fabric. However the boundary between the bundles
was not always distinct, and the lack of quantitative data for the third
dimension led us to examine histologically the straight portion of the vessel
segments for the purpose of identifying the basis of the pattern of muscle
bundles.

For the study of the straight portions, composite photomicrographs were
constructed using medium power optics. (The objective of the Nikon microscope
used was a 10x which had a suitable depth of focus to include the total 7.3μm
thickness of the section.) Subsequent detailed identification was done on each
nucleus to determine if it traversed the section and which end was down, this
being done under oil immersion. The digitized result of a sample composite
is shown in Figure 3 with only those nuclei shown which completely traversed
the section depth. The plot in Fig. 3 is a corrected true cross section,
corrected for the obliqueness of cutting (17.5°), cutting strain in the X-Y
plane, cutting strain in the Z plane which results in thickening of the section,
and rotation. The knobs are at the bottom ends. Attention is drawn to the
approximate uniformity of the nuclei in any region, to the nuclei being more
angled into the section at the sides (shorter projection) and being more
within the section at the top and bottom of the figure where they reverse
direction. In addition to analyzing actual histological sections we generated model oblique sections of cylindrical vessels and nuclei of uniform
helical pitch. We measured the dimensions of the nuclei for this study and
found their actual length to be 37 μm ± 6.4 (S.D.) and diameter 2.0 μm ± .83
(S.D.), a sample of 50 nuclei. We concluded that for the straight portions
of middle cerebral arteries (5 specimens from 2 autopsies) the smooth muscle
is arranged in a circular pattern in the tunica media, and that the pitch of
the helix is not different from 0°. Our finding, exampled in Figure 3, that
the projection of the nuclei in the section plane is not uniform around the
circumference is interpreted as an error in correcting for the obliqueness of
the section. In straight arterial segments the media was found to be 75%
smooth muscle.

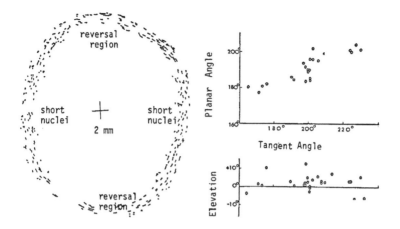

Fig. 3. Nuclear projections onto transformed cross-sectional plane.

Fig. 4. The planar and non planar components of collagen orientation.

## COLLAGEN

The orientation of collagen in the adventitia of straight vessel segments has been done using a Leitz polarizing microscope with a four axis rotating universal stage. After sectioning as above, the established birefringence of collagen (Schmitt, 1934) was intensified using silver impregnation stain (Mello & de Campos-Vidal, 1972) and Permount was used as an immersion medium.

The universal stage permits one to rotate the section in order to make a parallel fiber region aligned with, or perpendicular to the axis of the microscope. The results are obtained and recorded as two angles: a planar section angle and an elevation angle. Figure 4 shows data obtained from one section cut at 17° from a true cross section. Collagen bundles measured are of the order of 3 micrometers in any linear dimension. Attention is drawn to the correlation between the projected angle of the collagen bundle within the plane and the angle of the tangent to the vessel wall at that point. In contrast, the elevation angle (angle of the bundle out of the section plane) shows no correlation with tangent angle. The data in Fig. 4 is consistent with a circumferentially arranged outer collagen sheath around the cylindrical vessel.

Our plan is to complete the 3 dimensional orientation measurements on straight vessel sections, which to our knowledge is an application of the polarizing microscope and universal stage not previously reported. The method was learned from Professor John Starkey, a colleague in Physical Geology at the University.

Armed with the data on collagen and muscle in the straight segments we are returning to investigate quantitatively the arrangement in the region of vessel junctions.

## ACKNOWLEDGMENT

This work was supported by the Ontario Heart Foundation.

| BUNDLE NUMBER | Slide 72 θ | Slide 72 Ω | Slide 74 θ | Slide 74 Ω | Slide 76 θ | Slide 76 Ω |
|---|---|---|---|---|---|---|
| 1 | 80 | 53% | 76 | 45% | 83 | 45% |
| 2 | 190 | 64% | 176 | 80% | 163 | 62% |
| 3 | 20 | 69% | 21 | 59% | 33 | 51% |
| 4 | 120 | 50% | -- | -- | 133 | 56% |
| 5 | 170 | 39% | 141 | 92% | -- | -- |

Table 1. θ and Ω for bundled regions in front of apex (cf. Fig. 2).

## REFERENCES

Burston, W.R. & Thurley, K. (1957) A technique for the orientation of serial histological sections. *J. Anat.* 91, 409.

Hassler, O. (1961) Morphological studies on the large cerebral arteries with reference to the aetiology of subarachnoid haemorrhage. *Acta Psych. Neurol. Scand.* 36, Supp. 154, 1-140.

Macfarlane, T.W.R. (1975) *The Geometry of Cerebral Arterial Bifurcations and its Modification with Static Distending Pressure.* M.Sc. Thesis, University of Western Ontario.

Mello, M.L.S. & de Campos-Vidal, B. (1972) Evaluation of dichroism and anomalous dispersion of the birefringence on collagen subjected to metal impregnations. *Ann. Histochem.* 17, 333.

Schmitt, W.J. (1934) Polarisation optische Analyse des Submikroskopischen Baues von Zellen und Geweben. In: *Abderhalden's Handbuch der Biologischen Arberts methoden* (Ed. by Schmitt, W.J.), pp. 435-665. Springer, Berlin.

Stehbens, W.E. (1974) Changes in the cross-sectional area of the arterial fork. *Angiology* 25, 561.

Turner, F.J. & Weiss, L.E. (1963) *Structural Analysis of Metamorphic Tectonites*, pp. 197-199. McGraw-Hill, N.Y.

Underwood, E.E. (1970) *Quantitative Stereology*, Chapt. 3. Addison-Wesley, Reading, Mass.

National Bureau of Standards Special Publication 431
Proceedings of the FOURTH INTERNATIONAL CONGRESS FOR STEREOLOGY
held at NBS, Gaithersburg, Md., September 4-9, 1975 (Issued January 1976)

KARYO-INTERKARYOMETRY AS A PERIODIC CONTROL FOR CLINICAL PROGNOSIS

by E. C. Craciun and C. Tasca,
*Institute of Endocrinology, Bucharest, Romania*

Karyo-interkaryometry gradually developed as a quantitative branch of research in medicine and biology. Its aim is to give an exact description of cell nucleus in terms of quantitative and, therefore, well comparable data. These facts are correlated with the functional cellular aspects to allow a dynamic interpretation of cell physiology and physiopathology. Changes in cell function affect the cellular metabolism and consequently the volume of nucleus, nucleolus and cytoplasm. Nuclear size modifications gave rise to the principle of "clear intumescence" as described by Craciun, "nuclear functional oedema" (Benninghoff) and "working hypertrophy" (Eichner), each one of these cellular appearances with its definite nosological trends.

MATERIAL AND METHOD

One case is presented of benign hepatitis epidemica patient A.N., male 32 years, repeatedly studied by liver needle biopsy at 8, 50, and 180 days from the apparition of jaundice, on formalin fixed specimens.

Nuclear volumes, considered as rotation elipsoids were calculated by means of a manual method. The relation between statistical distribution of nuclear volumes and the distribution of nuclear surfaces was estimated. By calculating the error of nuclear volume starting from nuclear surface, the probability is that only in 50% we can determine the "true" nuclear volume. For karyometry the volumes of one thousand nuclei were calculated according to the formula of Puff: nuclear volume: $8S^2/3\pi A$ (S = surface of the section, A = major diameter). Secondly we measured the internuclear distance (interkaryometry) as an indirect proof to quantify the nucleo-plasmic ratio. The major reason is that such a parameter allows a quantitative expression of the intercellular and tissular milieu. Moreover it brings information on spatial distribution of nuclei. The logarithmic system was applied for calculating and grouping the observed values, which were then submitted to a statistical study.

RESULTS

The first needle biopsy, eight days after the beginning of jaundice, showed striking pictures of cytolysis with trabecular demolition. Hepatocytic chords can usually be identified by the neighboring blood capilaries. Hepatocytes are most irregular as to size of nuclei, their chromatin content, their reciprocal intervals and number of nuclei (Fig. 1). Many blood vessels are dilated without hemorrhage, as provided by the continuous reticulin net on Gomory silver impregnation and the absence of macrophages and siderophages. No necrotic elements and therefore no perinecrotic leucocytes. Histiocytic and lymphocytic polyblasts are present, but no polymorphonucleares.

Needle biopsy II (50 days) showed the same, but more accentuated features, especially from the point of view of cytolysis and trabecular disorganization. Needle biopsy III (180 days) presented a rather normal looking hepatocytes in as much nuclei and cytoplasma do not evoke any more the previous stages; some inflammatory changes still persist (Fig. 2). Clinical recovery followed, while hepatic tests and biochemical results were not quite normal six months after the beginning of icterus.

It is striking to compare lesions, figures, and graphics with clinical evolution.

419

From the point of view of the mathematical analysis it can be stated that the karyo-interkaryometric data correspond to a Gauss-Laplace curve as observed in biological processes. Bimodal and especially plurimodal curves point to a nonhomogenous cell population as observed in our histological slides at 8 and 50 days (Fig. 3). For the 180 days the graphic still showed a quite miscellaneous hepatocytic population as nuclear volumes are concerned (Fig. 4). Most frequent are the nuclear volumes of 132-250 $\mu^3$ and 70-95 $\mu^3$ at 8 and 50 days, while at 180 days the most frequent are those of 90-145 $\mu^3$. Accordingly the mean volume is reduced from 240 $\mu^3$ at 8 days to 182 $\mu^3$ at 50 days and finally to 110 $\mu^3$ at 180 days.

Internuclear distances reach a mean top value at 50 days of 10.6 $\mu$, as compared to 8 $\mu$ at 8 days and 9 $\mu$ at 180 days. A certain role is played by anisocytosis and by incomplete recovery of distrophic processes.

The graphical appearance of different stages of hepatitis epidemica corresponds to the histological picture. In the first needle biopsy a "feverish" sequence of high arrows with "saw teeth" appearance could be seen. Interkaryometry revealed a predominance of large intervals corresponding to cytolysis and trabecular demolition. In the second biopsy the irregular aspect still persists, but after 180 days the karyogram is almost completely normalized. A tendency exists to a reduction of the nuclear volumes and internuclear intervals with the elapsed time. Excepting for the 50 days stage the distribution of nuclear volumes is asymetric showing a dextrodeviation with the corresponding negative coefficient of asymmetry approaching null point. For the internuclear distances there is an asymmetry meaning a levodeviation. This reverse type of correlation between frequencies of nuclear volumes and intervals seems due to the favourable evolution of illness. This situation corresponds to an increased density of hepatocytic population in good biological condition. It means a positive situation concerning the chances of histological recovery.

DISCUSSION

Karyometry if looking for extrapolation is limited referring to the whole organ, given the relatively small number of measured cells. Such a situation cannot annihilate neither karyometry nor its current use. From our·data it results that nuclear dynamic parallels the anatomo-clinical picture. This synergism appeared in all stages of hepatitis we studied (Craciun; Craciun and Tasca). Quantitative variations of nuclear volumes could be therefore expressed as a morphopathological nuclear syndrome corresponding to a quantitative nuclear formula, typical for a certain stage of disease. From this point of view one could speak of a quantitative cytoprognosis (Tasca et al., Tasca). In the same way more than 1500 cases with various hepatic, endocrine, nervous, lymphoganglionary and neoplastic diseases were studied. A parallelism was always found between karyo-interkaryograms and clinical evolution. Karyometric methods could be used like leukograms or thermic curves as a periodic control for clinical prognosis. Moreover these methods need only routine laboratory equipment easily appliable to each hospital.

REFERENCES

BENNINGHOFF, A.: Funtionnelle Kernschwellung und Kernschrumpfung. Anat. Nachr., 1949, 1, 50-53.

CRACIUN, E. C.: Specificitatea leziunilor progressive si regresive din hepatita epidemica icterigena. An. Inst. V. Babes, 1943, 16, 178.

CRACIUN, E. C., TASCA, C.: La caryometrie et l'intercaryometrie dans l'hepatite epidemique evolutive. Rev. Intern. Hepatol., 1968, 18, 8, 967-986.

EICHNER, D.: Uber funtionelle Kernschwellung in den Nuclei supraoptici und paraventriculares des Hundes bei experimentellen Durstzustanden. Z. Zillforsch., 1952, 37, 406-414.

TASCA, C., ENACHESCU, D., CRACIUN, E. C.: A mathematical study of associated karyo-interkaryograms in Hepatitis Epidemica. Beitr. Path. Anat., 1968, 137, 316–329.

TASCA, C.: Cariocitometria ca detector cantitativ in diferite afectiuni. MD Thesis Bucuresti, 1967.

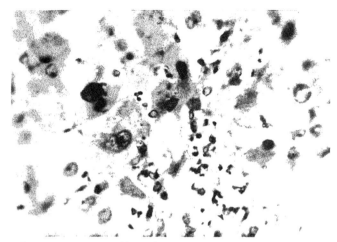

Fig. 1. Needle biopsy I (8 days). Distrophic cytoplasmic lesions and anisokarya. HE stain. x 200.

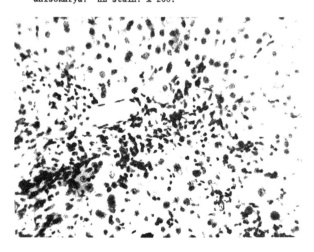

Fig. 2. Needle biopsy III (180 days). Some inflammatory changes still persist but the distrophic lesions are no more visible HE stain. x 100.

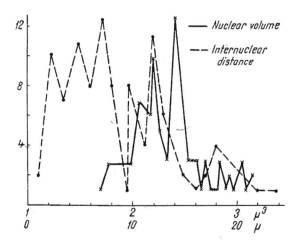

Fig. 3. Karyo-interkaryogram at 8 days shows a "feverish" sequence with high arrows.

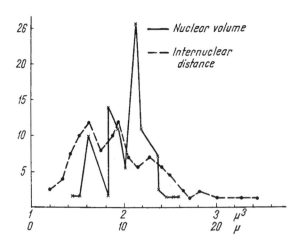

Fig. 4. Karyo-interkaryogram at 180 days shows a normal looking aspect with a single major peak.

National Bureau of Standards Special Publication 431
Proceedings of the FOURTH INTERNATIONAL CONGRESS FOR STEREOLOGY
held at NBS, Gaithersburg, Md., September 4-9, 1975 (Issued January 1976)

NUCLEIC ACIDS CONTENT-NUCLEAR VOLUME CORRELATION AS A QUANTITATIVE PARAMETER IN CELLULAR MALIGNANCY.

by C. Tasca and E. C. Craciun
*Institute of Endocrinology, Bucharest, Romania*

The histopathological diagnosis is still based on personal experience, the possibility to express it in quantitative terms being very limited. Cytophotometrical methods developed by Caspersson (1936) and later by Sandritter, in order to quantify by direct measurements the nucleic acids, brought interesting data on the normal and pathological nucleo-cytoplasmic metabolism. Unfortunately the cytophotometry alone was not able to confirm the malignant growth. As known, there are malignant tumors with either hyperchromatous or hypochromatous nuclei, i.e., with high or small quantities of nucleic acids. On the other side, the increase of nuclear volume is not an exclusive characteristic of neoplastic tissues because it may be found in many other etiological conditions. The correlation between nucleic acids content and nuclear volumes seemed to us to better quantify the "stages of malignancy."

MATERIAL AND METHODS

We selected two opposite kinds of tumors with respect to DNA content: skin hyperchromatous tumors induced in mice and liver adenoma in rats as hypochromatous tumors. The skin carcinoma were obtained in NMRI mice using 3,4 benzpyren; the hepatoma were induced in Sprague-Dawley rats fed for 8 months with 3 mg/kg weight di-ethyl-nitrosamine. As a model of not tumorous tissular growth we used the skin reversible hyperplasia obtained in mice by means of the co-carcinogen Phorbol-ester A1 (0.02μ Mol in 0.1 ml acetone).

DNA was stained by the Feulgen method. We also used the gallocyanin-chromalum staining method after treatment with 1% ribonuclease.

The cytophotometric measurements were accomplished by means of the Univeralmicrospectrophotometer (UMSP I) of the Firma Carl Zeiss Oberkochen with scanning table and rapid measurement device. The mean extinction and the projected surface of the nucleus were concomitantly measured. The Feulgen stained nuclei were measured in monochromatous light by 560 mμ; the gallocyanin-chromalum stained nuclei were measured by 500 mμ. By means of the computer RPC 4000 the nuclear volume and the content of DNA were calculated. To calculate the volume we used the formula: $V = 0.75\sqrt{S^3}$ (S = surface). The content (M) of DNA was calculated by the formula: $M = ES/\varepsilon'$ (S = nuclear surface, E = mean extinction, $\varepsilon'$ = specifical decadic extinction coefficient in $cm^2 g^{-1}$). The $\varepsilon'$ was calculated after the formula of Sandritter.

The correlation between the nuclear volume and the DNA content was expressed as a stochastic equation. The most probable dependence between DNA content and nuclear volume is represented as: DNA content = average DNA concentration × nuclear volume + const. In this equation the DNA concentration ($gcm^{-3}$) represents the direct equivalent of correlation.

RESULTS

*Correlation DNA-Nuclear Volume in Hypochromatous Heptacellular Tumors*

The nuclei of hepatoma were irregular with fine chromatin granules adherent to the nuclear membrane but with a clear center, i.e., with a low DNA concentration (Fig. 1). The results of our cytophotometric measurements are summarized in Fig. 2. The territories where most of the pair values (nuclear volumes and DNA content) were found, are represented as columns. The slopes of columns correspond to the regression equations. As compared to normal hepatic nuclei, the not yet tumorous nuclei around the neoplastic zones, showed a stronger correlation between DNA and nuclear volumes. The reason is that more nuclei contain a higher DNA content on one side and that the DNA synthesis is greater on the other side because an increased liver regeneration. The regression lines of the neoplastic liver nuclei showed a diminution of DNA concentration by an excessive

423

C. Tasca and E. C. Craciun

dyscorrelated nuclear volume increase. The correlation DNA-nuclear volume is therefore decrease in hepatoma.

*Correlation DNA-Nuclear Volume in Hyperchromatous Skin Tumors*
The results of cytophotometric measurements in hyperchromatic tumors induced by 3,4 benzpyren are summarized in Fig. 3. The surfaces where DNA concentrations and nuclear volumes were found, are represented as ellipses. The dotted lines correspond to the regression equations. By comparing the values of normal not treated mouse skin with those of the benzpyren treated skin but without morphological modifications, it results that DNA concentrations and nuclear volumes are similar. The regression lines showed however a smaller angle of the slope in to benzpyren exposed skin. The same diminished correlation DNA-nuclear volume appeared in all skin neoplasia.

*Correlation DNA-Nuclear Volume in the Skin Reversible Hyperplasia after the Co-Carcinogen Phorbol-ester Al*
In the reversible skin hyperplasia after Phorbol-ester Al treatment the regression line parallels the regression line of the normal skin (Fig. 4). On the contrary, the regression line of the with benzpyren-treated skin without morphological modifications showed a diminution of DNA-nuclear volume correlation.

DISCUSSION
From our data it results that the concentration of DNA is a sensible indicator for neoplastic mechanisms. In previous work (Tasca et al.) it was shown that the linear regression analysis disclosed for the nuclei of the experimentally induced papillomata a diminished correlation between nuclear volume and content of nucleic acids and histones. The diminution of the correlation could therefore be the first quantitative phenomenon in the course of cancerogenesis before other morphological modifications in routine histological methods are visible. Our results referred only to post-mitotic cell-collectives. The general value of our findings must be also further investigated.

REFERENCES
Caspersson, T.: Uber den chemischen Aufbau der Strukturen des Zellkerns. Scand. Arch.Physiol.73,Suppl.8,1936.
Sandritter, W.: Die Nachweismethoden der Naukleinsäuren. Z.Wiss.Mikr.Technik. 62,283-304,1954/1955.
Sandritter, W.: Ultravioletmikrospectrophotometrie. In: Graumann,W; Neumann, K. H.: Hdb.der Histochemie,pp.220-338, Stuttgart, Fischer,1958.
Tasca, C., Haag, D., Goerttler, Kl.: Cytophotometrische Befunde in der Trachealschleimhaut und in DANA-behandelten Trachealpapillomen beim syrischen Goldhamster. Z.Krebsforsch.,74,335-367,1970.

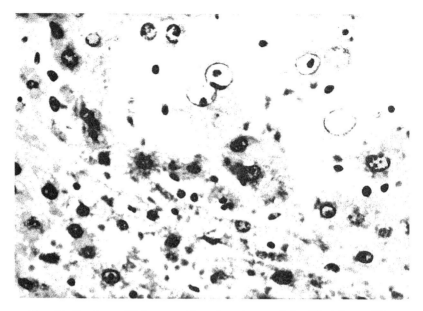

Fig. 1: Hepatoma with large and hypochromatous nuclei. HE stain. x500.

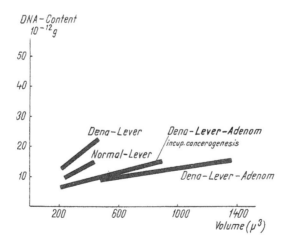

Fig. 2: Diagram of cytophotometric measurements in DENA-hepatoma Increase of nuclear volumes and diminution of DNA-concentration.

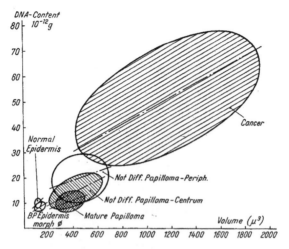

Fig. 3: DNA-concentration as a function of nuclear volumes. Diminution of
DNA-nuclear volume correlation of the skin without morphological
modifications and of all stages of malignancy.

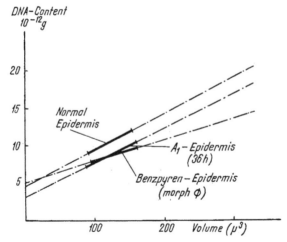

Fig. 4: Diminution of DNA-nuclear volume correlation in the mouse skin exposed
to benzpyren as a sign of incipient cancerogenesis.

426

*National Bureau of Standards Special Publication 431*
*Proceedings of the* FOURTH INTERNATIONAL CONGRESS FOR STEREOLOGY
*held at* NBS, Gaithersburg, Md., September 4-9, 1975 (Issued January 1976)

IS THERE AN UNRECOGNISED SYSTEMATIC ERROR IN THE ESTIMATION OF SURFACE DENSITY (SD) OF BIOMEMBRANES?

by Albrecht Reith and Tudor Barnard
*Norsk Hydro's Institute for Cancer Research, The Norwegian Radium Hospital, Oslo, Norway, and Wenner Gren Institute, Stockholm, Sweden*

A systematic error in the estimation of biomembrane SD is caused by the invisibility of membrane profiles oriented obliquely to the axis of the electron beam. This has been investigated quantitatively by Loud and a correction factor of 1.5 has been proposed for cristae membranes. SD estimates made by different groups on a model biomembrane system (membranes of liver mitochondria) vary by a factor of 2 from 23.5 to more than 53 $m^2/cm^3$ mitochondria. The theoretical SD of mitochondrial membranes (6 nm thick), tightly packed but including an intermembrane space (6 nm) between adjacent pairs of membrane sheets was calculated to be 117 $m^2/cm^3$. This means that liver mitochondria with a total membrane SD of 53 $m^2/cm^3$ would have a "membrane volume fraction" of nearly 0.5 $cm^3$, leaving only about 0.5 $cm^3$ for the matrix compartment. This would fit for types of mitochondria richly endowed with cristae such as heart and brown fat but not for liver. In this context the lower values are more reasonable. The fact that even SD estimates that are uncorrected for loss of oblique membrane profiles seem to be too high indicates that an as yet unrecognised systematic error seems to be affecting SD estimates, and therefore a more rigorous standardisation of the criteria used for defining membrane profile intersections would offer a first step towards decreasing the effects of this error.

It is agreed that there is a systematic error in the estimation of biomembrane SD by counting intercepts between membrane profiles and a test line. This is caused by the invisibility of membrane profiles oriented obliquely to the axis of the electron beam.

Loud, using the nuclear membrane of adult rat liver, estimated the critical angle of "50% visibility" to be about 60°. On this basis, correction factors of 1.5 and 1.25 have been applied to estimates of cristae and endoplasmic reticulum by several groups (2,3,4,6). Whether in fact estimates of biomembrane SD are routinely underestimated may be questionable, as we shall try to show by the following report:

The table shows the result of SD estimates obtained by several groups over a model system, mitochondria in adult rat livers.

The cristae values are corrected by 1.5 (uncorrected values are given in parenthesis): the outer membrane (and consequently the inner boundary membrane) values were not corrected, as the presence of these membranes can be assumed on the basis of other criteria, so that these intersection counts should be less sensitive to underestimation (3). The table shows that the estimates of total membrane SD falls into two groups, one of about 53 $m^2$ per $cm^3$, the other of 20-30 $m^2$ per $cm^3$.

In an attempt to decide which of the results was the most "reasonable", the theoretical SD of mitochondrial membranes (6 nm thickness) that could be accommodated in 1 $cm^3$ was calculated under two conditions: a) maximally tightly packed and b) tightly packed but with an intermembrane space (6 nm), intercalated between adjacent membrane pairs. The results were respectively 167 and 111 $m^2$ per $cm^3$.

Seen in this light, a total membrane SD of about 53 $m^2$ per 1 $cm^3$ would thus occupy nearly 0.5 of the mitochondrial volume, leaving only 0.5 for the matrix compartment. This might be expected for types of mitochondria richly endowed with cristae such as heart and brown fat, but not for liver (Fig.). Even if the uncorrected cristae SD estimates are used, the results still seem very high.

In Reith's study, only intersections with strictly cross-sectioned cristae profiles were recorded, and these corrected by 1.5. Under these conditions, the total membrane SD was 23.5 m$^2$ per cm$^3$, which gives a volume fraction of membranes and the space between adjacent pairs of membranes (intracristae and intermembrane space) of 0.14 or 0.07 cm$^3$/cm$^3$ respectively. This leaves 0.79 cm$^3$/cm$^3$ for the matrix volume fraction which is a more reasonable result for liver mitochondria. For comparison, estimates of total mitochondrial SD of heart and brown adipose tissue mitochondria gave values from 42 to 61 m$^2$/cm$^3$ (4,5,2). These figures appear to have correct relationships with the dense packing of cristae in such mitochondria compared to the relatively sparse cristae packing of liver mitochondria (see also Fig.).

A somewhat different approach, based on reconstructed particle size distribution, was used by the Louvain group(1), and their value of 32 m$^2$ per cm$^3$ is closer to 23 than to 53 m$^2$ per cm$^3$.

CONCLUSION: SD estimates made by different groups on a model biomembrane system vary by a factor of 2. The lower values obtained correspond better than the higher values with the matrix volume fraction as estimated by point-counting. The fact that even SD estimates that are uncorrected for loss of oblique membrane profiles can be unreasonable indicates that an as yet unrecognised systematic error seems to be affecting SD estimates. The error apparently counteracts the tendency to loss of membrane profile intersections and varies between laboratories. A more rigorous standardisation of the criteria used for defining membrane profile intersections offers a first step towards decreasing the effects of this error.

SURFACE DENSITIES OF MEMBRANES IN 1 CM$^3$ CUBE OF LIVER MITOCHONDRIA - m$^2$/cm$^3$
(in parenthesis values uncorrected for oblique sectioning).

from the rat:

liver

| | LOUD(3) | WEIBEL(6) | REITH(5) | LOUVAIN(1) |
|---|---|---|---|---|
| Outer membrane: | 8.96 | 8.07 | 7.0 | 7.3 |
| Cristae membrane: | 36.44(24.29) | 36.63(24.42) | 9.5 | 17.3 |
| Total inner membrane: | 45.40(33.25) | 44.70(32.49) | 16.5 | 24.7 |
| $\Sigma$ of all membranes | 54.35(42.20) | 52.77(40.56) | 23.5 | 32.0 |

heart

brown fat

REFERENCES

1. Baudhuin,P. & Berthet,J.(1967). Electron microscopic examination of subcellular fractions. J.Cell Biol. 35, 631.
2. Lindgren,C. & Barnard,T.(1972). Changes in interscapular brown adipose tissue of rat during perinatal and early postnatal development and after cold acclimation. Exptl.Cell Res.70,81.
3. Loud,A.V.(1968). A quantitative stereological description of the ultrastructure of normal rat liver parenchymal cells. J.Cell Biol.37,27.
4. Page,E. (1973). Morphometry of mitochondrial membranes and profile size distribution in rat

heart. J. Cell Biol. 59, 256a.
5. Reith,A.(1973). The influence of triiodothyronine and riboflavin deficiency on the rat liver with special reference to mitochondria. Lab. Invest. 29, 216.
Reith,A. & Fuchs,S.(1973). The heart muscle of the rat liver under influence of triiodothyronine and riboflavin deficiency with special reference to mitochondria. Lab.Invest. 29,229.
6. Weibel,E.R., Stäubli,W., Gnägi,H.R. & Hess, F.A.(1969). Correlated morphometric and biochemical studies of the liver cell.J.Cell Biol.42,68.

*National Bureau of Standards Special Publication 431*
*Proceedings of the* FOURTH INTERNATIONAL CONGRESS FOR STEREOLOGY
*held at* NBS, Gaithersburg, Md., September 4-9, 1975 (Issued January 1976)

QUANTITATIVE STUDIES WITH THE OPTICAL AND ELECTRONIC MICROSCOPE OF THE PATHOLOGICAL CELLS
OF THE ACUTE LYMPHOBLASTIC LEUKEMIA AND CHRONIC LYMPHOCYTIC LEUKEMIA

by J. L. Binet, P. Debré, P. D'Athis, D. Dighiero, and F. de Montaut.
*Département d'Hématologie, U.E.R. Pitié Solpêtrière, 91 boulevard de l'Hôpital, 75013 Paris, France*

ABSTRACT

 The stereological method gives new results in acute lymphoblastic leukemia
and in chronic lymphocytic leukemia. In A.L.L. there are two cell populations;
in C.L.L. there is only one volumetric population.

 After 2.5% glutaraldehyde fixation and further fixation in 2% osmium
tetroxide solution in phosphate buffer, the abnormal blood cells were dehydrated
and embedded in epon. Then they underwent a double treatment:
 First, they were cut in semi-fine (0.5 μ) sections, stained with Toluidine
blue, and the diameter of 1500 cells was measured on the Leitz Classimat. The
histogram constructed from these diameters was submitted to mathematical analysis in order to derive the volume according to the d'Athis method;
 Second, the same block was cut in fine sections and sixty pictures were
taken at random by the electron microscope. The volume density of the cytoplasm,
nucleus, and the two types of chromatin was also determined.
 Ten patients with acute lymphoblastic leukemia (ALL) were studied and the
volumetric distribution showed a double population. The first population included
the majority of cells (69 to 94%) and its mean volume ($V_1$) varied from 170 to
340 $\mu^3$. The mean volume was variable from one case to another. In 8 out of the
10 cases cytologists agreed with the sub-classification of the type of ALL. In
two cases there was disagreement. For 7 of the 8 patients, a good correlation
was found between morphological classification and volumetric determinations.
 The mean volume of the second population ($V_2$) was much higher (from 330
to 789 $\mu^3$) and the percentage of cells also varied considerably from one patient
to another (6 to 31%). The mean of the ratios $V_2/V_1$ was 2.1, and due to the
precision of the measurement, this ratio is not significantly different from 2.
 Autoradiographic studies of the same cells after incubation with tritriated
thymidine demonstrated a correlation between cells in the synthesis phase and
cells having a large volume.
 In 16 cases of chronic lymphocytic leukemia (CLL), the mean volume varied
from one patient to another, and the distribution demonstrated a single population. The ratios of the volume of the cytoplasm, the nucleus, the heterochroma-
tin, euchromatin and mitochondria varied little from one patient to another. A
sphericity index of the cytoplasmic membrane differentiated the smooth from the
villous cells and permitted one to place 12 patients into these two categories.

*National Bureau of Standards Special Publication 431*
*Proceedings of the* FOURTH INTERNATIONAL CONGRESS FOR STEREOLOGY
*held at* NBS, Gaithersburg, Md., September 4-9, 1975 (Issued January 1976)

SIMPLE DEVICES FOR STEREOLOGY AND MORPHOMETRY

by Hans Elias and Erich Botz
*Anatomic Institute, University of Heidelberg, Germany*

## Measuring and Point Counting Device

A light box with low voltage and a variable transformer is placed under the end of a focusable camera lucida (drawing tube ) of a light microscope (Fig. 1).

At first, the drawing tube is installed on the microscope body, immediately above it the photo attachment, while the binocular viewing piece takes the highest position. Using this arrangement, a luminous scale or grid consisting of bright lines on a black background placed on the light box is reflected, by the drawing tube, into the microscope so that its image appears simultaneously in the viewing eye pieces and on the film. Using scales or calibrated grids on the light box, direct measurements and point counts are independent of the choice of oculars, since the drawing tube produces an image of the scale or grid in the plane of the primary, real image.

The determination of a volume ratio

$$V/V_T = P/P_n$$

is independent of the dimensions of the point grid, since we are searching, in this case, for a dimensionless ratio. For all other stereological operations, such as

$$L/V_T = \frac{2P}{A_T} \ , \quad S/V_T = \frac{2P}{L_T} \ , \quad N/V_T = \frac{n}{A(D + t - 2h')} \ ,$$

the length of test lines and the size of the test area must be known.

Also for direct morphometric measurements such as the length of a longitudinally sectioned tubular gland or its diameter  or the thickness of a mucous membrane, the length of the luminous scale in relation to the object must be known.

A separate scale and a separate grid must be prepared for each specific objective. The scales or grids are calibrated by means of a stage micrometer. Figure 2 shows an example. The primary magnification of the objective is written on the left side and the optical length of the lines is written on the right side.

If photomicrographs are taken with this device, scales and grids automatically projected on the film partake in any secondary enlargement and remain invariant in regard to the image.(Fig. 4)

## The Particle Size Classifier

A perfected model of the optico-electrical particle size classifier, originally described by Schwartz and Elias (1970), will be demonstrated.

An illuminated iris diaphragm is reflected into the microscope by the focusable zoom drawing tube.

This device permits, together with the standard curves by
Hennig and Elias (1970), to determine rapidly size distribution
and the mean diameter of spheroid particles from their sections.

The observer searches, at first, for the largest section
through a particle.  The image of the fully open diaphragm is
superimposed on the image of that largest section (profile), so
that the area of the light circle (i.e. the image of the illuminated diaphragm) equals, as close as possible, the image of the
largest particle section.  This can be accomplished by the proper
choice of the objective, the correct position of the zoom device
and, if necessary, by varying the distance of the diaphragm from
the drawing tube.  If the particles to be measured are very small,
the distance of the diaphragm can be increased by inserting a
pair of mirrors between the camera lucida and the diaphragm.
Fig. 5 shows the open diaphragm superimposed on the largest
glomerular profile in a kidney.

The lever of the diaphragm (Fig. 3) is provided with two
contact points.  When depressed, one contact point closes a circuit with a continuous metal arc.  It activates the totaling
counter (on the right of the counter battery).

The other contact point closes a circuit by touching a
small metal disc and activates an individual counter for a
specific size class.

Ten positions of the iris diaphragm permit the classification
of sectional circles into ten size classes, a subdivision which
is sufficient and realistic.

On the counter battery one can read the number of sectional
circles for each size class, and from these numbers a histogram
is constructed.  Often, the two smallest size classes are empty or
deficient because polar or near polar "sections" through the particles are frequently unidentifiable, since they are too thin to
be visible, or they fall out of the slices.

These lowest size classes must be reconstituted by drawing a
smooth curve through the upper midpoints of the bars of the graph.
This curve, which often ends with size class 0.3 Dmax, must be
smoothly drawn through to origin of the coordinate system.

The next step is to match the resulting curve with one of our
standard curves (Hennig and Elias, 1970) for the size distribution
of sectional circles.

To each standard curve x for the distribution of sectional
circles there exists a coordinated curve $x_0$ which indicates the
size distribution of the spheres in space.  For each such curve
for the spheres there exists, in our curves, an indication of the
mean diameter of all spheres in the mixture.

If the observed, constructed curve does not exactly match one
of the standard curves, it will fall between two of them, and the
specific distribution can be found by interpolation.

References

Schwartz, E. and H. Elias, 1970, J. Microsc. 91: 57-59
Hennig, H. and H. Elias, 1970, J. Microsc. 93: 101-107

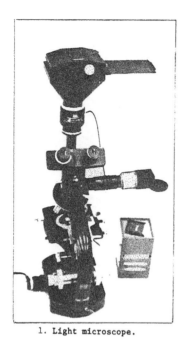

1. Light microscope.

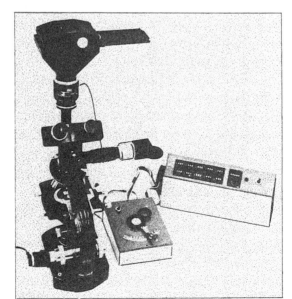

2. Sample grid.

3. Circuit arrangement.

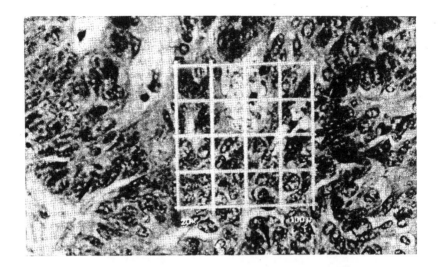

Fig. 4 : "Section" of colon carcinoma photographed with objective 20 x. (100 micron)$^2$ counting grid superimposed.

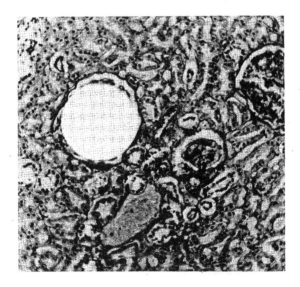

Fig. 5 : "Section" of kidney. Illuminated, open diaphragm superimposed on largest glomerular profile.

National Bureau of Standards Special Publication 431
Proceedings of the FOURTH INTERNATIONAL CONGRESS FOR STEREOLOGY
held at NBS, Gaithersburg, Md., September 4-9, 1975 (Issued January 1976)

THE EDEMA SYNDROME IN MOTION PICTURES

by Oscar C. Jaffee
Department of Biology, University of Dayton, Dayton, Ohio 45469, U.S.A.

The early embryo heart tube has a characteristic form
(Jaffee, '67; Patten, '68) which was illustrated in motion
pictures of 60 hr (incubation) chick embryo hearts. Moderate
hypoxia (Jaffee, '74) administered at this stage was followed
by an osmotic upset coupled with a depression of cardiac function
(chief features of the edema syndrome). Under these conditions
a swelling of the heart tube and some of the great vessels along
with a distortion of the form of the heart tube was noted.

A majority of the treated embryos revert to normal cardiac
function when returned to normal oxygen concentrations; more
than 25% of these survivors still display distorted heart tubes
even though normal function was restored. The arterial outflow
tract, a major site of cyanotic congenital heart disease was a
portion of the heart tube frequently malformed. In a following
series shown embryos had been treated as stated above but now
incubated until five days and then examined and photographed.
Heart tube form was still abnormal in many of those coupled
with abnormal blood flow patterns suggestive of ventricular
septal defects and double outlet right ventricles; both of these
malformations had been previously shown to follow hypoxic
exposure of embryos (Jaffee, '74).

The stereological form of the heart tube and of the
patterns of blood flow through the tube are major factors in
normal cardiogenesis (cf. Jaffee, '67). Changes in heart tube
form and flow patterns, produced by environmental changes, have
been traced into recognizable cardiac malformations, emphasizing
the morphogenetic role of these factors.

References

Jaffee, O.C., 1967. Anat. Rec. 158: 35-42.
Patten, B.M., 1968. Human Embryology, McGraw-Hill.
Jaffee, O.C., 1974. Teratology 10: 275-282.

PART II.  WORKSHOP SECTIONS

7.  MATHEMATICAL FOUNDATIONS OF STEREOLOGY

*Chairperson:*

R. T. De Hoff

*National Bureau of Standards Special Publication 431*
*Proceedings of the* FOURTH INTERNATIONAL CONGRESS FOR STEREOLOGY
*held at* NBS, Gaithersburg, Md., September 4-9, 1975 (Issued January 1976)

THE MATHEMATICAL FOUNDATIONS OF STEREOLOGY

The Proceedings of a Workshop held at the Fourth International Congress for Stereology

The inference of geometric information about three dimensional structures from observations made in lower dimensions (i.e., on plane sections, test lines, or point grids) involves the amalgamation of a number of components of the broad field of mathematics, including measure theory, topology, differential and integral geometry, probability theory, and statistical inference. In this workshop, four of the basic relations of stereology were presented in a series of derivations which began with the mathematical foundations and were developed into the simple, working relationships of stereology. A generalization of the intuitive notion of size was also presented. Some general comments on the use of statistics in the practical application of these simple results made the program for the workshop self-contained.

## I. OVERVIEW

by R. T. DeHoff, Department of Materials Science and Engineering, University of Florida, Gainesville, Florida, USA

A microstructure consists of points, lines, surfaces and volumes distributed in three dimensional space. Usually the surfaces are defined by the juxtaposition of two distinguishable kinds of volumes; lines by three, and points by four.* The microstructure may be defined descriptively, i.e., qualitatively; or it may be quantified by applying stereology, or some alternative techniques (DeHoff, (1975)). Thus, it is possible to define a

A. Qualitative Microstructural State.
In its most rudimentary form, this state is simply a list of all of the zero, one, two and three dimensional features that exist in the structure. If the number of distinguishable kinds of volumes in the structure exceeds two, such a listing will not be a trivial matter, as can be seen from Tables 1.1 through 1.3. The qualitative microstructural state may be further specified by descriptive statements about the features contained: e.g., "fine", or "coarse"; "small" or "large" amount; "equi-axed" or "elongated"; "simple" or "complex". If further specification of the structure is required, one may define its

B. Quantitative Microstructural State.
Each of the features listed in Tables 1.1 through 1.3 have associated with them geometric properties. The quantitative microstructural state of a structure is defined by assigning numerical values to one or more of these properties. The level of quantification is related to the number of different geometric properties evaluated for the structure. The properties themselves can be classified into two categories:
    1. The Quantitative Topological State is specified by evaluating the topological properties of the structure, Table 1.4. Topological properties include primarily (a) the number of

---

*Exceptions to these restrictions require a very high degree of homogeneity and order in the structure.

disconnected parts of the feature and (b) the connectivity, or number of redundant connections, of a connected feature (DeHoff, Aigeltinger and Craig, (1972)). These properties are listed in Table 1.4. Topological properties are not further developed in this workshop, because they are inaccessible to measurements made on sections, and can only be estimated from reconstructions of the three dimensional structure.

TABLE 1.1
Features that Exist in Single Phase Structures

| Dimension | Feature | Description |
|-----------|---------|-------------|
| 3 | Volumes (Grains) | α |
| 2 | Surfaces (Grain Boundaries) | αα |
| 1 | Curves (Grain Edges) | ααα |
| 0 | Points (Grain Corners) | αααα |

Total: 4 Features

TABLE 1.2
Features that May Exist in Two Phase Structures

| Dimension | Feature | Description | |
|-----------|---------|-------------|---|
| 3 | Volumes (Grains, Particles) | α β | |
| 2 | Surfaces (Grain Boundaries, Interfaces) | αα ββ αβ | |
| 1 | Curves (Grain Edges, Triple Lines) | ααα ααβ αββ βββ | On αβ Interface |
| 0 | Points (Quadruple Points) | ββββ αααβ ααββ αβββ ββββ | On αβ Interface |

Total: 14 Features

TABLE 1.3
Features that May Exist in Three Phase Structures

| Dimension | Feature | Description |
|---|---|---|
| 3 | Volumes (Grains, Particles) | α β γ |
| 2 | Surfaces (Grain Boundaries, Interfaces) | αα ββ γγ αβ βγ αγ |
| 1 | Curves (Grain Edges, Triple Lines) | ααα, βββ, γγγ ααβ, αββ ααγ, αγγ ββγ, βγγ αβγ |
| 0 | Points (Grain Corners, Quadruple Points) | αααα, ββββ, γγγγ αααβ, ααββ, αβββ αααγ, ααγγ, αγγγ βββγ, ββγγ, βγγγ ααβγ, αββγ, αβγγ |

Total: 34 Features

TABLE 1.4
Topological Properties of Features in Two Phase Structures

| Dimension | Feature | Designation | Property Number | Property Connectivity |
|---|---|---|---|---|
| 3 | Volumes | α | ✓ | ✓ |
|  |  | β | ✓ | ✓ |
| 2 | Surfaces | αα |  |  |
|  |  | ββ | * | - |
|  |  | αβ | ✓ | ✓ |
| 1 | Curves | ααα |  | ** |
|  |  | βββ |  | ** |
|  |  | ααβ |  | ** |
|  |  | αββ |  | ** |
| 0 | Points | αααα | ✓ |  |
|  |  | ββββ | ✓ |  |
|  |  | αααβ | ✓ |  |
|  |  | ααββ | ✓ |  |
|  |  | αβββ | ✓ |  |

Total Topological Properties: 21

*Number of individual elements and number of disconnected networks may be separately defined.
**Connectivity applies to network only, since the elements of the network are simply connected.

2. <u>The Quantitative Metric State</u> is specified by evaluating the metric properties of the structure, Table 1.5 (DeHoff and Rhines, (1968)).

.TABLE 1.5
Metric Properties of Features in Two Phase Structures

| Dimension | Feature | Designation | Property Symbol | Description |
|-----------|---------|-------------|--------|-------------|
| 3 | Volumes | $\alpha$<br>$\beta$ | $V_V{}^\alpha, V_V{}^\beta$ | Volume Fraction |
| 2 | Surfaces | $\alpha\alpha$<br>$\beta\beta$ | $S_V{}^{\alpha\alpha}$<br>$S_V{}^{\beta\beta}$ | Surface Area Per Unit Volume |
| | | $\alpha\beta$ | $S_V{}^{\alpha\beta}$<br>$K_V{}^{\alpha\beta}$ | Total Curvature Per Unit Volume |
| 1 | Curves | $\alpha\alpha\alpha$<br>$\beta\beta\beta$ | $L_V{}^{\alpha\alpha\alpha}$<br>$L_V{}^{\beta\beta\beta}$ | Line Length Per Unit Volume |
| | | $\alpha\alpha\beta$<br>$\alpha\beta\beta$ | $L_V{}^{\alpha\alpha\beta}$<br>$L_V{}^{\alpha\beta\beta}$<br>$k_V{}^{\alpha\alpha\beta}$ | Total Curvature Per Unit Volume |
| | | | $k_V{}^{\alpha\beta\beta}$ | |
| 0 | Points | - | | No Metric Properties |

Total Metric Properties: 11

The most obvious of these properties are those that report the extent of the feature:
(a) $V_V$, the <u>volume fraction</u>, reports the total volume of three dimensional features, per unit volume of structure;
(b) $S_V$, the <u>surface area per unit volume</u>, reports the total area of each specific kind of two dimensional feature in the unit volume;
(c) $L_V$, the <u>line length per unit volume</u>, reports the length of any lineal feature in unit volume of structure.
One additional parameter is defined and derived in this workshop;
(d) $K_V$, the <u>total curvature in unit volume</u>, is the integral of the local mean surface curvature over the area of surface in the volume, i.e.,

$$K_V = \iint_{S_V} \frac{1}{2}(\frac{1}{r_1} + \frac{1}{r_2})\, dS \qquad (1.1)$$

where $r_1$ and $r_2$ are the principal normal radii of curvature on the surface. Their reciprocals, $\kappa_1$ and $\kappa_2$, are the principal normal curvatures. These properties are called <u>metric</u> properties, because, in contrast to the topological properties, their values depend upon the dimensions or sizes of the features in the structure.

The essence of stereology is the demonstration that these geometric properties of features that exist in a three dimensional microstructure can be estimated from simple measurements, in fact from counts of statistical events that occur, made upon representative sections through the structure. The connections between the higher and lower dimensions are founded upon integral geometry, as is now developed.

## II.  PROBABILISTIC FOUNDATIONS OF STEREOLOGY

by R. E. Miles and Pamela Davy, Department of Statistics,
Institute of Advanced Studies, Australian National University,
P.O. Box 4, Canberra, A.C.T. 2600, Australia

Points $E_0$, lines $E_1$ and planes $E_2$ in space $E_3$ are parametrized by $(x, y, z)$, $(\theta, \phi; \xi, \eta)$ and $(\theta, \phi; p)$, respectively, where $(\theta, \phi)$ are spherical polar coordinates of the line or normal to the plane, $(\xi, \eta)$ are cartesian coordinates in the plane

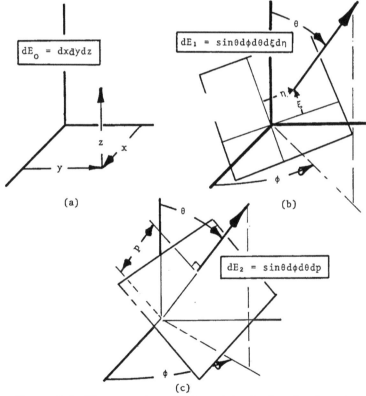

$$dE_0 = dxdydz$$

$$dE_1 = \sin\theta d\phi d\theta d\xi d\eta$$

$$dE_2 = \sin\theta d\phi d\theta dp$$

(a)          (b)

(c)

Figure 2.1  The parameters that must be assigned values to specify (a) a point, (b) a line, and (c) a plane.

through the origin perpendicular to the line, and p is the length of the perpendicular from the origin to the plane. The standard isotropic invariant volume elements

$$dE_0 = dxdydz, \qquad (2.1a)$$
$$dE_1 = \sin\theta \, d\theta d\phi d\xi d\eta, \qquad (2.1b)$$
$$dE_2 = \sin\theta \, d\theta d\phi dp \qquad (2.1c)$$

of integral Geometry for points, lines and planes are discussed, and the resulting measures

$$M_0(Y) = \int dE_0 = V(Y), \qquad (2.2a)$$

$$M_1(Y) = \int dE_1 = S(Y), \qquad (2.2b)$$

$$M_2(Y) = \int dE_2 = K(Y) \qquad (2.2c)$$

of the sets of points, lines and planes hitting a convex domain Y are derived; here V, S and K denote volume, surface area and integral of mean curvature, respectively.

The corresponding normalized (and therefore probability) elements

$$dE_0/M_0(Y), \quad dE_1/M_1(Y), \quad dE_2/M_2(Y)$$

govern the distribution of _uniform_ _random_ (UR) points in Y and _isotropic_ _uniform_ _random_ (IUR) lines and planes hitting Y. Next a classic result of Integral Geometry/Geometrical Probability, the

Crofton/Hostinsky Theorem. Let the distance between two independent UR points of Y be R, and let the length of the secant determined by an IUR line hitting Y by L. Then, with $\underline{E}$ denoting expectation,

$$(i+3)(i+4)V(Y)^2 \underline{E}(R^i) = \pi S(Y) \, \underline{E}(L^{i+4}) \quad (i \neq -3, -4). \qquad (2.3)$$

is presented, since it exhibits the essential elementary features of these two disciplines.

In the stereological context, the specimen X is taken as an opaque bounded connected (but not necessarily convex) domain which contains embedded sub-volumes, surfaces, curves and/or points.

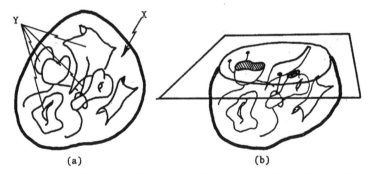

(a)                              (b)

Figure 2.2  The set X is the sample; the set Y consists of 0, 1, 2 and 3 dimensional features that make up the "internal structure" of X.

To these correspond the stereological 'fractions'

$$V_V = \frac{V}{V(X)}, \quad S_V = \frac{S}{V(X)}, \quad L_V = \frac{L}{V(X)}, \quad P_V = \frac{P}{V(X)} \qquad (2.4)$$

for volume, area, length and number, respectively. Also of importance for embedded surfaces are the integral of mean curvature and integral of gaussian curvature fractions

$$K_V = \frac{1}{2}\int(\kappa_1+\kappa_2)dS/V(X), \qquad (2.5a)$$

$$G_V = \int \kappa_1\kappa_2 \ dS/V(X). \qquad (2.5b)$$

It is shown that, for an IUR plane section of X,

$\underline{E}$ (area of section) = $2\pi \ V(X)/K(\bar{X})$           (2.6)
$\underline{E}$ (area of $\alpha$ phase sectioned) = $2\pi \ V(X_\alpha)/K(\bar{X})$   (2.7a)
$\underline{E}$ (length of intersection curve with embedded surface) =
    $(\pi^2/2)$ S(embedded surface)/K(X)         (2.7b)
$\underline{E}$ (total curvature of intersection curve with embedded
    surface) = $2\pi$ K(embedded surface)/K(X), and    (2.7c)
$\underline{E}$ (number of intersection points with embedded curve) =
    $\pi$ L(embedded curve)/K(X),           (2.7d)

where $\bar{X}$ denotes the convex hull of X. Note that each of (2.7a - d) is of the form

(2)   $\underline{E}(I) = 2\pi a$ J(phase, surface or curve)/$K(\bar{X})$.    (2.8)

It follows from (2.6) and (2.7) that

(3)   $\underline{E}(I_A) = aJ_V - \{K(\bar{X})/2\pi V(X)\} \ \underline{C}[I_A, A]$,      (2.9)

where $\underline{C}$ denotes covariance. That is, the customary stereological estimates for IUR plane sections are biased. The biases may be positive or negative, depending upon the spatial distribution of the phase, surface or curve within the specimen X.

However, the corresponding estimates $I_A^*$ for a different type of random plane section of X, viz. an IUR plane section weighted by its cross-sectional area with X or, equivalently, an isotropic plane through a UR point of X, are <u>unbiased</u>. Moreover, the mean square error of $I_A$ minus the variance of $I_A^*$ is

$$\{K(\bar{X})/2\pi V(X)\} \ \underline{C}[I_A(2aJ_V - I_A), A] \qquad (2.10)$$

which may be expected to be positive in general.

In fact, these results extend to specimens and sections of all dimensionalities. In this respect, this type of 'weighted' section possesses a further practical advantage not possessed by the corresponding IUR sections: such a line section of a spatial specimen is probabilistically equivalent to such a line section of such a plane section of the specimen.

## III. DERIVATIONS OF THE WORKING RELATIONSHIPS OF STEREOLOGY

### A. Volume Fraction Analysis*

by Hans Eckert Exner, Max-Planck,Institut für Metallforschung,
Stuttgart, W.-Germany

Volume fraction analysis is the oldest and, apart from grain size measurements, by far the most common procedure in practical stereology. It is particularly simple to obtain the volume fraction $V_V$ from plane sections, since volume fraction, area fraction, lineal fraction, and point fraction are all equal. The derivations of these equalities were first given by Delesse (1848), Rosiwal (1898) and Glagolev (1934), respectively, and have more recently been demonstrated by Hilliard (1961, 1968), Weibel (1963), and Underwood (1970).

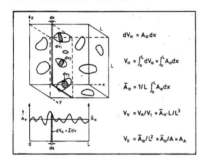
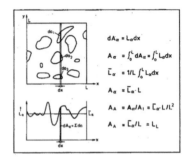

Figure 3.1 Derivation of $V_V=A_A$.  Figure 3.2 Derivation of $A_A=L_L$.

A simple integration of the area fraction over the volume of the sample yields the equality

$$V_V = A_A \qquad (3.1)$$

(Figure 3.1). Similarly, the equality

$$A_A = L_L \qquad (3.2)$$

is obtained (Figure 3.2). For proving the equality

$$V_V = P_P \qquad (3.3)$$

a different approach is taken relating the probability of hitting an α-feature by a point in the volume (Figure 3.3).
The experimental errors for a fixed number of section areas, linear intercepts, or hits on the α-phase are shown in Figure 3.4.

---

*This workshop paper consists entirely of information already available in the literature. Only random structures and plane random cross sections through opaque samples are considered. (Mathematical rigidity was thought to be less pertinent than simplicity in this context.)

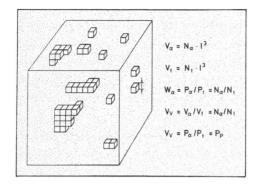

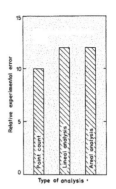

Figure 3.3 Schematic derivation of the equality $V_V = P_P$ ($V_\alpha$, $V_t$, $N_\alpha$, $N_t$: volume and number of elements of $\alpha$-features and the total cube. $W_\alpha$ = probability that $P_\alpha$ out of $P_t$, points fall on the $\alpha$-phase).

Figure 3.4 Experimental error due to statistical variations for 100 observations according to Hilliard (1961, 1968).

Systematic point counting (using a grid the spacing of which is large enough to prevent two points falling into the same $\alpha$-section) has the smallest error due to the fact that no variations in size and no measuring errors are involved, in contrast to lineal and areal analysis. Therefore, systematic point counting is usually the preferred method. The estimated range for the true volume fraction is

$$V_V \pm \Delta V_V = P_P \pm \sqrt{P_P \cdot (1 - P_P)}/\sqrt{P_t} \qquad (3.4)$$

where $P_t$ is the total number of points examined. Thus, the experimental error decreases in proportion to the inverse of the square root of the total number of points counted. It also can be seen from this useful relationship that a large total number of points must be counted for small volume fractions if a high relative accuracy is required. (For $\Delta V_V/V_V$ = 10% and volume fractions of 10 and 1%, 900 and 9 900 points must be counted, respectively.)

Figure 3.5 shows an example of a volume fraction analysis by

| Systematic point count with Weibel-grid | |
|---|---|
| One grid count | 100 grid counts |
| $P_t$ = 50 | $P_t$ = 5000 |
| $P_\alpha$ = 20 + 6/2 = 23 | $P_\alpha$ = 2138 + 642/2 = 2460 |
| $P_P$ = 23/50 = 0.46 | $P_P$ = 2460/5000 = 0.492 |
| $\pm P_P = \sqrt{0.46 \cdot 0.54/50}$ | $\pm P_P = \sqrt{0.492 \cdot 0.508/5000}$ |
| = 0.07 | = 0.0071 |
| $V_V \pm \Delta V_V$ = 46 ± 7 % | $V_V \pm \Delta V_V$ = 49.2 ± 0.7% |

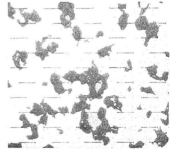

Figure 3.5 Volume fraction analysis by point counting. Points falling on phase boundaries are counted 1/2.

point counting. For one grid count, the result is too inaccurate to be useful. For 100 grid counts, the experimental error is reduced to 1 absolute pct or 2 relative pct.

If automatic instruments are used, the tedious counting and measuring procedures are speeded up. The experimental error, however, is reduced only if a larger number of α-sections is sampled than in point counting, and if no systematic errors are introduced due to insufficient phase detection. Therefore, there is no point in putting some hundred thousand points on a few features in a small field of the sample since these will characterize only the area fraction of this particular field. This will usually be a poor estimate of the average area fraction and, therefore, the volume fraction.

B.  Surface Area

by Ewald R. Weibel, Department of Anatomy, University of Bern, Bern, Switzerland

Consider the measurement of surface area in a simple structure made of two phases: an object phase embedded in a matrix. The interface between these two phases is measured by its surface area S, a stereologically measurable quantity being the surface density $S_V$ in the unit containing volume, i.e. in the volume V of the entire structure:

$$S_V = S/V \qquad (3.5)$$

Stereology offers two basic methods by which $S_V$ can be determined (Cornfield (1951), Smith (1953), Saltykov (1954), Tomkeieff (1945): (a) the structure is sectioned by a plane, $E_2$, (Figure 3.6a); the containing volume is represented on this section

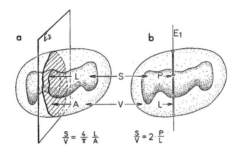

Figure 3.6 Fundamental relationships between surface, boundary length, and intersection densities.

by the area A of its profile, whereas the surface leaves a linear trace of length L. The following relationship holds:

$$S_V = \frac{S}{V} = \frac{4}{\pi} \cdot \frac{L}{A} = \frac{4}{\pi} \cdot L_A \qquad (3.6)$$

(b) The structure is intercepted by a test line, $E_1$, (Figure 3.6b); the containing volume is represented by the length of test line lying within the structure, L, and the surface by the number of point intersections, P, formed as the line traverses the surface, with the following relationship pertaining

$$S_V = \frac{S}{V} = \frac{2P}{L} = 2P_L \qquad (3.7)$$

These two relations shall now be derived, starting with a very simple case and then proceeding to more generally valid derivations. For reasons of simplicity we shall test the structure with a set of parallel test lines or planes; due to this we must require that the surface within the entire structure is isotropic, i.e. all orientations of surface elements must be equally likely. If this does not hold, the direction of the test lines must be suitably randomized.

1. <u>Line intersection measurement</u>
   a. <u>Isotropic model 1: sphere in cube</u>
   Let us consider first a simple structure of a cube of side $\ell$ containing a sphere of radius R (Figure 3.7). With $V = \ell^3$ and

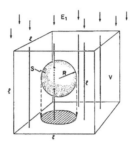

Figure 3.7   Isotropic model I.

$S = 4\pi R^2$ the surface density is

$$S_V = \frac{4\pi R^2}{\ell^3} \qquad (3.8)$$

This structure is now sampled with a "beam" of parallel vertical lines; some of these will pass through the cube, call their number n, and some will also intersect the sphere, call this $n_0$. What is the probability that such a line $E_1$ intersects the sphere? From Miles' presentation [Section II], it follows that this probability is the ratio of the measure of the enclosed object Y to the measure of the enclosing space X*:

$$p\{E_1 \uparrow S \subset V\} = \frac{M_1 |Y|}{M_1 |X|} \qquad (3.9)$$

These measures are the orthogonal projections of object and structure, if you want the "shadows" projected by the beam of lines onto a plane perpendicular to the lines. For the simple model of Figure 3.7 the projected area of the cube is $\ell^2$ and that of the sphere $\pi R^2$, so that

$$p\{E_1 \uparrow S \subset V\} = \frac{\pi R^2}{\ell^2} = \frac{n_0}{n} \qquad (3.10)$$

---

*The left side of equation (3.9) is read "the probability that a test line ($E_1$) intersects ($\uparrow$) a surface (S) contained in ($\subset$) a sample volume (V)".

This probability can evidently be empirically estimated by the ratio of the number of lines crossing the sphere to those crossing the cube.

Now notice the following relationships: the number of intersections formed between test lines and sphere surface is

$$P = 2 \cdot n_0 \qquad (3.11)$$

one for the entrance and one for the exit, and the total test line length within the structure is

$$L = \ell \cdot n \qquad (3.12)$$

The density of intersections per unit test line length is hence

$$P_L = \frac{P}{L} = \frac{2n_0}{\ell \, n} \qquad (3.13)$$

Substituting $n_0/n$ from eq. (3.10), we find

$$P_L = \frac{2\pi R^2}{\ell^3} \qquad (3.14)$$

Comparing this with the surface density defined by equation (3.8), we find that $P_L$ is just half the surface density; it follows immediately that

$$S_V = 2P_L$$

which is equation (3.7).

b. Isotropic model II: randomly oriented surface element

We now relax the restrictions on the shape of the objects (Figure 3.8); they may have any arbitrary shape, and we consider

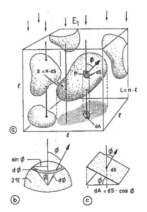

Figure 3.8 (a) Isotropic model II intersected by lines; (b) Orientation hemisphere; (c) Surface element and projected area.

their surface to be broken up into a very large number N of small surface elements of equal area dS so that the total object surface is

$$S = \int dS = N \cdot dS \qquad (3.15)$$

As containing space we retain, for reasons of simplicity, the cube of side $\ell$, so that the surface density is

$$S_V = \frac{\int dS}{\ell^3} = \frac{N \cdot dS}{\ell^3}$$

and we sample again with a beam of parallel vertical lines.

Isotropicity of the enclosed surface is ensured if all orientations of the surface elements, measured by the angle $\phi$ between the normal to the surface element and the direction of the test lines (Figure 3.8), are equally likely.

The derivation of the expected number of intersections follows the same reasoning as above: the probability that a line intersects one surface element is

$$p\{E_1 \uparrow dS \subset V\} = \frac{\overline{dA}}{\ell^2} \qquad (3.16)$$

where $\overline{dA}$ is the mean orthogonal projection of the surface element and $\ell^2$ is the projection of the cube, as above; $dA$ evidently depends on $dS$ and on the angle $\phi$, such that (Figure 3.8c)

$$dA(\phi) = dS \cdot \cos\phi \qquad (3.17)$$

Averaging must be performed over all orientations $\phi$, whereby we must note that not all angles $\phi$ are equally likely! Note that, in this formulation, we have already integrated out the angle $\theta$ measuring orientation with respect to one of the axes in the horizontal plane. Because of this, the probability of having an orientation between $\phi$ and $(\phi + d\phi)$ is

$$p\{\phi, d\phi\} = \sin\phi d\phi \qquad (3.18)$$

which is the area of a belt of width $d\phi$ and length $\sin\phi$ divided by the area of the orientation hemisphere $2\pi$ (assuming radius 1) (Weibel, (1963)). It is evident from Figure 3.8c that the area of this belt increases as $\phi$ increases. The average projection area of $dS$ is hence obtained by

$$\overline{dA} = \frac{\int_0^{\pi/2} dS \, \cos\phi \cdot \sin\phi \cdot d\phi}{\int_0^{\pi/2} \sin\phi \cdot d\phi} = \frac{dS}{2} \qquad (3.19)$$

By equation (3.16) this yields for the probability of obtaining an intersection with one surface element

$$p\{E_1 \uparrow dS \subset V\} = \frac{1}{2} \frac{dS}{\ell^2} \qquad (3.20)$$

If the object surface is made of $N$ surface elements and if $n$ test lines intersect the cube, the expected number of intersections in the entire structure is

$$\underline{E}(P) = N \cdot n \cdot p\{E_1 \uparrow dS \subset V\} \qquad (3.21)$$

Noting that $N = S/dS$ and $n \cdot \ell = L$, the total test line length, we obtain

$$\underline{E}(P) = \frac{S}{dS} \cdot \frac{L}{\ell} \cdot \frac{dS}{2\ell^2} = \frac{1}{2} \cdot \frac{S}{V} \cdot L \tag{3.22}$$

which finally leads us, by rearrangement, to

$$P_L = \underline{E}(P)/L = \frac{1}{2} S_V \tag{3.23}$$

a result identical to equation (3.7). Note that in this model the object surface can have any shape -- as long as it is isotropic on the whole -- and that the surface need not be closed; indeed, it can be fragmented into as small elements as is possible.

We could now also relax the condition on the shape of the containing space; it need not even be convex. In this general case the test line length $L_T$ enclosed in this space is the sum of all intercept lengths $\ell_i$:

$$L_T = \sum_i \ell_i \tag{3.24}$$

Eq. (3.23) would then read, after slight rearrangement

$$S_V = \frac{2 \cdot \Sigma P}{\Sigma \ell} = \frac{2\underline{E}(P)}{\underline{E}(L)} = 2P_L \tag{3.25}$$

It is important to note that in the case where the individual test line intercept with the structure has variable length, the unbiased estimator of $S_V$ is 2 times the ratio of the sum of intersection points to the sum of the test line lengths.

2. Boundary length measurement

To derive equation (3.6) we shall use the second model introduced in the preceding section (Figure 3.8a): a surface which can be fragmented into a large number of small surface elements dS (Figure 3.9).

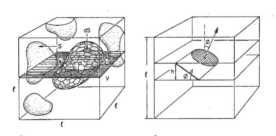

Figure 3.9 (a) Isotropic model II intersected by plane; (b) Caliper diameters of containing space and surface element.

The problem is now to find the probability that a test plane $E_2$ intersects a surface element. In this case the measure $M_2$ is the caliper diameter, i.e. the distance between the two tangent planes parallel to the section plane (= the orthogonal projection onto a line perpendicular to the section plane). For the cube this measure is $\ell$, and for the surface element (Figure 3.9b)

$$h = d \cdot \sin\phi \tag{3.26}$$

where d is the diameter of the element. Again we must average over all orientations:

$$\bar{h} = \frac{\int_0^{\pi/2} d \cdot \sin^2\phi \cdot d\phi}{\int_0^{\pi/2} \sin\phi \cdot d\phi} = \frac{\pi}{4} \cdot d \qquad (3.27)$$

so that

$$p\{E_2 \dagger dS \subset V\} = \frac{\bar{h}}{\ell} = \frac{\pi}{4} \cdot \frac{d}{\ell} \qquad (3.28)$$

The average length of line intercept between surface element and section plane is

$$\underline{E}(dL) = \pi \cdot \frac{A(dS)}{B(dS)} = \frac{\pi}{4} \cdot d \qquad (3.29)$$

where

$$A(dS) = \frac{\pi}{4} d^2 \qquad (3.30)$$

$$B(dS) = \pi d \qquad (3.31)$$

are area and circumference of the element, respectively [Section II].

Assuming the structure to contain a large number

$$N = \frac{S}{A(dS)} \qquad (3.32)$$

of such elements, the total length of intercepts on a test plane of area $A = \ell^2$ is

$$\Sigma \, dL = N \cdot \underline{E}(dL) \cdot p\{E_2 \dagger dS \subset V\}$$

$$= \{N \cdot \frac{\pi}{4} \cdot d^2\} \cdot \frac{\pi}{4} \cdot \frac{1}{\ell} \qquad (3.33)$$

where the term in square brackets is the total object surface. Dividing both sides by $\ell^2$ we obtain

$$\frac{\Sigma dL}{\ell^2} = \frac{\pi}{4} \frac{S}{\ell^3} \qquad (3.34)$$

This amounts to

$$L_A = \frac{\pi}{4} \cdot S_V$$

which is equal to equation (3.6).

Here again the containing space need not be a cube. In analogy to equation (3.25) we find as an unbiased estimator of surface density

$$S_V = \frac{4}{\pi} \cdot \frac{\Sigma L}{\Sigma A} = \frac{4}{\pi} \cdot \frac{E(L)}{E(A)} \qquad (3.35)$$

3. **Surface density estimated by two-stage procedure**

The usual practical procedure for estimating $S_V$ is to cut a section and then to count intersections formed by a set of test lines on the plane of section with the linear trace of the surface. The final question is, hence, whether this also yields the relationship $S_V = 2P_L$.

By a reasoning analogous to that used in Section 2 only reduced to a two-dimensional problem, one can show that

$$L_A = \frac{\pi}{2} \cdot P_L \qquad (3.36)$$

Break the linear (surface) trace into N short segments. (Figure 3.10),

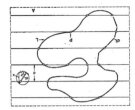

Figure 3.10   Length of line on plane.

so that

$$L_A = \int dL/A = N \cdot dL/A \qquad (3.37)$$

Deposit a test line grid of spacing d, so that

$$L_T = A/d \qquad (3.38)$$

The probability that such a segment is intersected by a test line is (Buffon (1777), Weibel (1967))

$$p\{E_1 \uparrow dL \subset A\} = \frac{2}{\pi} \cdot \frac{dL}{d} \qquad (3.39)$$

which is obtained again by allowing all orientations $\theta$ between the segment and the test lines to be equally likely (Weibel, (1967)) (Figure 3.10). The average number of intersections is

$$E(P) = N \cdot p\{E_1 \uparrow dL \subset A\} = \frac{2}{\pi} \cdot \frac{L}{d} \qquad (3.40)$$

or, by appropriate substitution for d from equation (3.38) and rearrangement

$$\frac{E(P)}{L_T} = \frac{2}{\pi} \cdot \frac{L}{A} \qquad (3.41)$$

which is equivalent to equation (3.36). Allowing again for irregular shape of the enclosing area, the unbiased estimator for $L_A$ is

$$L_A = \frac{\pi}{2} \frac{\bar{E}(P)}{\bar{E}(L_T)} \qquad (3.42)$$

In conclusion, we find that by substituting equation (3.36) into equation (3.6) we end up with equation (3.7):

$$S_V = \underbrace{\frac{4}{\pi} L_A \qquad\qquad\qquad L_A}_{S_V = 2\ P_L} = \frac{\pi}{2} P_L \qquad (3.43)$$

If we allow for variable containing space and section area and hence properly consider the ratio of means as given in equations (3.35) and (3.42), we find that we still obtain equation (3.25).

The final conclusion, therefore, is that the two-stage sampling procedure used in practice does not invalidate the fundamental relationship that twice the density of intersections on the length of test line is an unbiased estimator of surface density.

In equation (3.25) the ratio of surface area to volume is expressed as the ratio of mean intersection number to mean line intercept length. It can easily be shown that this formulation applies also to the estimation of the so-called surface to volume ratio of closed objects, as derived by Tomkeieff, (1945) and Cornfield and Chalkley, (1951), in which case the reference volume is taken to be the mean volume of the objects rather than the volume of an enclosing space.

C.  Length of Lineal Features in Space

by R. T. DeHoff, Department of Materials Science and Engineering, University of Florida, Gainesville, Florida, USA

Space curves exist in microstructures; e.g. dislocations in crystals observed in transmission electron microscopy are space curves, as are the edges (triple lines) in a cell structure. In addition, features that have one dimension very large in comparison to the other two (e.g., tubules) may be considered to be space curves. Such features possess a length. A stereological measure of the length of linear features (Y) in a three dimensional structure (X) is derived in this section.

Figure 3.11 shows such a linear feature Y in a sample volume, X. Figure 3.11b focusses upon an element of length, $d\lambda$, of the feature. According to equation (2.1c), the measure of the set of planes in space is:

$$dE_2 = dp\sin\theta d\theta d\phi \qquad (2.1c)$$

For the orientation shown in Figure 3.11b, event that a plane intersects the element, $d\lambda$, is satisfied by those planes with positions lying within the projection of dp on the direction normal to the sectioning plane. That is, the value of dp satisfying the event of interest is:

$$dp = d\lambda\cos\theta \qquad (3.44)$$

455

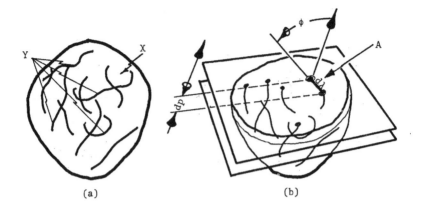

Figure 3.11  A sample containing a set of lineal features (space
curves) Y is shown in (a). The measure of the set of planes
that intersect an element of this set dλ is dp, the projection
of dλ on the direction normal to the test plane.

A count of the total number of such intersections divided by the
total area sampled by a set of test planes yields an estimate of
the total length of the set of lineal features in unit volume of
the structure, X (Saltykov, (1954), Smith, (1953)). This area
point count, $P_A$, is given by the ratio of the measure of the set of
planes that intersect lineal features to the measure of the area of
such planes. The numerator and denominator in this ratio are thus
integrals of the measure given in equation (2.1c):

$$P_A = \frac{\int\int\int_{\text{Intersections}} dE}{\int\int\int_{\text{All Planes}} AdE} = \frac{\int\int\int_{\text{Intersections}} dp \cdot \sin\theta \cdot d\phi d\theta}{\int\int\int_{\text{All Planes}} A \cdot dp \cdot \sin\theta \cdot d\phi d\theta} \qquad (3.45)$$

$$P_A = \frac{\int_0^{2\pi} \int_0^{\pi} \int_\lambda d\lambda \cdot \cos\theta \cdot \sin\theta \cdot d\phi d\theta}{\int_0^{2\pi} \int_0^{\pi} \int_p A \cdot dp \cdot \sin\theta \cdot d\phi d\theta}$$

$$= \frac{\int_\lambda d\lambda \cdot \int_0^{2\pi} d\phi \cdot 2\int_0^{\pi/2} \cos\theta \cdot \sin\theta \cdot d\theta}{\int_p A dp \cdot \int_0^{2\pi} d\phi \cdot 2\int_0^{\pi/2} \sin\theta \cdot d\theta}$$

Note that $\int d\lambda$ is the total length $\lambda$ of the feature in the volume, and that $\int Adp$ is the total volume V of the sample X, no matter what orientation is chosen. Carrying out the integration gives

$$P_A = \frac{\lambda \cdot 2\pi \cdot 2(1/2)}{V \cdot 2\pi \cdot 2 \cdot 1}$$

which simplifies to

$$P_A = \frac{1}{2} \cdot \frac{\lambda}{V} = \frac{1}{2} L_V \tag{3.46}$$

Thus, in a series of plane sections taken through a three dimensional structure, a count of the number of points of emergence of the lineal feature on the planes of observation, divided by the total area sampled, gives an unbiased estimate of the length of the lineal features per unit volume of microstructure.

## D.  The Area Tangent Count and the Total Curvature

by R. T. DeHoff, Department of Materials Science and Engineering, University of Florida, Gainesville, Florida, USA

Figure 3.12a shows a sample volume, X, that contains a set of embedded surfaces, Y.  If this volume is sampled with test planes, intersections with the surfaces will appear as linear traces on the plane of observation, Figure 3.12b.  Suppose the sectioning plane is sampled by sweeping lines across it, each sweep covering the plane area, A.  The event of interest in the present derivation is the formation of tangents between such sweeping test lines and the traces of the surfaces on the planar section, Cahn, (1967), DeHoff, (1967).  The tangent count records the number of times such sweeping lines form tangents in their sweep.  A count is registered as positive if the arc at tangency is convex; a negative count is registered if the arc is concave.  Note that whether an arc is "convex" or "concave" is a matter of convention, to be decided before counting begins.

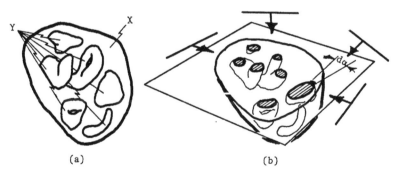

(a)                                          (b)

Figure 3.12  A sample X containing internal surfaces Y, (a), is cut by a sectioning plane of area A, (b).  Test lines swept across the plane from tangents with traces of the surfaces on the sectioning plane.

Recall that the measure of the set of planes in space is given by equation (2.1c).

$$dE_2 = dp \sin\theta d\phi d\theta \qquad (2.1c)$$

The measure of the set of sweeping lines in a plane is the same as the measure of the set of orientations in a plane, and is thus an angular interval, $d\alpha$. Thus, the measure of the set of sweeping test lines in space is

$$dT = d\alpha \ dE = d\alpha \ dp \ \sin\theta d\phi d\theta \qquad (3.47)$$

If A is the area of a sectioning plane, then the sweeping line traverses an area A for each sweep.

The $T_A$ count (number of tangents formed per unit area traversed) is thus given by

$$T_A = \frac{\underset{\text{Tangent Events}}{\iiiint} dT}{\underset{\text{All Values}}{\iiiint} AdT}$$

$$= \frac{\underset{\text{Tangent Events}}{\iiiint} d\alpha \cdot dp \cdot \sin\theta \cdot d\phi d\theta}{\underset{\text{All Values}}{\iiiint} Adp \cdot d\alpha \cdot \sin\theta \cdot d\phi d\theta}$$

For any plane, $\int d\alpha$ is or $\pi$; for any orientation, $\int Adp = V$, the volume of the sample X. Finally, $\iint \sin\theta d\theta d\phi$ is the area of the orientation sphere, $4\pi$. Hence, the denominator may be evaluated:

$$T_A = \frac{\underset{\text{Tangent Events}}{\iiiint} d\alpha \cdot dp \cdot \sin\theta \cdot d\phi d\theta}{4 \cdot \pi^2 V} \qquad (3.48)$$

It remains to evaluate the numerator in equation (3.48). Focus upon an element of length, $d\ell$, produced by intersecting a surface element dS with a sectioning plane. This element has a local curvature, k, which is related to principal normal curvatures of the surface being sectioned in the neighborhood of $d\ell$ by theorems of Meusnier and Euler (Struik, (1951)). This relationship depends upon the orientation of the sectioning plane, (defined by its normal), relative to the coordinate system on the surface defined by the surface normal and the principal directions:

$$k(\phi,\theta) = \frac{1}{\sin\theta} (\kappa_1 \cos^2\phi + \kappa_2 \sin^2\phi) \qquad (3.49)$$

A sweeping line in the plane will form a tangent with $d\ell$ if its direction of sweep lies within the interval of normal directions along $d\ell$, Figure 3.13a.

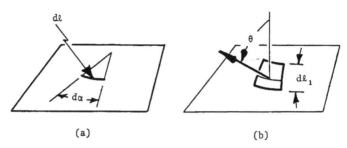

(a)                                    (b)

Figure 3.13  The measure of the set of sweeping test lines that
form tangents with a trace element $d\ell$ is related to the local
curvature of the element, (a).  The curvature of the element of
trace is in turn related to the principal normal curvatures
of the surface element from which it derives (b).

Thus, the event occurs if

$$d\alpha = kd\ell \qquad\qquad (3.50)$$

If $d\ell_1$ is an element of length in the surface being sectioned that
is perpendicular to $d\ell$ (Figure 3.13b), then the event that the
sectioning plane intersects the surface element defined by $d\ell$ and
$d\ell_1$ requires (Figure 3.12b)

$$dp = d\ell_1 \sin\theta \qquad\qquad (3.51)$$

The numerator in equation (3.48) may now be evaluated by combining
equations (3.40), (3.50) and (3.51):

$$T_A = \frac{\int\!\!\int\!\!\int\!\!\int \dfrac{1}{\sin\theta}(\kappa_1\cos^2\phi + \kappa_2\sin^2\phi)\cdot d\ell(d\ell_1\sin\theta)\cdot\sin\theta\cdot d\phi d\theta}{\text{Events}}{4\cdot\pi^2 V}$$

Carrying out the integration over $\theta$ and $\phi$ gives:

$$T_A = \frac{\int\!\!\int(1/2)(\kappa_1 + \kappa_2)\cdot dS}{\pi V}$$

The numerator is the integral of the mean curvature, $K$ [see
equation (1.1)]:

$$T_A = \frac{K}{\pi V} = \frac{1}{\pi} K_V \qquad\qquad (3.52)$$

In summary, sweeping test lines in a sectioning plane form
tangents with traces of surfaces in proportion to the angle
subtended by such traces.  This subtended angle is related to the
curvature of the trace.  The curvature on a trace is in turn
related to the principal normal curvatures on the surface from
which the trace is derived.  This combination of circumstances
makes the tangent count on a section related to the mean curvature
of the surfaces being sectioned, integrated over the surface area
in the system.

IV.   GRANULOMETRY FOR ISOLATED PARTICLES OR CONNECTED MEDIA

by Jean Serra, Center de Morphologie Mathematique,
Ecole de Mines de Paris, Fontainbleau, France

A.   Axiomatics

Starting from the idea that for unconnected, distinguishable grains, size dispersions are easily obtainable through the physical operation of sieving, and that more complex systems ought to be made amenable to similar procedures, G. Matheron has axiomatized the concept of granulometry in the following way.

Let Y be a population, or set, to be sieved by a family of sieves having mesh sizes $\lambda > 0$. Applying sieve $\lambda$ to Y will yield an oversize $\psi_\lambda$ (Y) depending upon the shape and dimensions of the sieve used as well as those of Y itself. It should thus be possible to use various kinds of sieve shapes yielding different results. If $\psi_\lambda$ (Y) is to define a granulometry, it must satisfy the following conditions:

$$\psi_\lambda \ (Y) \subset Y \text{ and } \psi_o \ (Y) = Y \tag{4.1}$$

$$\text{if } Z \subset Y \text{ then } \psi_\lambda \ (Z) \subset \psi_\lambda \ (Y) \tag{4.2}$$

i.e. the oversize from sieve $\lambda$ of any <u>subset</u> Y is contained in the oversize from sieve $\lambda$ of Y itself.

Successive sieving through various mesh sizes will yield a final result identical to that given by the sieve having the highest mesh number, i.e.:

$$\psi_\lambda \ o \ \psi_\mu \ (X) = \psi_\mu \ o \ \psi_\lambda \ (Y) = \psi_{\sup(\lambda,\mu)} \ (Y) \tag{4.3}$$

where o designates the combination of sieving operations.

B.   Examples

1.   Connected component volume (or area)
Let Y be a set of distinguishable grains or particles. If $\psi_\lambda$ (Y) is the subset consisting of all particles having a volume greater than $\lambda$, then $\psi_\lambda$ (Y) satisfies (4.1) to (4.3) above and thus defines a granulometry. Similar granulometries may be obtained with a cross section area in $E^2$.

2.   Maximum chord length
Again, let Y be a set of distinguishable grains. The collection $\psi_\lambda$ (Y) of all grains having a maximum chord length greater than $\lambda$ in any direction, defines a granulometry onto Y (i.e. when the chord direction is fixed, or normal to a fixed one).

These two examples assume the grains to be distinguishable, which is not always the case. They also ignore the stereological meaning of the granulometries on polished sections which will be considered later.

### 3. Openings by convex structuring elements

Let B be a convex compact set. The "opening of Y according to B" (noted $Y_B$) is defined as the volume of space swept by B under the condition that B remains entirely included in Y. Given B(o), let $B(\lambda)$ be the family of all sets homothetic to B(o). Then $Y_{B(\lambda)}$ defines a granulometry. There are thus as many different granulometries as basic shapes B(o). In planar measurements, two such shapes are of particular interest because of their morphological significance. They are the line segment and the circle, i.e. the least and the most isotropic convex bodies.

## C. Weights and Stereology

1. With each $\psi_\lambda$ (Y), we associate its volume (or area). The quantity

$$G(\lambda) = 1 - \frac{V[\psi_\lambda (Y)]}{V[Y]}$$ (4.4)

is then a distribution function.

2. Let $Z_1, \ldots Z_n$ be convex sets in $E^3$ and $Z_i'(\rho, \theta, \phi)$ be the section of $Z_i$ by an IUR plane $E_2$. Let us denote by $G_i(\lambda)$ the distribution function of the area $A[Z_i'(\rho, \theta, \phi)]$ when $\rho$, $\theta$ and $\phi$ vary (i.e. granulometry defined in B.1).

$$G(\lambda) = \frac{\Sigma K_i \, G_i(\lambda)}{\Sigma K_i}$$ (4.5)

where $G(\lambda)$ is the granulometry of $Z = \bigcup_i Z_i$ and $K_i$ the integral of the mean curvature of $Z_i$.

3. Let Y be the union of disjoint subsets $Y_i$ (the $Y_i$ not necessarily convex, nor connected), and $\psi_\lambda$ ($Y_i$) the granulometry according to the family of homothetic convex sets $B_\lambda$. $G(\lambda)$ is defined as formerly. Then:

$$G(\lambda) = \frac{\Sigma V_i \, G_i(\lambda)}{\Sigma V_i}$$ ($V_i$ = volume of $X_i$) (4.6)

REMARK: The openings may be spherical, circular, or linear intercepts, the weighting remains unchanged.

## V. STATISTICS OF MEASUREMENTS

by W. L. Nicholson, Pacific Northwest Laboratories,
A Division of BATTELLE MEMORIAL INSTITUTE,
Richland, Washington 99352, USA

## A. Introduction

The main topic of the workshop has been "what can be estimated?". Now we consider the topic "how many measurements must be taken?" or more generally "how precise is the estimate?". Basic stereological relationships describe global or average value geometric properties of a three-dimensional structure. Actual measurements will be distributed about such a global property. The precision of such a global property estimate depends upon the method of taking measurements and the complexity of the structure as well as the number of measurements. If the structure is spatially heterogeneous, precision is highly dependent on the sampling plan used to select the sections for examination.

## B. Theoretical Approach

A specimen is examined stereologically to estimate a three-dimensional global structure property $P_3$. A random section of the specimen is selected for examination. Observation is limited to a randomly selected field or subsection with a two-dimensional structure property $P_2$ which is an unbiased estimate of $P_3$. A scheme for making measurements / of the subsection produces an unbiased estimate $\hat{P}_2$ of $P_2$. The relative variance of $\hat{P}_2$ is expressible as

$$\sigma^2(\hat{P}_2)/P_3^2 = \underline{E}[\sigma^2(\hat{P}_2|P_2)]/P_3^2 + \sigma^2(P_2)/P_3^2 \qquad (5.1)$$

where $\sigma^2(\hat{P}_2|P_2)$ is the conditional variance of $\hat{P}_2$ given $P_2$ (i.e., the variance of $\hat{P}_2$ as an unbiased estimate of $P_2$) and $\underline{E}$ is the expectation operator over the randomization process which selected the field of observation. Formula (5.1) expresses relative variance as variability of $\hat{P}_2$ about $P_2$ plus variability of $P_2$ about $P_3$. In statistical jargon these variabilities are "within field" and "between field", respectively. Any investigation of precision of a global property estimate should consider both variabilities. The number of measurements should be thought of as a product (measurements/field) x fields. Formula (5.1) is primarily conceptual and of limited practical value without strong assumptions on the three-dimensional structure.

A simple structure for which working formulas have been developed is randomly distributed alpha phase particles in a three-dimensional beta phase matrix. Hilliard and Cahn, (1961) consider the relative advantages of areal analysis, random and systematic point counting, and lineal analysis for volume density estimation. They conclude that coarse grid systematic point counting is most economical. Formula (5.1) for coarse grid systematic point counting reduces to $1/N_\alpha$, where $N_\alpha$ is the expected number of points in the grid intersected by alpha phase particle profiles.

Nicholson and Merckx (1969) develop variance formulas for estimation of particle size distributions. While outside the realm of basic property estimation, this work shows the type of variance formulas that can be developed once sufficient assumptions are made about the geometry and spatial distribution of the three-dimensional structure. Scheaffer, (1975) determines an approximate within field variance formula for a line intersection count estimate $\pi N/2L$ of the boundary between a two phase Poisson field generated planar mosaic. A simple upper bound for this relative variance is $2/N$.

## C. Empirical Approach

When the structure of the specimen is not modeled, the number of measurements question must be determined empirically. Volume density data from rat lung specimens illustrates the empirical approach. The author is indebted to Professor Ewald Weibel for these data. To estimate volume density of alveoli in rat lung, six planar sections were prepared from each of seven lungs. On each section 12 fields were selected for analysis. For each field, a single random application of a regular triangular lattice of 168 test points gave a point count ratio estimate $P_a/168$ of area density in the alveoli phase. The 7 x 6 x 12 = 504 estimates range from 0.143 to 1.000 in a bimodal distribution, a composite of a smooth bell-shaped frequency function (mode of 0.70 and upper extreme of 0.86), and a spike at 1.000 consisting of 142 (28%) estimates. The bimodality suggests heterogeneity in alveoli structure. To learn more about the heterogeneity, the two

subdistributions of alveoli volume density estimates (partial
tissue fields and embedded in a single large alveolus cross-section
fields) were looked at separately. A nested analysis of variance
run on the 362 partial tissue fields is summarized below.

Table 5.1
Analysis of Variance Table for Alveoli Volume Density Data

| Source | Degrees of Freedom | Mean Square | Significance |
|---|---|---|---|
| Lungs | 6 | 0.07407 | 0.5% |
| Sections | 35 | 0.01778 | 10 % |
| Fields | 320 | 0.01235 | |

The between-fields mean square is too large to be purely within-
field sampling variability. Thus, real structural changes between
fields on the same section are indicated. The slightly significant
between-sections mean square suggests some global heterogeneity not
evident locally from field variability within a section. More data
is needed to establish the magnitude of such an effect. The highly
significant between-lung mean square is primarily due to a single
lung with a partial tissue alveolar volume density of 66% compared
to a median lung estimate of 73%.

An independent analysis was done of the frequency per section
of alveolus embedded fields. Comparison with theoretical binomial
frequencies suggests some degree of systematicness in the
positioning of large alveoli features on the section.

Summing the variabilities in the two distributions (assuming
no section effects, neither added variability nor systematicness)
gives 0.0240 as the variance of a single field alveoli density
estimate. Now a partial answer is available to the number of
measurements question. The basic sampling unit is the field and
the measurement a point count ratio from a single application of
the 168 point systematic grid. The 95% confidence level precision
of an averaged over K fields single lung alveoli volume density
estimate is

$$2\theta_a = 2\sqrt{0.0240/K} = 0.310/\sqrt{K}. \qquad (5.2)$$

For the present study K = 72 which gives a precision of 0.037.
With alveoli volume density of 0.8 this is a two sigma coefficient
of variation of about 4%. The precision of alternative sampling
plans follows directly from formula (5.2).

REFERENCES

1. Buffon, G. S. (1777): *Suppl. á L'Histoire Naturelle*, Vol. 4.

2. Cahn, J. W. (1967): *Trans. Met. Soc. AIME*, *239*, 610.

3. Cornfield, J. and Chalkley, H. W. (1951): *J. Wash. Acad. Sci.*, *41*, 226.

4. DeHoff, R. T. (1967): *Trans. Met. Soc. AIME*, *239*, 617.

5. DeHoff, R. T. and Rhines, F. N.: *Quantitative Microscopy*, McGraw Hill Co., Inc., New York.

6. DeHoff, R. T., Aigeltinger, E. H. and Craig, K. R. (1972): *J. Microscopy*, *95* Pt. 1, 69.

7. DeHoff, R. T. (1975): *Metallography,* *8*, 71.

8.  Hilliard, J. E. and Cahn, J. W. (1961): Trans. Met. Soc. AIME, 221, 334.

9.  Miles, R. E.: Probabilistic Foundations of Stereology, this volume.

10. Nicholson, W. L. and Merckx, K. R. (1969): Technometrics, 11, 707.

11. Saltykov, S. A. (1954, 1958 and 1972): Stereometric Metallography, Moscow, State Publishing House for Metals Sciences.

12. Scheaffer, R. L. (1975): J. Microscopy, in press.

13. Smith, C. S. and Guttman, L. (1953): Trans. AIME, 197, 81.

14. Struik, D. J. (1950): Lectures on Classical Differential Geometry, Addison-Wesley Press, Cambridge, Mass., p. 76 and 81.

15. Tomkeieff, S. I. (1945): Nature, 155, 24.

16. Weibel, E. R. (1963): Lab. Investigation, 12, 2.

17. Weibel, E. R. and Elias, H. (1967); Quantitative Methods in Morphology, Berlin, Springer Verlag, p. 12.

# 8. PARTICLE SCIENCE

*Chairperson:*

B. H. Kaye

National Bureau of Standards Special Publication 431
Proceedings of the FOURTH INTERNATIONAL CONGRESS FOR STEREOLOGY
held at NBS, Gaithersburg, Md., September 4-9, 1975 (Issued January 1976)

WORKSHOP SESSION– PARTICLE SCIENCE

by Brian H. Kaye, *Chairman*
*Institute for Fineparticle Research, Laurentian University, Sudbury, Ontario, Canada*

INTRODUCTION REMARKS

Fineparticle science embraces those systems in which at least one component exists in a finely divided state such that the physical properties and behaviour of that component are determined by the competition between surface and microscopic properties. (Note: Classical physics deals in general with systems in which surface forces can be neglected, colloid science is concerned with those systems in which surface forces are dominant).

Stereology has been concerned historically with those fineparticle systems in which the dispersed particles are trapped in an inaccessible matrix such as particle of ore in a piece of rock. When characterizing fineparticles in a rock matrix the analyst has availed himself of the classic work of Rosiwal and Chayes.

From a fundamental scientific point of view pores in a sintered ceramic or a piece of coke also constitute a dispersed phase of zero density. Again the analyst in this situation has used stereological technology to measure the inaccessible pores of this type of system by making a section through the system and conducting an examination of the two dimensional system exposed in the section. One paper of this session deals with a fineparticle problem which up until now has remained intractable, that of determining the size distribution of pigment particles as actually dispersed in a paint film.

Wherever possible the fineparticle analyst has chosen to free particles from an inaccessible matrix and study them in an alternative medium because of the difficulty of obtaining enough measurement from sections to gain sufficiently high levels of confidence in the measured parameters. The advent of the new automated iconometric systems such as the Bausch and Lomb Omnicon, the Leitz Texture Analyzer and the Imanco systems, each of which can be interfaced directly to large digital computers, open up the possibility that the analyst will turn more and more to the technology of stereology in order to characterize his fineparticle systems.

In general when discussing a fineparticle one talks about its external morphology, that is its gross shape characteristics, its surface topography which concern itself with the humps and hollows that are visible on the surface of the particle, the particle texture that is any obvious regular structure discernable as being superimposed on the topography of the particle and the topology of the particle, that is the relationship of holes within a particle to its total structure. Thus a particle which contains no pores is said to be Genus 0 one topologically speaking, any particle which contains a pore which does not reach the surface of the fineparticle is termed a Genus 1 solid and so on. The fineparticle analyst does not have good techniques for determining external morphology of very fineparticles and one of the areas of progress may well be that of developing techniques for placing fineparticles in a cast resin and giving them a preferred orientation by the application of electrostatic forces.

In summary, in the past the fineparticle analyst has avoided stereological measurement wherever possible because of the difficulty of making enough measurements on a section. As we will probably find out in this workshop session the use of new automated iconometrics technology will probably shift the balance of preference of the fineparticle analyst in his choice of techniques to include more often stereological examination.

National Bureau of Standards Special Publication 431
Proceedings of the FOURTH INTERNATIONAL CONGRESS FOR STEREOLOGY
held at NBS, Gaithersburg, Md., September 4-9, 1975 (Issued January 1976)

STEREOLOGY AND PARTICLE TECHNOLOGY

by Hans Eckart Exner
Max-Planck-Institut für Metallforschung, Stuttgart, Germany

Quantitative applications of microscopy are common features in the evaluation of microstructures and in the testing of particulate matter. Much of the progress in one field would be useful in the other as well.

Figure 1 shows, for example, the evaluation of size distributions. Graticule sizing as developed for particle technology (P/T) is now used in stereology (ST) in many other fields of research (e.g. biology, anatomy, geology, materials technology) as well. Lineal analysis, on the other hand, has been developed in ST and, though infrequently, is adopted for P/T. Mathematical evaluation by special functions (e.g. the RRS distribution) is more common for powder particles but is often applied to particle and grain size distributions as well. Television microscopy is utilized in both fields.

Figure 2 gives a review of some geometric quantities and their importance in particle and materials technology, respectively. Some of them are very important in both fields (++), e.g. specific surface, others cannot be used (-), e.g. the degree of patterness for characterizing a powder. As in Fig. 1, the arrows show in which direction parameters were adopted.

| Field | P/T | ST |
|---|---|---|
| Graticule Sizing | + ⟶ | (+) |
| Television Microscopy | + ⟵ | + |
| Lineal Analysis | - ⟵-- | ++ |
| Mathematical Treatment | ++ ⟶ | + |

| Field → / Features → | Particle Technology / Particles or Pores | | Materials Technology / Grains or Particles | |
|---|---|---|---|---|
| **Parameters** | | | | |
| $V_V$ | - | ++ | - | ++ |
| $S_V$ | ++ | ++ | ++ | ++ |
| $\bar{D}$ | + | + ⟶ | (+) | (+) |
| $f(D)$ | ++ | + ⟶ | + | |
| Shape | (+) | (+) ⇌ | (+) | + |
| $D_1/D_2$ | + | - | + | + |
| Orientation | - | + ⟵ | + | + |
| Patterness | - | ? | - | + |

Fig. 1 Size distribution analysis in ST and P/T.

Fig. 2 Geometric quantities in ST and P/T.

Powder particles are usually studied in projection. The first part of Figure 3 shows some relationships derived in mathematical stereology between quantities which can be measured on projected images and some useful parameters for describing the geometric nature of powder particles. Mean surface curvature is not yet used for powder characterization. Even more interesting is the adoption of parameters calculated from quantities measured from random cross sections (Fig. 3). Random cross sectioning of a powder is possible by means of dispersion in high-viscosity, fast hardening resins. A concise characterization of specific surface and shape is possible in this way.

For determining particle size distributions, stereological relationships are only available for spherical particles (see Fig. 3) For discrete, insoluble particles in opaque materials, etching away the matrix would allow the use of all the methods developed for particle size distribution analysis in P/T. In Figure 4, some of these

469

methods are evaluated with respect to sensitivity, correlation
with physical properties and accuracy according to an analysis of
tungsten carbide powders (Exner and Fischmeister, 1966). Microscopic analysis using a graticule proved to be among the most sensitive and accurate methods. Figure 5 shows a way to characterize
the accuracy of a size distribution analysis developed for P/T
(Fischmeister and co-workers, 1961, 1963, and 1966). Deviations of
the calculated and measured curves for a 1:1 blend show that
the method is not useful in this size range.

Projection

1. $S = 4A'$

2. $\bar{D}' = N'_L/N'_A$

3. $L'_P = \pi N'_L/N'_A$

4. $\bar{K}_m = \pi \bar{D}'/2\bar{A}' = \frac{\pi}{2}/\bar{L}'$

5. $N(D_i) = \{A_o N(L'_i) - \sum_{j=1}^{k} A N(L'_{i+j})\}/L_i$

6. $N(D_i) = \sum_{j=1}^{k} N(L'_j) A_j/\Delta$

Section

1. $S_V = 4N/L_T \cdot V_V$

2. $\bar{L} = 4/S_V = L_T \cdot V_V/N$

3. $f_1 = \frac{2}{3\pi} \cdot P_L^2 \cdot M_2^2/V_V \cdot N_A \cdot M_1 M_3$

4. $f_2 = 12 S^{3/2}/\pi V = M_L (L)/\bar{L}^L$

5. $N(D_{i,1/2}) = \frac{4}{\pi}\{2N(L_i) - N(L_{i+1})\}/L_i^2$

6. $N(D_{i+1/2}) = \frac{4}{\pi}[N(L_i)]/(2L_i-\Delta) - N(L_{i+1})]/(2L_i+\Delta)]/\Delta$

* for spheres only

**Fig. 3** Useful stereological relationships for P/T.

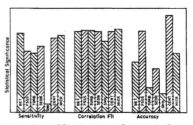

**Fig. 4** Efficiency of particle
sizing methods.

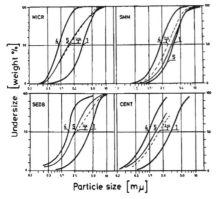

**Fig. 5** Additivity test for measured
size distributions (dashed curves calculated, curve 5 measured for 1:1 blend
of powders of size distributions 1 and
4).

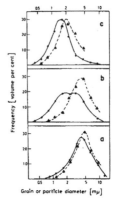

**Fig. 6** Comparison of
size distributions of
WC particles before and
after sintering (a) coarse
powder, (b) blend 1:1 of a
and c, (c) fine powder.

An example of the application of ST to P/T is shown in Figure 6.
Here, the size distributions of particles of WC-powders are compared
with those of the WC-grains in cemented carbide samples sintered
from these powders (Fischmeister, Exner, Lindelöf, 1966). Useful conclusions on sintering kinetics can be drawn if the methods used for
the analysis are comparable.

In conclusion, stereology can be usefully applied to particle
technology. However, more effort is needed to make stereological methods more widely known in all fields of possible application.

National Bureau of Standards Special Publication 431
Proceedings of the FOURTH INTERNATIONAL CONGRESS FOR STEREOLOGY
held at NBS, Gaithersburg, Md., September 4-9, 1975 (Issued January 1976)

REPRESENTATION OF PARTICLE SIZE AND SHAPE

by B. Scarlett
Department of Chemical Engineering, University of Technology, Loughborough, Leicestershire, England

## 1. Introduction

The subject of stereology is concerned with the measurement and quantification of the geometry of a inhomogeneous medium and it is fitting that, at a
conference on stereology, we should consider the purpose of making such measurements. The ultimate objective of making such measurements must be to relate
them in a quantitative manner to the behaviour of the medium. In a three dimensional world it is possible to measure only length, area or volume. All of
these parameters are scalar parameters and thus, in that form, they can only be
related to other scalar parameters. On the other hand, it is known that many of
the physical properties of particles are tensor quantities of varying rank. For
example, the viscous resistance to motion of an irregular particle has been shown
by Brenner (1) to be described by three tensors of the second rank. The question to be asked is whether the scalar measurements of length, area and volume
can ever be related to such complicated parameters. The answer to this question
lies in the realisation that, in making the scalar measurements of length,
information has already been discarded about the relative position and orientation of the extremities of the length. Another way of stating the same fact is
that, in making the scalar measurements, the associated unit vectors have been
disregarded.

In this paper the philosophy is proposed that, in all problems which
attempt to relate the geometry and the behaviour of a set of particles, it
should be assumed that the complete vector locus of the particle array is known.
Although the paper is addressed to particulate systems, the formulation proposed
is equally applicable to other studies.

If the complete vector locus is assumed, it can then be reduced to the
form necessary for the solution of a particular problem. Only at that stage, is
it necessary to consider how to actually make the measurements. It should be
emphasised that each physical problem requires a different parameter of varying
complexity and that there is no universal particle parameter, there are only the
basic building blocks of length, area and volume. Furthermore, it should be
recognised that, in regarding these measurements as the basic building blocks,
a set of Cartesian axes has been subconsiously imposed on the problem. The
basic vector locus is completely invariant of axes and it would be possible to
devise a different stereological system such as the polar coordinate approach
as proposed by Cheng and Sutton (2).

## 2. Particle Vector Locus

A particle shape is itself a vector locus. The particle outline can be
considered, with respect to any arbitrary origin to be the three dimensional
graph of its own particle size and shape. This is illustrated in Figure 1,
where the vector $\underline{r}$ denotes the position vector of any element of surface, dS.
An array of particles is similarly a unique vector locus. The vector locus is
independent of axes or co-ordinate systems. It can, however, be changed by
reorienting the particle or moving the particle. Thus, it is immediately apparent that the locus contains more information than simply the size and shape of
the particle, it also contains information about the position and orientation of
the particle. It is clear that this total amount of information can be discarded

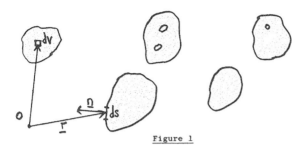

Figure 1

and reduced to parameters which are independent of these factors if it is
required to isolate the particle parameters from those of the array.

In considering the vector locus of a single particle, or of an array of
particles, as shown in Figure 1 we could consider that, in principle, the position vector of each element of the volume is recorded. Alternatively, the outline is recorded if each element of the surface of the particle is known. In
fact, neither of these sets of data is sufficient to generate meaningful parameters about the particle. Not only must the position of each element of the
surface be known, but also the direction of some of the vector quantities
associated with that element of surface. The most important of these is the
unit·vector, $\underline{n}$, which lies at right angles to the surface. The other vectors
are those associated with the curvature of the surface. These vectors correspond to first and second differentials of the position vector and it would be
possible to use also even higher orders whose physical significance is less
apparent. The assumption that these vectors are known, therefore, is quite ·
compatible with total knowledge of the vector locus, it means that the locus can
also be differentiated. The use of the volume position vector would require a
similar autocorrelation to be useful but is not necessary for particles which,
although they may be hollow, have a homogeneous solid texture.

### 3. Particle Parameters

From a set of vector quantities, tensor quantities of any order may be
formed. Thus, two vectors form a dyadic which is a completely general form of a
second rank tensor. This form of tensor algebra is well described by Gibbs and
Wilson ( 3 ). It should be emphasised again that the purpose of·stereological
measurement is to determine a strategy of·breaking the particle vectors into
constituent scalars which can be measured. However, from the particle vectors,
tensors of any rank can be formed.

The tensors may then be integrated or averaged over the whole particle
surface to form particle parameters. Alternatively, the limits of integration
may be restricted, for example to the exterior surface of a hollow· particle.
In this way, it may be possible to describe mathematically the most complex
physical behaviour.

For the purpose of illustration, it seems worthwhile to list all the possible combinations of vectors and dyadics, integrated over· the whole particle
surface, which can be formed from. the position vector, $\underline{r}$, and the unit normal,
$\underline{n}$. The properties of each are described although the description of physical ·
properties usually require more complex combinations of the basic vectors.

(a)      $\int_s ds\ \underline{r}$ is a vector quantity

It expresses only the position of the centre of gravity of a thin shell which
has the same shape as the body with respect to the origin O. This is easily
proved. If $\underline{r}_{oc}$ is the position of the centre of gravity with respect to the
arbitrary origin, O:-

$$\int_s ds\ \underline{r}_{op} = \int_s ds\ \underline{r}_{oc} + \int_s ds\ \underline{r}_{cp}$$

$$\therefore \int_s ds\ \underline{r}_{op} = S\ \underline{r}_{oc} + 0$$

where S is the total surface.

The second term is zero by definition.

(b)      $\int_s ds\ \underline{n}$ is a vector quantity and is zero by definition.

Thus, the projection of the surface in any direction, is also zero:

i.e.      $\int_s ds\ \underline{n} \cdot \underline{i} = 0$

where $\underline{i}$ is the unit vector in some fixed direction.

However, the integral of the modulus of the projected surface is, of course,
a commonly used stereological property.

$\int_s ds\ |\underline{n} \cdot \underline{i}|$ = twice the total projection in direction $\underline{i}$ and is the
quantity relating the mean chord and the volume.

$$\frac{V}{L} = \frac{1}{2} \int_s ds\ |\underline{n} \cdot \underline{i}|$$

where $\underline{i}$ can be any direction

It should be emphasised that this parameter, total projection, is not a
resolvable quantity. If a mutually perpendicular set of axes, $\underline{i}_1$ , $\underline{i}_2$ , $\underline{i}_3$ are
formed and the projection measured in each direction, the result-
ant is three scalar values of the projection. Because of the modulus sign
around the direction cosines, it is impossible to rotate these values to another
set of axes. It is not, therefore, in any sense a reliable measurement of
anisotropy. Consequently, the mean chord is also not a means of detecting
anisotropy.

(c)      $\int_s ds\ \underline{r}\ \underline{r}$ is a second rank tensor.

It is of limited interest. Its scalar invariant represents the moment of inertia of a thin shell of the same shape.

Thus:-      $\int_s ds\ \underline{r}.\underline{r} = S.\ M.$

(d)      $\int_s ds\ \underline{n}\ \underline{n}$ is a second rank tensor.

This tensor is a most convenient means of expressing anisotropy in a particle
array. The tensor can be resolved in any direction, $\underline{i}$.

Thus:-      $\int_s ds\ \underline{n}\ \underline{n}.\underline{i}$ is a vector quantity which may lie in any direction.

473

It expresses the direction of the normal to the surface averaged with respect to the projected surface in direction $\underline{i}$.

If the particles are all randomly orientated the tensor reduces to a simple form:-

$$\int_s ds\ \underline{n}\ \underline{n}\ =\ \frac{1}{3}\ S\ \underline{\underline{I}}$$

where $\underline{\underline{I}}$ is the idemfactor (unit tensor).

Even if the system is not randomly orientated, the scalar invariant of the tensor is the total surface, S. The tensor can thus be made dimensionless by dividing by S.

It should be emphasised that this tensor expresses anisotropy due to particle orientation, as illustrated in Figure 2a, not that due to particle position as illustrated in Figure 2b.

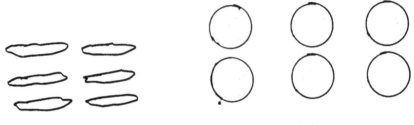

Figure 2a                              Figure 2b

This tensor is resolvable in the sense that if it is determined in one set of axes it can be rotated into an alternative set. This statement is really an expression of the fact that anisotropy is a tensor quantity and cannot, as shown in section (b) be expressed as a vector.

(e)      $\int_s ds\ \underline{r}\ \underline{n}$  reduces to a very simple parameter.

Thus:-    $\int_s ds\ \underline{r}\ \underline{n}\ =\ \underline{\underline{I}}\ V$

where V is the volume of the particle or particles.

This is easily proved by considering the particle locus in Cartesian form and multiplying by the idemfactor.

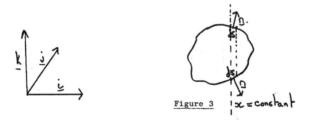

Figure 3    $x = $ constant

$$\int_s ds \; \underline{r} \; \underline{n} = \int_s ds \; \underline{r} \; \underline{n} \cdot (\underline{i} \; \underline{i} + \underline{j} \; \underline{j} + \underline{k} \; \underline{k})$$

$$\therefore \quad \int_s ds \; \underline{r} \; \underline{n} = \int_s (\underline{i} \; x + \underline{j} \; y + \underline{k} \; z) \{ (\underline{n}.\underline{i}) \; ds \; \underline{i} + (\underline{n}.\underline{j}) \; ds \; \underline{j}$$
$$+ (\underline{n}.\underline{k}) \; ds \; \underline{k} \}$$

The diagonal elements of the tensor are each seen to be equal to the volume of the particle i.e.

$$\int_s x \; ds \; (\underline{n}.\underline{i}) = \int_s y \; ds \; (\underline{n}.\underline{j}) = \int_s z \; ds \; (\underline{n}.\underline{k}) = V$$

The other elements of the tensor are zero because the projected surface at a fixed coordinate forms a thin slice, as shown in Figure 3, and its value is zero parallel to the coordinate. Thus, each of the other six terms in the tensor is zero.

Thus, the scalar invariant of this tensor is also related to the volume.

$$\int_s ds \; \underline{r}.\underline{n} = 3V$$

This can be most easily illustrated by considering Figure 4 . The volume of the element of particles is the difference in volume of the two cones which have O as the apex.

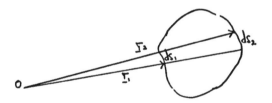

Figure 4

Thus:-    $dV = \dfrac{1}{3} (\underline{r}_2 \cdot \underline{n} \; ds_2 - \underline{r}_1 \cdot \underline{n} \; ds_1)$

$\therefore \quad 3V = \int_s \underline{r}.\underline{n} \; ds$

It follows also that the vector invariant of the tensor is zero.

Thus:- $\quad \int_s ds\ \underline{r} \times \underline{n} = 0$

It also follows that the tensor is symmetric.

Thus:- $\quad \int_s ds\ \underline{r}\ \underline{n} = \int_s ds\ \underline{n}\ \underline{r}$

These generalised expressions of the particle size and shape are particularly convenient because they can be manipulated mathematically. This is in contrast to the statistical distributions often produced by measurements such as chord lengths. Furthermore, although the various moments of these distributions can be used in a similar manner, their relationship to a physical problem is not obvious. In contrast, the generalised expressions can arise naturally as the solution to a problem. The ease of manipulation of the expressions can easily be appreciated by the well known transformations for translation and rotation of the axes.

a) The particle can be moved with respect to the origin, O, to a new origin, p, by subtraction of the transposition vector:-

$$\underline{r}^1 = \underline{r} - \underline{r}_{op}$$

b) The particle may be rotated from axes $\underline{i}$, $\underline{j}$, $\underline{k}$ to axes $\underline{i}^1$, $\underline{j}^1$, $\underline{k}^1$ by multiplying by the idemfactor.

$$\underline{r}^1 = \underline{r} \cdot (\underline{i}^1 \underline{i}^1 + \underline{j}^1 \underline{j}^1 + \underline{k}^1 \underline{k}^1)$$

## Conclusions

In considering the characterisation of a particle array, the number of possible parameters is infinite. The choice of the appropriate parameter for a particular problem is most logically made by assuming a knowledge of the complete vector locus and then mathematically reducing that data to the minimum amount necessary for the problem. At that stage, the measurement strategy should be considered. This philosophy is in contrast to postulating a parameter and attempting to relate or correlate it with any convenient property of the system.

Complex parameters can best be formed from the position vector of an element of surface and its various differentials juxtaposed to constitute tensors of varying rank. The tensor quantity is usually integrated over part or the whole of the surface.

## References

1. Happel J., Brenner H., Low Reynolds Number Hydrodynamics (Noordhoff) 1973

2. Cheng D.C.H., Sutton H.M., Nature (Physical Science) 232 No.35, p.192 (1971)
   Sutton H.M., Bunalli M., Proc. Soc. Anal. Chem. p.13, 1973

3. Gibbs J.W., Wilson E.B., Vector Analysis, (Dover)(1960)

## Acknowledgements:

The author is grateful to the British Steel Corporation for the award of a Fellowship during the academic year 1970/71. The ideas contained in this paper were formulated at that time.

National Bureau of Standards Special Publication 431
Proceedings of the FOURTH INTERNATIONAL CONGRESS FOR STEREOLOGY
held at NBS, Gaithersburg, Md., September 4-9, 1975 (Issued January 1976)

## THE APPLICATION OF STEREOLOGY TO PARTICLE TECHNOLOGY

by B. Scarlett and P. J. Lloyd
*Department of Chemical Engineering, University of Technology, Loughborough,-Leicestershire, England*

### Introduction

One of the major problems facing the Particle Technologist in
formulating the basic science of particulate systems which are found
in the form of powders, slurries, pastes, emulsions or aerosols is
how to describe the discreteness of the particles. With the exception
of spheres, no particle can be described by a single parameter and
throughout the development of the subject use has been made of the concept
of an equivalent sphere, that is to say the diameter of a sphere which would
have the same physical property as the true particle. Thus an irregular
particle may be described by a number of equivalent spheres diameters with
respect to its volume, its surface area, its settling velocity, its ability
to pass through a sieve mesh etc. This situation is in itself difficult
but it is made more complex by the fact that real particulate systems consist
of distributions of particle diameters and possible also of particle shapes.
This problem has not been tackled by many investigators who prefer to ignore
the problem and use a mean size which is often dependent both on the
analytical technique used and the material used in the study. This is not
very satisfactory and there is a need for a method to characterise
distributions of particles     both uniquely and in a way such that the
characterisations can be used to describe the physical properties of the
particle system.

In 1966 Rumpf & Debbas (1) used many of the principles of stereology
in a study of the packings of size distributed spheres and irregular
particles. They did not use any of the stereology standard references but
did refer to a paper by an astronomer Wicksell (2) who used stereology
principles in the analysis of photographs of stars taken through a
telescope. Rumpf & Debbas were interested in discovering if size
segregation of spheres occurred in the packing and analysed the size of the
exposed areas at a series of sections through the packing. The size
distribution of the areas were related back to the original sphere
distributions by relatively simple formulae. They also attempted to use
the same method with irregular particles but the result was not as elegant
or convincing. Other workers notably Dullien & co-workers (3) (4) have
followed a similar approach but in each case an attempt is made to relate
back the distributions of the areas or laminae or of the chords, or
filaments, back to the original particle size distribution. This method
is severely limited since this relationship is only valid if the shape of
the particles is known and is the same for all particles. In 1970,
Scarlett (3) proposed that the sub-particle characteristics of laminae and
filaments together with the volume of the particle were a sufficient
description of a system of particles and moreover could be used to describe
some of the physical properties of particle systems.

### Sub-Particle Characteristics

A particle may be considered as being cut into a large number of laminae
each with an identical but small finite thickness. The area of each of the
laminae is clearly well defined and easily measured. If the same particle
was sliced in all possible directions, the resulting distribution of the areas

would uniquely characterise the particle.  It may not be possible to reconstruct the particle but the distribution will contain all the information
concerning the particle diameter and shape.  The distribution is also easily
measured.  A large number of identical particles are scattered onto the
surface of a cold setting resin close to setting point.  More resin is poured
on top and after the resin has set the block is cut through the plane of
particles.  After polishing the plane the areas of the particles are measured
using an automatic microscope.  Each of the laminae can now be considered cut
into filaments of small but uniform cross-section.  The resulting distribution
of the lengths will also be characteristic of the particle.  This procedure
is shown in figure 1.

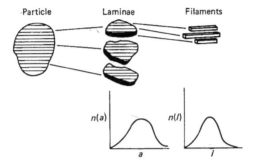

Fig. 1    Distribution function of a number of laminae against area

Of course it is not necessary to consider just a single particle as a distribution of particles of different sizes and shapes can be analysed in
exactly the same way.  In fact the filaments having only a length dimension
can be considered as the primary sub-particle characteristic.  The filaments
can be linked together to form laminae and a distribution of areas, or to
form complete particles and a distribution of volumes.  This is illustrated
in figure 2.

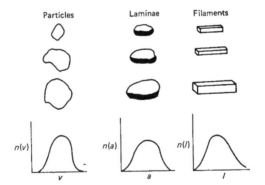

Fig. 2    Distributions that uniquely characterise a set of particles

The distributions of filament lengths will not however uniquely define a
particle system since they can be linked together in many different ways.
A similar argument applies to the distribution of areas, but the three
distributions together will contain sufficient information to define a

system of particles. The difficulty of defining shape has been avoided but the definition will be implicit in the three distributions.

Fortunately these three distributions can be measured relatively easily. The distributions of the filament lengths and laminae areas are very easily measured on an automatic microscope of the Quantimet or πmc type. Indeed these distributions are easier to measure than the conventional microscope equivalent sphere diameters such as Feret's or Martin's. The distribution of volumes can easily be measured with a Coulter Counter since it is the distribution that is always measured but which is always converted using an assumption of constant shape factor into a weight versus the diameter of a sphere of equivalent volume. Thus techniques are available to measure the distributions. Fortunately, however, as the following sections indicate, only one of the distributions may be needed to describe a physical property of the particle system.


Critical Porosity

A bed of particles can only be sheared easily when the porosity has reached a value known as the critical porosity. If the density of the packed bed is too great the bed must dilate to the critical porosity before any relative movement of the particles is possible. This property is peculiar to particulate systems and was named by Osborne Reynolds as the property of dilatancy. This critical porosity is a function of the size, size distribution and shape of the particles. Scarlett and Todd (6) measured the critical porosity of the sheared plane in an annular shear cell (7) by scanning the cell with an x-ray beam. The attenuation of the x-ray beam is a function of the solid material in the beam and was calibrated directly against each material used.

Using the concept of sub-particle characteristics, Scarlett et al (8) are able to predict the critical porosity. Two assumptions are made initially:

(1)  The bed is completely random (i.e. any particle has an equal chance of being at any point in the bed).

(2)  The bed is initially isotropic (i.e. any line drawn at random through the bed will indicate both the same filament size distribution and the same ratio of void filaments to the total length of line (i.e. the same porosity).

When the bed is sheared the motion of the particles is very complex but three distinct components can be detected:

(a)  the particles translate perpendicular to plane of shear in order to pass over one another;

(b)  the particles may have motion relative to one another in the plane of shear because they may overlap;

(c)  the particles may rotate in the plane of shear to present a smaller aspect perpendicular to the plane of shear;

A random distribution of particles (Figure 3) can be represented by filaments selected as those perpendicular to the plane of shear. Movement of the particles relative one to another is now represented by the relative movements of the filaments. Two planes AA[1] and BB[1] can be drawn through the filaments such that AA[1] moves relative to BB[1]. These planes are drawn parallel

479

Fig. 3     Random chord distribution

Fig. 4     Representation of a bed of particles by random chords

Fig. 5     Representation of sectioned chords contained in a
           critical volume

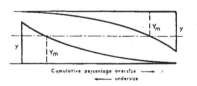

Fig. 6     Condition for flow

but in fact must be tortuous and will cut each filament at some random position. The separation between AA' and BB' will be the critical value when movement occurs and will depend on the randomly sectioned filament distribution g(y) which is related to the filament distribution f (x) by the formula

$$g(y) = \int_{x=y}^{x=x_{max}} \frac{f(x)}{x} \, dx$$

Clearly if the separation is greater than x max flow can occur but the critical porosity is less than this maximum value. By considering conditions (b) and (c) Scarlett & Todd concluded that the separation should equal twice the medium value of the sectioned filament distribution Ym. This is demonstrated in Figure 6, and the critical porosity εc found to be given by the relation

$$1 - \varepsilon c = \frac{2 \int_{0}^{x_{max}} y \, g(y) \, dy}{2 Y_m \int_{0}^{x_{max}} g(y) \, dy} = \frac{\int_{0}^{x_{max}} y \, g(y) \, dy}{Y_m}$$

Experimental results are compared with the predicted for three samples of sand and the results shown in Table 1. As is seen the predicted results are all low but the agreement is close.

Table 1          Critical Porosity Measurements

| Mesh Size B.S.S. | Experimental Critical Porosities | | | Predicted Critical Porosity |
|---|---|---|---|---|
| 14-18 | .481 | .475 | .472 | .452 |
| 36-52 | .455 | .452 | .446 | .444 |
| 60-85 | .459 | .458 | .454 | .443 |

## Structure of Packed Beds

In a recent investigation, Ward et al (9) examined the structure of packed beds of sand particles formed in by number of different methods:- by filtration, by dry compaction and by sedimentation. In each case the final compact was set in resin and sectioned. After careful polishing the compacts were examined with an Quantimet image analysing system. The purpose of the investigation was to find out if there were detectable differences in the structure. Since the study was primarily concerned with filtration attention was focussed on the void space and the particles selected were rounded sand particles which would show little orientation effects. The compacts were analysed in two directions both in the direction of flow or applied force and perpendicular to that direction. Each time the filament length distribution of both the solids and the voids was determined. A one way analysis of variance was used to compare the variance of filament distribution in the direction of scan with that perpendicular to the scan. The results presented showed the value of the technique and showed that the structure of the beds of particles depended on the method of construction.

## Transmission of Force through a Powder Compact

In 1972 De Silva (10) explored the possibility of using a Monte Carlo technique to explore the transmission of force through a bed of particles. The basic concept was to consider the equilibrium of each particle of the bed and the transmission of force through the points of contact. To do this it was necessary to be able to predict the positions of point contacts on each particle. The maximum distance in the direction of the applied force will be one of the Feret diameters. If the distribution of Feret diameters is sectioned randomly, as described previously, a point A is obtained. A second sectioned filament at right angles will give B. The position r and direction $\Theta$ of the contact B from the input of force O is now known. In the azimuthal direction $\phi$ is assumed to be another random variable.

By representing the sectioned Feret diameters and the filament lengths as cumulative fractional number undersize v. length, a representative length/ diameter could be obtained by selecting a random number from 0 to 1 and referring this fraction to the distribution.

De Silva traced a unit input force through the bed of particles by considering the equilibrium on a large number of particles. Only one force was traced to the container wall. The process was repeated many times until the distribution of force on the container was obtained. The predicted values agreed well with those measured with small pressure transducers.

### Conclusion

This paper has put forward the suggestion that the distributions of the sub particle characteristics, filaments and laminae together with the distribution of volumes are a complete description of a particle system which are easily measured and avoid the difficulties of defining shape since such information is implicit in the differences in the distributions. Furthermore it has been shown in three cases that these sub-particle characteristics can be used in describing and/or predicting some physical properties of a system of particles. Clearly more research into this concept is required but it is believed to have sufficient promise for an extended research programme.

## BIBLIOGRAPHY

1.    Debbas, S., & Rumpf, H., (1966) Chem. Eng. Sci., 21, 583.

2.    Wicksell, S.D., (1925) Biometrika, 17, 84.

3.    Dullien, F.A.L, & Mehta P.N., (1971/72) Powdr Technol, 5, 179..

4.    Dullien, F.A.L., & Dhawan, E.G.K., (1973) Powdr Technol, 1, 305.

5.    Scarlett, B., Particle Size Analysis 1970
            (Pub. Soc. for Anal. Chem., London).

6.    Scarlett, B., & Todd A.C. (1969) Trans. ASME 91 478

7.    Scarlett, B, & Todd A.C. (1968) J. Sci. Inst. 1 655

8.    Scarlett, B., Akers, R.J., Parkinson, J.S., Todd, A.C., (1969/70
            Powdr Technol 3 299.

9.    Ward, A.S., Lloyd P.J., Smith I., (1973) Chemical Engineer. 584

10.   De Silva (1972) Ph.D. Thesis Loughborough University of Technology

*National Bureau of Standards Special Publication 431*
*Proceedings of the* FOURTH INTERNATIONAL CONGRESS FOR STEREOLOGY
*held at* NBS, Gaithersburg, Md., September 4-9, 1975 (Issued January 1976)

TECHNIQUES FOR PARTICLE MEASUREMENT USING IMAGE ANALYSIS

by Roger R. A. Morton
*Bausch & Lomb, Rochester, New York 14625, U.S.A.*

Particle analysis has long been recognized as one of the important fields of application for image analysis. Accurate measurements of size distributions for a wide variety of particulates and powders are routinely obtained using image analysis equipment. Certain precautions must, however, be taken to ensure that the desired accuracy of measurement is attained. It is a purpose of this paper to discuss some of the precautions which should be taken.

Sampling – The first step, one of the most important in obtaining data which accurately represents a powder sample, is to ensure that the image presented to the instrumentation accurately represents the parent sample from which it was derived. Very rarely in particle measurements is it possible to measure every particle in a sample to be characterized. Thus, it is generally inferred that the portion of the sample actually measured does indeed represent the entire sample from which it was derived. To ensure that this inference is indeed valid, care must be taken in selecting and preparing a specimen from the parent sample. (See, for example, Herdan, Ch. 5.)

The optimum sampling technique will, of course, depend on the nature of the particulates, whether, for example, the parent sample is in suspension in a gas or liquid, or whether it is only accessible as it passes along a conveyor belt, or is held in a storage bin.

A broad range of sampling techniques are available for use in these various situations. To be successful, such techniques must be able to independently select every particle which is to form the selected sample. Furthermore, in general, this sample must be selected from throughout the entire parent sample.

Overlap – The methods of mounting a selected sample on a microscope slide are numerous and well documented. A major requirement is that individual particles are made to lie on the microscope slide at random positions, uniformly across the slide, and that the position of each particle be independent of the position of other particles. These precautions will, of course, not ensure that overlap does not occur but they do permit us to predict the degree of overlap which we can expect. Today's image analysis instrumentation, such as the Omnicon Pattern Analysis System, provides the user with a number of alternatives and solutions to the overlap problem.

1. Based on a prior knowledge of the expected shapes of the particles to be measured, it is possible, using shape and size discrimination to simultaneously identify agglomerates from individual features and to simply measure the individual objects. If the particular distribution has a large spread of size ranges, statistical correction can be applied to correct for the bias arising from excluding the agglomerates. If the distribution of particle size has a narrow spread, no such correction need be applied.

2.  Based on prior knowledge of the shape of the particles, agglomerates may be identified as above and separated into their component
particles using boundary fitting techniques based on the prior shape
information.

3.  In cases where a more advanced image analysis instrumentation
is not justified, or where the shape of the particles is quite
arbitrary so that agglomerated particles may look just like a
single particle, a sample may be prepared with a sufficiently low
concentration (expressed as fraction of area of the image covered
by particles $A_A$) and the effect of any of the residual overlap
upon results may be predicted. Thus, for example, the reduction
of count, due to overlap for low concentrations will for circular
particles of identical size be given by Armitage as $4A_A$.

If the sample has large spread in size, then this reduction in
count may increase to as high as $8A_A$ and if the sample has both
a large spread in size and irregular shapes the multiplier may go
higher. Similarly, the total area of such a sample can be expected to be reduced by the fraction $A_A$ and the mean size increased
by $3A_A$ or more.

If the sample is prepared so that the value of $A_A$ is between 0.5%
and 0.2% and stage motion together with automatic focus options
employed in the image analysis system to provide rapid field to
field analysis; accurate results may be obtained quickly without
significant overlap error.

How Many Particles? - Another question frequently asked when using
image analysis for particle sizing is, "How many particles must be
counted before the results of the specimen can be taken as accurately corresponding to the results of the parent sample?" The
answer to such a question, of course, depends on the type of results to be extracted. As a general guideline, the accuracy with
which a count within a certain classification can be expected to
correlate, on a normalized basis, to the count for the entire
parent sample is given by (with 95% confidence)

$$\frac{2}{\sqrt{N}} \times 100\%$$

where N is the number of particles counted in the classification.
A classification may be a particular size range classification or
a classification based on shape, density or other criterion. Thus
classification applies, for example, to size ranges in a size
histogram, where the number in each interval is independent of the
number in any other subinterval.

In an oversize or undersize cumulative distribution, however, this
independence between each classification does not hold and a different expression is required for the estimated error. In a oversize or undersize distribution normalized to 100%, the error (with
95% confidence) is given by Kolmogorov (see Walsh, Ch. 10) as

$$\frac{1.36}{\sqrt{N}} \times 100\%$$

where N is the total number of particles in the distribution.
Other results, of course, apply to measurements such as means,
standard deviations, and so on. Unfortunately, space does not
permit their inclusion at this time.

Selection of Measurements - A wide range of measurements is now available to the image analysis user from systems such as the Omnicon Pattern Analysis System. The choice of the appropriate measurement to be used for particular situations is an important one. It depends largely on characteristics which it is desired to analyze. Again, space does not permit a detailed coverage of the correlation of different measurements to the characteristics. However, it should be noted that measurements may be correlated to surface area, equivalent sedimentation diameter, equivalent filter of blocking diameter, equivalent weight distribution as well as correlated to a wide range of shape factors, and orientation measures.

Data Processing - Measurement data extracted from particulate sample may be presented in many different ways. A list of the measurements made for each particle is the most extensive and probably least informative of these. On the other hand, presenting the data as a size distribution is both compact and yet highly informative. Many different statistical parameters may be extracted from the distribution including mean, standard deviation, various higher order moments, median, mode, and percentile measurements, which may be used to indicate the fraction of particles in the sample outside specified bounds.

Figure 1 shows display of size distribution as a sideways plot on a television monitor. The upper part of the display shows the commands used to obtain the measurements and resulting size distribution. Next comes the statistical parameters computed from the distribution, and finally, the distribution presented as a size distribution plot rotated through 90°. The left column represents the size intervals, while the right column represents the number of particles in each size interval.

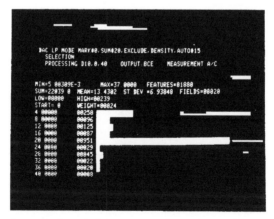

Figure 1: An example of data processing output as presented on the Omnicon PAS display. The measurement area/longest dimension (A/C) was measured for 1880 features over 20 fields of view. The distribution is shown over 10 intervals in the range 0 to 40 microns.

REFERENCES

ARMITAGE, P.

An overlap problem arising in particle counting.
Biometrika, V.36, pp. 256-266.  (1949).

HERDAN, G.

Small particle statistics.  2nd ed.
Academic Press Inc., New York, N.Y.  (1960).

WALSH, J.E.

Handbook of non-parametric statistics,
Van Nostrand, New Jersey, (1962).

*National Bureau of Standards Special Publication 431*
*Proceedings of the* FOURTH INTERNATIONAL CONGRESS FOR STEREOLOGY
*held at* NBS, Gaithersburg, Md., September 4-9, 1975 (Issued January 1976)

MICROSCOPIC CHARACTERIZATION OF FINE POWDERS

by P. Ramakrishnan
*Department of Metallurgical Engineering, Indian Institute of Technology, Bombay-400076, India*

ABSTRACT

Problems associated with the characterization of fine flaky powders in the
subsieve range have been considered. Although water covering area can be used
to characterize the mass of powder, special attention is focused on scanning
electron microscopy for the study of agglomeration, particle morphology and surface characteristics. Since the chemical reactivity of the particles is also
related to the internal structure, transmission electron microscopy can provide
very valuable information. The importance of microscopic observations in the
characterization of these particles has also been discussed.

INTRODUCTION

Although characterization of powders is of fundamental importance, the
factors which characterize an individual particle and a mass of powder are
difficult to determine. Further different test methods used to determine a
single property, say particle size of a powder, do not have identical
results.[1,2] At the same time these characteristics are very important in the
compacting and sintering of metallic and ceramic powders, the hiding power of
paints and pigments or the sensitivity of a chemical reaction. Though the
shape of a powder is an important characteristic, it is difficult to define.[3]
By using the surface contour concept as a guide line it may be possible to
distinguish rounded, irregular, angular, etc., particles. Many attempts have
also been made for shape characterization on a quantitative basis by the use
of various mathematical treatments including the dynamic shape factor of
particles.[4] But these are based on simplified models and are far away from
the real characteristics of the particles. When the particle size of the powder
is reduced the powder becomes more cohesive and sticking and the flow is
restricted. The problems related to handling and dispersion become more acute.
At the same time large quantities of these powders are required in many industrial applications. In this study the characterization of fine flaky aluminum
powders used in the paint, pigment and explosive industries have been attempted.

EXPERIMENTAL WORK

The powders were prepared by the atomization of a thin stream of molten
metal passing through a nozzle with compressed air under pressure.[5] Since
the characteristics of powders are different for different applications[6]
the atomized powders were wet ball milled to achieve the required characteristics. The powders were filtered, dried and the -325 mesh fraction (44
microns) was used for subsequent investigation. A simple sieve analysis is
not sufficient to characterize the powders and hence other methods have been
tried. For fine flaky powders, covering area on water can give reasonably good
reproducible results. But this again does not provide much information about
the size, shape and surface topography of the particles. One of the best
approaches will be the direct observation of the particles microscopically by
using suitable techniques. Hence subsequent characterization studies were made
by using optical microscopy, transmission electron microscopy and scanning
electron microscopy. Only few micrographs are presented to illustrate the
salient features.

RESULTS AND DISCUSSION

Optical microscopy is simple and has been of considerable value in the characterization of certain powders. But there are many limitations when applied to fine particles. This is because of the limitation in the resolution at the highest available magnifications. Fig. 1 shows the optical micrograph of the powders. Although one can observe a comparatively large number of particles which may give some idea about the distribution, this micrograph does not give much information about the shape and other surface characteristics. Scanning electron microscopy offers many advantages[7,8,9] because of its better resolution, depth of focus and range of magnifications (about 20 to 50,000). In the scanning electron microscope one can achieve resolutions approaching 200 A and a depth of focus of about 300 times the optical microscope. The scanning electron micrograph fig. 2 can provide considerable information about the aggregation, particle morphology and surface characteristics. Fig. 3 is the micrograph of the same material at a higher magnification showing more details about particle morphology and surface topography. Particles which appear to be single at a low magnification really consist of an aggregate of smaller particles. Another interesting feature is that the information obtained from scanning electron microscopy regarding the particle characteristics can be used for the interpretation of the results from the water covering area of the powders. Fig. 4 and fig. 5 are the scanning electron micrographs of the two different powders with water covering area of 12,000 $cm^2$/gm and 9000 $cm^2$/gm respectively. The lower water covering area can be attributed to the segregation and the associated morphology of the particles as shown in fig. 5 which is not favorable for spreading, while fig. 4 shows the well developed particle morphology which can cover a larger area.

The reactivity of these powders are also related to their internal structure. Transmission electron microscopy can provide considerable information about the internal structure.[10] The powders are flaky in nature and many particles have thicknesses less than 1000 A, favorable for direct electron transmission. Fig. 6 is a direct electron transmission micrograph showing a considerable amount of dislocations, while fig. 7 is comparatively free from such defects. Powders shown in fig. 7 are found to be less sensitive in a chemical reaction than the powders shown in fig. 6.

CONCLUSIONS

Microscopic observations are playing a predominent role in the characterization of fine flaky powders. By selecting suitable techniques such as optical, electron scanning and electron transmission microscopy, considerable information could be collected regarding the distribution, agglomeration, particle morphology, surface topography and internal structure of these powders. These are valuable tools for the characterization of powders where conventional methods are inadequate and do not provide the necessary information.

REFERENCES

1. Kaye, B. H., Chemical Engineering 7, 239 (1966).

2. Fischmeister, H. F., et al., Powder Met. 7, 82 (1961).

3. Hausner, H. H., Planseeber. Pulvermet. 14, 2, 75 (1966).

4. Medalia, A. I., Powder Technology 4, 3, 117 (1970/71).

5. Ramakrishnan, P., et al., Trans. Ind. Inst. Metals 22, 3, 48 (1969).

6. Ramakrishnan, P., Compaction (1973), ed. A. S. Goldberg, Powder Advisory centre, 23.

7. Johari, O., Metal Progress 94, 147 (1968).

8. Hung-Chi Chao, Modern developments in powder metallurgy 5, 369 (1971), ed. H. H. Hausner, Plenum Press.

9. Nixon, W. C., J. Sei. Instr. 7, 8, 685 (1974).

10. Brimhall, J. L., et al., A.S.T.M., S.T.P., 48, 97 (1970).

Fig. 1. Optical micrograph.

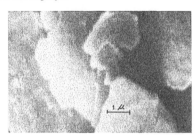

Fig. 2. Scanning electron micrograph.    Fig. 3. Scanning electron micrograph.

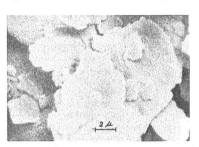

Fig. 4. Scanning electron micrograph.    Fig. 5. Scanning electron micrograph.

Fig. 6.  Direct transmission          Fig. 7.  Direct transmission
        Electron micrograph.                   electron micrograph.

National Bureau of Standards Special Publication 431
Proceedings of the FOURTH INTERNATIONAL CONGRESS FOR STEREOLOGY
held at NBS, Gaithersburg, Md., September 4-9, 1975 (Issued January 1976)

## STEREOLOGY OF PAINT FILMS

by Brian H. Kaye and Ian Robb
Institute for Fineparticle Research, Laurentian University, Sudbury, Ontario, Canada

Classical studies of the performance of pigment particles in an organic film matrix have concerned themselves with the particle size of pigment as measured in very dilute systems before being incorporated in the paint film. There is a growing realization that this approach has severe limitations since any section through a real paint film soon demonstrates the highly agglomerated structure of the pigment dispersed in the paint film. A stereological examination of paint particles set in the paint film poses severe problems for the analyst. Not the least of his problem is the development of an adequate sectioning technique. Since the organic matrix is orders of magnitudes softer than the fineparticles any attempt to cut through the film usually ends in the distortion of the film or the popping out of pigment particles in the sectioning technique. Determination of the average pigment content in any exposed section is relatively simple however the determination of the agglomeration state is very complex. The complexity of the field of view is such that automated iconometric systems are not able to deal with them without extensive interference from the executive analyst with his light pen. And even then the analyst is taking decisions as to what constitutes a homogeneous or an existential agglomerate (i.e. one that exists from the point of view of the physical process being studied in the experiment). Two approaches are being considered at Laurentian University to overcome these problems on analysing any section through a paint film. The first technique involves a new optical information processing technology known as "spatial filtering". In this technology a diffraction pattern is generated by passing a laser beam through a transparency of the whole field of view. This diffraction pattern is filtered in the Fourier transform plane, the spatial filter used being constructed by studying a single particle with laser light. It is then possible to reconstruct a field of view in which all the individual particles which are the same size as the particles used to construct the spatial filter can be studied as single points of light in the plane of reconstruction. There is some difficulty in reconstructing images when the particles are non-symmetrical and there are many technical problems to be overcome, however the reconstruction of the field as a set of bright spots enables its size and number information of the pigment to be collected very easily by iconometric devices. Also this process is a parallel processing technology.

In general it should be noticed that stereology workers have tended to ignore the tremendous power of optical parallel processing for image analysis and probably one of the coming developments in stereology will be the further use of optical image processing rather than sequential iconometric analysis based on a linescan logic.

REFERENCES

1. Kaye, B.H., Naylor, A.G.: "Spatial Filtering for Detection of Particles of a Particular Shape". Journal of Colloid and Interface Science, Vol. 39, no. 1, 1972.

2. Kaye, B.H., Naylor, A.G.: "Optical Information Processing in Characterization by Microscopy". American Laboratory, April, 1974.

*National Bureau of Standards Special Publication 431*
*Proceedings of the* FOURTH INTERNATIONAL CONGRESS FOR STEREOLOGY
*held at* NBS, Gaithersburg, Md., September 4-9, 1975 (Issued January 1976)

WORKSHOP SESSION— PARTICLE SCIENCE

by Brian H. Kaye, *Chairman*
*Institute for Fineparticle Research, Laurentian University, Sudbury, Ontario, Canada*

CONCLUDING REMARKS

From the papers that have been presented in this workshop we can see the
development of three lines of endeavour. Dr. Exner has focussed attention
on the fact that we can obtain information on fineparticle systems by embedding particles in a matrix followed by section of the matrix. It is probably
generally true that the fineparticle analyst is not aware of the power and the
number of relationships that have been developed within stereology for abstract-
ing information from sectioned fields of view. One of the values of a
conference such as this is the bringing together of fineparticle analysts with
people in the field of stereology. I am sure that as a result of this conference we will see a wider diffusion of stereological relationships into the
technology of the fineparticle analyst. We also see developments of mathematical relationships for evaluating fineparticles as viewed through iconometric
systems. In the past the tremendous information level required has tended to
deter the analyst from developing these relationships. Now that we have the
automated iconometric technology available we can anticipate increasing
development of studies such as those presented by Drs. Scarlett and Lloyd
and Dr. Dullien.

What is missing from iconometric technology however is a realization that often
iconometric devices work in the opposite direction from the point of view of
experimental efficiency. They gather all the information in a field of view
and then discard, intelligently, information not required. More efficient
iconometric devices must work the other way; they must be able to discriminate
in a field of view what they need to know and abstract that information for
future use. Perhaps a name for this approach to iconometrics device would be
"minimal iconometrics" as distinct from present forms of automated iconometrics.
So far in these remarks we have focussed on what the stereologist has to offer
to the fineparticle analyst. There is the other side of the story. It is
obvious that the art of optical information processing of images is more
advanced in fineparticle science than in stereology. The powerful methodology
for the parallel processing of images using holographs, spatial filtering and
Fourier transforms will obviously be a great interest to the stereologists as
they become more widely known. In conclusion one can say that this workshop
has been a fruitful encounter between two groups, both of which are interdisciplinary and it should stimulate the diffusion of exciting technology into
both areas of endeavour.

## 9.  SIZE DISTRIBUTIONS

*Chairperson:*

J. H. Steele, Jr.

National Bureau of Standards Special Publication 431
Proceedings of the FOURTH INTERNATIONAL CONGRESS FOR STEREOLOGY
held at NBS, Gaithersburg, Md., September 4-9, 1975 (Issued January 1976)

ESTIMATION OF LINEAR PROPERTIES OF SPHERICAL BODIES IN THIN FOILS FROM THEIR PROJECTIONS

by Frank Piefke
Institut B für Mathematik, Technische Universität Braunschweig, 33 Braunschweig, Pockelsstr. 14, Germany

We investigate the size distribution of spheres in a given
unit volume V=1. Consider $N_V$ opaque spherical bodies with the radii
$x_i$ (i=1,2,...,$N_V$), which are distributed at random in a transparent
matrix; the centers of them may be distributed homogeneously.
A linear property $\Theta$ is defined by (NICHOLSON 1970):

$$\Theta = \sum_{i=1}^{N_V} H(x_i) \qquad (1)$$

where H(x) is a given real function. Two examples of linear
properties are:

$$N_V \text{ with } H(x) = 1 \text{ and } S_V \text{ with } H(x) = 4\pi x^2.$$

We take a slice of thickness T and area A from the structure.
This slice contains a certain number $AN_A$ of sections of the
spheres. We measure the radii of the $A$ circular projections of
these sections. These radii may be $y_i$ (i=1,2,...,$AN_A$). We suppose
that a positive resolution point $y_0$ exists; $A$ particles
with radii less than $y_e$ cannot be seen.

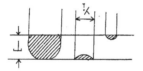
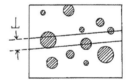

Now we can formulate the estimators of linear properties using the
theory of NICHOLSON. An unbiassed estimator of the linear property
$\Theta$ given by (1) is:

$$\hat{\Theta} = \sum_{i=1}^{AN_A} h(y_i) \qquad (2)$$

where h(y) is the solution of a certain integral-equation. In our
case h(y) is given by:

$$(3)$$

$$h(y) = \frac{V}{AT} \sqrt{\frac{2}{\pi}} \{ f(\frac{\sqrt{2\pi}}{T}\sqrt{y^2-y_0^2})H(y_0) + \int_{y_0}^{y} f(\frac{\sqrt{2\pi}}{T}\sqrt{y^2-\eta^2})H'(\eta)\,d\eta \}$$

H(y) must be differentiable. The function f is given by:

497

$$f(w) = \exp(\ w^2/2\ )\ \int_w^\infty \exp(\ -t^2/2\ )\ dt \qquad (4)$$

For example an unbiaséd estimator of the number $N_V$ is:

$$\hat{N}_V = \frac{V}{A\ T}\sqrt{\frac{2}{\pi}}\ \sum_{i=1}^{AN_A} f(\frac{\sqrt{2\pi}}{T}\ \sqrt{y_i^2 - y_0^2}\ ) \qquad (5)$$

This is an old result of my teacher G. BACH. I have calculated estimators of $L_V$ the sum over all radii, $S_V$ the total surface of the particles, $V_V$ the total volume of the particles and of the variances of thése estimators. An explicit formula for h(y) is known if the function H(x) is of the form:

$$H(x) = M\ x^r \quad \text{with } r=0,1,2,\ldots$$

Now I have tested the estimation theory by a Monte-Carlo-Simulation on a computer. A cube of volume 1 is constructed in which the centers of the spheres are distributed homogeneously. Then a theoretical density function g(x) of the radii of the spheres is chosen; a radius given by a random number corresponds to each center. A real model of a distribution of spheres in space is constructed by this method. For the function g(x) I have chosen three cases:
1) triangle-distribution, 2) $g(x) = B\ x\ \exp(\ -Bx^2/2\ )$ and
3) two-point-distribution.
Secondly the slice of thickness T is taken from the cube and the radii $y_i$ are measured. Now the estimators can be calculated and be compared with the true values of the linear properties.
I have chosen $N_V$ = 1000 spheres in the cube. A few hundred of them dependent on T and $y_0$ are seen in the slice. Table 1 shows some results of my calculations.

### Table 1

| T | $y_0$ | $N_V$ | $\hat{N}_V$ | $S_V$ | $\hat{S}_V$ | distribution |
|---|---|---|---|---|---|---|
| 5 | 0 | 1000 | 1005 | $2.26\ 10^5$ | $2.28\ 10^5$ | 1) |
| 5 | 1 | 1000 | 971 | $2.26\ 10^5$ | $2.28\ 10^5$ | 1) |
| 2 | 1 | 1000 | 1012 | $2.26\ 10^5$ | $2.29\ 10^5$ | 1) |
| 5 | 0 | 1000 | 1000 | $4.05\ 10^5$ | $3.99\ 10^5$ | 2) |
| 5 | 1 | 1000 | 986 | $4.05\ 10^5$ | $3.99\ 10^5$ | 2) |
| 2 | 1 | 1000 | 992 | $4.05\ 10^5$ | $3.93\ 10^5$ | 2) |
| 5 | 0 | 1000 | 1062 | $2.38\ 10^5$ | $2.42\ 10^5$ | 3) |
| 5 | 1 | 1000 | 1051 | $2.38\ 10^5$ | $2.41\ 10^5$ | 3) |
| 2 | 1 | 1000 | 1085 | $2.38\ 10^5$ | $2.42\ 10^5$ | 3) |

REFERENCES   NICHOLSON, G.S., Biometrika 57, 2, 273 (1970)
             BACH, G., Dissertation, Selbstverlag des Mathematischen Seminars
             Giessen (1959)
             PIEFKE, F., Dissertation, Technische Universität Braunschweig (1975)

*National Bureau of Standards Special Publication 431*
*Proceedings of the* FOURTH INTERNATIONAL CONGRESS FOR STEREOLOGY
*held at* NBS, Gaithersburg, Md., September 4-9, 1975 (Issued January 1976)

ON SIZE DISTRIBUTION METHODS

by F. A. L. Dullien and K. S. Chang,
*University of Waterloo, Waterloo, Ontario,. Canada*

I. NOTE ON SECTION DIAMETER THEORIES

Kendall and Moran[1] in their little book on "Geometrical Probability" presented a derivation of the integral equation for the probability distribution $\phi(\delta)d\delta$ of the diameters of the circle of intersection when it is known that the sphere does intersect the plane. The final expression is

$$\phi(\delta)d\delta = \frac{\delta}{D_o} \int_\delta^\infty \frac{F(D)}{\sqrt{D^2 - \delta^2}} \, dD d\delta \tag{1}$$

They commented that an incorrect theory, which was apparently widely used, neglected the fact that larger spheres have a greater probability of being included in the section. In an attempt to clarify matters it will be shown below that a different widely used theory (e.g. Scheil[2]) in which there is no explicit reference to this fact, is equivalent to (1). First the derivation in Kendall and Moran is given.

The expected number of spheres whose diameters lie in the range $D \to D + dD$ and which intersect an arbitrary plane in unit area of the latter is $N_V DF(D)dD$ where $N_V$ is the mean number of centers of spheres in unit volume and $F(D)dD$ is the probability of a sphere chosen at random to have diameter in the range $D \to D + dD$. The probability distribution of diameters of spheres which intersect the plane will be

$$f(D)dD = \frac{DF(D)dD}{\int_0^\infty DF(D)dD} = \frac{DF(D)dD}{D_o} \tag{2}$$

where $D_o$ is the mean sphere diameter. Note that $N_V D_o$ is the number of spheres intersected by unit area of the plane.

If a sphere of diameter D does intersect the plane, the probability that its center lies at a distance $X \to X + dX$ from the plane is $2D^{-1}dX$ and the diameter of the circle of intersection is $(D^2 - 4X^2)^{1/2}$.

Hence the probability of intersecting a sphere of diameter D and obtaining a circle of diameter $\delta$ is

$$f(D)dD \frac{D^{-1} \delta \, d\delta}{\sqrt{D^2 - \delta^2}} = \frac{\delta F(D) dD d\delta}{D_o \sqrt{D^2 - \delta^2}} \tag{3}$$

where (2) has been used.

The probability distribution $\phi(\delta)d\delta$ of the diameters of the circle of intersection when it is known that the sphere does intersect the plane is therefore given by (1).

According to the other theory, the expected number of spheres of diameter $D \to D + dD$ and which intersect an arbitrary plane in a unit area of the latter at a distance $\overline{X} \to X + dX$ from the plane is $2N_V F(D)dDdX$. Notice that $N_V F(D)dD$ is equal to $N(D)dD$, the number of spheres with diameters $D \to D + dD$ in unit volume. Hence the expected number of circles of diameter $\delta$ in unit area of plane, obtained by intersecting spheres of diameter D is

$$N_V \delta \; \frac{F(D) \; dDd\delta}{\sqrt{D^2 - \delta^2}} \tag{4}$$

The expected number of circles of diameter $\delta$ in unit area of plane, $n(\delta)d\delta$, obtained by the intersecting spheres of any diameter $(D \geq \delta)$ is

$$n(\delta) d\delta = N_V \delta d\delta \int_\delta^\infty \frac{F(D) \; dD}{\sqrt{D^2 - \delta^2}} \tag{5}$$

The total expected number of circles in unit area of plane is equal to the number of spheres intersected by unit area of plane

$$N_V \int_0^\infty D \; F(D) \; dD = N_V D_o \tag{6}$$

Hence the probability distribution of the diameters of the circle of intersection also by this theory is

$$\frac{n(\delta) d\delta}{N_V D_o} = \frac{\delta}{D_o} \int_\delta^\infty \frac{F(D) \; dDd\delta}{\sqrt{D^2 - \delta^2}} \tag{7}$$

The lefthand side of (7) is nothing but $\phi(\delta)d\delta$ and thus (5) agrees with (1). Therefore the two theories are equivalent.


II.  NOTE ON THE SECTION - DIAMETER METHOD FOR NON-SPHERICAL OBJECTS

Dullien and co-workers[3-5] have devised stereological methods for non-spherical objects which they have used successfully for void size distribution determination in sandstones and bead packs.  In the original presentation of the theory there was no clear line drawn between rigorous mathematical theory and the simplified procedure used.  An attempt is made here to clarify some of these problems.

The size of an object is characterized by the magnitude of the longest chord, or intercept length, $D_{2MAX}$.  It is convenient to choose this chord as axis A of the object.  The objects may have any shape, but for increasingly non-convex objects the longest chords will represent the true size of the object with decreasing accuracy.  All objects are assumed to have the same shape and they may differ only in size.

The orientation of an object in a cartesian coordinate system is given by the polar angle $\theta$ enclosed by the axis A with the positive z-direction, and the angle $\phi$ enclosed by the projection of A onto the x-y plane with the positive x-direction.

Each object is thought to be placed in the coordinate system in such a way that the x-y plane is exactly half way the distance $D_1$ between the tangent planes to the two extremities of the object which are parallel to the x-y plane.

The expected number of objects of diameter $D_{2MAX} \rightarrow D_{2MAX} + dD_{2MAX}$ of any given orientation which are intersected by a plane parallel to the x-y plane in unit area of the latter at a distance $z \rightarrow z + dz$ from the x-y plane, giving rise to two-dimensional features of diameter $\delta \rightarrow \delta + d\delta$ in the intersecting plane, is

$$2N_V F(D_{2MAX}) \; dD_{2MAX} dz \; (\sin\theta d\theta d\phi)/4\pi = 2N_V F(D_{2MAX}) \; dD_{2MAX} \left| \frac{\partial z}{\partial \delta} \right| \; d\delta (\sin\theta d\theta d\phi)/4\pi \tag{8}$$

The diameter $\delta$ of a two-dimensional feature is defined as the longest chord of intersection with the feature in the x-direction.

The total number of two-dimensional features $\beta(\delta)\,d\delta\,dD_{2MAX}$ with diameter $\delta$, obtained by intersecting objects of diameter $D_{2MAX}$, which has been assumed random, is therefore

$$\beta(\delta)\,d\delta\ dD_{2MAX} = 2N_V F(D_{2MAX})\,d\delta\,dD_{2MAX} \int_{\phi=0}^{\pi} \int_{\theta=0}^{\pi} \left|\frac{\partial z}{\partial \delta}\right| \sin\theta\,d\theta\,d\phi/4\pi =$$

$$= 2N_V F(D_{2MAX}) \overline{\left|\frac{\partial z}{\partial \delta}\right|}\ dD_{2MAX}\,d\delta \qquad (9)$$

The operation in (9) presumes that $z = g(\delta)$ is known as a function of the orientation of the object as given by the pair of numbers $(\theta,\phi)$.

Determination of $z = g(\delta)$ by direct observation would involve making successive parallel sections of individual objects in various random, but known orientations and measuring $\delta$ in each section as a function of z. Could the derivative $\overline{\left|\partial z/\partial \delta\right|}$ be obtained by direct observation then this would be some function of $D_{2MAX}$ and which would take the place of the expression $(1/2)(\delta/\sqrt{D^2-\delta^2})$, used for spherical objects. The expected number of two-dimensional features of diameter $\delta$, obtained in unit area of sectioning plane, regardless of $D_{2MAX}$ would then be

$$n(\delta)\,d\delta = 2N_V\,d\delta \int_{\delta}^{\infty} \overline{\left|\frac{\partial z}{\partial \delta}\right|}\ F(D_{2MAX})\,dD_{2MAX} \qquad (10)$$

The question arises what happens when the object is not convex and $g(\delta)$ is, in general, not unique and $\partial z/\partial \delta$ has singularities. Evidently the probability of an event cannot be infinite and therefore at the points where $\partial z/\partial \delta$ has singularity $d\delta$ must diminish to zero. It is worthwhile to note that even for the case of spherical objects $\partial z/\partial \delta$ is infinite at the equator. Therefore singularities which may arise in the case of non-convex objects do not present a real problem. $\left|\partial z/\partial \delta\right|\,d\delta$ merely provides a measure of the number of two-dimensional features with diameter $\delta \rightarrow \delta + d\delta$ obtained when sectioning an object of a specified orientation with equidistant parallel planes. It is immaterial whether features of a given diameter $\delta$ originate when sectioning the object only at one or at several different values of z. Therefore lack of convexity of the objects does not invalidate the theory.

For practical purposes the function $g(\delta)$ must be determined from observations made on sections through many objects of random orientation which are imbedded in a solid matrix. Moreover the various objects never have exactly the same shape. In view of these complications and in order to simplify the procedure y vs. $\delta$ has been measured in the two-dimensional features obtained by sectioning the objects of random orientation imbedded in the solid matrix. The assumption has been made that the average of $y = g(\delta)$ taken over two-dimensional features of various sizes and orientations in a sample is the same as the average of $z = G(\delta)$ taken over the corresponding three-dimensional objects.

The dimensionless form has been introduced

$$\left(\frac{y}{D_1/2}\right) = \psi\left(\frac{\delta}{D_2}\right) \qquad (11)$$

where $D_2$ is the largest value of $\delta$ for a given feature of specified orientation.

Differentiating (11) gives

$$\left|\frac{\partial y}{\partial \delta}\right| = \frac{D_1}{2D_2} \left| \psi'\left(\frac{\delta}{D_2}\right) \right| \tag{12}$$

whence, by averaging over a large number of two-dimensional features of various sizes, shapes, and orientations

$$\overline{\left|\frac{\partial y}{\partial \delta}\right|} = \left(\frac{1}{2}\right)\left(\overline{\frac{D_1}{D_2}}\right) \overline{\left|\psi'\left(\frac{\delta}{D_2}\right)\right|} \tag{13}$$

Using (13) and the assumption $\overline{|\partial y/\partial \delta|} = \overline{|\partial z/\partial \delta|}$ in (9), the following approximate expression is obtained for the expected number of two-dimensional features of diameter $\delta$ in unit area of sectioning plane, regardless of $D_{2MAX}$ and orientation of the objects

$$n(\delta)\,d\delta \cong \left(\overline{\frac{D_1}{D_2}}\right) N_V d\delta \int_{D_{2MAX}}^{\infty} F(D_{2MAX}) \, \overline{|\psi'(\delta/D_2)|} \, dD_{2MAX} \tag{14}$$

This is the integral equation to solve for the unknown $N_V F(D_{2MAX}) dD_{2MAX}$ in the case of non-spherical objects. The solution has been worked out by Dhawan[6] and it has been successfully applied to the void size distribution determination of various sandstones and bead packs. In certain cases where the relationship

$$\overline{|\psi'(\delta/D_2)|} = |\overline{\psi}'(\delta/D_2)| \tag{15}$$

holds $\overline{|\psi'(\delta/D_2)|}$ may be replaced by $|\overline{\psi}'(\delta/D_2)|$ in (14).

REFERENCES

1. Kendall, M.G., and Moran, P.A.P.: "Geometrical Probability". Hafner Publishing Co., New York, 1963.

2. Scheil, E.: Die Berechnung der Anzahl und Grössenverteilung kugelförmiger Körper mit Hilfe der durch ebenen Schnitt erhaltenen Schnittkreise. Zeit. anorg. allgem. Chem., 201 (1931), 259-264.

3. Dullien, F.A.L. and Dhawan, G.K.: "Photomicrographic Size Distribution Determination of Non-Spherical Objects," Powder Tech. 7 (1973), 305-313.

4. Dullien, F.A.L. and Dhawan, G.K.: "Characterization of Pore Structure by a Combination of Quantitative Photomicrography and Mercury Porosimetry," J. Coll. Int. Sci., 47 (1974), 337-349.

5. Dullien, F.A.L.: "Photomicrographic Pore Size Distributions Using Quantitative Stereology and Application of Results in Tertiary Petroleum Recovery," in "Pore Structure and Properties of Materials" RILEM-IUPAC Proceedings of the International Symposium, Prague, September 18-21, 1973, Preliminary Report - Part I. Academia Prague 1973.

6. Dhawan, G.K.: "Photomicrographic Investigation of the Structure of Sandstones and Other Porous Media by the Methods of Quantitative Stereology: With Application to Oil Recovery," Ph.D. Thesis, University of Waterloo, 1972.

National Bureau of Standards Special Publication 431
Proceedings of the FOURTH INTERNATIONAL CONGRESS FOR-STEREOLOGY
held at NBS, Gaithersburg, Md., September 4-9, 1975 (Issued January 1976)

SIZE DISTRIBUTION OF RETAINED AUSTENITE PHASE IN A QUENCHED STAINLESS STEEL

by Anant V. Samudra and Om Johari
IIT Research Institute, Chicago, Illinois 60616, U.S.A.

INTRODUCTION

Retained austenite is a metastable phase in quenched steel. The properties of the steel and their behavior toward operating stress and temperature variations depend on the resistance of the retained austenite to martensite formation. This resistance, called the austenite stability, has been quantitatively correlated to the microstructural features of retained austenite.[1] The portion of work on characterizing the size distribution of retained austenite is presented here.

MATERIAL AND PROCESSING

A high-purity Fe-28Ni-0.1C alloy was austenitized and quenched to a range of temperatures from $+10^\circ C$ to $-196^\circ C$ to achieve different volume fractions of martensite. The specimens were prepared and studied by optical metallography.

RESULTS

As martensitic transformation progresses, the austenite gets partitioned into compartments. The austenite and martensite distribution is nonuniform, and the austenite compartment formation is not well developed until the transformation exceeds 80 percent. At higher extents of transformation the retained austenite is finely distributed in the martensite matrix. The data on two-dimensional size distribution in retained austenite in six samples are presented in Table 1.

DISCUSSION

Quantitative characterization of three-dimensional size distribution in retained austenite requires a quantitative metallography method. At present, most methods depend on the assumption that particles approach a spherical or spheroidal shape.[2-5] In the present study, the Johnson-Saltykov method based on size distribution of two-dimensional areas was employed. Retained austenite was assumed as dispersed uniformly as discrete, spherical particles in a martensite matrix. The original Saltykov table had to be extended to smaller sizes. There were great practical difficulties in exact counting and classifying of very small austenite areas, in spite of the 1000X magnification employed.[2,3,6] Also, the apparent zero count in size group (-1) resulted in negative numbers because of the sequential dependence of the derivation. The analyses therefore exclude data beyond size group (0). The analyses summarized in Tables 2 and 3, allow the following conclusions:

1. A size distribution in the retained austenite phase exists for each sample.

2. The maximum size and the mean size of retained austenite particles decrease with lower volume fraction of retained austenite.

3. The number of large size austenite particles decreases with lower volume fraction of retained austenite.

The particle size distributions are approximate, and no special significance is to be attached to secondary peaks and valleys. The metallurgical interpretations of these size distributions and their correlation with the stability of retained austenite will be published elsewhere.

As mentioned earlier, the accuracy of area counts and their classification is limited in the small areas. In the analyses presented, the very small areas (size group (-2)) were excluded. To estimate how the results would be affected if these data were included, the counts from size group (-1) and (-2) were lumped together to facilitate the sequential analysis. The new estimates for sample B are presented in parentheses in Tables 2 and 3.

Clearly, the new particle counts in these small size groups are very large and, therefore, the average particle size has decreased to half its value. Also, the relative proportion of the particles in different size groups changes significantly (see Table 3).

It is well known that the mean value of the particle volume depends on the magnification employed and that the estimate of mean particle volume decreases as magnification increases.[2,3] Thus, comparisons of mean particle volume must be made at the same magnification. It should be realized that the relative contribution to particle counts by very small size areas increases and there is a greater and greater error in the estimate of mean austenite particle volume, as the volume fraction of austenite decreases.

For improving the accuracy of measurement in this type of study, techniques that can give higher magnifications and better resolution than optical microscopy are necessary. Scanning electron microscopy is not suitable for studying smooth surfaces and requires a somewhat deeper etch (for metallographic samples), and the attendant broadening of the phase boundaries makes quantitative measurements difficult. Transmission electron microscopy would require large sample areas of uniform thickness to obtain statistically usable data.

Underwood[3] has described the new simplified method suggested by Saltykov for deriving particle size from size distribution of areas. If the data from Table 1 are analyzed by this new formulation, estimates of particle counts result are higher by 15-20 percent than those by the Johnson-Saltykov method. Table 4 compares the analyses for the two methods. The reason for the discrepancy is not clear.

CONCLUSION

The particle size distribution in a practical case of austenite-martensite mixture is not easily determined. With the currently available experimental techniques and mathematical formulations, the chances of improving these results appear small.

REFERENCES

1.  A.V. Samudra, "Stability and Structure of Retained Austenite in a Quenched Stainless Steel," Ph.D. Thesis, IIT, Chicago, 1975.

2.  E.E. Underwood, "Particle Size Distribution," in Quantitative Microscopy, R.T. DeHoff and F.N. Rhines (eds.), McGraw-Hill, New York, 1968, pp. 149-200.

3.  E.E. Underwood, "The Mathematical Foundations of Quantitative Stereology," in Stereology and Quantitative Metallography, ASTM STP 504, 1972, pp. 3-38.

4.  J.W. Cahn and R.L. Fullman, "On the Use of Lineal Analysis for Obtaining Particle Size Distribution Function in Opaque Samples," Trans. AIME, 1956, vol. 206, no. 5, pp. 610-612.

5.  R.T. DeHoff, "The Determination of the Size Distribution of Ellipsoidal Particle from Measurements Made on Random Plane Sections," Trans. AIME, 1962, vol. 224, no. 6, pp. 474-477.

6.  J.E. Hilliard, "Measurement of Volume in Volume," in Quantitative Microscopy, R.T. DeHoff and F.N. Rhines (eds.), McGraw-Hill, New York, 1968, pp. 45-74.

Table 1 - Distribution of Section Areas of Austenite Phase

| Group No. | Section Area Limits, $\mu m^2$ | A (17% Aust.) | B (8.6% Aust.) | C (3.4% Aust.) | D (1.8% Aust.) | E (1.5% Aust.) | F (1.1% Aust.) |
|---|---|---|---|---|---|---|---|
| | | | Number of Section Areas | | | | |
| 12 | 124-197 | | 1 | | | | |
| 11 | 78.5-124 | 2 | 3 | 1 | | | |
| 10 | 49.6-78.5 | 9 | 8 | 2 | 1 | | |
| 9 | 31.3-49.6 | 23 | 15 | 12 | 2 | | |
| 8 | 19.7-31.3 | 43 | 26 | 22 | 9 | | |
| 7 | 12.4-19.7 | 49 | 33 | 35 | 15 | | |
| 6 | 7.85-12.4 | 76 | 81 | 57 | 38 | 8 | 4 |
| 5 | 4.96-7.85 | 74 | 80 | 68 | 67 | 8 | 9 |
| 4 | 3.13-4.96 | 60 | 59 | 44 | 55 | 16 | 10 |
| 3 | 1.97-3.13 | 83 | 141 | 102 | 108 | 42 | 41 |
| 2 | 1.24-1.97 | 33 | 52 | 31 | 36 | 26 | 45 |
| 1 | 0.785-1.24 | 24 | 59 | 50 | 45 | 26 | 33 |
| 0 | 0.496-0.785 | 15 | 41 | 46 | 25 | 10 | 12 |
| -1 | 0.313-0.496 | 0** | 0 | 0 | 0 | 0 | 0 |
| -2 | 0.197-0.313 | 87** | 215 | NC* | NC* | NC* | NC* |
| Total area examined, $mm^2$ | | 0.0301 | 0.0507 | 0.0507 | 0.0608 | 0.0608 | 0.0608 |

*NC stands for not counted.

**The reason for zero count in group (-1) and a large count in group (-2) is due to practical difficulties in exact measurement of small areas and in their accurate classification.

Table 2 - Number of Austenite Particles per $mm^3$

| Group No. | A (17% Aust.) | B (8.6% Aust.) | C (3.4% Aust.) | D (1.8% Aust.) | E (1.5% Aust.) | F (1.1% Aust.) |
|---|---|---|---|---|---|---|
| | Number of Austenite Particles Per $mm^3$ in Samples $(x\ 10^3)$ | | | | | |
| 12 | | 1.9 | | | | |
| 11 | 8.0 | 6.5 | 2.4 | | | |
| 10 | 43.3 | 21.4 | 5.0 | 2.5 | | |
| 9 | 107.9 | 47.7 | 42.7 | 5.5 | | |
| 8 | 295.8 | 101.8 | 86.0 | 33.3 | | |
| 7 | 371.3 | 149.9 | 168.4 | 62.2 | | |
| 6 | 778.7 | 525 | 346 | 208.5 | 50.1 | 25.0 |
| 5 | 779.3 | 529 | 473 | 441 | 46.7 | 62.9 |
| 4 | 620 | 374 | 251.5 | 341 | 131.2 | 73.3 |
| 3 | 1378 | 1767 | 1254 | 1100 | 520 | 470 |
| 2 | -- | 87.2 | -- | 26.5 | 195.3 | 519 |
| 1 | -- | 802 | 516 | 511 | 308.5 | 366.5 |
| 0 | -- | 515 (7070)* | -- | 173 | | |
| Total | 4382.3 | 4928.4 (11998)* | 3145.0 | 2904.7 | 1251.8 | 1516.7 |
| Avg Vol/ Particle, $x\ 10^{-8}\ mm^3$ | 3.93 | 1.94 (0.81)* | 2.10 | 1.15 | 0.51 | 0.39 |

*The numbers in parentheses show the new values after including the counts of very small particles in the analysis.

Table 3 - Relative Percentages of Austenite Particles

| Group No. | Austenite Particles per mm$^3$ in Samples, Relative Percentages | | | | | |
|---|---|---|---|---|---|---|
| | A (17% Aust.) | B (8.6% Aust.) | C (3.4% Aust.) | D (1.8% Aust.) | E (1.5% Aust.) | F (1.1% Aust.) |
| 12 | | 0.04 | | | | |
| 11 | 0.18 | 0.13 | | | | |
| 10 | 0.99 | 0.43(0.18)* | 0.07 | 0.08 | | |
| 9 | 2.46 | 0.97(0.39) | 0.16 | 0.19 | | |
| 8 | 6.75 | 2.07(0.85) | 1.36 | 1.15 | | |
| 7 | 8.47 | 3.04(1.24) | 2.73 | 2.14 | | |
| 6 | 17.77 | 10.65(4.37) | 5.36 | 7.15 | 4.0 | 1.65 |
| 5 | 17.78 | 10.74(4.41) | 11.0 | 15.2 | 3.73 | 4.15 |
| 4 | 14.15 | 7.6(3.11) | 15.0 | 11.73 | 10.45 | 4.84 |
| 3 | 31.44 | 35.9(14.73) | 7.99 | 37.9 | 41.5 | 31.0 |
| 2 | -- | 1.77(0.73) | 39.9 | 0.91 | 15.6 | 34.25 |
| 1 | -- | 16.3(6.68) | -- | 17.6 | 24.6 | 24.2 |
| 0 | -- | 10.5(4.29) | 16.4 | 5.96 | -- | -- |
| -1 | | (58.93) | -- | | | |
| Total | 100 | 100  (100.0) | 100 | 100 | 100 | 100 |

*The numbers in parentheses show the new values after including the counts of very small particles in the analysis.

Table 4 - Comparison of Particle Distributions According to Two Methods

| Group No. | Sample A | | Sample B | | Sample C | |
|---|---|---|---|---|---|---|
| | New Saltykov | J-S* | New Saltykov | J-S | New Saltykov | J-S |
| 12 | | | 2.3 | 1.9 | | |
| 11 | 9.7 | 8.0 | 7.8 | 6.5 | 2.9 | 2.4 |
| 10 | 51.5 | 43.3 | 25.7 | 21.5 | 6.2 | 5.0 |
| 9 | 156.3 | 107.9 | 57.2 | 47.7 | 51.7 | 42.7 |
| 8 | 347.3 | 295.8 | 121.4 | 101.8 | 105.8 | 86.0 |
| 7 | 428.4 | 371.3 | 176.9 | 149.0 | 201.6 | 168.4 |
| 6 | 899.1 | 778.7 | 629.5 | 525.0 | 412.1 | 346.0 |
| 5 | 927.5 | 779.3 | 623.o | 529.0 | 559.8 | 473.0 |
| 4 | 784.4 | 620 | 430.8 | 374.0 | 287.0 | 251.5 |
| 3 | 1786.3 | 1378 | 2111.4 | 1767.0 | 1502.0 | 1254.0 |
| 2 | 105.3 | Neg | 127.8 | 87.2 | Neg | Neg |
| 1 | 298.7 | Neg | 914.1 | 802.2 | 917.6 | 516.0 |
| 0 | 163.5 | Neg | 584.9 | 515.0 | 953.1 | Neg |

*J-S refers to the Johnson-Saltykov Method.

*National Bureau of Standards Special Publication 431*
*Proceedings of the* FOURTH INTERNATIONAL CONGRESS FOR STEREOLOGY
*held at* NBS, Gaithersburg, Md., September 4-9, 1975 (Issued January 1976)

ANALYSIS OF A SET OF SPHERICAL CELLS RELATIVE TO THEIR VOLUME

by Philippe d'Athis
*Département d'Hématologie, U.E.R. Pitié Salpêtrière, 91 boulevard de l'Hôpital, 75013 Paris, France*

DEFINITIONS

Given a set E of cells, let g be the density of frequency for the cellular volume in this set; g is a positive function such as:

$$\int_a^b g(v)\,dv = \text{frequency of cells, the volume of which lies between the values a and b,}$$

and consequently gives the equality:

$$\int_0^{+\infty} g(v)\,dv = 1$$

Let $(t \to r_t)$ a family of positive functions such as:

$$\int_0^{+\infty} r_t(v)\,dv = 1$$

(Each function $r_t$ may be looked at as a density of frequency.)
We say the set E is <u>homogeneous</u> relative to the family $(t \to r_t)$ if a value t can be found such as:

$$r_t = g$$

and is heterogeneous relative to that family if values $(t_1,p_1)$, $(t_2,p_2)$, . . . $(t_n,p_n)$ (where $n \geq 2$) can be found, such as:

1) $p_1 > 0$, $p_2 > 0$, . . . $p_n > 0$

2) $p_1 + p_2 + . . . + p_n = 1$

3) $p_1 r_{t_1} + p_2 r_{t_n} = g$

(Demonstrating a set of blood cells to be heterogeneous leads to presuming it to contain physiological different cells.)

PROBLEM

A sample being randomly extracted from the set E, every cell of this sample is cut randomly and therefore gives a plane section.

The problem is to analyze E, that is to say, to appreciate with the help of measures on sections whether the set E is homogeneous or heterogeneous relative to a given family $(t \to r_t)$ of densities.

METHOD OF ANALYSIS

This abstract only considers a set of SPHERICAL cells; then a section is a spherical zone, the thickness of which is equal to the thickness h of the slice and the function of repartition F of the APPARENT (= larger) diameter can be defined:

$F(z)$ = frequency of sections, the apparent diameter of which is lower than the value z.

When it can be assumed that

1) the random sampling uniformly probabilizes the set E; and

2) every cell from the sample is cut uniformly, relative to one of its diameters;

it is then demonstrated that:

$$F(z) = 1 - \frac{\pi}{2} \int_z^{+\infty} x^2 \left(\frac{h + \sqrt{x^2 - z^2}}{h + x}\right) g \left(\frac{\pi}{2}x^3\right) dx$$

Then, with the theoretical equality

$$g = p_1 r_{t_1} + p_2 r_{t_2} + \ldots + p_n r_{t_n}$$

and the experimental function $F_1$, we can estimate the parameters $t_1, t_2, \ldots t_n$; $p_1, p_2, \ldots p_n$ by the "least square method;" that is, by minimizing the function:

$$Q(t,p) = \left|\left| F(z) - F^* z \right|\right|^2$$

where

$$t = (t_1, t_2, \ldots t_n) \text{ and } p = (p_1, p_2, \ldots p_n)$$

$$F^*(z) = 1 - \frac{\pi}{2} \int_z^{+\infty} x^2 \left(\frac{h + \sqrt{x^2 - z^2}}{h + x}\right) g^* \left(\frac{\pi}{2}x^3\right) dx$$

$$g^* = p_1 r_{t_1} + p_2 r_{t_2} + \ldots + p_n r_{t_n}.$$

APPLICATION

The two most frequently used families $(t \rightarrow r_t)$ are

1) the normal family:

$$r_t(v) = \frac{1}{s\sqrt{2\pi}} \cdot \exp\left(-\frac{1}{2}\{\frac{v-m}{s}\}^2\right)$$

$$t = (m,s)$$

which can only be an approximation because

$$\int_0^{+\infty} r_t(v) dv = 1$$

The quality of this approximation improves with the ratio $\frac{m}{s}$ and may be considered admissible from the value $\frac{m}{s} = 3$.

2) The log-normal family:

$$\begin{cases} r_t(v) = \frac{1}{s\sqrt{2\pi}} \cdot \exp\left(-\frac{1}{2}\{\frac{\ln(v)-m}{s}\}^2\right) \cdot \frac{1}{v} \\ t = (m,s) \end{cases}$$

*National Bureau of Standards Special Publication 431*
*Proceedings of the* FOURTH INTERNATIONAL CONGRESS FOR STEREOLOGY
*held at* NBS, Gaithersburg, Md., September 4-9, 1975 (Issued January 1976)

## APPENDIX: BASIC STEREOLOGY

by Ervin E. Underwood
*President of the International Society for Stereology*

Since its inception in 1961, the International Society for Stereology has been active in promoting a uniform system of nomenclature for the new science of stereology. Because of the broad scope of stereology, no attempt is made to define all possible terms. Rather, only the basic geometrical quantities are defined, affording instant recognition to scientists anywhere, regardless of language. In this way, stereological symbols, like music and mathematics, have attained an equally international acceptance.

The following brief review has been prepared with a dual purpose in mind. One is to familiarize readers of stereological papers with this system of nomenclature. The other objective is to encourage those that publish stereological papers to use these basic symbols, thereby broadening their audience. Thus, we present the basic system of symbols, measurements, definitions and equations commonly used in the stereological analysis of microstructures.

### Table 1. List of Basic Symbols and Their Definitions

| Symbol | | Definitions |
|---|---|---|
| $P$ | | Number of point elements. |
| $P_P$ | | Point fraction. Number of points (in areal features) per test point |
| $P_L$ | $m^{-1}$ | Number of point intersections per unit length of test line |
| $P_A$ | $m^{-2}$ | Number of points per unit test area |
| $P_V$ | $m^{-3}$ | Number of points per unit test volume |
| $L$ | $m$ | Length of lineal elements. Mean length: $\bar{L} = L_L/N_L$ |
| $L_L$ | $m/m$ | Lineal fraction. Length of intercepts per unit length of test line |
| $L_A$ | $m/m^2$ | Length of lineal elements per unit test area |
| $L_V$ | $m/m^3$ | Length of lineal elements per unit test volume |
| $A$ | $m^2$ | Planar area of intercepted features. Mean area: $\bar{A} = A_A/N_A$ |
| $S$ | $m^2$ | Surface or interface area (not necessarily planar). $\bar{S} = S_V/N_V$ |
| $A_A$ | $m^2/m^2$ | Area fraction. Area of intercepted features per unit test area |
| $S_V$ | $m^2/m^3$ | Surface area per unit test volume |
| $V$ | $m^3$ | Volume of three-dimensional features. Mean volume: $\bar{V} = V_V/N_V$ |
| $V_V$ | $m^3/m^3$ | Volume fraction. Volume of features per unit test volume |
| $N$ | | Number of features (as opposed to points) |
| $N_L$ | $m^{-1}$ | Number of interceptions of features per unit length of test line |
| $N_A$ | $m^{-2}$ | Number of interceptions of features per unit test area |
| $N_V$ | $m^{-3}$ | Number of features per unit test volume |

Dimensions arbitrarily shown in meters.

509

Basically, P is associated with points, L with lines, S with surfaces, and V with volumes. The quantity A represents the flat area cut by the random section plane through a volume element, and thus varies from zero to a maximum value. S, on the other hand, is the external surface or interface (usually curved) and has a fixed value for any particular microstructural feature. N is reserved for the number of <u>objects</u> or <u>features</u>, as opposed to the number of <u>points</u> (P). The compound symbols represent fractions, and are ratios of a microstructural quantity to a test quantity. Thus, for example, $S_V$, or $S/V_T$, is the surface area per unit volume of the microstructure, where $V_T$ is the test volume. Similarly, $A_T$, $L_T$ and $P_T$ are the test area, length of test line, and number of test points.

These combined symbols are merely a convenient way of writing the fraction. Each symbol has a definite geometrical meaning and an associated dimension. Thus the dimensionality of the combined terms is readily apparent, as well as the dimensional consistency of the equations. Note that descriptive names are avoided because the symbol changes for each name. For example, if B is used for boundary, or P for perimeter, or I for interface, etc., this instantly creates confusion. Basically, they are all lines, L, in the section plane. Appropriate identifying subscripts can be used as required.

The triangular matrix of Table 2 is arranged to show the interrelationship of measured quantities on the section plane to the calculated spatial quantities. The equations listed in Table 2 represent the basic relationships of stereology. The top line reveals the direct equality of volume ratio, areal ratio, lineal ratio and point ratio of an adequately sampled and measured second phase. The second line includes two equations, $S_V = 2\ R_L$ and $L_A = (\pi/2)\ P_L$, in which the surface area per unit volume and the line length per unit area are both related to the $P_L$ measurement. The equation $L_V = 2\ P_A$ is for a system of lines in space where $P_A$ represents the points per unit area created by the random section plane cutting the lines. $P_V$ in the last line cannot be obtained except for special cases. If the points can be approximated by small particles, then $N_V$ is the quantity sought.

Table 2. Relationship of Measured (○) and Calculated (□) Quantities

| Microstructural Feature | Dimensions of Symbols | | | |
|---|---|---|---|---|
| | $mm^0$ | $mm^{-1}$ | $mm^{-3}$ | $mm^{-3}$ |
| Points | $P_P$ | $P_L$ | $P_A$ | $P_V$ |
| Lines | $L_L$ | $L_A$ | $L_V$ | |
| Surfaces | $A_A$ | $S_V$ | | |
| Volumes | $V_V$ | | | |

Basic Equations

$$V_V = A_A = L_L = P_P \qquad mm^0$$
$$S_V = (4/\pi)\ L_A = 2P_L \qquad mm^{-1}$$
$$L_V = 2P_A \qquad mm^{-2}$$
$$P_V \qquad mm^{-3}$$

Although this is not the place to go into the derivations of the basic equations, it may be of interest to indicate the approach adopted. Figure 1 shows a model used to derive the relationship between $V_V$ and $A_A$. (1) Randomly oriented and positioned volume elements (e.g., $\alpha$-phase) in a test volume are cut by a thin test slice of thickness $\delta x$. By letting $\delta x$ become sufficiently small, we can express the volume of the $\alpha$-phase in terms of the area fraction. Then, in the limit, the total volume of $\alpha$-phase is obtaining by integrating from 0 to $\ell$. Dividing through by the test volume $V_T$ gives $V_V = \overline{A_A}$, which states that the average value of $A_A$ represents an estimate of the volume fraction, $V_V$.

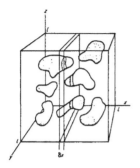

Figure 1. Model for Deriving the Relationship $V_V = A_A$.

Figure 2 is used in deriving the relationship $S_V = 2P_L$. (1) A cube containing randomly-oriented surfaces is penetrated by many randomly-positioned vertical test lines. An expression can be set up for the density of intersections with elementary surface areas having all orientations in space. Those elementary surface elements oriented <u>normal</u> to the test lines will have a larger number of intersections than elements oriented <u>parallel</u> to the test lines. Then, by adding the individual contributions, we obtain the total number of intersections with the total surface area.

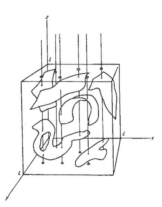

Figure 2. Model for Deriving the Relationship $S_V = 2P_L$.

A similar idea is employed in Figure 3 to obtain the relationship $L_V = 2P_A$. (1) A cube containing a randomly-oriented system of lines is cut by a large number of vertical and parallel test planes. Since the probability of intersection of the test planes with the elementary line segments is proportional to their projected lengths normal to the sectioning planes, it can be shown that the proportionality constant is equal to 2.

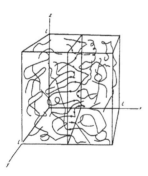

Figure 3. Model for Deriving the Relationship $L_V = 2P_A$.

511

Basic measurements are shown graphically in Figure 4. The six combined symbols in the top line all represent simple counting measurements. Examples of each are given in the microstructures below. $P_P$ represents points that fall within a selected phase divided by the total number of test points; $P_L$ is the number of points of intersection with linear elements in the microstructure per unit length of test line; and $P_A$ means the number of point features in the microstructure per unit test area. Note that the use of P in all three cases involves no ambiguity when used in the combined term. In any case, if additional specification is necessary, the basic terms are modified by appropriate subscripts; for example, $(P_P)\alpha$ would mean the point fraction for the $\alpha$- phase. Note that the $T_A$, or tangent count measurement, is made with a sweeping test line instead of a stationary test line. The $N_L$ and $N_A$ counting measurements are made on objects (such as particles or second-phase regions) rather than points. They represent the number of objects per unit length and per unit area, respectively. The three remaining combined terms require measurements of area and length. Note that $P_L = N_L$ for $V_V = 1$, and $P_L = 2N_L$ for $V_V < 1$.

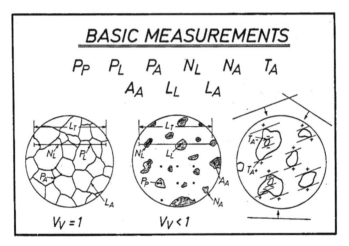

Figure 4. Basic measurements in stereology.

It should be emphasized that adequate sampling and sufficient measurements on the samples must be made to achieve the desired accuracy. These questions are discussed in the literature of stereology. For a selection of source books on stereology see the references (1-5) listed below.

REFERENCES

1.  Quantitative Stereology by E. E. Underwood, Addison-Wesley Publ. Co., Reading, Mass. (1970).
2.  Stereometric Metallography by S. A. Saltykov, Metallurgizdat, Moscow (1970), 3rd Ed.
3.  Stereology 3, Proceedings of the Third International Congress for Stereology, Berne, Switzerland (1972) edited by E. R. Weibel, et al., Blackwell Scientific Publs., Oxford.
4.  Quantitative Microscopy edited by R. T. DeHoff and F. N. Rhines, McGraw-Hill Book Co., New York (1968).
5.  Stereology, Proceedings of the Second International Congress for Stereology, Chicago, Ill. (1967), edited by H. Elias, Springer-Verlag, New York.

In order to assist the reader with the symbolism in some of the more mathematical papers, we have assembled this glossary of equivalent symbols. Note that all terms are not included (see p. 509 for a complete list of basic terms). Here we show only those terms that differ from or duplicate others.

The Editors

| Description of Symbol | Equivalent I.S.S. Symbol | Miles (p.3) | DeHoff & Gehl (p.29) | Rhines (p.233) |
|---|---|---|---|---|
| **BASIC TERMS** | | | | |
| Number of objects, particles, etc. | N | $n^{(a)}, m^{(a)}$ | $M^{(a)}$ | -- |
| per unit length | $N_L$ | -- | -- | $N_L$ |
| Number of points | P | -- | -- | -- |
| per unit length, area or volume | $P_L, P_A, P_V$ | -- | -- | $N_L, N_A, N_V$ |
| Number of particle centroids per unit volume | $P_V$ | D | -- | -- |
| Surface area per unit volume | $S_V$ | $A_V$ | -- | -- |
| Mean caliper, or tangent, diameter | $\bar{D}$ | $\bar{M}$ | -- | $\bar{D}$ |
| Number of tangent points | T | -- | T | -- |
| per unit area or volume | $T_A, T_V$ | -- | $T_A, T_V$ | $T_A, T_V$ |
| Length of test line | $L_T$ | $T^{(a)}$ | -- | -- |
| test area | $A_T$ | -- | -- | -- |
| test volume | $V_T$ | -- | $V_o$ | -- |
| Length of line segment | $L, \ell, d\ell$ | L | $d\lambda, dp$ | -- |
| Foil thickness | t | t | t | -- |
| **SPECIAL TERMS** | | | | |
| Curvature at a point of a curve, $= d\theta/d\ell$ | $\kappa$ | $\kappa$ | k | -- |
| Integrated curvature of curves, $= \int \kappa d\ell$ | k | -- | $\int k(\lambda) d\lambda$ | -- |
| per unit length, $= k/\int d\ell$ | $k_L$ | -- | $\bar{k}$ | -- |
| per unit volume, $= k/\int\int\int dv$ | $k_V$ | -- | -- | $C_V$ |
| Integrated mean curvature of surfaces $^{(c)}$ | K | K | M | -- |
| per unit surface area, $= K/\int\int ds$ | $K_S$ | -- | -- | -- |
| per unit volume, $= K/\int\int\int dv$ | $K_V$ | $K_V$ | $M_V$ | $M_V$ |
| Integrated Gaussian curvature of surfaces $^{(d)}$ | G | G | -- | -- |
| per unit surface area, $= G/\int\int ds$ | $G_S$ | -- | -- | -- |
| per unit volume, $= G/\int\int\int dv$ | $G_V$ | $G_V$ | -- | -- |
| Number of inflection points | $(P)_{infl}$ | -- | I | -- |
| per unit volume | $(P_V)_{infl}$ | -- | $I_V$ | -- |
| Number of intersection points per unit volume | $(P_V)_{I\text{-points}}$ | $N_V$ | -- | -- |
| Length of intersection curves per unit volume | $(L_V)_{I\text{-curves}}$ | $L_V$ | -- | -- |
| Number of branches per unit volume | $--^{(b)}$ | -- | -- | $B_V$ |
| Number of nodes per unit volume | $--^{(b)}$ | -- | -- | $K_V$ |
| Connectivity per unit volume | $--^{(b)}$ | -- | -- | $G_V$ |
| Constant of proportionality | -- | -- | -- | $k^{(a)}$ |
| Random convex set | -- | $K^{(a)}$ | -- | -- |
| Complementary cone at I-point | -- | $C^{(a)}$ | -- | -- |
| **PROJECTED TERMS** | | | | |
| (All quantities refer to the projection plane) | | | | |
| Number of points | P' | N' | $N_{proj}$ | -- |
| per unit length or area | $P_L', P_A'$ | $N_L', N_A'$ | $N_{L proj}$ | -- |
| Number of inflection points | $(P')_{infl}$ | -- | $I_{proj}$ | -- |
| per unit area | $(P_A')_{infl}$ | -- | $I_{A proj}$ | -- |
| Number of tangent points | T' | -- | $T_{proj}$ | -- |
| per unit area | $T_A'$ | -- | $T_{A proj}$ | -- |
| Curvature at a point of a curve | $\kappa'$ | $\kappa$ | -- | -- |
| Integrated curvature of curve | k' | C' | -- | -- |
| per unit area | $k_A'$ | $C_A'$ | -- | -- |
| Length of line segment | $d\ell'$ | -- | dp | -- |

$^{(a)}$Ad hoc notation, $^{(b)}$Notation not well-established, $^{(c)}\frac{1}{2}\int\int(\kappa_1+\kappa_2)ds$, $^{(d)}\int\int\kappa_1\kappa_2 ds$

513

September 1975

C = Conferee            Cia = Conferee in absentia
Ch = Chairperson        A = Author

Dr. Shora Agajanyan                     Cia     Heinz Barthel                           Cia
Academy of Science of the USSR                  Veitscher Magnesitwerke A.-G.
Pushchino.                                      Forschunsinstitut Leoben
Moscow Region, U.S.S.R.                         Magnesitstrasse 2
                                                A-8707 Leoben-Goss, Austria


Dr. R. V. Ambartzumian                  Cia     Ian Bartky                              C
Institute of Mathematics                        Scientific Assistant
Academy of Science of the Armenian              Institute for Applied Technology
S.S.R.                                          National Bureau of Standards
Erevan, USSR                                    Washington, DC  20234
                                                U.S.A.
Mevr. Drs. W.H.M. Amesz-Voorhoeve  C
Department of Anatomy                           Dr. Rolf Baur
69 Oostersingel                                 Anatomisches Institut
Groningen                                       Pestalozzistr. 20
THE NETHERLANDS                                 CH 4056 Basel
                                                Switzerland
Prof. Dr. G. C. Amstutz, Direktor
Mineralog.-Petrographischen                     Marie Bellemare
Institutes der Universität Heidelberg           Department of Anatomy
Postfach 840, Berlinerstrasse 19                University of Montreal, CP 6128
D69 Heidelberg, Germany                         Montreal, PQ
                                                Canada
Dr. R. S. Anderssen                     C,
Australian National University          A,      Mr. G. Bernroider                       C,
Computer Centre, P. O. Box 4            Ch      Universität Salzburg                    Ch,
Canberra ACT 2600                               Zoologisches Institut                   A
Australia                                       Akademiestrasse 26
                                                A-5020 Salzburg, Austria
Mr. R. H. Atkinson                      Cia
861 Tice Place                                  Dr. Martin Berry                        Cia,
Westfield, NJ    07090                          Anatomy Dept., Medical School           A
U.S.A.                                          University of Birmingham
                                                Vincent Drive
                                                Birmingham B15 2TJ, England
J. Baak
Vrije Universiteit de Boelelaan                 Dr. Felix D. Bertalanffy
Amsterdam                                       Professor of Anatomy, Medical College
The Netherlands                                 750 Bannatyne Avenue
                                                Winnipeg 3, Manitoba
                                                Canada
Dr. Gunter Bach
Oppelnstrasse 28                                Prof. Michael B. Bever                  Cia
33, Braunschweig                                Massachusetts Institute of Technology
Germany                                         Room 13-5066
                                                Cambridge, MA   02139
                                                U.S.A.
Prof. Lida K. Barrett                   Cia
Dept. of Mathematics                            Dr. Indra Bhargava                      C,
Ayres Hall                                      P. O. Box 1059                          A
University of Tennessee                         Ahwaz
Knoxville, TN    37916                          Iran
U.S.A.

Prof. Jacques-Louis Binet  C,
Faculté Pitié-Salpêtrière  Ch,
91 Boulevard de L'Hôpital  A
75013 Paris
France

Dr. Edwin S. Boatman  C
F561 Health Sciences SC-34
University of Washington
Seattle, WA  98195
U.S.A.

Dr. Robert P. Bolender  Cia
University of Washington
Seattle, WA  98195
U.S.A.

Dr. Jose Bosch-Figueroa
Dept. of Crystallography & Mineralogy
University of Barcelona
Av. Jose Antonio 585
Barcelona 7, Spain

Dr. E. A. Bossen  C
Department of Pathology
Duke University Medical Center
Durham, NC  27710
U.S.A.

Dr. Mary Blair Bowers  C
Bldg. 3, Room 318
National Institutes of Health
Bethesda, MD  20014
U.S.A.

Dr. S. Bradbury  Cia
Department of Human Anatomy
South Parks Road
Oxford OX1 3QX
England

Dr. A. Braun
Laboratoire Suisse de Res. Horlogeres
Rue A. L. Breguet 2
CH-2000 Neuchatel
Switzerland

Dr. L. G. Briarty  C,
Botany Department  A
Nottingham University
University Park
Nottingham NG7 2RD
England

David Brown  C
A-205 Chemistry Bldg.
National Bureau of Standards
Washington, DC  20234
U.S.A.

Mr. R. D. Buchheit
Battelle Memorial Institute
505 King Avenue
Columbus, OH  43201
U.S.A.

Dr. G. Colin Budd  Cia
Medical College of Ohio at Toledo
P. O. Box 6190
Toledo, OH  43614
U.S.A.

H. C. Burnett  C
B-264, Bldg. 223
National Bureau of Standards
Washington, DC  20234
U.S.A.

Peter H. Burri, M.D.
Dept. of Anatomy
University of Berne
Buhlstrasse 26, CH-3012 Berne
Switzerland

Dr. Robert J. Buschmann
Veterans Administration West Side
Hospital
MP 113, P. O. Box 8195
Chicago, IL  60680, U.S.A.

Dr. Romulo Luís Cabrini
Departamente de Radiobiologia
Comison Nacional de Energia Atomila
Avda. Libertador 8250
Buenos Aires, Argentina

Massimo Calabresi, M.D.
300 Ogden St.
New Haven, CT  06511
U.S.A.

Dr. César Cánepa  Cia
Division de Exploraciones
MINERO-PERU
1649, Av. Arcquipa
Lima, Peru

Dr. Peter B. Canham  Cia,
Department of Biophysics  A
The University of Western Ontario
London N6A 5C1
Canada

Nancy V. Carlson
New Eng. Deaconess Hospital
Cancer Research Institute
Electron Microscopy
194 Pilgrim Road
Boston, MA  02215
U.S.A.

Dr. Anna-Mary Carpenter C,
University of Minnesota Ch
Dept. of Anatomy
254 Jackson Hall
Minneapolis, MN 55455, U.S.A.

Dr. Wayne A. Cassatt C
A-121, Bldg. 222
National Bureau of Standards
Washington, DC 20234, U.S.A.

Dr. Philippe R. L. Cauwe C,
Wild Leitz-France A
86, avenue du 18 juin 1940, B.P. 107
92504 Rueil-Malmaison
France

Dr. James P. Chalcroft Cia
Meat Industry Res. Inst. of
New Zealand
P. O. Box 9002
Hamilton North, New Zealand

Dr. Harold W. Chalkley
Bethesda, MD, U.S.A.
Deceased, September 25, 1975

Dr. Chi-Hau Chen C,
Electrical Engineering Department Ch,
Southeastern Mass. University A
N. Dartmouth, MA 02747, U.S.A.

George C. Cheng Ch,
Rogers Road, M-102 A
Athens, GA 30601
U.S.A.

Dr. J.-L. Chermant Cia
Université de Caen A
Laboratoire de Chimie Minérale
Industrielle
CAEN-Cedex 14032
France

Dr. Robert B. Chiasson
Dept. of Biology Sciences
University of Arizona
Tucson, AZ 85721
U.S.A.

Dr. Antonio Chizzola
Instituto de Clinica Med. Gen.
dell Universita de Parma
Ospedali Riuniti
43100 Parma
Italy

Dr. William P. Clancy
Materials Research Lab.
Army Mat. & Mech. Research Center
Watertown, MA 02172
U.S.A.

Charles I. Cohen C
Owens Corning Fiberglass Corp.
Technical Center, P. O. Box 415
Granville, OH 43023, U.S.A.

Dr. Morris Cohen
Room 13-5046
Mass. Institute of Technology
Cambridge, MA 02139
U.S.A.

Docent Yrjö Collan, M.D. C,
Second Dept. of Pathology A
University of Helsinki
00290 Helsinki 29
FINLAND

J. Stanley Cook C
Roy C. Ingersoll Research Center
Wolf & Algonquin Roads
Des Plaines, IL 60018, U.S.A.

Dr. M. Coster C,
Laboratoire de Chimie A
Minérale Industrielle
Université de Caen
CAEN Cedex 14032, France

Prof. Emil Craciun, M.D. Cia,
Inst. of Endrocrinology A
Str. Dionisie Lupu 62
Bucarest I (22) Rumania

Dr. Ian Crain Cia
Rm. 226, Health Protection Bldg.
Dept. of Health & Welfare
Tunney's Pasture
Ottawa, Ontario
Canada

Dr. Luis-M. Cruz-Orive C,
Dept. of Probability & Statistics A
The University
Sheffield, S3 7RH
England

Prof. Robin L. Curtis
Department of Anatomy
Marquette School of Medicine
561 North 15th Street
Milwaukee, WI 53233
U.S.A.

517

*Directory*

Mr. Ronald K. DeFord
University Station
Box 7609
Austin, TX  78712, U.S.A.

Robert M. Doerr                        C,
U.S. Bureau of Mines                   Ch,
P. O. Box 280                          A
Rolla, MO  65401, U.S.A.

Dr. Robert T. DeHoff                   C,
Dept. of Materials Sci. & Eng.         Ch,
University of Florida                  A
Gainesville,. FL  32611, U.S.A.

C. G. Dodd                             Cia
Schick Safety Razor Company
Division of Warner-Lambert Company
Milford, CT  06460, U.S.A.

Dr. Felix A. de la Iglesia
Warner-Lambert Research Institute
 of Canada
Sheridan Park
Mississauga L5K 1B4, Ontario
Canada

Prof. F. A. L. Dullien                 C,
University of Waterloo                 A
Waterloo
Ontario N2L 3G1
Canada

Prof. Dr. Andre G. DeWilde             C
Anat. Emb. Lab.
Oostersingel 69
Groningen
Netherlands

Dr. Sven O. E. Ebbesson
Dept. of Neurological Surgery,
University of Virginia
School of Medicine
Charlottesville, VA  22901, U.S.A.

Dr. Roland deWit                       C,
National Bureau of Standards           Ch,
B-120, Bldg. 223                       A
Washington, DC. 20234, U.S.A.

Cammander George T. Eden, DC, USN  C,
A-145, Polymers Bldg. (311.04)     A
National Bureau of Standards
Washington, DC  20234, U.S.A.

J. Leland Daniel                       C
Battelle - Northwest
PSL-1610
Richland, WA  99352
U.S.A.

Dr. Siegfried Eins                     Cia
Max-Planck-Institut. für
 Biophysikalische Chemie
D 3400 Göttingen- Nikolausberg
Am Fassberg, Germany

Prof. Per E. Danielsson                C
Dept. of Electrical Engineering
Linköping University
Linköping S-58183
Sweden

Dr. Brenda Eisenberg                   C
Department of Medicine
U.C.L.A. Medical School
Los Angeles, CA  90024
U.S.A.

Dr. Paul Davies                        C
Beth Israel Hospital
330 Brookline Avenue
Boston, MA   02215, U.S.A.

Dr. Farouk El-Baz
National Air & Space Museum
Smithsonian Institution
Washington, DC  20560, U.S.A.

Dr. L. Delaey
Dept. of Metallurgy
De Croylaan 2
B 3030 Heverlee
Belgium

Prof. Dr. phil. Hans Elias             C,
463 Marietta Drive                     CH,
San Francisco, CA  94127               A
U.S.A.

Dr. Russell L. Deter                   C,
Dept. of Cell Biology                  A
Baylor College of Medicine
1200 Moursund Blvd.
Houston, TX  77025
U.S.A.

Dr. Peter M. Elias
Dept. of Pathology
School of Medicine
University of California
San Francisco, CA  94143
U.S.A.

518

Dr. David Bertram Ellson                    Cia
Division of Tribophysics, CSIRO
Box 4331
Melbourne 3001
Australia

Dr. Colin Fisher
Metals Research Ltd.
Melbourn, Royston
Hertfordshire, SG8 6EJ
England

Sami M. El-Soudani                          Cia,
Dept. of Metallurgy & Mat. Sci.              A
University of Cambridge
Cambridge CB2 3QZ
England

D. E. Fleet                                 C
Dept. of Metallurgy & Materials Sci.
University of Cambridge
Pembroke Street
Cambridge CB2 3QZ
England

Dr. Edgar S. Etz                            C
Rm. A-121, Bldg. 222
National Bureau of Standards (312.02)
Washington, DC  20234
U.S.A.

Mr. Peter G. Fleming, Admin.
The Royal Microscopical Society
Clarendon House
Corn Market Street
Oxford, OX1 3HA
England

Dr. Hans Eckart Exner          C, Ch, A
Max-Planck-Institut für Metallforschung
175 Buesnauer Str.
7000 Stuttgart
Germany

Dr. Eva Foh
Anatomisches Institut der Universitat
Olshausenstrasse 40-60
23 Kiel
Germany

Dr. D. G. Fagan                             Cia
Hospital for Sick Children
555 University Avenue
Toronto, Ontario M5G 1X8
Canada

J. B. Forrest                               C
McMaster University Medical Centre
1200 Main Street West
Hamilton 16, Ontario L8S 4J9
Canada

J. K. Fawell                                Cia
General Toxicology Division
Inveresk Research International
Musselburgh, Midlothian, EH21 7UB
Scotland

Mr. Richard H. Foster                       Cia
The City University
St. John Street
London EC1R 4TR
England

Dr. Micheline Federman                      Cia
Cancer Research Institute
N.E. Deaconess Hospital
185 Pilgrim Road
Boston, MA  02215, U.S.A.

Dr. William G. Fricke, Jr.
Physical Metallurgy Division
Alcoa Technical Center
Alcoa Center, PA   15069
U.S.A.

Dr. Emmanuel Feinermann                     C,A
Centre d'Etudes Nucléaires de Saclay
SCP, BP No. 2
91190 GIF/Yvette
France

Dr. med. habil. Siegfried Fritsch  Cia
Zentralinstitut für Mikrobiologie
   und Experimentelle Therapie
Beutenbergstr. 11, 69 Jena
Germany

R. E. Filipy                                C
Battelle-Northwest
Box 999 (LSL II, 3000)
Richland, WA   99352
U.S.A.

Yasumichi Fujimoto, Lecturer         C,
Dept. of Anatomy                     A
Osaka Dental University
Osaka, 540
Japan

Dr. Vernon W. Fischer
St. Louis University
Med. School, Anatomy Department
1402 S. Grand Avenue
St. Louis, MO  63104
U.S.A.

Dr. R. Buckminster Fuller
3500 Market Street
Philadelphia, PA   19104
U.S.A.

Dr. Ralph Gander
Wild Heerbrugg Co.
9435 Heerbrugg
Switzerland

Dr. James Q. Gant, Jr.                    C
1835 Eye Street, N.W., Suite 201
Washington, DC  20006
U.S.A.

Dr. Peter Gehr                           C,
Department of Anatomy                     A
University of Bern
Bühlstrasse 26
3012 Bern, Switzerland

Dr. H. D. Geissinger
University of Guelph
Dept. of Biomedical Science
O.V.C. Research Station
Guelph N1G 2W1
Ontario, Canada

Dr. Juan Gil
University of Berne
Anatomisches Institut
Bühlstrasse 26,
3012 Berne, Switzerland

Dr. Pierre Gilles
Ingenieur Service Chimie Physique
Commissariat Energie Atom. CEN/Saclay
BP #2, 91190 GIF-sur-Yvette
France

Dr. Walter Good                          C,
Swiss Avalanche Research Institute       A
CH-7260 Weissfluhjoch/Davos
Switzerland

Dr. Richard Gordon                       Ch,
Image Processing Unit                     A
Bldg. 36, Room 4D-28
National Institutes of Health
Bethesda, MD   20014, U.S.A.

Prof. Earl E. Gose                       Cia
University of Illinois, Chicago
Box 4348
Chicago, IL  60680
U.S.A.

Dr. Tony Greday
Centre de Recherches Metallurgiques
Rue du Val-Benoit, 69
B-4000 Liege
Belgium

Dr. Hilmar Grimm
Inst. Mikrobiologie u. exp. Therapie
Beuthenbergstrasse 11
Jena
Germany

Dr. Sverker Griph                        C
Dept. of Anatomy
University of Uppsala, Box 571
S-75123 Uppsala
Sweden

Dr. Miguel Guirao-Perez
Director Instituto "Foloriz"
Facultad de Medicinia de Granada
Granada
Spain

Prof. J. Gurland                         C,
Brown University                         Ch,
Division of Engineering                   A
Providence, RI   02912
U.S.A.

Prof. Ernest L. Hall                     C,
Univ. of Southern California             A
Dept. of Electrical Engineering
Los Angeles, CA  90007
U.S.A.

Dr. Gilbert F. Hamilton
Department of Anatomy
Marshal College University
Aberdeen
Scotland

Dr. Paul Hamosh                          C
Dept. of Physiology & Biophysics
Georgetown University School
   of Medicine
Washington, DC  20007, U.S.A.

Prof. John W. Harman
Diagnostic Laboratory
Department of Pathology
University College
Dublin 2, Ireland

Dr. Bruce S. Hass                        C
Argonne National Laboratory
Bio-Medical Division
7900 Cass Avenue
Argonne, IL   60439, U.S.A.

Prof. Dr. med. H. Haug                   C,
Medizinischen Akademie Lübeck            Ch,
Ratzeburger Allee-160                     A
D-2400. Lübeck
Germany

Dr. P. W. Hawkes      Cia
Laboratoire d'optique Electronique
29 Rue Jeanne Marrig
31055 Toulouse Cedex
France

Dr. August Hennig
Department of Anatomy
University of Munich
Pettenkoferstrasse 15 D-8
München 11
Germany

Dr. Jean-Pierre Herveg      C
Rm. 4B-47, Bldg. 10
National Institutes of Health
Bethesda, MD    20014, U.S.A.

Dr. Peter Hill      C
Carl Zeiss, Inc.
444 Fifth Avenue
New York, NY    10018, U.S.A.

Prof. John E. Hilliard      C,
Dept. of Materials Science      Ch,
   and Engineering      A
Northwestern University
Evanston, IL   60201, U.S.A.

Jan Hinsch      C
E. Leitz, Inc.
3930 Knowles Avenue
Kensington, MD   20795, U.S.A.

Robert W. Hinton      C
Homer Research Lab
Bethlehem Steel Corp.
Bethlehem, PA    18016
U.S.A.

Miss Clara S. Hires
Mistaire Laboratories
152 Glen Avenue
Millburn, NJ    07041, U.S.A.

Dr. John D. Hoffman, Director    C,
Institute for Materials Research    A
National Bureau of Standards
Washington, DC   20234
U.S.A.

Ing. Vratislav Horalek, Dr. Sc.    Cia
Mechenicka 14/2558
141 00 Praha 4 - Sporilov II
Czechoslovakia

Dr. Emanuel Horowitz      C
Deputy Director
Institute for Materials Research
National Bureau of Standards
Washington, DC   20234, U.S.A.

Dr. Hans P. Hougardy      Ch, C, A
Max-Planck-Inst. für Eisenforschung
Max-Planck Strasse 1
4000 Düsseldorf
Germany

Werner Hunn      C,
E. Leitz, Inc.      A
Link Drive
Rockleigh, NJ    07647, U.S.A.

Dr. Otto Hunziker      C,
Basic Medical Research Dept.    A
Sandoz Ltd.
CH-4002 Basel, Switzerland

Dal Hyde      C,
School of Veterinary Medicine    A
University of California
Davis, CA   95616
U.S.A.

Dr. Geza Ifju      C,
Forestry Products Dept.      A
V.P.I. and S.U.
Blacksburg, VA    24061, U.S.A.

M. Indra      Cia,
Inst. of Physiology      A
Czechoslovak Academy of Sciences
Praha 4 - Krc
Budejovická 1083
Czechoslovakia

Oscar A. Iseri, M.D.      C
V. A. Hospital (Univ. of Maryland)
3900 Loch Raven Blvd.
Baltimore, MD    21218, U.S.A.

Dr. M. I. Ismail      Cia
Chemical Engineering Dept.
Alexandria University
Alexandria
Egypt

Prof. Genrich R. Ivanitsky, D.Sc. Cia
Biophysical Inst. of Ac. Sci. USSR
Dept. of Applied Mathematics, I42292
Pushchino
Moscow Region
U.S.S.R.

Prof. Oscar C. Jaffee    C,
University of Dayton, Biology    A
300 College Park Avenue
Dayton, OH 45469, U.S.A.

Dr. A. J. Jakeman    C,
Computer Center    A
Australian National University
P. O. Box 4
Canberra, A.C.T. 2600
Australia

Dr. Nigel T. James    Cia,
Dept. of Human Biology and Anatomy    A
University of Sheffield
Sheffield S10 2LA
England

Dominique Jeulin    C, A
Institut de Recherches de la Sidérurgie
IRSID
Maizières-lès-Metz 57210
France

Mr. Stig Johansson    C,
Sandvik AB    A
TMP Process Metallurgy
S-81101 Sandviken 1
Sweden.

Dr. Jay A. Johnson    C,
Forestry Products Division    A
V.P.I. and S.U.
Blacksburg, VA 24061
U.S.A.

K. A. Johnson    Cia
P. O. Box 1663, MS-328
Los Alamos, NM 87544
U.S.A.

Ronald B. Johnson    C
Executive Officer
Institute for Materials Research
National Bureau of Standards
Washington, DC 20234, U.S.A.

Meurig P. Jones, Senior Lecturer    Cia,
Dept. of Mining and Mineral Tech.    A
Imperial College
London SW7
England

Prof. dr. Miroslav Kalisnik
Institute for Hist.-Embryol., Med. Fac.
University of Ljubljana
Zaloska 4/I
Ljubljana 61105
Yugoslavia

Prof. Laveen Kanal    Cia
Dept. of Computer Science
University of Maryland
College Park, MD 20742, U.S.A.

Dr. Shirley L. Kauffman    C
State University of New York
450 Clarkson Avenue
Brooklyn, NY 11203
U.S.A.

Dr. Brian H. Kaye, Director    Ch,
Inst. for Fine Particle Research    A
Laurentian University    C
Sudbury, Ontario P3E 2C6
Canada

Dr. Hausjorg Keller    C,
Department of Anatomy    A
University of Bern
Bühlstrasse 26
3012 Bern, Switzerland

Thomas Kelly    C,
Image Analysing Computers    A
40 Robert Pitt Drive
Monsey, NY 10952
U.S.A.

Dr. M. D. Kendal    C
St. Thomas's Hospital Medical School
Lambeth Palace Road
London, SE1 7EH
England

Dr. Charles R. Key    Cia
Dept. of Pathology
University of New Mexico
915 Stanford N.E.
Albuquerque, NM 87131, U.S.A.

Russell A. Kirsch    C,
National Bureau of Standards    Ch
A-317, Administration Bldg.
Washington, DC 20234
U.S.A.

Dr. Gonzague S. Kistler
Anat. Institut der Universitat
Gloriastrasse 19
8006 Zurich
Switzerland

Prof. Victor Klee
University of Washington
Department of Mathematics
Seattle, WA 98105
U.S.A.

Dr. Andres J. P. Klein-Szanto
Abteilung fur Orale Strukturbiologie
Plattenstrasse 11
8028 Zurich
Switzerland

Prof. Andre G. Laurent
Wayne State University
Department of Mathematics
Detroit, MI  48202
U.S.A.

Prof. Dr. Ekkehard Kleiss
Head, Dept. of Embrology
University de Los Andes
Apartado 38, Merida
Venezuela

Dr. Katheryn E. Lawson          C,
Organization 5833               A
Sandia Laboratories
Albuquerque, NM  87115
U.S.A.

Prof. Dr. Wayne Kraft          Cia
Dept. of Met. and Mat. Science
Lehigh University
Bethlehem, PA  18015
U.S.A.

Priv. Doz. Dr. med. Ortwin Leder
Anatomisches Inst. d.
 Albert-Ludwigs-U.
Albertstrasse 17
Freiburg im Breisgau
Germany

John Krc, Jr.                   C
26669 Hunting Road
Huntington Woods, MI  48070
U.S.A.

Vladimír Levický, M.D., Ph.D.   C,
Inst. of Normal & Pathological  A
 Physiology
Slovak Academy of Sciences
Sienkiewiczova 1
884 23 Bratislava, Czechoslovakia

Ing. Ivan Krekule               C,
Inst. of Physiology             A
Czechoslovak Academy of Sciences
Praha 4 - Krc
Budejovická 1083
Czechoslovakia

Dr. Karl G. Lickfeld
Institute of Med. Microbiology
55 Hufelandstr.
D-43 Essen 1
Germany

Dr. David E. Kuhl
Dept. of Radiology
Hospital of the Univ. of Pa.
Philadelphia, PA  19104
U.S.A.

Ass. Prof. Lars Göran Lindberg  C,
Cytodiagnostic Department       A
University Hospital of Lund
Lund S-22185
Sweden

Dr. Tine Kuiper-Goodman         C,
National Health & Welfare       A
Health Protection Branch
Tunney's Pasture
Ottawa K1A OL2
Canada

Prof. Dr. Willi Lindemann
Lehrstuhl fur Kristallstrukturlehre
Konigsbergerstrasse
87 Würzburg
Germany

Mr. P. Lamb                     Cia
ICI Plastics Division
Bessemer Road
Welwyn Garden City
Herts AL7 1HD, England

Dr. Lewis E. Lipkin
National Institutes of Health
8000 Rockville Pike
Bldg. 36, Rm. 4D24
Bethesda, MD  20014, U.S.A.

P. D. Lambert                   Cia,
Coal Preparation Branch         A
National Coal Board, South Notts Area
Bestwood, Nottingham N96 8UE
England

Mr. P. J. Lloyd                 C,
Dept. of Chemical Engineering   A
University of Technology
Loughborough, Leicestershire
England

Dipl. Math. Reinhard Lang       Cia
Universität Bielefeld
Fak. Mathematik
Kurt-Schumacher-Str. 6
D-48 Bielefeld, West Germany

Dr. Arthur L. Loeb
Dept. of Visual and Environmental
 Studies
Harvard University
Cambridge, MA  02138, U.S.A.

523

Gerald H. Lolmaugh
7721 Bristow Drive
Annandale, VA   22003
U.S.A.

Prof. Alden V. Loud                        C
Dept. of Pathology
New York Medical College
Walhalla, NY   10595
U.S.A.

John Ludbrook
Department of Surgery
Royal Adelaide Hospital
Adelaide 5000
South Australia

Mr. C. J. McCarthy                         C
Bausch & Lomb
820 Linden Avenue
Rochester, NY   14625
U.S.A.

Walter C. McCrone
McCrone Associates, Inc.
2820 S. Michigan Avenue
Chicago, IL   60616, U.S.A.

Bruce McCollough, D.V.M., Ph.D.            C
Southwest Foundation for Res.
   and Education
P. O. Box 28147
San Antonio, TX   78284, U.S.A.

Prof. Paul J. McMillan                     C
Dept. of Anatomy
School of Medicine
Loma Linda University
Loma Linda, CA   92354, U.S.A.

Dr. Manuel Miranda Magalhaes
Lab. de Histologia
Faculdade de Medicina de Porto
Porto
Portugal

Philippe Maire
CEA
Cen-Sacley
91 Gif-yvette BP No. 2
France

Dr. Sasha Malamed                          C,
Rutgers Medical School                     A
College of Medicine and Dentistry
   of New Jersey
Piscataway, NJ   08854
U.S.A.

Dr. J. R. Manning                          C
B-264, Bldg. 223
National Bureau of Standards
Washington, DC  20234, U.S.A.

Jacques Marilleau                          C
CEA Commissariat a L'Energie
   Atomique
Boite Postale N° 7
93270 Sevran, France

Dr. R. Bruce Martin                        C,
Division of Orthopedic Surgery            A
West Virginia Univ. Medical Ctr.
Morgantown, WV   26506
U.S.A.

Robert Martin                              C
Administrative Officer
Institute for Materials Research
National Bureau of Standards
Washington, DC  20234, U.S.A.

Prof. Luigi Martino
No. Via Carulli
5 Bari
Italy

Dr. Tadeusz B. Massalski
Head, Met. Phys. Group
Carnegie-Mellon University
4400 Fifth Avenue
Pittsburgh, PA   15213, U.S.A.

Gloria D. Massaro, M.D.                     C
Veterans Administration Hospital
50 Irving Street, N.W.
Washington, DC  20422
U.S.A.

E. D. Massey                                C
Medical Research Council Laboratories
Woodmansterne Road
Carshalton, Surrey SM5 4EF
England

Odile Mathieu                              C,
Dept. d'Anatomie, Faculté de Méd. A
Univ. de Montréal, C.P. 6128
Montréal 101, P.Q.
Canada

Dr. T. M. Mayhew
Dept. of Human Biology & Anatomy
Univ. of Sheffield, Western Bank
Sheffield S10 2TN
England

Dr. Geoffrey A. Meek C, Ch,
University of Sheffield
Dept. of Human Biology and Anatomy A
Sheffield S10 2TN
England

R. Moy C,
Nat'l. Environmental A
 Satellite Service
NOAA, FOB #4, S1124
Washington, DC 20233, U.S.A.

Dr. Wolfgang J. Mergner C
Dept. of Pathology
University of Maryland
22 South Green Street
Baltimore, MD 21201, U.S.A.

Prof. Pieter Muije
Georgia Institute of Technology
School of Chemical Engineering
Atlanta, GA 30332
U.S.A.

Dr. Paul-Emil Messier C,
Dept. of Anatomy A
University of Montreal
B. 0. 6128 Montreal
Quebec, Canada

Dr. Timo Nevalainen
Dept. of Pathology
University of Turku
Kiinamyllynkatu 10
Turku, Finland

Dr. Roger E. Miles C,
Dept. of Statistics Ch,
Australian National University A
P. O. Box 4
Canberra A.C.T. 2600
Australia

Dr. Wesley L. Nicholson C,
Battelle - Northwest Ch,
P. O. Box 999 A
Richland, WA 99352
U.S.A.

Bert Mobley
1234 19th Street, #13
Santa Monica, CA 90404
U.S.A.

Dr. Gastone G. Nussdorfer
Istituto Anatomia Umana Normale
Lab. Micr. Electr.
Via Gabelli, 37
35100 Padova, Italy

Fernand Moliexe
Inst. de Rech. de la Sidérurgie Franc.
185 rue de President Roosevelt
Saint Germain-en-Laye, 78104
France

Dr. Yoshikuni Ohta A
Dept. of Anatomy
Osaka Dental University
Osaka, 540
Japan

Dr. George A. Moore C,
B-118, Bldg. 223 Ch,
National Bureau of Standards A
Washington, DC 20234
U.S.A.

Dr. Daniel G. Oldfield C,
Dept. of Biological Sciences A
DePaul University
1036 West Belden Avenue
Chicago, IL 60614, U.S.A.

Prof. Richard Moore C,
Box 701, Mayo Memorial Bldg. Ch,
University of Minnesota A
Minneapolis, MN 55455
U.S.A.

Dr. Gerhard Ondracek C,
Institut für Material Ch,
u. Festkörperforschung A
Kernforschungszentrum
D-7500 Karlsruhe, Postfach 947
Germany

Mr. Richard E. M. Moore Cia
Anatomy Dept.
Guy's Hospital
London SE1 9RT
England

Prof. P. Darrell Ownby Cia,
Dept. of Ceramic Engineering A
School of Mines and Metallurgy
Rolla, MO 65401
U.S.A.

Dr. Roger R. A. Morton C,
Analytical Systems Division A
Bausch & Lomb
820 Linden Avenue
Rochester, NY 14625
U.S.A.

Prof. Monique Pavel Cia
145, Avenue de Malakoff A
75116 - Paris
France

Paul Parniere     C
I.R.S.I.D.
185 rue President Roosevelt
St. Germain-en-Laye, 78104
France

Prof. Lee D. Peachey     C,
Department of Biology     Ch,
University of Pennsylvania     A
Philadelphia, PA   19174, U.S.A.

E. C. Pearson     C
Alcan International Ltd.
Research Centre
Box 8400
Kingston, Ontario
Canada

Prof. Dr. med Tomas Pexieder     Cia
Inst. D'Histologie et D'Embryologie
9 Rue du Bugnon
1011 Lausanne/Suisse
Switzerland

Dr. Ulrich Pfeifer     C,
Pathologisches Institut
Universität Würzburg
Luitpoldkrankenhaus
87 Würzburg, Germany

Dr. Victor A. Phillips
Martin Marietta Laboratories
1450 S. Rolling Road
Baltimore, MD   21227
U.S.A.

William G. Pichel     Cia,
National Environmental Satellite     A
 Service
NOAA, S11212
Washington, DC   20233, U.S.A.

Dipl. Phys. Frank Piefke     C,
Institut B für Mathematik     A
TU Braunschweig
Pockelstrasse 14
33 Braunschweig, Germany

Dr. Ruggero Pierantoni     Cia
Laboratorio de Cibernetica
Biofisica del C.N.T.
Cammogli 16032
Italy

Dr. Hermann Pinkus     Cia
Dept. of Dermatology
Wayne State University
540 E. Canfield
Detroit, MI   48201
U.S.A.

Dr. Jan Pitha     Cia
Laboratory Service
VA Hospital
921 N.E. 13th Street
Oklahoma City, OK   73104, U.S.A.

Floyd W. Preston     Cia
Dept. of Chem. and Petr. Eng.
University of Kansas
Lawrence, KS   66046' U.S.A.

Dr. J. M. Prévosteau     C,
B.R.G.M.     A
B.P. 6009
45018 - Orléans Cédex
France

Prof. T. Radil-Weiss     Cia,
Czechoslovak Academy of Sciences     A
Institute of Physiology
Budejovická 1083,
Prague 4, Czechoslovakia

Edward Rae     C
E. Leitz, Inc.
1123 Grand View Drive
S. San Francisco, CA   94080
U.S.A.

Dr. Brian Ralph     C,
Dept. of Met. & Materials Sci.     Ch,
University of Cambridge     A
Pembroke Street
Cambridge CB2 3QZ
England

Dr. P. Ramakrishnan     C,
Indian Institute of Technology     A
Dept. of Metallurgical Engineering
Powai, Bombay-400076
India

Dr. J. Rathje,     Cia
c/o Bayer AG
Postfach 1140, Geb. Da 5
4047 Dormagen
West Germany

Dr. William Reichel
Franklin Square Hospital
9000 Franklin Square Drive
Baltimore, MD   21237
U.S.A.

Dr. Curt W. Reimann     C
Scientific Assistant
Institute for Materials Research
National Bureau of Standards
Washington, DC   20234
U.S.A.

Dr. Albrecht Reith                          Cia,    Prof. Alvin H. Rothman
The Norwegian Radium Hospital               A .     California State College
Montebello, Oslo-3                                  Department of Biology
Norway                                              Fullerton, CA  92631, U.S.A.

Dr. F. N. Rhines                            C,      Mr. Richard R. Rowand
Dept. of Materials Sci. and Eng.            Ch,     Tech. Mgr. for NDT
University of Florida                       A       Air Force Material Lab. (MAMN)
Gainesville, FL  32611, U.S.A.                      Wright-Patterson AFB, OH  45433, U.S.A.

Dr. Manfred Rink                            C,      Dr. Letty Salentijn
Technical University Clausthal              A       Dept. of Anatomy
Institut für Geophysik                              College of Phys. & Surgeons
Adolf Römer-Str. 2A                                 Columbia University
D-3392 Clausthal-Zellerfeld                         630 West 168th Street
Germany                                             New York, NY   10032, U.S.A.

Ian Robb                                    C,      Prof. Dr. S. A. Saltikov        Cia
Inst. for Fine Particle Research            A       October Ave. 27, Apt. 16
Laurentian University                               Erevan 18, Armenia
Sudbury, Ontario P3E 2C6                            U.S.S.R.
Canada

Max Robinowitz, M.D.                        C       Dr. Anant V. Samudra            Cia
Georgetown University Medical School                IIT Research Institute          A
3900 Reservoir Rd., N.W.                            10 West 35th Street
Washington, DC  20007, U.S.A.                       Chicago, IL  60616, U.S.A.

Dr. Otto Röhm                                       Prof. Dr. Masao Satake          C
Ch. des Mouetts 16                                  Dept. of Civil Engineering
CH-1007 Lausanne                                    Tohoku University
Switzerland                                         Aoba, Aramaki
                                                    Sendai 980, Japan

Prof. Dr. med. Hanspeter Rohr               C,      Mr. Brian Scarlett              C,
Pathologisches Institut                     A       Dept. of Chemical Engineering   A
University of Basel                                 University of Technology
Schönbeinstrasse 40                                 Loughborough, Leicestershire
CH-4056 Basel, Switzerland                          England

Dr. Cesar Romero-Sierra                             Robert J. Schaefer              C,
Anatomy Dept.                                       Naval Research Lab.             A
Queens University                                   Code 6350
Kingston, Ontario K7L 3N6                           Washington, DC  20375
Canada                                              U.S.A.

Mark Rosenblum                              C       Axel Schleicher                 C,
University of Cincinnati                            Medizinische Hochschule         A
339 Brookside Dr.                                   3000 Hannover 61
Dayton, OH  45406                                   Karl-Wiechert Allee 9
U.S.A.                                              Germany

Azriel Rosenfeld                            Cia     Dr. Holger Schmeisser           C,
Computer Science Center                             IMANCO GmbH                     A
University of Maryland                               D605 Offenbach
College Park, MD  20742                             Schreberstrasse 18
U.S.A.                                              Germany

Dr. Renate Schmidt-Koşchel
Bundesanstalt für Materialprüfung
Dienststelle 0.12
Unter den Eichen 87
Berlin 45, Germany

Dr. Charles C. Schock     C
Henry Ford Hospital
2799 W. Grand Blvd.
Detroit, MI  48202
U.S.A.

Dr. Jürgen R. Schopper    C,
Institut für Geophysik     A
Technische University
Adolf-Römer-Strasse 2A
D-3392 Clausthal-Zellerfeld
Germany

Prof. Dr. H. E. Schroeder   Cia
Department Oral Structural Biology
Dental Institute
University of Zurich
11 Plattenstrasse
Zurich 8028, Switzerland

Dr. Bernard Schwartz     Cia
Tufts - New England Med. Center
171 Harrison Avenue
Boston, MA  02111, U.S.A.

Prof. Jean Serra      C,
Centre de Morphologie Mathématique Ch,
Ecole des Mines      A
Fountainebleau, Paris
France

Dr. Joseph Sherrick
1128 Jeffrey Court West
Northbrook, IL  60062
U.S.A.

Dr. Leon Simar      C,
Lab. d'Anatomie Pathologique   A
Inst. de Pathologie
Université de Liège
1, rue des Bonnes Villes
B-4000 Liège, Belgium

Dr. Jack Sklansky     Cia
School of Engineering
University of California
Irvine, CA  92664
U.S.A.

Dr. Arne Sollberger
Life Science I
Southern Ill. Univ., School of Med.
Carbondale, IL  62901
U.S.A.

Dr. G. C. Sornberger     C
Beth Israel Hospital, KB-23
330 Brookline Avenue
Boston, MA  02215
U.S.A.

Dr. Ian J. Spark     Cia
Gippsland Inst. of Advanced Ed.
Box 42, Churchill
Victoria, 3842
Australia

Dr. Michael B. Sporn     C
Bldg. 37, Rm. 3C09
National Institutes of Health
Bethesda, MD  20014
U.S.A.

Dr. James H. Steele, Jr.    C,
Armco Steel Research Center   Ch,
703 Curtis Street     A
Middletown, OH  45043
U.S.A.

Dr. Sherman F. Stinson    C
Bldg. 37, Room 3C02
National Cancer Inst.
Bethesda, MD  20014, U.S.A.

Dr. Piet Stroeven   Ch, C, A
Dept. of Civil Eng., Mat. Sci. Group
Delft Univ. of Technology
Stevinweg 4, Delft
Netherlands

Dr. Jen Sturgess     Cia
Dept. of Biochemistry
Hospital for Sick Children
555 University Avenue
Toronto 2, Ontario, Canada

Mr. Louis Sutro     Cia
Mail Station 20, DL-15, Room 218
Draper Laboratories
68 Albany Street
Cambridge, MA  02139
U.S.A.

C. Tasca, M.D.     Cia,
Inst. of Endocrinology   A
Str. Dionisie Lupu 62
Bucuresti, I(22)
Romania

Dr. Maurice M. Taylor
Dept. Nat. Defence, Perception Group
D.C.I.E.M., Box 2000
Downsview, Ontario
Canada

A. C. Terrell      C,
Metals Research Limited   A
Moat House, Melbourn
Royston, Herts. SG8 6EJ
England

Dr. Thomas Thorsen
Division of Math. & Eng.
El Camino College
16007 Crenshaw Blvd.
Torrance, CA 90506, U.S.A.

Dr. Géza M. Timčák    Cia,
Mineral Research Lab    A
Faculty of Mining
Košice Technical University
Svermova 5c
Košice, Czechoslovakia

Mr. Herold A. Treibs    C
Westinghouse Hanford Co.
Box 1970
Richland, WA 99352
U.S.A.

Dr. Oleh J. Tretiak
Biomedical Engineering & Science
Drexel University
Philadelphia, PA 19104, U.S.A.

Dr. Heinrich E. Tuchschmid
Metallurgist
Franzosenweg 10A Ch 850
Frauenfeld
Switzerland

Dr. Donald Ruomi     Cia
Roy C. Ingersoll Research Ctr.
Borg-Warner
Wolf and Algonquin Roads
Des Plaines, IL 60018, U.S.A.

Mr. Allan F. Turpin    C
Bausch & Lomb
847 Azalea Drive
Rockville, MD 20850
U.S.A.

Walter S. Tyler, D.V.M., Ph.D. C,
Director, Calif. Primate Res. Ctr. Ch,
University of California   A
Davis, CA 95616
U.S.A.

Dr. Ervin E. Underwood   C,
School of Chemical Engr. (Met.) A
Georgia Institute of Technology
Atlanta, GA 30332
U.S.A.

George A. Uriano     C
Scientific Assistant to the Director
Institute for Materials Research
National Bureau of Standards
Washington, DC 20234, U.S.A.

H. B. M. Uylings     Cia
Central Institute for
 Brain Research
Amsterdam
The Netherlands

Prof. dr. ir. c. J. D. M. Verhagen
Delft University of Technology Cia
Lorentzweg 1
Delft 2208
The Netherlands

Dr. Werner Villiger
Lab. für Elektronenmikrospie
Pestalozzistrasse 20
4000 Basel
Switzerland

Dr. Robin T. Vollmer    C
Department of Pathology
Duke University
Durham, NC 27705, U.S.A.

Mr. James G. Walmsley   C,
Dept. of Biophysics    A
University of Western Ontario
London, Ontario
Canada

Francis J. Warmuth    Cia
Special Metals Corporation
Middle Settlement Road, Rt 5B
New Hartford, NY 13413
U.S.A.

Dr. Nancy E. Warner
Dept. of Pathology
Univ. of Southern California
2025 Zonal Avenue
Los Angeles, CA 90033, U.S.A.

Dr. Bruce A. Warren
Associate Prof., Dept. of Pathology
The University of Western Ontario
London, Ontario
Canada

Dr. Richard Warren    Cia,
Dept. Engineering Metals   A
Chalmers Univ. of Technology
S-402 20 Gothenburg
Sweden

Prof. Geoffrey S. Watson          C
Dept. of Statistics
Fine Hall
Princeton University
Princeton, NJ   08540, U.S.A.

Dr. John H. L. Watson            C,
Dept. of Physics & Biophysics     A
Edsel B. Ford Inst. for Med. Research
2799 W. Grand Blvd.
Detroit, MI  48202, U.S.A.

LCDR Robert A. Waugh
U.S. Navy
Box 129, Naval Hospital
Bethesda, MD   20014, U.S.A.

Prof. Dr. Ewald R. Weibel        C,
Dept. of Anatomy                 Ch,
University of Bern                 A
Bühlstrasse 26
3012 Bern, Switzerland

Robert F. Whitman                 C
IMANCO
5209 Edmondson Avenue
Baltimore, MD   21229
U.S.A.

Robert L. Willes                 C,
Image Analysing Computers, Inc.   A
40 Robert Pitt Drive
Monsey, NY   10952
U.S.A.

D. G. Williams
Joyce, Loebel & Co. Ltd.
Princes Way
Team Valley-Gateshead NE11 0UJ
England

Norman J. Wilsman                 C
University of Minnesota
1342 Keston Street
St. Paul, MN   55108
U.S.A.

Dr. Robert Wlodawer              Cia,
Sonnenblickstrasse 8              A
CH-8404 Winterthur
Switzerland

Dr. Lowell A. Woodbury            C
Radiobiology Lab
University of Utah
Salt Lake City, UT   84132
U.S.A.

Dr. Ch. Zaminer                  Cia
Bundesansalt für Materialprufung
Unter den Eichen 87
1 Berlin 45
Germany

Dr. Johan Zwaan                  Cia
The Children's Hospital Medical
   Center, Dept. Ophthalmology
300 Longwood Avenue
Boston, MA  02115
U.S.A.

*Page number for Chairmen in italics*

---

* Paper presented at Congress but not printed in Proceedings.

## A

Abnormal distributions, 197
ACTH, 379
Acute leukemia, 401
Adenocarcinoma, 321
Adrenocortical cells, 379
Aerodynamic equivalent diameters, 123
Agglomerations of inclusions, 261
Aggregate size, 507
Ahlberg method, 225
Air pollutants, 355
Aluminum powders, 487
Amyloplasts, 385
Anisotropic materials, 55, 367
Architecture, of blood vessel wall, 415
Area distribution, of sections, 291
Area frequency, of cube sections, 291
Arterial wall, composition, 337
  Elastic tissue, 337
  Smooth muscle walls, 337
Aspect ratio factors, 117
Autofocus, 181
Automated image analyzing computer, 123
Automated measurements, 197
  Analysis, 355
Automatic comparative feature selection, 127
Automatic feature analysis, 185
  Linear analysis, 189
Automatic image analysis, 177, 327
  Accuracy, 141
  Central nervous tissue, 327
  Instruments, 141
Autoradiographic studies, 375, 429
Axons, 331

## B

Ball models, contactivity, 41, 45
  Hard spheres, 41, 45
  Soft spheres, 41, 45
Banded structure, 163
Basic equations, 509
  Derivations, 509
  Measurements, 509
Biased samples, 79
Biases, 59
Bifurcation ratios, 49
Bifurcations, cerebral arteries, 415
Bimodal distribution, 401, 429, 463
Biological research, 341
Biomembranes, 427
Blastocyst formation, 397
Blood vessels, 337
Bone, human, 55

Boolean scheme, 83
Boundary length in a plane, 452
  Derivation, 454
Brain, human, 167, 389
  Structure analysis, 327
Branching patterns, 49
Brinell hardness, 233
  vs. grain-boundary area, 233

## C

Cancer, in human intestine, 321
Cancerous transformation, 363
Canonical patterns, 133
Cardiac disease, 225
Cardiogenesis, 435
Cardiovascular radiology, 225
Cast iron, 315
Cat, brain cortex, 203
  Cerebral cortex, 203
Categorical setting, 133
Category theory, 133
  Oriented approach, 107
Cell constituents, identification 341
  Biology, 341
Cell Organelles, 359
  Volumes, 359
Cellular orientation, 415
  Malignancy, 423
  Structure, 211
Central nervous tissue, 327, 331
  Automatic image analysis, 327
  Heterogeneous composition of, 327
  Sampling, 327
Cerebral cortex, 203, 331
  Synaptic density distribution, 331
Characterization of powders, 487
Chemical resolution, 211
Chick embryo, 393, 435
  Heart, 435
Cholesterol, 379
Chord sizing, 163
Cineangiography, 225
Classifier - collector, 123
Cleanness assessment, of steel, 261
Closing, 151, 157
Clustering, network model of, 49
Clusters, of stringers, 265
Coefficient of variation, 59
Comminution, 299
Composite materials, 287, 295
Computational methods, 69
Computer-assisted-by-man, 197
Computer input and processing, 121
Computer-plotted projection, 93
Concentric spheroids, 79
Concrete, crack structure, 281
  Grain structure, 281

Lightning Source UK Ltd.
Milton Keynes UK
UKHW041149260219
338006UK00012B/575/P